Stochastic Finance

An Introduction with Market Examples

CHAPMAN & HALL/CRC
Financial Mathematics Series

Aims and scope:
The field of financial mathematics forms an ever-expanding slice of the financial sector. This series aims to capture new developments and summarize what is known over the whole spectrum of this field. It will include a broad range of textbooks, reference works and handbooks that are meant to appeal to both academics and practitioners. The inclusion of numerical code and concrete real-world examples is highly encouraged.

Published Titles

American-Style Derivatives; Valuation and Computation, *Jerome Detemple*
Analysis, Geometry, and Modeling in Finance: Advanced Methods in Option Pricing, *Pierre Henry-Labordère*
Computational Methods in Finance, *Ali Hirsa*
Credit Risk: Models, Derivatives, and Management, *Niklas Wagner*
Engineering BGM, *Alan Brace*
Financial Modelling with Jump Processes, *Rama Cont and Peter Tankov*
Interest Rate Modeling: Theory and Practice, *Lixin Wu*
Introduction to Credit Risk Modeling, Second Edition, *Christian Bluhm, Ludger Overbeck, and Christoph Wagner*
An Introduction to Exotic Option Pricing, *Peter Buchen*
Introduction to Risk Parity and Budgeting, *Thierry Roncalli*
Introduction to Stochastic Calculus Applied to Finance, Second Edition, *Damien Lamberton and Bernard Lapeyre*
Monte Carlo Methods and Models in Finance and Insurance, *Ralf Korn, Elke Korn, and Gerald Kroisandt*
Monte Carlo Simulation with Applications to Finance, *Hui Wang*
Nonlinear Option Pricing, *Julien Guyon and Pierre Henry-Labordère*
Numerical Methods for Finance, *John A. D. Appleby, David C. Edelman, and John J. H. Miller*
Option Valuation: A First Course in Financial Mathematics, *Hugo D. Junghenn*
Portfolio Optimization and Performance Analysis, *Jean-Luc Prigent*
Quantitative Finance: An Object-Oriented Approach in C++, *Erik Schlögl*

Quantitative Fund Management, *M. A. H. Dempster, Georg Pflug, and Gautam Mitra*

Risk Analysis in Finance and Insurance, Second Edition, *Alexander Melnikov*

Robust Libor Modelling and Pricing of Derivative Products, *John Schoenmakers*

Stochastic Finance: An Introduction with Market Examples, *Nicolas Privault*

Stochastic Finance: A Numeraire Approach, *Jan Vecer*

Stochastic Financial Models, *Douglas Kennedy*

Stochastic Processes with Applications to Finance, Second Edition, *Masaaki Kijima*

Structured Credit Portfolio Analysis, Baskets & CDOs, *Christian Bluhm and Ludger Overbeck*

Understanding Risk: The Theory and Practice of Financial Risk Management, *David Murphy*

Unravelling the Credit Crunch, *David Murphy*

Proposals for the series should be submitted to one of the series editors above or directly to:
CRC Press, Taylor & Francis Group
3 Park Square, Milton Park
Abingdon, Oxfordshire OX14 4RN
UK

Chapman & Hall/CRC FINANCIAL MATHEMATICS SERIES

Stochastic Finance

An Introduction with Market Examples

Nicolas Privault

CRC Press
Taylor & Francis Group
Boca Raton London New York

CRC Press is an imprint of the
Taylor & Francis Group, an **informa** business
A CHAPMAN & HALL BOOK

CRC Press
Taylor & Francis Group
6000 Broken Sound Parkway NW, Suite 300
Boca Raton, FL 33487-2742

Printed on acid-free paper
Version Date: 20131029

International Standard Book Number-13: 978-1-4665-9402-9 (Hardback)

Visit the Taylor & Francis Web site at
http://www.taylorandfrancis.com

and the CRC Press Web site at
http://www.crcpress.com

Contents

List of Figures

Preface

This text is an introduction to pricing and hedging in discrete and continuous time financial models without friction (i.e., without transaction costs), with an emphasis on the complementarity between analytical and probabilistic methods. Its contents are mostly mathematical, and aim at making the reader aware of both the power and limitations of mathematical models in finance, by taking into account their conditions of applicability. The book covers a wide range of classical topics including Black–Scholes pricing, exotic and American options, term structure modeling and change of numéraire, as well as models with jumps. It is targeted at the advanced undergraduate and graduate level in applied mathematics, financial engineering, and economics. The point of view adopted is that of mainstream mathematical finance in which the computation of fair prices is based on the absence of arbitrage hypothesis, therefore excluding riskless profit based on arbitrage opportunities and basic (buying low/selling high) trading. Similarly, this document is not concerned with any "prediction" of stock price behaviors that belong to other domains such as technical analysis, which should not be confused with the statistical modeling of asset prices. The text also includes 104 figures and simulations, along with about 20 examples based on actual market data.

The descriptions of the asset model, self-financing portfolios, arbitrage and market completeness, are given first in Chapter 1 in a simple two time-step setting. These notions are then reformulated in discrete time in Chapter 2. Here, the impossibility of accessing future information is formulated using the notion of adapted processes, which will play a central role in the construction of stochastic calculus in continuous time.

In order to trade efficiently it would be useful to have a formula to estimate the "fair price" of a given risky asset, helping for example to determine whether the asset is undervalued or overvalued at a given time. Although such a formula is not available, we can instead derive formulas for the pricing of options that can act as insurance contracts to protect their holders against adverse changes in the prices of risky assets. The pricing and hedging of options in discrete time, particularly in the fundamental example of the Cox–Ross–Rubinstein model, are considered in Chapter 3, with a description of the passage from discrete to continuous time that prepares the transition to the subsequent chapters.

A simplified presentation of Brownian motion, stochastic integrals, and the associated Itô formula, is given in Chapter 4. The Black–Scholes model is presented from the angle of partial differential equation (PDE) methods in Chapter 5, with the derivation of the Black–Scholes formula by transforming the Black–Scholes PDE into the standard heat equation which is then solved by a heat kernel argument. The martingale approach to pricing and hedging is then presented in Chapter 6, and complements the PDE approach of Chapter 5 by recovering the Black–Scholes formula via a probabilistic argument. An introduction to volatility estimation is given in Chapter 7, including historical, local, and implied volatilities. This chapter also contains a comparison of the prices obtained by the Black–Scholes formula with option price market data.

Exotic options such as barrier, lookback, and Asian options in continuous asset models are treated in Chapter 8. Optimal stopping and exercise, with application to the pricing of American options, are considered in Chapter 9. The construction of forward measures by change of numéraire is given in Chapter 10 and is applied to the pricing of interest rate derivatives in Chapter 12, after an introduction to the modeling of forward rates in Chapter 11, based on material from [56]. The pricing of defaultable bonds is considered in Chapter 13.

Stochastic calculus with jumps is dealt with in Chapter 14 and is restricted to compound Poisson processes which have only a finite number of jumps on any bounded interval. Those processes are used for option pricing and hedging in jump models in Chapter 15, in which we focus mostly on risk minimizing strategies as markets with jumps are generally incomplete. Chapter 16 contains an elementary introduction to finite difference methods for the numerical solution of PDEs and stochastic differential equations, dealing with the explicit and implicit finite difference schemes for the heat equations and the Black–Scholes PDE, as well as the Euler and Milshtein schemes for SDEs. The text is completed with an appendix containing the needed probabilistic background.

The material in this book has been used for teaching in the masters of science in financial engineering program at the City University of Hong Kong and at the Nanyang Technological University in Singapore. The author thanks Ju-Yi Yen (University of Cincinnati), the editor and the reviewers for many corrections, suggestions and improvements of the manuscript.

Singapore Nicolas Privault

Introduction

Modern mathematical finance and quantitative analysis require a strong background in fields such as stochastic calculus, optimization, partial differential equations (PDEs) and numerical methods, or even infinite dimensional analysis. In addition, the emergence of new complex financial instruments on the markets makes it necessary to rely on increasingly sophisticated mathematical tools. Not all readers of this book will eventually work in quantitative financial analysis, nevertheless they may have to interact with quantitative analysts, and becoming familiar with the tools they employ will be an advantage. In addition, despite the availability of ready-made financial calculators it still makes sense to be able oneself to understand, design and implement such financial algorithms. This can be particularly useful under different types of conditions, including an eventual lack of trust in financial indicators, possible unreliability of expert advice such as buy/sell recommendations, or other factors such as market manipulation. To some extent we would like to have some form of control on the future behavior of random (risky) assets; however, since knowledge of the future is not possible, the time evolution of the prices of risky assets will be modeled by random variables and stochastic processes.

Historical Sketch

We start with a description of some of the main steps, ideas and individuals that have played an important role in the development of the field over the last century.

Robert Brown, botanist, 1827

Brown observed the movement of pollen particles as described in his paper "A brief account of microscopical observations made in the months of June, July and August, 1827, on the particles contained in the pollen of plants; and on the general existence of active molecules in organic and inorganic bodies." *Phil. Mag.* 4, 161–173, 1828.

Philosophical Magazine, first published in 1798, is a journal that "publishes articles in the field of condensed matter describing original results, theories and concepts relating to the structure and properties of crystalline materials, ceramics, polymers, glasses, amorphous films, composites and soft matter."

Louis Bachelier, mathematician, PhD 1900

Bachelier used Brownian motion for the modeling of stock prices in his PhD thesis "Théorie de la spéculation," *Annales Scientifiques de l'Ecole Normale Supérieure* 3 (17): 21-86, 1900.

Albert Einstein, physicist

Einstein received his 1921 Nobel Prize in part for investigations on the theory of Brownian motion: "... in 1905 Einstein founded a kinetic theory to account for this movement," presentation speech by S. Arrhenius, Chairman of the Nobel Committee, Dec. 10, 1922.

Albert Einstein, Über die von der molekularkinetischen Theorie der Wärme geforderte Bewegung von in ruhenden Flüssigkeiten suspendierten Teilchen," *Annalen der Physik* 17 (1905) 223.

Norbert Wiener, mathematician, founder of cybernetics

Wiener is credited, among other fundamental contributions, for the mathematical foundation of Brownian motion, published in 1923. In particular he constructed the Wiener space and Wiener measure on $\mathcal{C}_0([0,1])$ (the space of continuous functions from $[0,1]$ to $\mathbb{R}$ vanishing at 0).

Norbert Wiener, "Differential space," *Journal of Mathematics and Physics of the Massachusetts Institute of Technology*, 2, 131-174, 1923.

Kiyoshi Itô, mathematician, Gauss Prize 2006

Itô constructed the Itô integral with respect to Brownian motion, cf. Itô, Kiyosi, Stochastic integral. *Proc. Imp. Acad. Tokyo* 20, (1944). 519–524. He also constructed the stochastic calculus with respect to Brownian motion, which laid the foundation for the development of calculus for random processes, cf. Itô, Kiyoshi, "On stochastic differential equations," *Mem. Amer. Math. Soc.* 1951, (1951).

"Renowned math wiz Itô, 93, dies." (*The Japan Times*, Saturday, Nov. 15, 2008).

> Kiyoshi Itô, an internationally renowned mathematician and professor emeritus at Kyoto University died Monday of respiratory failure at a Kyoto hospital, the university said Friday. He was 93. Itô was once dubbed "the most famous Japanese in Wall Street" thanks to his contribution to

the founding of financial derivatives theory. He is known for his work on stochastic differential equations and the "Itô Formula," which laid the foundation for the Black–Scholes model, a key tool for financial engineering. His theory is also widely used in fields like physics and biology.

Paul Samuelson, economist, Nobel Prize 1970

In 1965, Samuelson rediscovered Bachelier's ideas and proposed geometric Brownian motion as a model for stock prices. In an interview he stated, "In the early 1950s I was able to locate by chance this unknown [Bachelier's] book, rotting in the library of the University of Paris, and when I opened it up it was as if a whole new world was laid out before me." We refer to "Rational theory of warrant pricing" by Paul Samuelson, *Industrial Management Review*, p. 13–32, 1965.

In recognition of Bachelier's contribution, the Bachelier Finance Society was started in 1996 and now holds the World Bachelier Finance Congress every 2 years.

Robert Merton, Myron Scholes, economists

Robert Merton and Myron Scholes shared the 1997 Nobel Prize in economics: "In collaboration with Fisher Black, developed a pioneering formula for the valuation of stock options ... paved the way for economic valuations in many areas ... generated new types of financial instruments and facilitated more efficient risk management in society."

Black, Fischer; Myron Scholes (1973). "The Pricing of Options and Corporate Liabilities." *Journal of Political Economy* 81 (3): 637-654.

The development of options pricing tools contributed greatly to the expansion of option markets and led to development several ventures such as the "Long Term Capital Management" (LTCM), founded in 1994. The fund yielded annualized returns of over 40% in its first years, but registered a loss in the US of \$4.6 billion in less than four months in 1998, which resulted in its closure in early 2000.

Oldrich Vasiček, economist, 1977

Interest rates behave differently from stock prices, notably due to the phenomenon of mean reversion, and for this reason they are difficult to model using geometric Brownian motion. Vasiček was the first to suggest a mean-reverting model for stochastic interest rates, based on the Ornstein-Uhlenbeck process, in "An equilibrium characterisation of the term structure," *Journal of Financial Economics* 5: 177–188.

David Heath, Robert Jarrow, A. Morton

These authors proposed in 1987 a general framework to model the evolution of (forward) interest rates, known as the HJM model, see their joint paper "Bond Pricing and the Term Structure of Interest Rates: A New Methodology for Contingent Claims Valuation," *Econometrica*, (January 1992), Vol. 60, No. 1, pp 77–105.

Alan Brace, Dariusz Gatarek, Marek Musiela (BGM)

The BGM model is actually based on geometric Brownian motion, and it is especially useful for the pricing of interest rate derivatives such as caps and swaptions on the LIBOR market, see "The Market Model of Interest Rate Dynamics." *Mathematical Finance* Vol. 7, page 127. Blackwell 1997, by Alan Brace, Dariusz Gatarek, Marek Musiela.

European Call and Put Options

We close this introduction with a description of European call and put options, which are at the basis of risk management. As mentioned above, an important concern for the buyer of a stock at time t is whether its price S_T can fall down at some future date T. The buyer of the stock may seek protection from a market crash by purchasing a contract that allows him to sell his asset at time T at a guaranteed price K fixed at time t. This contract is called a put option with strike price K and exercise date T.

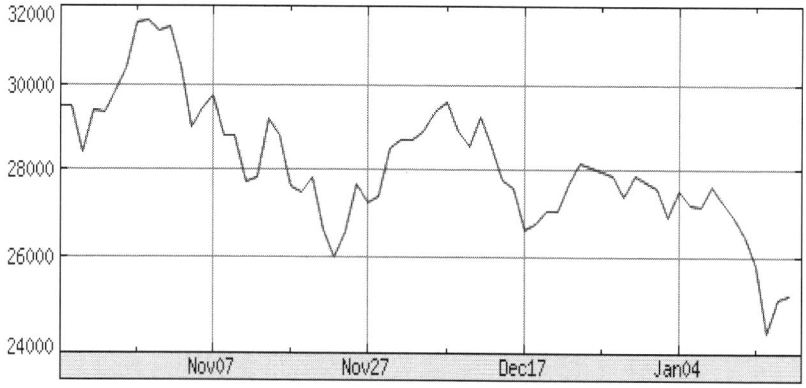

FIGURE 0.1: Graph of the Hang Seng index – holding a put option might be useful here.

Definition 0.1 *A (European)* put *option is a contract that gives its holder* *the right (but not the obligation) to* sell *a quantity of assets at a predefined price K called the strike and at a predefined date T called the maturity.*

In case the price S_T falls down below the level K, exercising the contract will give the holder of the option a gain equal to $K - S_T$ in comparison to those who did not subscribe to the option and sell the asset at the market price S_T. In turn, the issuer of the option will register a loss also equal to $K - S_T$ (in the absence of transaction costs and other fees).

If S_T is above K then the holder of the option will not exercise the option as he may choose to sell at the price S_T. In this case the profit derived from the option is 0.

In general, the payoff of a (so-called European) put option will be of the form

$$\phi(S_T) = (K - S_T)^+ = \begin{cases} K - S_T, & S_T \leq K, \\ \\ 0, & S_T \geq K. \end{cases}$$

Two possible scenarios (S_T finishing above K or below K) are illustrated in Figure 0.2.

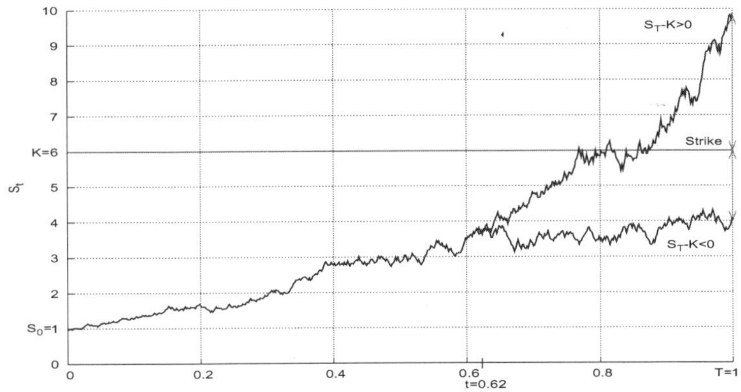

FIGURE 0.2: Sample price processes simulated by a geometric Brownian motion.

On the other hand, if the trader aims at buying some stock or commodity, his interest will be in prices not going up and he might want to purchase a call option, which is a contract allowing him to buy the considered asset at time T at a price not higher than a level K fixed at time t.

Here, in the event that S_T goes above K, the buyer of the option will register a potential gain equal to $S_T - K$ in comparison to an agent who did not subscribe to the call option.

Definition 0.2 *A (European)* call *option is a contract that gives its holder the right (but not the obligation) to* buy *a quantity of assets at a predefined price K called the strike and at a predefined date T called the maturity.*

In general, a (European) call option is an option with payoff function

$$\phi(S_T) = (S_T - K)^+ = \begin{cases} S_T - K, & S_T \geq K, \\ 0, & S_T \leq K. \end{cases}$$

In market practice, options are often divided into a certain number n of *warrants*, the (possibly fractional) quantity n being called the *entitlement ratio.*

In order for an option contract to be fair, the buyer of the option should pay a fee (similar to an insurance fee) at the signing of the contract. The computation of this fee is an important issue, which is known as option *pricing*.

The second important issue is that of *hedging*, i.e., how to manage a given portfolio in such a way that it contains the required random payoff $(K - S_T)^+$ (for a put option) or $(S_T - K)^+$ (for a call option) at the maturity date T.

The next figure illustrates a sharp increase and sharp drop in asset price, making it valuable to hold a call option during the first half of the graph, whereas holding a put option would be recommended during the second half.

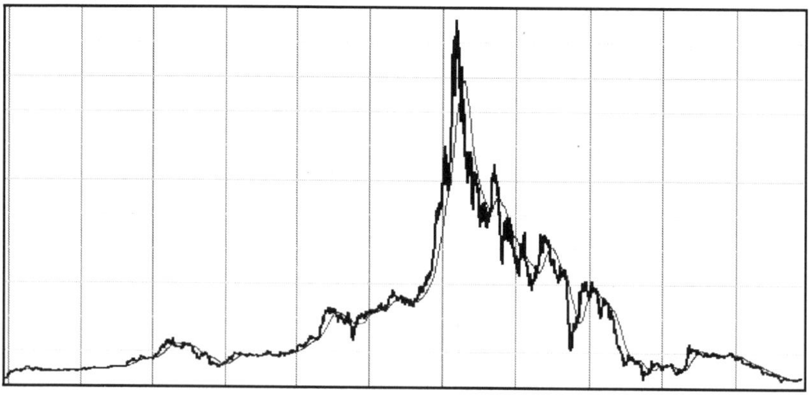

FIGURE 0.3: "Infogrames" stock price curve.

Chapter 1

Assets, Portfolios and Arbitrage

We consider a simplified financial model with only two time instants $t = 0$ and $t = 1$. In this simple setting we introduce the notions of portfolio, arbitrage, completeness, pricing and hedging using the notation of [24]. A binary asset price model is considered as an example in Section 1.7.

1.1 Definitions and Formalism

We will use the following notation. An element $\bar{x}$ of $\mathbb{R}^{d+1}$ is a vector

$$\bar{x} = (x_0, x_1, \ldots, x_d)$$

made of $d+1$ components. The scalar product $\bar{x} \cdot \bar{y}$ of two vectors $\bar{x}$, $\bar{y} \in \mathbb{R}^{d+1}$ is defined by

$$\bar{x} \cdot \bar{y} = x_0 y_0 + x_1 y_1 + \cdots x_d y_d.$$

The vector

$$\bar{\pi} = \left(\pi^{(0)}, \pi^{(1)}, \ldots, \pi^{(d)}\right)$$

denotes the prices $\pi^{(i)} > 0$ at time $t = 0$ of $d+1$ assets numbered $i = 0, 1, \ldots, d$.

The values $S^{(i)} > 0$ at time $t = 1$ of assets $i = 1, \ldots, d$ are represented by the random vector

$$\bar{S} = \left(S^{(0)}, S^{(1)}, \ldots, S^{(d)}\right)$$

defined on a probability space $(\Omega, \mathcal{F}, \mathbb{P})$.

In addition we will assume that asset n° 0 is a riskless asset (of savings account type) that yields an interest rate $r > 0$, i.e., we have

$$S^{(0)} = (1 + r)\pi^{(0)}.$$

1.2 Portfolio Allocation and Short-Selling

A *portfolio* based on the assets $0, 1, 2, \ldots, d$ is viewed as a vector $\bar{\xi} \in \mathbb{R}^{d+1}$ in which $\xi^{(i)}$ represents the (possibly fractional) quantity of asset n^o i owned by an investor, $i = 0, 1, \ldots, d$. The *price* of such a portfolio is given by

$$\bar{\xi} \cdot \bar{\pi} = \sum_{i=0}^{d} \xi^{(i)} \pi^{(i)}$$

at time $t = 0$.

At time $t = 1$ the *value* of the portfolio has evolved into

$$\bar{\xi} \cdot \bar{S} = \sum_{i=0}^{d} \xi^{(i)} S^{(i)}.$$

If $\xi^{(0)} > 0$, the investor puts the amount $\xi^{(0)} \pi^{(0)} > 0$ in a savings account with interest rate r, while if $\xi^{(0)} < 0$ he borrows the amount $-\xi^{(0)} \pi^{(0)} > 0$ with the same interest rate.

For $i = 1, \ldots, d$, if $\xi^{(i)} > 0$ then the investor buys a (possibly fractional) quantity $\xi^{(i)} > 0$ of the asset n^o i, while if $\xi^{(i)} < 0$ he borrows a quantity $-\xi^{(i)} > 0$ of asset i and sells it to obtain the amount $-\xi^{(i)} \pi^{(i)} > 0$. In the latter case one says that the investor *short sells* a quantity $-\xi^{(i)} > 0$ of the asset n^o i.

Usually, profits are made by first buying at a lower price and then selling at a higher price. Short-sellers apply the same rule but in the reverse time order: first sell high, and then buy low if possible, by applying the following procedure.

1. Borrow the asset n^o i.

2. At time $t = 0$, sell the asset n^o i on the market at the price $\pi^{(i)}$ and invest the amount $\pi^{(i)}$ at the interest rate $r > 0$.

3. Buy back the asset n^o i at time $t = 1$ at the price $S^{(i)}$, with hopefully $S^{(i)} < (1 + r)\pi^{(i)}$.

4. Return the asset to its owner, with possibly a (small) fee $p > 0$.[1]

[1]The cost p of short-selling will not be taken into account in later calculations.

At the end of the operation the profit made on share n^o i equals

$$(1+r)\pi^{(i)} - S^{(i)} - p > 0,$$

which is positive provided $S^{(i)} < (1+r)\pi^{(i)}$ and $p > 0$ is sufficiently small.

1.3 Arbitrage

As stated in the next definition, an arbitrage opportunity is the possibility to make a strictly positive amount of money starting from 0 or even from a negative amount. In a sense, an arbitrage opportunity can be seen as a way to "beat" the market.

The short-selling procedure described in Section 1.2 represents a way to realize an arbitrage opportunity. One can proceed similarly by simply buying an asset instead of short-selling it.

1. Borrow the amount $-\xi^{(0)}\pi^{(0)} > 0$ on the riskless asset n° 0.

2. Use the amount $-\xi^{(0)}\pi^{(0)} > 0$ to buy the risky asset n^o i at time $t = 0$ and price $\pi^{(i)}$, for a quantity $\xi^{(i)} = -\xi^{(0)}\pi^{(0)}/\pi^{(i)}$, $i = 1, \ldots, d$.

3. At time $t = 1$, sell the risky asset n^o i at the price $S^{(i)}$, with hopefully $S^{(i)} > \pi^{(i)}$.

4. Refund the amount $-(1+r)\xi^{(0)}\pi^{(0)} > 0$ with interest rate $r > 0$.

At the end of the operation the profit made is

$$S^{(i)} - (1+r)\pi^{(i)} > 0,$$

which is positive provided $S^{(i)} > \pi^{(i)}$ and r is sufficiently small.

Next we state a mathematical formulation of the concept of arbitrage.

Definition 1.1 *A portfolio $\bar{\xi} \in \mathbb{R}^{d+1}$ constitutes an arbitrage opportunity if the three following conditions are satisfied:*

i) $\bar{\xi} \cdot \bar{\pi} \leq 0$, *[start from 0 or even with a debt] ii)* $\bar{\xi} \cdot \bar{S} \geq 0$, *[finish with a non-negative amount]*

iii) $\mathbb{P}(\bar{\xi} \cdot \bar{S} > 0) > 0$. *[a profit is made with non-zero probability]*

The are many real-life examples of situations where arbitrage opportunities can occur, such as:

– assets with different returns (finance),

– servers with different speeds (queueing, networking, computing),

– highway lanes with different speeds (driving).

In the latter two examples, the absence of arbitrage is a consequence of the fact that switching to a faster lane or server may result in congestion, thus annihilating the potential benefit of the shift.

六合彩投注換算表

MARK SIX INVESTMENT TABLE

複式 **Multiple**		一胆拖 **One Banker with**		兩胆拖 **Two Bankers with**		三胆拖 **Three Bankers with**		四胆拖 **Four Bankers with**		五胆拖 **Five Bankers with**	
所選號碼總數 No. of Selections	HK$	配腳數目 No. of Legs	HK$	配腳數目 No. of Legs	HK$	配腳數目 No. of Legs	HK$	配腳數目 No. of Legs	HK$	配腳數目 No. of Legs	HK$
7	35	6	30	5	25	4	20	3	15	2	10
8	140	7	105	6	75	5	50	4	30	3	15
9	420	8	280	7	175	6	100	5	50	4	20
10	1,050	9	630	8	350	7	175	6	75	5	25
11	2,310	10	1,260	9	630	8	280	7	105	6	30
12	4,620	11	2,310	10	1,050	9	420	8	140	7	35
13	8,580	12	3,960	11	1,650	10	600	9	180	8	40
14	15,015	13	6,435	12	2,475	11	825	10	225	9	45
15	25,025	14	10,010	13	3,575	12	1,100	11	275	10	50
49	69,919,080	48	8,561,520	47	891,825	46	75,900	45	4,950	44	220

FIGURE 1.1: Another example of absence of arbitrage.

In the table of Figure 1.1 the absence of arbitrage opportunities is materialized by the fact that the price of each combination is found to be proportional to its probability, thus making the game fair and disallowing any opportunity or arbitrage that would result from betting on a more profitable combination.

In the sequel we will work under the assumption that arbitrage opportunities do not occur and we will rely on this hypothesis for the pricing of financial instruments.

Let us give a market example of pricing by absence of arbitrage.

From March 24 to 31, 2009, HSBC issued *rights* to buy shares at the price of \$28. This *right* actually behaves like a call option since it gives the right (with no obligation) to buy the stock at $K = \$28$. On March 24 the HSBC stock price finished at \$41.70.

The question is: how to value the price $\$R$ of the right to buy one share ? This question can be answered by looking for arbitrage opportunities. Indeed, there are two ways to buy the stock:

1. directly buy the stock on the market at the price of \$41.70. Cost: \$41.70,

or:

2. first purchase the right at price $\$R$ and then the stock at price \$28. Total cost: \$R+\$28.

For an investor who owns no stock and no rights, arbitrage would be possible in case $\$R + \$28 < \$41.70$ by buying the right at a price $\$R$, then the stock at price \$28, and finally selling the stock at the market price of \$41.70. The profit made by the investor would equal

$$\$41.70 - (\$R + \$28) > 0.$$

On the other hand, for an investor who owns the rights, in case $\$R + \$28 > \$41.70$, arbitrage would be possible by first selling the right at price $\$R$, and then buying the stock on the market at \$41.70. At time $t = 1$ the stock could be sold at around \$28, and profit would equal

$$\$R + \$28 - \$41.70 > 0.$$

In the absence of arbitrage opportunities, the above argument implies that $\$R$ should satisfy

$$\$R + \$28 - \$41.70 = 0,$$

i.e., the *arbitrage* price of the right is given by the equation

$$\$R = \$41.70 - \$28 = \$13.70. \tag{1.1}$$

Interestingly, the *market* price of the right was \$13.20 at the close of the session on March 24. The difference of \$0.50 can be explained by the presence of various market factors such as transaction costs, the time value of money, or simply by the fact that asset prices are constantly fluctuating over time. It may also represent a small arbitrage opportunity, which cannot be at all excluded. Nevertheless, the absence of arbitrage argument (1.1) prices the right at \$13.70, which is quite close to its market value. Thus the absence of arbitrage hypothesis appears as an accurate tool for pricing.

1.4 Risk-Neutral Measures

In order to use absence of arbitrage in the general context of pricing financial derivatives, we will need the notion of *risk-neutral measure.*

The next definition says that under a risk-neutral (probability) measure, the risky assets n^o $1,\ldots,d$ have same *average* rate of return as the riskless asset n° 0.

Definition 1.2 *A probability measure* $\mathbb{P}^*$ *on* Ω *is called a risk-neutral measure if*

$$\mathbb{E}^*[S^{(i)}] = (1+r)\pi^{(i)}, \qquad i = 0, 1, \ldots, d. \tag{1.2}$$

Here, $\mathbb{E}^*$ denotes the expectation under the probability measure $\mathbb{P}^*$.

In other words, $\mathbb{P}^*$ is called "risk neutral" because taking risks under $\mathbb{P}^*$ by buying a stock $S^{(i)}$ has a neutral effect: on average the expected yield of the risky asset equals the riskless rate obtained by investing in the savings account with interest rate r.

On the other hand, under a "risk premium" probability measure $\mathbb{P}^\#$, the expected return of the risky asset $S^{(i)}$ would be higher than r, i.e., we would have

$$\mathbb{E}^\#[S^{(i)}] > (1+r)\pi^{(i)}, \qquad i = 1, \ldots, d.$$

The following result can be used to check for the existence of arbitrage opportunities, and is known as the first fundamental theorem of mathematical finance. In the sequel we will only consider probability measures $\mathbb{P}^*$ that are *equivalent* to $\mathbb{P}$ in the sense that $\mathbb{P}^*(A) = 0$ if and only if $\mathbb{P}(A) = 0$ for all $A \in \mathcal{F}$.

Theorem 1.1 *A market is* without *arbitrage opportunity if and only if it admits at least one equivalent risk-neutral measure* $\mathbb{P}^*$.

Proof. For the sufficiency, given $\mathbb{P}^*$ a risk-neutral measure we have

$$\bar{\xi} \cdot \bar{\pi} = \sum_{i=0}^{d} \xi^{(i)} \pi^{(i)} = \frac{1}{1+r} \sum_{i=0}^{d} \xi^{(i)}\, \mathbb{E}^*[S^{(i)}] = \frac{1}{1+r}\, \mathbb{E}^*[\bar{\xi} \cdot \bar{S}] > 0,$$

because $\mathbb{P}^*(\bar{\xi} \cdot \bar{S} > 0) > 0$ as $\mathbb{P}(\bar{\xi} \cdot \bar{S} > 0) > 0$ and $\mathbb{P}^*$ is equivalent to $\mathbb{P}$, and this contradicts Definition 1.1-(i). The proof of necessity relies on the theorem of separation of convex sets by hyperplanes, cf. Theorem 1.6 of [24]. □

1.5 Hedging of Contingent Claims

In this section we consider the notion of contingent claim, according to the following broad definition.

Definition 1.3 *A contingent claim is any non-negative random variable $C \geq 0$.*

In practice the random variable C represents the payoff of an (option) contract at time $t = 1$.

Referring to Definition 0.2, a European call option with maturity $t = 1$ on the asset n^o i is a contingent claim whose the payoff C is given by

$$C = (S^{(i)} - K)^+ = \begin{cases} S^{(i)} - K & \text{if } S^{(i)} \geq K, \\ \\ 0 & \text{if } S^{(i)} < K, \end{cases}$$

where K is called the *strike price*. The claim C is called "contingent" because its value may depend on various market conditions, such as $S^{(i)} > K$. A contingent claim is also called a "derivative" for the same reason.

Similarly, referring to Definition 0.1, a European put option with maturity $t = 1$ on the asset n^o i is a contingent claim with payoff

$$C = (K - S^{(i)})^+ = \begin{cases} K - S^{(i)} & \text{if } S^{(i)} \leq K, \\ \\ 0 & \text{if } S^{(i)} > K, \end{cases}$$

Definition 1.4 *A contingent claim with payoff C is said to be attainable if there exists a portfolio strategy $\bar{\xi}$ such that*

$$C = \bar{\xi} \cdot \bar{S}.$$

When a contingent claim C is attainable, a trader will be able to:

1. at time $t = 0$, build a portfolio allocation $\bar{\xi} = (\xi^{(0)}, \xi^{(1)}, \ldots, \xi^{(d)}) \in \mathbb{R}^{d+1}$,

2. invest the amount

$$\bar{\xi} \cdot \bar{\pi} = \sum_{i=0}^{d} \xi^{(i)} \pi^{(i)}$$

in this portfolio at time $t = 0$,

3. at time $t = 1$, pay the claim amount C using the value $\bar{\xi} \cdot \bar{S}$ of the portfolio.

The above shows that in order to attain the claim, an initial investment $\bar{\xi} \cdot \bar{\pi}$ is needed at time $t = 0$. This amount, to be paid by the buyer to the issuer of the option (the option writer), is also called the *arbitrage price* of the contingent claim C, and denoted by

$$\pi(C) := \bar{\xi} \cdot \bar{\pi}. \tag{1.3}$$

The action of allocating a portfolio $\bar{\xi}$ such that

$$C = \bar{\xi} \cdot \bar{S} \tag{1.4}$$

is called *hedging*, or *replication*, of the contingent claim C.

As a rough illustration of the principle of hedging, one may buy oil-related stocks in order to hedge oneself against a potential price rise of gasoline. In this case, any increase in the price of gasoline that would result in a higher value of the derivative C would be correlated to an increase in the underlying stock value, so that the equality (1.4) would be maintained.

In case the value $\bar{\xi} \cdot \bar{S}$ exceeds the amount of the claim, i.e., if

$$\bar{\xi} \cdot \bar{S} \geq C,$$

we talk about *super-hedging*.

In this book we focus on hedging (i.e., *replication* of the contingent claim C) and we will not consider super-hedging.

1.6 Market Completeness

Market completeness is a strong property saying that any contingent claim can be perfectly hedged.

Definition 1.5 *A market model is said to be* complete *if every contingent claim C is attainable.*

The next result is the second fundamental theorem of mathematical finance.

Theorem 1.2 *A market model without arbitrage is complete if and only if it admits only one risk-neutral measure.*

Proof. cf. Theorem 1.40 of [24]. □

Theorem 1.2 will give us a concrete way to verify market completeness by searching for a unique solution $\mathbb{P}^*$ to Equation (1.2).

1.7 Example

In this section we work out a simple example that allows us to apply Theorem 1.1 and Theorem 1.2.

We take $d = 1$, i.e., there is only a riskless asset n° 0 and a risky asset $S^{(1)}$. In addition we choose

$$\Omega = \{\omega^-, \omega^+\},$$

which is the simplest possible non-trivial choice of a probability space, made of only two possible outcomes with

$$\mathbb{P}(\{\omega^-\}) > 0 \quad \text{and} \quad \mathbb{P}(\{\omega^+\}) > 0,$$

in order for the setting to be non-trivial. In other words the behavior of the market is subject to only two possible outcomes, for example, one is expecting an important binary decision of yes/no type, which can lead to two distinct scenarios called ω^- and ω^+.

In this context, the asset price $S^{(1)}$ is given by a random variable

$$S^{(1)} : \Omega \longrightarrow \mathbb{R}$$

whose value depends whether the scenario ω^-, resp. ω^+, occurs.

Precisely, we set

$$S^{(1)}(\omega^-) = a, \qquad \text{and} \qquad S^{(1)}(\omega^+) = b,$$

i.e., the value of $S^{(1)}$ becomes equal a under the scenario ω^-, and equal to b under the scenario ω^+, where $0 < a \leq b$.

The first natural question we ask is:

– Are there arbitrage opportunities in such a market ?

We will answer this question using Theorem 1.1, which amounts to searching for a risk-neutral measure $\mathbb{P}^*$. In this simple framework, any measure $\mathbb{P}^*$ on $\Omega = \{\omega^-, \omega^+\}$ is characterized by the data of two numbers $\mathbb{P}^*(\{\omega^-\}) \in [0, 1]$ and $\mathbb{P}^*(\{\omega^+\}) \in [0, 1]$, such that

$$\mathbb{P}^*(\Omega) = \mathbb{P}^*(\{\omega^-\}) + \mathbb{P}^*(\{\omega^+\}) = 1. \tag{1.5}$$

Here, saying that $\mathbb{P}^*$ is *equivalent* to $\mathbb{P}$ simply means that

$$\mathbb{P}^*(\{\omega^-\}) > 0 \quad \text{and} \quad \mathbb{P}^*(\{\omega^+\}) > 0.$$

In addition, according to Definition 1.2 a risk-neutral measure $\mathbb{P}^*$ should satisfy

$$\mathbb{E}^*[S^{(1)}] = (1+r)\pi^{(1)}. \tag{1.6}$$

Although we should solve this equation for $\mathbb{P}^*$, at this stage it is not yet clear how $\mathbb{P}^*$ appears in (1.6).

In order to make (1.6) more explicit we write the expectation as

$$\mathbb{E}^*[S^{(1)}] = a\mathbb{P}^*(S^{(1)} = a) + b\mathbb{P}^*(S^{(1)} = b),$$

hence Condition (1.6) for the existence of a risk-neutral measure $\mathbb{P}^*$ reads

$$a\mathbb{P}^*(S^{(1)} = a) + b\mathbb{P}^*(S^{(1)} = b) = (1+r)\pi^{(1)}.$$

Using the Condition (1.5) we obtain the system of two equations

$$\begin{cases} a\mathbb{P}^*(\{\omega^-\}) + b\mathbb{P}^*(\{\omega^+\}) = (1+r)\pi^{(1)} \\ \\ \mathbb{P}^*(\{\omega^-\}) + \mathbb{P}^*(\{\omega^+\}) = 1, \end{cases} \tag{1.7}$$

with solution

$$\mathbb{P}^*(\{\omega^-\}) = \frac{b-(1+r)\pi^{(1)}}{b-a} \quad \text{and} \quad \mathbb{P}^*(\{\omega^+\}) = \frac{(1+r)\pi^{(1)}-a}{b-a}.$$

In order for a solution $\mathbb{P}^*$ to exist as a probability measure, the numbers $\mathbb{P}^*(\{\omega^-\})$ and $\mathbb{P}^*(\{\omega^+\})$ must be non-negative.

We deduce that if a, b and r satisfy the condition

$$a < (1+r)\pi^{(1)} < b, \tag{1.8}$$

then there exists a risk-neutral measure $\mathbb{P}^*$ which is unique, hence by Theorems 1.1 and 1.2 the market is without arbitrage and complete.

If $a = b = (1+r)\pi^{(1)}$ then (1.2) admits an infinity of solutions, hence the market is without arbitrage but it is not complete. More precisely, in this case both the riskless and risky assets yield a deterministic return rate r and the value of the portfolio becomes

$$\bar{\xi} \cdot \bar{S} = (1+r)\bar{\xi} \cdot \bar{\pi},$$

at time $t = 1$, hence the terminal value $\bar{\xi} \cdot \bar{S}$ is deterministic and this *single* value can not always match the value of a random contingent claim C that would be allowed to take *two* distinct values $C(\omega^-)$ and $C(\omega^+)$. Therefore, market completeness does not hold when $a = b = (1+r)\pi^{(1)}$.

Note that if $a = (1+r)\pi^{(1)}$, resp. $b = (1+r)\pi^{(1)}$, then $\mathbb{P}^*(\{\omega^+\}) = 0$, resp. $\mathbb{P}^*(\{\omega^-\}) = 0$, and $\mathbb{P}^*$ is not equivalent to $\mathbb{P}$.

On the other hand, under the conditions

$$a < b < (1+r)\pi^{(1)} \qquad \text{or} \qquad (1+r)\pi^{(1)} < a < b, \tag{1.9}$$

no risk neutral measure exists and as a consequence there exist arbitrage opportunities in the market.

Let us give a financial interpretation of Conditions (1.9).

1. If $(1+r)\pi^{(1)} < a < b$, let $\xi^{(1)} = 1$ and choose $\xi^{(0)}$ such that $\xi^{(0)}\pi^{(0)} + \xi^{(1)}\pi^{(1)} = 0$, i.e.,
$$\xi^{(0)} = -\xi^{(1)}\pi^{(1)}/\pi^{(0)} < 0.$$
This means that the investor borrows the amount $-\xi^{(0)}\pi^{(0)} > 0$ on the riskless asset and uses it to buy one unit $\xi^{(1)} = 1$ of the risky asset. At time $t = 1$ she sells the risky asset $S^{(1)}$ at a price at least equal to a and refunds the amount $-(1+r)\xi^{(0)}\pi^{(0)} > 0$ she borrowed, with interest. Her profit is
$$\begin{aligned}\bar{\xi}\cdot\bar{S} &= (1+r)\xi^{(0)}\pi^{(0)} + \xi^{(1)}S^{(1)} \\ &\geq (1+r)\xi^{(0)}\pi^{(0)} + \xi^{(1)}a \\ &= -(1+r)\xi^{(1)}\pi^{(1)} + \xi^{(1)}a \\ &= \xi^{(1)}(-(1+r)\pi^{(1)} + a) \\ &> 0. \qquad \smile\end{aligned}$$

2. If $a < b < (1+r)\pi^{(1)}$, let $\xi^{(0)} > 0$ and choose $\xi^{(1)}$ such that $\xi^{(0)}\pi^{(0)} + \xi^{(1)}\pi^{(1)} = 0$, i.e.,
$$\xi^{(1)} = -\xi^{(0)}\pi^{(0)}/\pi^{(1)} < 0.$$
This means that the investor borrows a (possibly fractional) quantity $-\xi^{(1)} > 0$ of the risky asset, sells it for the amount $-\xi^{(1)}\pi^{(1)}$, and invests this money on the riskless account for the amount $\xi^{(0)}\pi^{(0)} > 0$. As mentioned in Section 1.2, in this case one says that the investor *short-sells* the risky asset. At time $t = 1$ she obtains $(1+r)\xi^{(0)}\pi^{(0)} > 0$ from the riskless asset and she spends at most b to buy the risky asset and return it to its original owner. Her profit is
$$\begin{aligned}\bar{\xi}\cdot\bar{S} &= (1+r)\xi^{(0)}\pi^{(0)} + \xi^{(1)}S^{(1)} \\ &\geq (1+r)\xi^{(0)}\pi^{(0)} + \xi^{(1)}b \\ &= -(1+r)\xi^{(1)}\pi^{(1)} + \xi^{(1)}b \\ &= \xi^{(1)}(-(1+r)\pi^{(1)} + b) \\ &> 0, \qquad \smile\end{aligned}$$

since $\xi^{(1)} < 0$. Note that here, $a \leq S^{(1)} \leq b$ became

$$\xi^{(1)} b \leq \xi^{(1)} S^{(1)} \leq \xi^{(1)} a$$

because $\xi^{(1)} < 0$.

Under Condition (1.8) there is absence of arbitrage and Theorem 1.1 shows that no portfolio strategy can yield $\bar{\xi} \cdot \bar{S} \geq 0$ and $\mathbb{P}(\bar{\xi} \cdot \bar{S} > 0) > 0$ starting from $\xi^{(0)}\pi^{(0)} + \xi^{(1)}\pi^{(1)} \leq 0$, although this is less simple to show directly.

Finally if $a = b \neq (1+r)\pi^{(1)}$ then (1.2) admits no solution as a probability measure $\mathbb{P}^*$ hence arbitrage opportunities exist and can be constructed by the same method as above.

The second natural question is:

– Is the market complete, i.e., are all contingent claims attainable ?

In the sequel we work under the condition

$$a < (1+r)\pi^{(1)} < b,$$

under which Theorems 1.1 and 1.2 show that the market is without arbitrage and complete since the risk-neutral measure $\mathbb{P}^*$ exists and is unique.

Let us recover this fact by elementary calculations. For any contingent claim C we need to show that there exists a portfolio $\bar{\xi} = (\xi^{(0)}, \xi^{(1)})$ such that $C = \bar{\xi} \cdot \bar{S}$, i.e.,

$$\begin{cases} \xi^{(0)}(1+r)\pi^{(0)} + \xi^{(1)} a = C(\omega^-) \\ \\ \xi^{(0)}(1+r)\pi^{(0)} + \xi^{(1)} b = C(\omega^+). \end{cases} \tag{1.10}$$

These equations can be solved as

$$\xi^{(0)} = \frac{bC(\omega^-) - aC(\omega^+)}{\pi^{(0)}(1+r)(b-a)} \quad \text{and} \quad \xi^{(1)} = \frac{C(\omega^+) - C(\omega^-)}{b-a}. \tag{1.11}$$

In this case we say that the portfolio $(\xi^{(0)}, \xi^{(1)})$ *hedges* the contingent claim C. In other words, any contingent claim C is attainable and the market is indeed complete. Here, the quantity

$$\xi^{(0)}\pi^{(0)} = \frac{bC(\omega^-) - aC(\omega^+)}{(1+r)(b-a)}$$

represents the amount invested on the riskless asset.

Note that if $C(\omega^+) \geq C(\omega^-)$ then $\xi^{(1)} \geq 0$ and there is not short selling.

This occurs in particular if C has the form $C = h(S^{(1)})$ with $x \longmapsto h(x)$ a non-decreasing function, since

$$\begin{aligned}\xi^{(1)} &= \frac{C(\omega^+) - C(\omega^-)}{b-a} \\ &= \frac{h(S^{(1)}(\omega^+)) - h(S^{(1)}(\omega^-))}{b-a} \\ &= \frac{h(b) - h(a)}{b-a} \\ &\geq 0,\end{aligned}$$

thus there is no short-selling.

The *arbitrage price* $\pi(C)$ of the contingent claim C is defined in (1.3) as the initial value at $t = 0$ of the portfolio hedging C, i.e.,

$$\pi(C) = \bar{\xi} \cdot \bar{\pi}, \tag{1.12}$$

where $(\xi^{(0)}, \xi^{(1)})$ are given by (1.11). Note that $\pi(C)$ cannot be 0 since this would entail the existence of an arbitrage opportunity according to Definition 1.1.

The next proposition shows that the arbitrage price $\pi(C)$ of the claim can be computed as the expected value of its payoff C under the risk-neutral measure, after discounting at the rate $1 + r$ for the time value of money.

Proposition 1.1 *The arbitrage price $\pi(C) = \bar{\xi} \cdot \bar{\pi}$ of the contingent claim C is given by*

$$\pi(C) = \frac{1}{1+r} \mathbb{E}^*[C]. \tag{1.13}$$

Proof. We have

$$\begin{aligned}\pi(C) &= \bar{\xi} \cdot \bar{\pi} \\ &= \xi^{(0)}\pi^{(0)} + \xi^{(1)}\pi^{(1)} \\ &= \frac{bC(\omega^-) - aC(\omega^+)}{(1+r)(b-a)} + \pi^{(1)}\frac{C(\omega^+) - C(\omega^-)}{b-a} \\ &= \frac{1}{1+r}\left(C(\omega^-)\frac{b - \pi^{(1)}(1+r)}{b-a} + C(\omega^+)\frac{(1+r)\pi^{(1)} - a}{b-a}\right) \\ &= \frac{1}{1+r}\Big(C(\omega^-)\mathbb{P}^*(S^{(1)} = a) + C(\omega^+)\mathbb{P}^*(S^{(1)} = b)\Big) \\ &= \frac{1}{1+r}\mathbb{E}^*[C].\end{aligned}$$

□

In the case of a European call option with strike $K \in [a, b]$ we have $C = (S^{(1)} - K)^+$ and

$$\pi((S^{(1)} - K)^+) = \pi^{(1)} \frac{b - K}{b - a} - \frac{(b - K)a}{(1 + r)(b - a)}.$$

Here, $(\pi^{(1)} - K)^+$ is called the *intrinsic value* at time 0 of the call option.

The simple setting described in this chapter raises several questions and remarks.

Remarks

1. The fact that $\pi(C)$ can be obtained by two different methods, i.e., an algebraic method via (1.11) and (1.12) and a probabilistic method from (1.13) is not a simple coincidence. It is actually a simple example of the deep connection that exists between probability and analysis.

   In a continuous time setting, (1.11) will be replaced with a *partial differential equation* (PDE) and (1.13) will be computed via the *Monte Carlo* method. In practice, the quantitative analysis departments of major financial institutions can be split into the *PDE team* and the *Monte Carlo* team, often trying to determine the same option prices by two different methods.

2. What if we have three possible scenarios, i.e., $\Omega = \{\omega^-, \omega^o, \omega^+\}$ and the random asset $S^{(1)}$ is allowed to take more than two values, e.g., $S^{(1)} \in \{a, b, c\}$ according to each scenario ? In this case the system (1.7) would be rewritten as

$$\begin{cases} a\mathbb{P}^*(\{\omega^-\}) + b\mathbb{P}^*(\{\omega^o\}) + c\mathbb{P}^*(\{\omega^+\}) = (1 + r)\pi^{(1)} \\ \mathbb{P}^*(\{\omega^-\}) + \mathbb{P}^*(\{\omega^o\}) + \mathbb{P}^*(\{\omega^+\}) = 1, \end{cases}$$

   and this system of two equations for three unknowns does not have a unique solution, hence the market can be without arbitrage but it cannot be complete. Completeness can be reached by adding a second risky asset, i.e., taking $d = 2$, in which case we will get three equations and three unknowns. More generally, when Ω has $n \geq 2$ elements, completeness of the market can be reached provided we consider d risky assets with $d+1 \geq n$. This is related to the Meta-Theorem 8.3.1 of [4] in which the number d of traded underlying risky assets is linked to the number of random sources through arbitrage and completeness.

Exercises

Exercise 1.1 Consider a financial model with two instants $t = 0$ and $t = 1$ and two assets:

- a riskless asset π with price π_0 at time $t = 0$ and value $\pi_1 = \pi_0(1 + r)$ at time $t = 1$,
- a risky asset S with price S_0 at time $t = 0$ and random value S_1 at time $t = 1$.

We assume that S_1 can take only the values $S_0(1 + a)$ and $S_0(1 + b)$, where $-1 < a < r < b$. The *return* of the risky asset is defined as

$$R = \frac{S_1 - S_0}{S_0}.$$

1. What are the possible values of R ?
2. Show that under the probability measure $\mathbb{P}^*$ defined by

$$\mathbb{P}^*(R = a) = \frac{b - r}{b - a}, \qquad \mathbb{P}^*(R = b) = \frac{r - a}{b - a},$$

   the expected return $\mathbb{E}^*[R]$ of S is equal to the return r of the riskless asset.

3. Do there exist arbitrage opportunities in this model ? Explain why.
4. Is this market model complete ? Explain why.
5. Consider a contingent claim with payoff C given by

$$C = \begin{cases} \alpha & \text{if } R = a, \\ \\ \beta & \text{if } R = b. \end{cases}$$

   Show that the portfolio (η, ξ) defined[2] by

$$\eta = \frac{\alpha(1 + b) - \beta(1 + a)}{\pi_0(1 + r)(b - a)} \quad \text{and} \quad \xi = \frac{\beta - \alpha}{S_0(b - a)},$$

   hedges the contingent claim C, i.e., show that at time $t = 1$ we have

$$\eta\pi_1 + \xi S_1 = C.$$

   Hint: distinguish two cases $R = a$ and $R = b$.

[2]Here, η is the (possibly fractional) quantity of asset π and ξ is the quantity held of asset S.

6. Compute the arbitrage price $\pi(C)$ of the contingent claim C using η, π_0, ξ, and S_0.

7. Compute $\mathbb{E}^*[C]$ in terms of a, b, r, α, β.

8. Show that the arbitrage price $\pi(C)$ of the contingent claim C satisfies

$$\pi(C) = \frac{1}{1+r}\mathbb{E}^*[C]. \tag{1.14}$$

9. What is the interpretation of Relation (1.14) above ?

10. Let C denote the payoff at time $t = 1$ of a put option with strike $K = \$11$ on the risky asset. Give the expression of C as a function of S_1 and K.

11. Letting $\pi_0 = S_0 = 1$ and $a = 8$, $b = 11$, compute the portfolio (ξ, η) hedging the contingent claim C.

12. Compute the arbitrage price $\pi(C)$ of the claim C.

Chapter 2

Discrete-Time Model

A basic limitation of the two time-step model considered in Chapter 1 is that it does not allow for trading until the end of the time period is reached. In order to be able to re-allocate the portfolio over time we need to consider a discrete-time financial model with $N+1$ time instants $t = 0, 1, \ldots, N$. The practical importance of this model lies also in its direct computer implementability.

2.1 Stochastic Processes

A *stochastic process* on a probability space $(\Omega, \mathcal{F}, \mathbb{P})$ is a family $(X_t)_{t\in T}$ of random variables $X_t : \Omega \to \mathbb{R}$ indexed by a set T. Examples include:

- the two-instant model: $T = \{0, 1\}$,
- the discrete-time model with finite horizon: $T = \{0, 1, 2, \ldots, N\}$,
- the discrete-time model with infinite horizon: $T = \mathbb{N}$,
- the continuous-time model: $T = \mathbb{R}_+$.

For real-world examples of stochastic processes one can mention:

- the time evolution of a risky asset – in this case X_t represents the price of the asset at time $t \in T$.
- the time evolution of a physical parameter – for example, X_t represents a temperature observed at time $t \in T$.

In this chapter we will focus on the finite horizon discrete-time model with $T = \{0, 1, 2, \ldots, N\}$.

Here the vector

$$\bar{\pi} = (\pi^{(0)}, \pi^{(1)}, \ldots, \pi^{(d)})$$

denotes the prices at time $t = 0$ of $d + 1$ assets numbered $0, 1, \ldots, d$.

The *random* vector

$$\bar{S}_t = (S_t^{(0)}, S_t^{(1)}, \ldots, S_t^{(d)})$$

on Ω denotes the values at time $t = 1, \ldots, N$ of assets $0, 1, \ldots, d$, and forms a stochastic process $(\bar{S}_t)_{t=0,1,\ldots,N}$ with $\bar{S}_0 = \bar{\pi}$.

Here we still assume that asset 0 is a riskless asset (of savings account type) yielding an interest rate r, i.e., we have

$$S_t^{(0)} = (1+r)^t \pi^{(0)}, \qquad t = 0, 1, \ldots, N.$$

2.2 Portfolio Strategies

A portfolio strategy is a stochastic process $(\bar{\xi}_t)_{t=1,\ldots,N} \subset \mathbb{R}^{d+1}$ where $\xi_t^{(i)}$ denotes the (possibly fractional) quantity of asset i held in the portfolio over the period $(t-1, t]$, $t = 1, \ldots, N$.

Note that the portfolio allocation

$$\xi_t = (\xi_t^{(0)}, \xi_t^{(1)}, \ldots, \xi_t^{(d)})$$

remains constant over the period $(t-1, t]$ while the stock price changes from S_{t-1} to S_t over this period.

In other terms,

$$\xi_t^{(i)} S_{t-1}^{(i)}$$

represents the amount invested in asset i at the beginning of the time period $(t-1, t]$, and

$$\xi_t^{(i)} S_t^{(i)}$$

represents the value of this investment at the end of $(t-1, t]$, $t = 1, \ldots, N$.

The value of the porfolio at the beginning of the time period $(t-1, t]$ is

$$\bar{\xi}_t \cdot \bar{S}_{t-1} = \sum_{i=0}^{d} \xi_t^{(i)} S_{t-1}^{(i)},$$

when the market "opens" at time $t-1$, and becomes

$$\bar{\xi}_t \cdot \bar{S}_t = \sum_{i=0}^{d} \xi_t^{(i)} S_t^{(i)} \tag{2.1}$$

at the end of $(t-1, t]$, i.e., when the market "closes", $t = 1, \ldots, N$.

At the beginning of the next trading period $(t, t+1]$ the value of the portfolio becomes

$$\bar{\xi}_{t+1} \cdot \bar{S}_t = \sum_{i=0}^{d} \xi_{t+1}^{(i)} S_t^{(i)}. \tag{2.2}$$

Note that the stock price $\bar{S}_t$ is assumed to remain constant "overnight," i.e., from the end of $(t-1, t]$ to the beginning of $(t, t+1]$.

Obviously the question arises whether (2.1) should be identical to (2.2). In the sequel we will need such a consistency hypothesis, called self-financing, on the portfolio strategy $\bar{\xi}_t$.

Definition 2.1 *We say that the portfolio strategy* $(\bar{\xi}_t)_{t=1,\ldots,N}$ *is* self-financing *if*

$$\bar{\xi}_t \cdot \bar{S}_t = \bar{\xi}_{t+1} \cdot \bar{S}_t, \qquad t = 1, \ldots, N-1. \tag{2.3}$$

The meaning of the self-financing condition (2.3) is simply that one cannot take any money in or out of the portfolio during the "overnight" transition period at time t. In other words, at the beginning of the new trading period $(t, t+1]$ one should re-invest the totality of the portfolio value obtained at the end of period $(t-1, t]$. The next figure is an illustration of the self-financing condition.

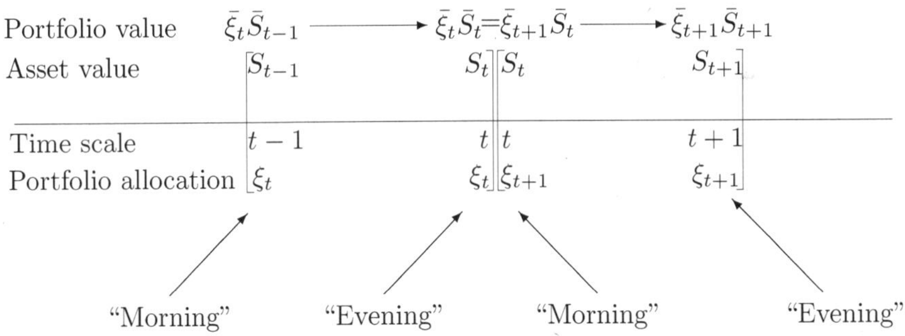

FIGURE 2.1: Illustration of the self-financing condition (2.3).

Note that portfolio re-allocation happens "overnight" during which time the portfolio global value remains the same due to the self-financing condition. The portfolio allocation ξ_t remains the same throughout the day, however the portfolio value changes from morning to evening due to a change in the stock price. Also, $\bar{\xi}_0$ is not defined and its value is actually not needed in this framework.

Of course the chosen unit of time may not be the day, and it can be replaced by weeks, hours, minutes, or even fractions of seconds in high-frequency trading.

We will denote by

$$V_t := \bar{\xi}_t \cdot \bar{S}_t$$

the value of the portfolio at time $t = 1, \ldots, N$, with

$$V_t = \bar{\xi}_{t+1} \cdot \bar{S}_t, \qquad t = 0, \ldots, N-1,$$

by the self-financing condition (2.3), and in particular

$$V_0 = \bar{\xi}_1 \cdot \bar{S}_0.$$

Let also

$$\bar{X}_t := (X_t^{(0)}, X_t^{(1)}, \ldots, X_t^{(d)})$$

denote the vector of discounted asset prices defined as:

$$X_t^{(i)} = \frac{1}{(1+r)^t} S_t^{(i)}, \qquad i = 0, 1, \ldots, d, \quad t = 0, 1, \ldots, N,$$

or

$$\bar{X}_t := \frac{1}{(1+r)^t} \bar{S}_t, \qquad t = 0, 1, \ldots, N.$$

The *discounted* value at time 0 of the portfolio is defined by

$$\widetilde{V}_t = \frac{1}{(1+r)^t} V_t, \qquad t = 0, 1, \ldots, N.$$

We have

$$\begin{aligned} \widetilde{V}_t &= \frac{1}{(1+r)^t} \bar{\xi}_t \cdot \bar{S}_t \\ &= \frac{1}{(1+r)^t} \sum_{i=0}^{d} \xi_t^{(i)} S_t^{(i)} \\ &= \sum_{i=0}^{d} \xi_t^{(i)} X_t^{(i)} \\ &= \bar{\xi}_t \cdot \bar{X}_t, \qquad t = 1, \ldots, N, \end{aligned}$$

and

$$\widetilde{V}_0 = \bar{\xi}_1 \cdot X_0 = \bar{\xi}_1 \cdot S_0.$$

The effect of discounting from time t to time 0 is to divide prices by $(1+r)^t$, making all prices comparable at time 0.

2.3 Arbitrage

The definition of arbitrage in discrete time follows the lines of its analog in the two-step model.

Definition 2.2 *A portfolio strategy* $(\bar{\xi}_t)_{t=1,\ldots,N}$ *constitutes an arbitrage opportunity if all* three *following conditions are satisfied:*

(i) $V_0 \leq 0$, *[start from* 0 *or even with a debt]*

(ii) $V_N \geq 0$, *[finish with a non-negative amount]*

(iii) $\mathbb{P}(V_N > 0) > 0$. *[a profit is made with non-zero probability]*

2.4 Contingent Claims

Recall that from Definition 1.3, a contingent claim is given by the non-negative random payoff C of an option contract at time $t = N$. For example, in the case of the European call of Definition 0.2, the payoff C is given by $X = (S_N - K)^+$ where K is called the strike price.

In a discrete-time setting we are able to consider path-dependent options in addition to European type options. One can distinguish between vanilla options whose payoff depends on the terminal value of the underlying asset, such as simple European contracts, and exotic or path-dependent options such as Asian, barrier, or lookback options, whose payoff may depend on the whole path of the underlying asset price until expiration time.

The list provided below is actually very restricted and there exist many more option types, with new ones appearing constantly on the markets.

European options

The payoff of a European call on the underlying asset no i with maturity N and strike K is

$$C = (S_N^{(i)} - K)^+.$$

The payoff of a European put on the underlying asset no i with exercise date N and strike K is

$$C = (K - S_N^{(i)})^+.$$

Let us mention also the existence of binary, or digital options, also called cash-or-nothing options, whose payoffs are

$$C = \mathbf{1}_{[K,\infty)}(S_N^{(i)}) = \begin{cases} \$1 & \text{if } S_N^{(i)} \geq K, \\ 0 & \text{if } S_N^{(i)} < K, \end{cases}$$

for call options, and

$$C = \mathbf{1}_{(-\infty,K]}(S_N^{(i)}) = \begin{cases} \$1 & \text{if } S_N^{(i)} \leq K, \\ 0 & \text{if } S_N^{(i)} > K, \end{cases}$$

for put options.

Asian options

The payoff of an Asian call option (also called average value option) on the underlying asset no i with exercise date N and strike K is

$$C = \left(\frac{1}{N+1} \sum_{t=0}^{N} S_t^{(i)} - K \right)^+ .$$

The payoff of an Asian put option on the underlying asset no i with exercise date N and strike K is

$$C = \left(K - \frac{1}{N+1} \sum_{t=0}^{N} S_t^{(i)} \right)^+ .$$

We refer to Section 8.5 for the pricing of Asian options in continuous time.

Barrier options

The payoff of a down-an-out barrier call option on the underlying asset no i with exercise date N, strike K and barrier B is

$$C = \left(S_N^{(i)} - K \right)^+ \mathbf{1}_{\left\{ \min\limits_{t=0,1,\ldots,N} S_t^{(i)} > B \right\}} = \begin{cases} S_N^{(i)} - K & \text{if } \min\limits_{t=0,1,\ldots,N} S_t^{(i)} > B, \\ 0 & \text{if } \min\limits_{t=0,1,\ldots,N} S_t^{(i)} \leq B. \end{cases}$$

This option is also called a Callable Bull Contract with no residual value, in which B denotes the call price $B \geq K$. It is also called a turbo warrant with no rebate.

The payoff of an up-and-out barrier put option on the underlying asset no i with exercise date N, strike K and barrier B is

$$C = \left(K - S_N^{(i)}\right)^+ \mathbf{1}_{\left\{\max\limits_{t=0,1,\ldots,N} S_t^{(i)} < B\right\}} = \begin{cases} K - S_N^{(i)} & \text{if } \max\limits_{t=0,1,\ldots,N} S_t^{(i)} < B, \\ \\ 0 & \text{if } \max\limits_{t=0,1,\ldots,N} S_t^{(i)} \geq B. \end{cases}$$

This option is also called a Callable Bear Contract with no residual value, in which the call price B usually satisfies $B \leq K$. See [22], [74] for recent results on the pricing of CBBCs, also called turbo warrants. We refer the reader to Chapter 8 for the pricing and hedging of similar exotic options in continuous time. Barrier options in continuous time are priced in Section 8.3.

Lookback options

The payoff of a floating strike lookback call option on the underlying asset no i with exercise date N is

$$C = S_N^{(i)} - \min_{t=0,1,\ldots,N} S_t^{(i)}.$$

The payoff of a floating strike lookback put option on the underlying asset no i with exercise date N is

$$C = \left(\max_{t=0,1,\ldots,N} S_t^{(i)}\right) - S_N^{(i)}.$$

We refer to Section 8.4 for the pricing of lookback options in continuous time.

2.5 Martingales and Conditional Expectation

Before proceeding to the definition of risk-neutral probability measures in discrete time we need to introduce more mathematical tools such as conditional expectations, filtrations, and martingales.

Conditional expectations

Clearly, the expected value of any risky asset or random variable is dependent on the amount of available information. For example, the expected return on a real estate investment typically depends on the location of this investment.

In the probabilistic framework the available information is formalized as a collection $\mathcal{G}$ of events, which may be smaller than the collection $\mathcal{F}$ of all available events, i.e., $\mathcal{G} \subset \mathcal{F}$.[1]

The notation $\mathbb{E}[F|\mathcal{G}]$ represents the expected value of a random variable F given (or conditionally to) the information contained in $\mathcal{G}$, and it is read "the conditional expectation of F given $\mathcal{G}$." In a certain sense, $\mathbb{E}[F|\mathcal{G}]$ represents the best possible estimate of F in mean square sense, given the information contained in $\mathcal{G}$.

The conditional expectation satisfies the following five properties, cf. Section A for details and proofs.

(i) $\mathbb{E}[FG \mid \mathcal{G}] = G\,\mathbb{E}[F \mid \mathcal{G}]$ if G depends only on the information contained in $\mathcal{G}$.

(ii) $\mathbb{E}[G \mid \mathcal{G}] = G$ when G depends only on the information contained in $\mathcal{G}$.

(iii) $\mathbb{E}[\mathbb{E}[F \mid \mathcal{H}] \mid \mathcal{G}] = \mathbb{E}[F \mid \mathcal{G}]$ if $\mathcal{G} \subset \mathcal{H}$, called the *tower property*, cf. also Relation (A.20).

(iv) $\mathbb{E}[F \mid \mathcal{G}] = \mathbb{E}[F]$ when F "does not depend" on the information contained in $\mathcal{G}$ or, more precisely stated, when the random variable F is *independent* of the σ-algebra $\mathcal{G}$.

(v) If G depends only on $\mathcal{G}$ and F is independent of $\mathcal{G}$, then

$$\mathbb{E}[h(F,G) \mid \mathcal{G}] = \mathbb{E}[h(x,F)]_{x=G}.$$

When $\mathcal{H} = \{\emptyset, \Omega\}$ is the trivial σ-algebra we have $\mathbb{E}[F \mid \mathcal{H}] = \mathbb{E}[F]$, $F \in L^1(\Omega)$. See (A.20) and (A.21) for illustrations of the tower property by conditioning with respect to discrete and continuous random variables.

Filtrations

The total amount of "information" present in the market at time $t = 0,1,\ldots,N$ is denoted by $\mathcal{F}_t$. We assume that

$$\mathcal{F}_t \subset \mathcal{F}_{t+1}, \qquad t = 0,1,\ldots,N-1,$$

which means that the amount of information available on the market increases over time.

[1]The collection $\mathcal{G}$ is also called a σ-algebra, cf. Section A.

Usually, $\mathcal{F}_t$ corresponds to the knowledge of the values $S_0^{(i)}, \ldots, S_t^{(i)}$, $i = 1, \ldots, d$, of the risky assets up to time t. In mathematical notation we say that $\mathcal{F}_t$ is generated by $S_0^{(i)}, \ldots, S_t^{(i)}$, and we usually write

$$\mathcal{F}_t = \sigma\left(S_0^{(i)}, \ldots, S_t^{(i)}\right), \qquad t = 0, 1, \ldots, N.$$

The notation $\mathcal{F}_t$ is useful to represent a quantity of information available at time t. Note that different agents or traders may work with distinct filtration. For example, an insider will have access to a filtration $(\mathcal{G}_t)_{t=0,1,\ldots,N}$ larger than the filtration $(\mathcal{F}_t)_{t=0,1,\ldots,N}$ available to an ordinary agent, in the sense that

$$\mathcal{F}_t \subset \mathcal{G}_t, \qquad t = 0, 1, \ldots, N.$$

The notation $\mathbb{E}[F|\mathcal{F}_t]$ represents the expected value of a random variable F given (or conditionally to) the information contained in $\mathcal{F}_t$. Again, $\mathbb{E}[F|\mathcal{F}_t]$ denotes the best possible estimate of F in mean square sense, given the information known up to time t.

We will assume that no information is available at time $t = 0$, which translates as

$$\mathbb{E}[F \mid \mathcal{F}_0] = \mathbb{E}[F]$$

for any integrable random variable F.

As above, the conditional expectation with respect to $\mathcal{F}_t$ satisfies the following five properties:

(i) $\mathbb{E}[FG \mid \mathcal{F}_t] = F\,\mathbb{E}[G \mid \mathcal{F}_t]$ if F depends only on the information contained in $\mathcal{F}_t$.

(ii) $\mathbb{E}[F \mid \mathcal{F}_t] = F$ when F depends only on the information known at time t and contained in $\mathcal{F}_t$.

(iii) $\mathbb{E}[\mathbb{E}[F \mid \mathcal{F}_{t+1}] \mid \mathcal{F}_t] = \mathbb{E}[F \mid \mathcal{F}_t]$ if $\mathcal{F}_t \subset \mathcal{F}_{t+1}$ (by the *tower property*, cf. also Relation (6.1) below).

(iv) $\mathbb{E}[F \mid \mathcal{F}_t] = \mathbb{E}[F]$ when F does not depend on the information contained in $\mathcal{F}_t$.

(v) If F depends only on $\mathcal{F}_t$ and G is independent of $\mathcal{F}_t$, then

$$\mathbb{E}[h(F,G) \mid \mathcal{F}_t] = \mathbb{E}[h(x,G)]_{x=F}.$$

Note that by the tower property *(iii)* the process $t \mapsto \mathbb{E}[F \mid \mathcal{F}_t]$ is a martingale, cf. e.g., Relation (6.1) for details.

Martingales

A martingale is a stochastic process whose value at time $t+1$ can be estimated using conditional expectation given its value at time t. Recall that a process $(M_t)_{t=0,1,\ldots,N}$ is said to be $\mathcal{F}_t$-adapted if the value of M_t depends only on the information available at time t in $\mathcal{F}_t$, $t = 0, 1, \ldots, N$.

Definition 2.3 *A stochastic process* $(M_t)_{t=0,1,\ldots,N}$ *is called a discrete time* martingale *with respect to the filtration* $(\mathcal{F}_t)_{t=0,1,\ldots,N}$ *if* $(M_t)_{t=0,1,\ldots,N}$ *is* $\mathcal{F}_t$*-adapted and satisfies the property*

$$\mathbb{E}[M_{t+1}|\mathcal{F}_t] = M_t, \qquad t = 0, 1, \ldots, N-1.$$

Note that the above definition implies that $M_t \in \mathcal{F}_t$, $t = 0, 1, \ldots, N$. In other words, a random process $(M_t)_{t=0,1,\ldots,N}$ is a martingale if the best possible prediction of M_{t+1} in the mean square sense given $\mathcal{F}_t$ is simply M_t.

As an example of the use of martingales we can mention weather forecasting. If M_t denotes the random temperature observed at time t, this process is a martingale when the best possible forecast of tomorrow's temperature M_{t+1} given information known up to time t is just today's temperature M_t, $t = 0, 1, \ldots, N-1$.

In the sequel we will say that a stochastic process $(\xi_k)_{k\geq 0}$ is *predictable* if ξ_k depends only on the information in $\mathcal{F}_{k-1}$, $k \geq 1$. In particular, ξ_0 is a constant.

An important property of martingales is that the martingale transform (2.4) of a predictable process is itself a martingale, see also Proposition 6.1 for the continuous-time analog of the following proposition.

Proposition 2.1 *Given* $(X_t)_{t\in\mathbb{N}}$ *a martingale and* $(\xi_k)_{k\in\mathbb{N}}$ *a square-summable predictable process, the discrete-time process* $(M_t)_{t\in\mathbb{N}}$ *defined by*

$$M_t = \sum_{k=1}^{t} \xi_k (X_k - X_{k-1}), \qquad t \in \mathbb{N}, \tag{2.4}$$

is a martingale.

Proof. Given $n \geq 0$ we have

$$\begin{aligned} \mathbb{E}[M_n \mid \mathcal{F}_t] &= \mathbb{E}\left[\sum_{k=1}^{n} \xi_k (X_k - X_{k-1}) \Big| \mathcal{F}_t\right] \\ &= \mathbb{E}\left[\sum_{k=1}^{n} \mathbb{E}[\xi_k (X_k - X_{k-1}) \mid \mathcal{F}_t]\right] \end{aligned}$$

$$= \sum_{k=1}^{t} \mathbb{E}[\xi_k(X_k - X_{k-1}) \mid \mathcal{F}_t] + \sum_{k=t+1}^{n} \mathbb{E}[\xi_k(X_k - X_{k-1}) \mid \mathcal{F}_t]$$

$$= \sum_{k=1}^{t} \xi_k(X_k - X_{k-1}) + \sum_{k=t+1}^{n} \mathbb{E}[\xi_k(X_k - X_{k-1}) \mid \mathcal{F}_t]$$

$$= M_t + \sum_{k=t+1}^{n} \mathbb{E}[\xi_k(X_k - X_{k-1}) \mid \mathcal{F}_t].$$

To conclude we need to show that

$$\mathbb{E}[\xi_k(X_k - X_{k-1}) \mid \mathcal{F}_t] = 0, \qquad t+1 \leq k \leq n.$$

We note that when $0 \leq t \leq k-1$ we have $\mathcal{F}_t \subset \mathcal{F}_{k-1}$, and by the "tower property" of conditional expectations we get

$$\mathbb{E}[\xi_k(X_k - X_{k-1}) \mid \mathcal{F}_t] = \mathbb{E}[\mathbb{E}[\xi_k(X_k - X_{k-1}) \mid \mathcal{F}_{k-1}] \mid \mathcal{F}_t].$$

In addition, since the process $(\xi_k)_{k\in\mathbb{N}}$ is predictable, ξ_k depends only on the information in $\mathcal{F}_{k-1}$, and using Property (ii) of conditional expectations we may pull out ξ_k out of the expectation since it behaves as a constant parameter given $\mathcal{F}_{k-1}$, $k = 1, \ldots, n$, hence

$$\mathbb{E}[\xi_k(X_k - X_{k-1}) \mid \mathcal{F}_{k-1}] = \xi_k \mathbb{E}[X_k - X_{k-1} \mid \mathcal{F}_{k-1}] = 0$$

because $(X_t)_{t\in\mathbb{N}}$ is a martingale, and more generally,

$$\mathbb{E}[\xi_k(X_k - X_{k-1}) \mid \mathcal{F}_t] = 0,$$

for all $k = t+1, \ldots, n$. This yields

$$\begin{aligned}\mathbb{E}[X_k - X_{k-1} \mid \mathcal{F}_{k-1}] &= \mathbb{E}[X_k \mid \mathcal{F}_{k-1}] - \mathbb{E}[X_{k-1} \mid \mathcal{F}_{k-1}] \\ &= \mathbb{E}[X_k \mid \mathcal{F}_{k-1}] - X_{k-1} \\ &= 0, \qquad k = 1, \ldots, n,\end{aligned}$$

because $(X_t)_{t\in\mathbb{N}}$ is a martingale. □

2.6 Risk-Neutral Probability Measures

As in the two time-step model, the concept of risk neutral measures will be used to price financial claims under the absence of arbitrage hypothesis.

Definition 2.4 *A probability measure* $\mathbb{P}^*$ *on* Ω *is called a risk-neutral measure if under* $\mathbb{P}^*$*, the expected return of each risky asset equals the return* r *of the riskless asset, that is*

$$\mathbb{E}^*[S_{t+1}^{(i)} \mid \mathcal{F}_t] = (1+r)S_t^{(i)}, \qquad t = 0, 1, \ldots, N-1, \tag{2.5}$$

$i = 0, 1, \ldots, d$. *Here,* $\mathbb{E}^*$ *denotes the expectation under* $\mathbb{P}^*$.

Since $S_t^{(i)} \in \mathcal{F}_t$, Relation (2.5) can be rewritten in terms of asset returns as

$$\mathbb{E}^*\left[\frac{S_{t+1}^{(i)} - S_t^{(i)}}{S_t^{(i)}} \Big| \mathcal{F}_t\right] = r, \qquad t = 0, 1, \ldots, N-1.$$

In other words, taking risks under $\mathbb{P}^*$ by buying the risky asset n^o i has a neutral effect, as the expected return is that of the riskless asset. The measure $\mathbb{P}^*$ would represent a risk premium if we had

$$\mathbb{E}^*[S_{t+1}^{(i)} \mid \mathcal{F}_t] = (1+\tilde{r})S_t^{(i)}, \qquad t = 0, 1, \ldots, N-1,$$

with $\tilde{r} > r$.

The definition of risk-neutral probability measure can be reformulated using the notion of martingale.

Proposition 2.2 *A probability measure* $\mathbb{P}^*$ *on* Ω *is a risk-neutral measure if and only if the discounted price process* $X_t^{(i)}$ *is a martingale* under $\mathbb{P}^*$, i.e.,

$$\mathbb{E}^*[X_{t+1}^{(i)} \mid \mathcal{F}_t] = X_t^{(i)}, \qquad t = 0, 1, \ldots, N-1, \tag{2.6}$$

$i = 0, 1, \ldots, d$.

Proof. It suffices to check that Conditions (2.5) and (2.6) are equivalent since

$$\mathbb{E}^*[S_{t+1}^{(i)} \mid \mathcal{F}_t] = (1+r)^{t+1}\,\mathbb{E}^*[X_{t+1}^{(i)} \mid \mathcal{F}_t] \qquad \text{and} \qquad S_t^{(i)} = (1+r)^t X_t^{(i)},$$

$t = 0, 1, \ldots, N-1$, $i = 1, \ldots, d$. □

Next we restate the first fundamental theorem of mathematical finance in discrete time, which can be used to check for the existence of arbitrage opportunities.

Theorem 2.1 *A market is* without *arbitrage opportunity if and only if it admits at least one risk-neutral measure.*

Proof. cf. Theorem 5.17 of [24]. □

2.7 Market Completeness

Definition 2.5 *A contingent claim with payoff C is said to be attainable (at time N) if there exists a portfolio strategy $(\bar{\xi}_t)_{t=1,\ldots,N}$ such that*

$$C = \bar{\xi}_N \cdot \bar{S}_N. \tag{2.7}$$

In case $(\bar{\xi}_t)_{t=1,\ldots,N}$ is a portfolio that attains the claim C at time N, i.e., if (2.7) is satisfied, we also say that $(\bar{\xi}_t)_{t=1,\ldots,N}$ *hedges* the claim C. In case (2.7) is replaced by the condition

$$\bar{\xi}_N \cdot \bar{S}_N \geq C,$$

we talk of super-hedging. When $(\bar{\xi}_t)_{t=1,\ldots,N}$ hedges C, the arbitrage price $\pi_t(C)$ of the claim at time t will be given by the value

$$\pi_t(C) = \bar{\xi}_t \cdot \bar{S}_t$$

of the portfolio at time $t = 0, 1, \ldots, N$. Note that at time $t = N$ we have

$$\pi_N(C) = \bar{\xi}_N \cdot \bar{S}_N = C,$$

i.e., since exercise of the claim occurs at time N, the price $\pi_N(C)$ of the claim equals the value C of the payoff.

Definition 2.6 *A market model is said to be* complete *if every contingent claim is attainable.*

The next result can be viewed as the second fundamental theorem of mathematical finance.

Theorem 2.2 *A market model without arbitrage is complete if and only if it admits only one risk-neutral measure.*

Proof. cf. Theorem 5.38 of [24]. □

2.8 Cox–Ross–Rubinstein (CRR) Market Model

We consider the discrete time Cox–Ross–Rubinstein model [14] with $N + 1$ time instants $t = 0, 1, \ldots, N$ and $d = 1$ risky asset, also called the *binomial model*. The price $S_t^{(0)}$ of the riskless asset evolves as

$$S_t^{(0)} = \pi^{(0)}(1 + r)^t, \qquad t = 0, 1, \ldots, N.$$

Let the *return* of the risky asset $S = S^{(1)}$ be defined as

$$R_t := \frac{S_t - S_{t-1}}{S_{t-1}}, \qquad t = 1, \ldots, N.$$

In the CRR model the return R_t is random and allowed to take only two values a and b at each time step, i.e.,

$$R_t \in \{a, b\}, \qquad t = 1, \ldots, N,$$

with $-1 < a < b$. That means, the evolution of S_{t-1} to S_t is random and given by

$$S_t = \left\{ \begin{array}{ll} (1+b)S_{t-1} & \text{if } R_t = b \\ \\ (1+a)S_{t-1} & \text{if } R_t = a \end{array} \right\} = (1 + R_t)S_{t-1}, \qquad t = 1, \ldots, N,$$

and

$$S_t = S_0 \prod_{j=1}^{t} (1 + R_j), \qquad t = 0, 1, \ldots, N.$$

Note that the price process $(S_t)_{t=0,1,\ldots,N}$ evolves on a binary recombining (or binomial) tree. The discounted asset price is

$$X_t = \frac{S_t}{(1+r)^t}, \qquad t = 0, 1, \ldots, N,$$

with

$$X_t = \left\{ \begin{array}{ll} \dfrac{1+b}{1+r} X_{t-1} & \text{if } R_t = b \\ \\ \dfrac{1+a}{1+r} X_{t-1} & \text{if } R_t = a \end{array} \right\} = \frac{1 + R_t}{1 + r} X_{t-1}, \qquad t = 1, \ldots, N,$$

and

$$X_t = \frac{S_0}{(1+r)^t} \prod_{j=1}^{t} (1 + R_j) = X_0 \prod_{j=1}^{t} \frac{1 + R_j}{1 + r}.$$

In this model the discounted value at time t of the portfolio is given by

$$\bar{\xi}_t \cdot \bar{X}_t = \xi_t^{(0)} \pi_0 + \xi_t^{(1)} X_t, \qquad t = 1, \ldots, N.$$

The information $\mathcal{F}_t$ known in the market up to time t is given by the knowledge of $S_1, \ldots, S_t$, which is equivalent to the knowledge of $X_1, \ldots, X_t$ or $R_1, \ldots, R_t$, i.e., we write

$$\mathcal{F}_t = \sigma(S_1, \ldots, S_t) = \sigma(X_1, \ldots, X_t) = \sigma(R_1, \ldots, R_t), \qquad t = 0, 1, \ldots, N,$$

where as a convention $\mathcal{F}_0 = \{\emptyset, \Omega\}$ contains no information.

Theorem 2.3 *The CRR model is without arbitrage if and only if $a < r < b$. In this case the market is complete.*

Proof. In order to check for arbitrage opportunities we may use Theorem 2.1 and look for a risk-neutral measure $\mathbb{P}^*$. According to the definition of a risk-neutral measure this probability $\mathbb{P}^*$ should satisfy Condition (2.5), i.e.,

$$\mathbb{E}^*[S_{t+1}^{(i)} \mid \mathcal{F}_t] = (1+r)S_t^{(i)}, \qquad t = 0, 1, \ldots, N-1.$$

Rewriting $\mathbb{E}^*[S_{t+1} \mid \mathcal{F}_t]$ as

$$\mathbb{E}^*[S_{t+1} \mid \mathcal{F}_t] = (1+a)S_t\mathbb{P}^*(R_{t+1} = a \mid \mathcal{F}_t) + (1+b)S_t\mathbb{P}^*(R_{t+1} = b \mid \mathcal{F}_t),$$

it follows that any risk-neutral measure $\mathbb{P}^*$ should satisfy the equations

$$\begin{cases} (1+b)S_t\mathbb{P}^*(R_{t+1} = b \mid \mathcal{F}_t) + (1+a)S_t\mathbb{P}^*(R_{t+1} = a \mid \mathcal{F}_t) = (1+r)S_t \\ \mathbb{P}^*(R_{t+1} = b \mid \mathcal{F}_t) + \mathbb{P}^*(R_{t+1} = a \mid \mathcal{F}_t) = 1, \end{cases}$$

i.e.,

$$\begin{cases} b\mathbb{P}^*(R_{t+1} = b \mid \mathcal{F}_t) + a\mathbb{P}^*(R_{t+1} = a \mid \mathcal{F}_t) = r \\ \mathbb{P}^*(R_{t+1} = b \mid \mathcal{F}_t) + \mathbb{P}^*(R_{t+1} = a \mid \mathcal{F}_t) = 1, \end{cases}$$

with solution

$$\mathbb{P}^*(R_{t+1} = b \mid \mathcal{F}_t) = \frac{r-a}{b-a} \qquad \text{and} \qquad \mathbb{P}^*(R_{t+1} = a \mid \mathcal{F}_t) = \frac{b-r}{b-a}. \tag{2.8}$$

Clearly, $\mathbb{P}^*$ can be a non singular probability measure only if $r - a > 0$ and $b - r > 0$. In this case the solution $\mathbb{P}^*$ of the problem is unique hence the market is complete by Theorem 2.2. □

Note that the values of $\mathbb{P}^*(R_{t+1} = b \mid \mathcal{F}_t)$ and $\mathbb{P}^*(R_{t+1} = a \mid \mathcal{F}_t)$ computed in (2.8) are non random, hence they are independent of the information contained in $\mathcal{F}_t$. As a consequence, under $\mathbb{P}^*$, the random variable R_{t+1} is independent of the information $\mathcal{F}_t$ up to time t, which is generated by $R_1, \ldots, R_t$. We deduce that $(R_1, \ldots, R_N)$ form a sequence of independent and identically distributed (i.i.d.) random variables.

In other words, R_{t+1} is independent of $R_1, \ldots, R_t$ for all $t = 1, \ldots, N-1$, the random variables $R_1, \ldots, R_N$ are *independent* under $\mathbb{P}^*$, and by (2.8) we have

$$\mathbb{P}^*(R_{t+1} = b) = \frac{r-a}{b-a} \qquad \text{and} \qquad \mathbb{P}^*(R_{t+1} = a) = \frac{b-r}{b-a}.$$

As a consequence, letting $p^* := (r-a)/(b-a)$, when $(k_1, \ldots, k_n) \in \{a, b\}^{N+1}$ we have

$$\mathbb{P}^*(R_1 = k_1, \ldots, R_N = k_n) = (p^*)^l(1-p^*)^{N-l},$$

where l, resp. $N - l$, denotes the number of times the term "a", resp. "b", appears in the sequence $\{k_1, \ldots, k_N\}$.

Exercises

Exercise 2.1 We consider the discrete-time Cox–Ross–Rubinstein model with $N+1$ time instants $t = 0, 1, \ldots, N$, and the price π_t of the riskless asset evolves as $\pi_t = \pi_0(1+r)^t$, $t = 0, 1, \ldots, N$. The evolution of S_{t-1} to S_t is given by

$$S_t = \begin{cases} (1+b)S_{t-1} \\ \\ (1+a)S_{t-1} \end{cases}$$

with $-1 < a < r < b$. The *return* of the risky asset S is defined as

$$R_t := \frac{S_t - S_{t-1}}{S_{t-1}}, \qquad t = 1, \ldots, N,$$

and $\mathcal{F}_t$ is generated by $R_1, \ldots, R_t$, $t = 1, \ldots, N$.

1. What are the possible values of R_t ?

2. Show that under the probability measure P^* defined by

$$P^*(R_{t+1} = a \mid \mathcal{F}_t) = \frac{b-r}{b-a}, \qquad P^*(R_{t+1} = b \mid \mathcal{F}_t) = \frac{r-a}{b-a},$$

   $t = 0, 1, \ldots, N-1$, the expected return $\mathbb{E}^*[R_{t+1} \mid \mathcal{F}_t]$ of S is equal to the return r of the riskless asset.

3. Show that under P^* the process $(S_t)_{t=0,\ldots,N}$ satisfies

$$\mathbb{E}^*[S_{t+k} \mid \mathcal{F}_t] = (1+r)^k S_t, \qquad t = 0, \ldots, N-k, \quad k = 0, \ldots, N.$$

Chapter 3

Pricing and Hedging in Discrete Time

We consider the pricing and hedging of options in a discrete time financial model with $N+1$ time instants $t = 0, 1, \ldots, N$. Vanilla options are treated using backward induction, and exotic options with arbitrary payoff functions are considered using the Clark–Ocone formula in discrete time.

3.1 Pricing of Contingent Claims

Let us consider an attainable contingent claim with payoff $C \geq 0$ and maturity N. Recall that by the Definition 2.5 of attainability there exists a hedging portfolio strategy $(\xi_t)_{t=1,2,\ldots,N}$ such that

$$\bar{\xi}_N \cdot \bar{S}_N = C \tag{3.1}$$

at time N. Clearly, if (3.1) holds, then investing the amount

$$V_0 = \bar{\xi}_1 \cdot \bar{S}_0 \tag{3.2}$$

at time $t = 0$, resp.

$$V_t = \bar{\xi}_t \cdot \bar{S}_t \tag{3.3}$$

at time $t = 1, \ldots, N$, into a self-financing hedging portfolio will allow one to hedge the option and to obtain the perfect replication (3.1) at time N.

The value (3.2)–(3.3) at time t of a self-financing portfolio strategy $(\xi_t)_{t=1,2,\ldots,N}$ hedging an attainable claim C will be called an *arbitrage price* of the claim C at time t and denoted by $\pi_t(C)$, $t = 0, 1, \ldots, N$.

Next we develop a second approach to the pricing of contingent claims, based on conditional expectations and martingale arguments. We will need the following lemma.

Lemma 3.1 *The following statements are equivalent:*

(i) *The portfolio strategy* $(\bar{\xi}_t)_{t=1,\ldots,N}$ *is self-financing.*

(ii) $\bar{\xi}_t \cdot \bar{X}_t = \bar{\xi}_{t+1} \cdot \bar{X}_t$ *for all* $t = 1, \ldots, N-1$.

(iii) *We have*

$$\widetilde{V}_t = \widetilde{V}_0 + \sum_{j=1}^{t} \bar{\xi}_j \cdot (\bar{X}_j - \bar{X}_{j-1}), \qquad t = 0, 1, \ldots, N. \tag{3.4}$$

Proof. First, the self-financing condition (i)

$$\bar{\xi}_{t-1} \cdot \bar{S}_{t-1} = \bar{\xi}_t \cdot \bar{S}_{t-1}, \qquad t = 1, \ldots, N,$$

is clearly equivalent to (ii) by division of both sides by $(1+r)^{t-1}$.

Next, assuming that (ii) holds we have

$$\begin{aligned} \widetilde{V}_t &= \widetilde{V}_0 + \sum_{j=1}^{t} (\widetilde{V}_j - \widetilde{V}_{j-1}) \\ &= \widetilde{V}_0 + \sum_{j=1}^{t} \bar{\xi}_j \cdot \bar{X}_j - \bar{\xi}_{j-1} \cdot \bar{X}_{j-1} \\ &= \widetilde{V}_0 + \sum_{j=1}^{t} \bar{\xi}_j \cdot \bar{X}_j - \bar{\xi}_j \cdot \bar{X}_{j-1} \\ &= \widetilde{V}_0 + \sum_{j=1}^{t} \bar{\xi}_j \cdot (\bar{X}_j - \bar{X}_{j-1}), \qquad t = 1, \ldots, N. \end{aligned}$$

Finally, assuming that (iii) holds we get

$$\widetilde{V}_t - \widetilde{V}_{t-1} = \bar{\xi}_t \cdot (\bar{X}_t - \bar{X}_{t-1}),$$

hence

$$\bar{\xi}_t \cdot \bar{X}_t - \bar{\xi}_{t-1} \cdot \bar{X}_{t-1} = \bar{\xi}_t \cdot (\bar{X}_t - \bar{X}_{t-1}),$$

and

$$\bar{\xi}_{t-1} \cdot \bar{X}_{t-1} = \bar{\xi}_t \cdot \bar{X}_{t-1}, \qquad t = 1, \ldots, N.$$

□

In Relation (3.4), the term $\bar{\xi}_t \cdot (\bar{X}_t - \bar{X}_{t-1})$ represents the profit and loss

$$\widetilde{V}_t - \widetilde{V}_{t-1} = \bar{\xi}_t \cdot (\bar{X}_t - \bar{X}_{t-1}),$$

of the self-financing portfolio strategy $(\bar{\xi}_j)_{j=1,\ldots,N}$ over the time period $[t-1, t]$, computed by multiplication of the portfolio allocation $\bar{\xi}_t$ with the change of price $\bar{X}_t - \bar{X}_{t-1}$, $t = 1, \ldots, N$.

Relation (3.4) admits a natural interpretation by saying that when a portfolio is self-financing the value $\widetilde{V}_t$ of the (discounted) portfolio at time t is given by summing up the (discounted) profits and losses registered over all time periods from time 0 to time t.

The sum (3.4) is also referred to as a discrete time *stochastic integral* of the portfolio process $(\bar{\xi}_t)_{t=1,\ldots,N}$ with respect to the random process $(\bar{X}_t)_{t=0,1,\ldots,N}$. In particular, it can be shown from (3.4) that $(\widetilde{V}_t)_{t=0,1,\ldots,N}$ is a martingale under $\mathbb{P}^*$ by the martingale transform argument of Proposition 2.1, as in the proof of Theorem 3.1 below.

As a consequence of the above Lemma 3.1, if a contingent claim C with discounted payoff

$$\widetilde{C} := \frac{C}{(1+r)^N}$$

is attainable by a self-financing portfolio strategy $(\bar{\xi}_t)_{t=1,\ldots,N}$ then we have

$$\widetilde{C} = \bar{\xi}_N \cdot \bar{X}_N = \widetilde{V}_N = \widetilde{V}_0 + \sum_{t=1}^{N} \bar{\xi}_t \cdot (\bar{X}_t - \bar{X}_{t-1}). \tag{3.5}$$

Note that in the above formula it is the use of discounted asset price $\bar{X}_t$ that allows us to add up the profits and losses $\bar{\xi}_t \cdot (\bar{X}_t - \bar{X}_{t-1})$ since they are expressed in units of currency "at time 0." In general, \$1 at time $t = 0$ and \$1 at time $t = 1$ cannot be added without proper discounting.

Theorem 3.1 *The arbitrage price $\pi_t(C)$ of a contingent claim C is given by*

$$\pi_t(C) = \frac{1}{(1+r)^{N-t}} \mathbb{E}^*[C \mid \mathcal{F}_t], \qquad t = 0, 1, \ldots, N, \tag{3.6}$$

where $\mathbb{P}^*$ denotes any risk-neutral probability measure.

Proof. Let $\widetilde{C} = C/(1+r)^N$ denote the discounted payoff of the claim C. We will show that under any risk-neutral measure $\mathbb{P}^*$ the discounted value of any self-financing portfolio hedging C is given by

$$\widetilde{V}_t = \mathbb{E}^*\left[\widetilde{C} \mid \mathcal{F}_t\right], \qquad t = 0, 1, \ldots, N, \tag{3.7}$$

which shows that

$$V_t = \frac{1}{(1+r)^{N-t}} \mathbb{E}^*[C \mid \mathcal{F}_t]$$

after multiplication of both sides by $(1+r)^t$. To conclude we will note that the arbitrage price $\pi_t(C)$ of the claim at any time t is by definition equal to the value V_t of the corresponding self-financing portfolio.

We now need to prove (3.7), and for this we will use the martingale transform argument of Proposition 2.1. Since the portfolio strategy $(\xi_t)_{t=1,2,\ldots,N}$ is self-financing, from Lemma 3.1 we have

$$\begin{aligned}
\mathbb{E}^*\left[\widetilde{C} \mid \mathcal{F}_t\right] &= \mathbb{E}^*\left[\widetilde{V}_N \mid \mathcal{F}_t\right] \\
&= \mathbb{E}^*\left[\widetilde{V}_0 + \sum_{j=1}^{N} \bar{\xi}_j \cdot (\bar{X}_j - \bar{X}_{j-1}) \Big| \mathcal{F}_t\right] \\
&= \mathbb{E}^*\left[\widetilde{V}_0 \mid \mathcal{F}_t\right] + \sum_{j=1}^{N} \mathbb{E}^*\left[\bar{\xi}_j \cdot (\bar{X}_j - \bar{X}_{j-1}) \mid \mathcal{F}_t\right] \\
&= \widetilde{V}_0 + \sum_{j=1}^{t} \mathbb{E}^*\left[\bar{\xi}_j \cdot (\bar{X}_j - \bar{X}_{j-1}) \mid \mathcal{F}_t\right] + \sum_{j=t+1}^{N} \mathbb{E}^*\left[\bar{\xi}_j \cdot (\bar{X}_j - \bar{X}_{j-1}) \mid \mathcal{F}_t\right] \\
&= \widetilde{V}_0 + \sum_{j=1}^{t} \bar{\xi}_j \cdot (\bar{X}_j - \bar{X}_{j-1}) + \sum_{j=t+1}^{N} \mathbb{E}^*\left[\bar{\xi}_j \cdot (\bar{X}_j - \bar{X}_{j-1}) \mid \mathcal{F}_t\right] \\
&= \widetilde{V}_t + \sum_{j=t+1}^{N} \mathbb{E}^*\left[\bar{\xi}_j \cdot (\bar{X}_j - \bar{X}_{j-1}) \mid \mathcal{F}_t\right],
\end{aligned}$$

where we used Relation (3.4) of Lemma 3.1. In order to obtain (3.7) we need to show that

$$\sum_{j=t+1}^{N} \mathbb{E}^*\left[\bar{\xi}_j \cdot (\bar{X}_j - \bar{X}_{j-1}) \mid \mathcal{F}_t\right] = 0.$$

Let us show that

$$\mathbb{E}^*\left[\bar{\xi}_j \cdot (\bar{X}_j - \bar{X}_{j-1}) \mid \mathcal{F}_t\right] = 0,$$

for all $j = t+1, \ldots, N$. We have $0 \le t \le j-1$ hence $\mathcal{F}_t \subset \mathcal{F}_{j-1}$, and by the "tower property" of conditional expectations we get

$$\mathbb{E}^*\left[\bar{\xi}_j \cdot (\bar{X}_j - \bar{X}_{j-1}) \mid \mathcal{F}_t\right] = \mathbb{E}^*\left[\mathbb{E}^*\left[\bar{\xi}_j \cdot (\bar{X}_j - \bar{X}_{j-1}) \mid \mathcal{F}_{j-1}\right] \mid \mathcal{F}_t\right],$$

therefore it suffices to show that

$$\mathbb{E}^*\left[\bar{\xi}_j \cdot (\bar{X}_j - \bar{X}_{j-1}) \mid \mathcal{F}_{j-1}\right] = 0.$$

We note that the porfolio allocation $\bar{\xi}_j$ over the time period $[j-1, j]$ is predictable, i.e., it is decided at time $j-1$ and it thus depends only on the information $\mathcal{F}_{j-1}$ known up to time $j-1$, hence

$$\mathbb{E}^*\left[\bar{\xi}_j \cdot (\bar{X}_j - \bar{X}_{j-1}) \mid \mathcal{F}_{j-1}\right] = \bar{\xi}_j \cdot \mathbb{E}^*\left[\bar{X}_j - \bar{X}_{j-1} \mid \mathcal{F}_{j-1}\right].$$

Finally we note that

$$\mathbb{E}^*\left[\bar{X}_j - \bar{X}_{j-1} \mid \mathcal{F}_{j-1}\right] = \mathbb{E}^*\left[\bar{X}_j \mid \mathcal{F}_{j-1}\right] - \mathbb{E}^*\left[\bar{X}_{j-1} \mid \mathcal{F}_{j-1}\right]$$

$$\begin{aligned} &= \mathbb{E}^*\left[\bar{X}_j \mid \mathcal{F}_{j-1}\right] - \bar{X}_{j-1} \\ &= 0, \qquad j = 1, \ldots, N, \end{aligned}$$

because $(\bar{X}_t)_{t=0,1,\ldots,N}$ is a martingale under the risk-neutral measure $\mathbb{P}^*$, and this concludes the proof. □

Note that (3.6) admits an interpretation in an insurance framework, in which $\pi_t(C)$ represents an insurance premium and C represents the random value of an insurance claim made by a subscriber. In this context, the premium of the insurance contract reads as the average of the values (3.6) of the random claims after time discounting. In addition, the discounted price process $((1+r)^{-t}\pi_t(C))_{t=0,1,\ldots,N}$ is a martingale under $\mathbb{P}^*$.

As a consequence of Theorem 3.1, the discounted portfolio process $(\widetilde{V}_t)_{t=0,1,\ldots,N}$ is a martingale under $\mathbb{P}^*$, since

$$\begin{aligned} \mathbb{E}^*\left[\widetilde{V}_{t+1} \mid \mathcal{F}_t\right] &= \mathbb{E}^*\left[\mathbb{E}^*\left[\widetilde{C} \mid \mathcal{F}_{t+1}\right] \mid \mathcal{F}_t\right] \\ &= \mathbb{E}^*\left[\widetilde{C} \mid \mathcal{F}_t\right] \\ &= \widetilde{V}_t, \qquad t = 0, \ldots, N-1, \end{aligned}$$

from the "tower property" of conditional expectations.

In particular for $t = 0$ we obtain the price of the contingent claim C at time 0:

$$\pi_0(C) = \mathbb{E}^*\left[\widetilde{C} \mid \mathcal{F}_0\right] = \mathbb{E}^*\left[\widetilde{C}\right] = \frac{1}{(1+r)^N}\mathbb{E}^*[C].$$

3.2 Hedging of Contingent Claims – Backward Induction

The basic idea of hedging is to allocate assets in a portfolio in order to protect oneself from a given risk. For example, a risk of increasing oil prices can be hedged by buying oil-related stocks, whose value should be positively correlated with the oil price. In this way, a loss connected to increasing oil prices could be compensated by an increase in the value of the corresponding portfolio.

In the setting of this chapter, hedging an attainable contingent claim C means computing a self-financing portfolio strategy $(\bar{\xi}_t)_{t=1,\ldots,N}$ such that $\bar{\xi}_N \cdot \bar{S}_N = C$, i.e.,

$$\bar{\xi}_N \cdot \bar{X}_N = \widetilde{C}, \tag{3.8}$$

by first solving Equation (3.8) for $\bar{\xi}_N$. The idea is then to work by *backward induction* and to compute successively $\bar{\xi}_{N-1}$, $\bar{\xi}_{N-2}$, ..., $\bar{\xi}_4$, down to $\bar{\xi}_3$, $\bar{\xi}_2$, and finally $\bar{\xi}_1$.

In order to implement this algorithm we may use the self-financing condition which yields $N-1$ equations

$$\bar{\xi}_t \cdot \bar{X}_t = \bar{\xi}_{t+1} \cdot \bar{X}_t, \qquad t = 1, \ldots, N-1, \tag{3.9}$$

and allows us in principle to compute the portfolio strategy $(\bar{\xi}_t)_{t=1,\ldots,N}$.

After solving (3.8) for $\bar{\xi}_N$, we then use $\bar{\xi}_N$ to solve the self-financing condition

$$\bar{\xi}_{N-1} \cdot \bar{S}_{N-1} = \bar{\xi}_N \cdot \bar{S}_{N-1}$$

for $\bar{\xi}_{N-1}$, then

$$\bar{\xi}_{N-2} \cdot \bar{S}_{N-2} = \bar{\xi}_{N-1} \cdot \bar{S}_{N-2}$$

for $\bar{\xi}_{N-2}$, and successively $\bar{\xi}_2$ down to $\bar{\xi}_1$.

Then the discounted value $\widetilde{V}_t$ at time t of the portfolio claim can be obtained from

$$\widetilde{V}_0 = \bar{\xi}_1 \cdot \bar{X}_0 \qquad \text{and} \qquad \widetilde{V}_t = \bar{\xi}_t \cdot \bar{X}_t, \qquad t = 1, \ldots, N. \tag{3.10}$$

In the proof of Theorem 3.1 we actually showed that the price $\pi_t(C)$ of the claim at time t coincides with the value V_t of any self-financing portfolio hedging the claim C, i.e.,

$$\pi_t(C) = V_t, \qquad t = 0, 1, \ldots, N,$$

as given by (3.10). In addition, (3.6) shows that

$$V_t = \frac{1}{(1+r)^{N-t}} \mathbb{E}^*[C \mid \mathcal{F}_t], \qquad t = 0, 1, \ldots, N, \tag{3.11}$$

hence the price of the claim can be computed either algebraically by solving (3.8) and (3.9) and then using (3.10), or by a probabilistic method by evaluating the expectation (3.11).

3.3 Pricing of Vanilla Options in the CRR Model

In this section we consider the pricing of contingent claims in the discrete time Cox-Ross-Rubinstein model, with $d = 1$. More precisely we are concerned with vanilla options whose payoffs depend on the terminal value of the underlying asset, as opposed to exotic options whose payoffs may depend on the whole path of the underlying asset price until expiration time.

Recall that the portfolio value process $(V_t)_{t=0,1,\ldots N}$ and the discounted portfolio value process respectively satisfy

$$V_t = \bar{\xi}_t \cdot \bar{S}_t \qquad \text{and} \qquad \widetilde{V}_t = \frac{1}{(1+r)^t} V_t = \bar{\xi}_t \cdot \bar{X}_t, \qquad t = 1, 2, \ldots, N.$$

Here we will be concerned with the pricing of vanilla options with payoffs of the form

$$C = f(S_N),$$

e.g., $f(x) = (x - K)^+$ in the case of a European call. Equivalently, the discounted claim

$$\widetilde{C} = \frac{C}{(1+r)^N}$$

satisfies $\widetilde{C} = \widetilde{f}(S_N)$ with $\widetilde{f}(x) = f(x)/(1+r)^N$, i.e., $\widetilde{f}(x) = \frac{1}{(1+r)^N}(x - K)^+$ in the case of a European call with strike K.

From Theorem 3.1, the discounted value of a portfolio hedging the attainable (discounted) claim $\widetilde{C}$ is given by

$$\widetilde{V}_t = \mathbb{E}^*\left[\widetilde{f}(S_N) \mid \mathcal{F}_t\right], \qquad t = 0, 1, \ldots, N,$$

under the risk-neutral measure $\mathbb{P}^*$. Equivalently, the arbitrage price $\pi_t(C)$ of the contingent claim $C = f(S_N)$ is given by

$$\pi_t(C) = \frac{1}{(1+r)^{N-t}} \mathbb{E}^*[f(S_N) \mid \mathcal{F}_t], \qquad t = 0, 1, \ldots, N. \tag{3.12}$$

In the next proposition we implement the calculation of (3.12).

Proposition 3.1 *The price $\pi_t(C)$ of the contingent claim $C = f(S_N)$ satisfies*

$$\pi_t(C) = v(t, S_t), \qquad t = 0, 1, \ldots, N,$$

where the function $v(t, x)$ is given by

$$v(t,x) = \frac{1}{(1+r)^{N-t}} \mathbb{E}^*\left[f\left(x \prod_{j=t+1}^{N}(1+R_j)\right)\right] \tag{3.13}$$

$$= \frac{1}{(1+r)^{N-t}} \sum_{j=0}^{N-t} \binom{N-t}{j} (p^*)^j (1-p^*)^{N-t-j} f\left(x\,(1+b)^j\,(1+a)^{N-t-j}\right).$$

Proof. From the relations

$$S_N = S_t \prod_{j=t+1}^{N} (1 + R_j),$$

and (3.12) we have, using Property (v) of the conditional expectation and the independence of the returns $\{R_1, \ldots, R_t\}$ and $\{R_{t+1}, \ldots, R_N\}$,

$$\begin{aligned}\pi_t(C) &= \frac{1}{(1+r)^{N-t}} \mathbb{E}^*[f(S_N) \mid \mathcal{F}_t] \\ &= \frac{1}{(1+r)^{N-t}} \mathbb{E}^*\left[f\left(S_t \prod_{j=t+1}^{N} (1+R_j) \right) \Big| \mathcal{F}_t \right] \\ &= \frac{1}{(1+r)^{N-t}} \mathbb{E}^*\left[f\left(x \prod_{j=t+1}^{N} (1+R_j) \right) \right]_{x=S_t}.\end{aligned}$$

Next we note that the number of times R_j is equal to b for $j \in \{t+1, \ldots, N\}$, has a binomial distribution with parameter $(N-t, p^*)$, where

$$p^* = \frac{r-a}{b-a} \quad \text{and} \quad 1 - p^* = \frac{b-r}{b-a}, \tag{3.14}$$

since the set of paths from time $t+1$ to time N containing j times "$(1+b)$" has cardinal $\binom{N-t}{j}$ and each such path has the probability $(p^*)^j (1-p^*)^{N-t-j}$, $j = 0, \ldots, N-t$. Hence we have

$$\begin{aligned}\pi_t(C) &= \frac{1}{(1+r)^{N-t}} \mathbb{E}^*[f(S_N) \mid \mathcal{F}_t] \\ &= \frac{1}{(1+r)^{N-t}} \sum_{j=0}^{N-t} \binom{N-t}{j} (p^*)^j (1-p^*)^{N-t-j} f\left(S_t (1+b)^j (1+a)^{N-t-j} \right).\end{aligned}$$

□

In the above proof we have also shown that $\pi_t(C)$ is given by the conditional expectation

$$\pi_t(C) = \frac{1}{(1+r)^{N-t}} \mathbb{E}^*[f(S_N) \mid S_t]$$

given the value of S_t at time $t = 0, 1, \ldots, N$, i.e., the price of the claim C is written as the average (path integral) of the values of the contingent claim over all possible paths starting from S_t.

The discounted price $\widetilde{V}_t$ of the portfolio can also be computed via a *backward induction* procedure. Namely, by the "tower property" of conditional expectations we have

$$\widetilde{V}_t = \widetilde{v}(t, S_t)$$

$$
\begin{aligned}
&= \mathbb{E}^*\left[\widetilde{f}(S_N) \mid \mathcal{F}_t\right] \\
&= \mathbb{E}^*\left[\mathbb{E}^*\left[\widetilde{f}(S_N) \mid \mathcal{F}_{t+1}\right] \mid \mathcal{F}_t\right] \\
&= \mathbb{E}^*\left[\widetilde{V}_{t+1} \mid \mathcal{F}_t\right] \\
&= \mathbb{E}^*\left[\widetilde{v}(t+1, S_{t+1}) \mid \mathcal{F}_t\right] \\
&= \widetilde{v}\,(t+1,(1+a)S_t)\,\mathbb{P}^*(R_{t+1}=a) + \widetilde{v}\,(t+1,(1+b)S_t)\,\mathbb{P}^*(R_{t+1}=b) \\
&= (1-p^*)\widetilde{v}\,(t+1,(1+a)S_t) + p^*\widetilde{v}\,(t+1,(1+b)S_t)\,,
\end{aligned}
$$

which shows that $\widetilde{v}(t,x)$ satisfies the induction relation

$$
\widetilde{v}(t,x) = (1-p^*)\widetilde{v}\,(t+1, x(1+a)) + p^*\widetilde{v}\,(t+1, x(1+b))\,,
$$

while the terminal condition $\widetilde{V}_N = \widetilde{f}(S_N)$ implies

$$
\widetilde{v}(N,x) = \widetilde{f}(x).
$$

3.4 Hedging of Vanilla Options in the CRR model

In this section we consider the hedging of contingent claims in the discrete time Cox–Ross–Rubinstein model. Our aim is to compute a self-financing portfolio strategy hedging a vanilla option with payoff of the form

$$
C = f(S_N).
$$

Proposition 3.2 *The replicating portfolio $(\xi_t^{(0)}, \xi_t^{(1)})_{t=1,\ldots,N}$ hedging the contingent claim $C = f(S_N)$ is given by*

$$
\xi_t^{(1)} = \frac{\widetilde{v}\,(t,(1+b)S_{t-1}) - \widetilde{v}\,(t,(1+a)S_{t-1})}{X_{t-1}(b-a)/(1+r)}, \qquad t = 1,\ldots,N,
$$

and

$$
\xi_t^{(0)} = \frac{\widetilde{v}(t-1, S_{t-1}) - \xi_t^{(1)} X_{t-1}}{\pi^{(0)}}, \qquad t = 1,\ldots,N,
$$

where the function $\widetilde{v}(t,x) = (1+r)^{-t} v(t,x)$ is given by (3.13).

Proof. Recall that by Lemma 3.1 the following statements are equivalent:

(i) The portfolio strategy $(\bar{\xi}_t)_{t=1,\ldots,N}$ is self-financing.

(ii) We have

$$
\widetilde{V}_t = \widetilde{V}_0 + \sum_{j=1}^{t} \bar{\xi}_j \cdot (\bar{X}_j - \bar{X}_{j-1}), \qquad t = 1,\ldots,N.
$$

As a consequence, any self-financing hedging strategy $(\bar{\xi}_t)_{t=1,\ldots,N}$ should satisfy

$$\widetilde{v}(t, S_t) - \widetilde{v}(t-1, S_{t-1}) = \widetilde{V}_t - \widetilde{V}_{t-1} = \bar{\xi}_t \cdot (\bar{X}_t - \bar{X}_{t-1}).$$

Note that since the discounted price $X_t^{(0)}$ of the riskless asset satisfies

$$X_t^{(0)} = (1+r)^{-t} S_t^{(0)} = \pi^{(0)},$$

we have

$$\begin{aligned}
\bar{\xi}_t \cdot (\bar{X}_t - \bar{X}_{t-1}) &= \xi_t^{(0)}(X_t^{(0)} - X_{t-1}^{(0)}) + \xi_t^{(1)}(X_t^{(1)} - X_{t-1}^{(1)}) \\
&= \xi_t^{(0)}(\pi^{(0)} - \pi^{(0)}) + \xi_t^{(1)}(X_t^{(1)} - X_{t-1}^{(1)}) \\
&= \xi_t^{(1)}(X_t^{(1)} - X_{t-1}^{(1)}) \\
&= \xi_t^{(1)}(X_t - X_{t-1}), \qquad t = 1, \ldots, N.
\end{aligned}$$

Hence we have

$$\widetilde{v}(t, S_t) - \widetilde{v}(t-1, S_{t-1}) = \xi_t^{(1)}(X_t - X_{t-1}), \qquad t = 1, \ldots, N,$$

and from this we deduce the two equations

$$\begin{cases}
\widetilde{v}\,(t, (1+a)S_{t-1}) - \widetilde{v}(t-1, S_{t-1}) = \xi_t^{(1)} \left(\dfrac{1+a}{1+r} X_{t-1} - X_{t-1} \right), \\
\\
\widetilde{v}\,(t, (1+b)S_{t-1}) - \widetilde{v}(t-1, S_{t-1}) = \xi_t^{(1)} \left(\dfrac{1+b}{1+r} X_{t-1} - X_{t-1} \right),
\end{cases}$$

$t = 1, \ldots, N$, i.e.,

$$\begin{cases}
\widetilde{v}\,(t, (1+a)S_{t-1}) - \widetilde{v}(t-1, S_{t-1}) = \xi_t^{(1)} \dfrac{a-r}{1+r} X_{t-1}, \\
\\
\widetilde{v}\,(t, (1+b)S_{t-1}) - \widetilde{v}(t-1, S_{t-1}) = \xi_t^{(1)} \dfrac{b-r}{1+r} X_{t-1}, \qquad t = 1, \ldots, N,
\end{cases}$$

hence

$$\xi_t^{(1)} = \frac{\widetilde{v}\,(t, (1+a)S_{t-1}) - \widetilde{v}\,(t-1, S_{t-1})}{X_{t-1}(a-r)/(1+r)}, \qquad t = 1, \ldots, N,$$

and

$$\xi_t^{(1)} = \frac{\widetilde{v}\,(t, (1+b)S_{t-1}) - \widetilde{v}\,(t-1, S_{t-1})}{X_{t-1}(b-r)/(1+r)}, \qquad t = 1, \ldots, N.$$

From the obvious relation

$$\xi_t^{(1)} = \frac{b-r}{b-a} \xi_t^{(1)} - \frac{a-r}{b-a} \xi_t^{(1)},$$

we get

$$\xi_t^{(1)} = \frac{\widetilde{v}\,(t,(1+b)S_{t-1}) - \widetilde{v}\,(t,(1+a)S_{t-1})}{X_{t-1}(b-a)/(1+r)}, \qquad t = 1,\ldots,N,$$

which only depends on S_{t-1} as expected. This is consistent with the fact that $\xi_t^{(1)}$ represents the (possibly fractional) quantity of the risky asset to be present in the portfolio over the time period $[t-1,t]$ in order to hedge the claim C at time N, and is decided at time $t-1$.

Concerning the quantity $\xi_t^{(0)}$ of the riskless asset in the portfolio at time t, recall that we have

$$\widetilde{V}_t = \bar{\xi}_t \cdot \bar{X}_t = \xi_t^{(0)} X_t^{(0)} + \xi_t^{(1)} X_t^{(1)}, \qquad t = 1,\ldots,N,$$

hence

$$\begin{aligned}
\xi_t^{(0)} &= \frac{\widetilde{V}_t - \xi_t^{(1)} X_t^{(1)}}{X_t^{(0)}} \\
&= \frac{\widetilde{V}_t - \xi_t^{(1)} X_t^{(1)}}{\pi^{(0)}} \\
&= \frac{\widetilde{v}(t,S_t) - \xi_t^{(1)} X_t^{(1)}}{\pi^{(0)}}, \qquad t = 1,\ldots,N.
\end{aligned}$$

Note that we have

$$\begin{aligned}
\xi_t^{(0)} &= \frac{\widetilde{v}(t-1,S_{t-1}) + (\widetilde{v}(t,S_t) - \widetilde{v}(t-1,S_{t-1})) - \xi_t^{(1)} X_t^{(1)}}{\pi^{(0)}} \\
&= \frac{\widetilde{v}(t-1,S_{t-1}) + \xi_t^{(1)}(X_t - X_{t-1}) - \xi_t^{(1)} X_t^{(1)}}{\pi^{(0)}} \\
&= \frac{\widetilde{v}(t-1,S_{t-1}) - \xi_t^{(1)} X_{t-1}}{\pi^{(0)}}, \qquad t = 1,\ldots,N.
\end{aligned}$$

□

Hence the discounted amount $\xi_t^{(0)}\pi^{(0)}$ invested on the riskless asset is

$$\xi_t^{(0)}\pi^{(0)} = \widetilde{v}(t-1,S_{t-1}) - \frac{\widetilde{v}\,(t,(1+b)S_{t-1}) - \widetilde{v}\,(t,(1+a)S_{t-1})}{(b-a)/(1+r)}, \qquad (3.15)$$

$t = 1,\ldots,N$, and we recover the fact that $\xi_t^{(0)}$ depends only on S_{t-1} and not on S_t.

Using the relation

$$v(t-1,S_{t-1}) = (1+r)^{t-1}\widetilde{v}(t-1,S_{t-1})$$

the amount (3.15) can be rewritten without discount as

$$\begin{aligned}\xi_t^{(0)} S_t^{(0)} &= (1+r)^t \xi_t^{(0)} \pi^{(0)} \\ &= (1+r)^t \widetilde{v}(t-1, S_{t-1}) - (1+r)^t \frac{\widetilde{v}(t,(1+b)S_{t-1}) - \widetilde{v}(t,(1+a)S_{t-1})}{(b-a)/(1+r)} \\ &= (1+r)v(t-1, S_{t-1}) - (1+r)\frac{v(t,(1+b)S_{t-1}) - v(t,(1+a)S_{t-1})}{b-a} \\ &= \frac{1+r}{b-a}\left((b-a)v(t-1, S_{t-1}) - v(t,(1+b)S_{t-1}) + v(t,(1+a)S_{t-1})\right),\end{aligned}$$

$t = 1, \ldots, N$. Recall that this portfolio strategy $(\xi_t^{(0)}, \xi_t^{(1)})_{t=1,\ldots,N}$ hedges the claim $C = f(S_N)$, i.e., at time N we have

$$V_N = f(S_N),$$

and it is self-financing by Lemma 3.1.

3.5 Hedging of Exotic Options in the CRR Model

In this section we take $p = p^*$ given by (3.14) and we consider the hedging of path dependent options. Here we choose to use the finite difference gradient and the discrete Clark–Ocone formula of stochastic analysis, see also [24], [44], [54], Chapter 1 of [55], [64], or §15-1 of [73]. See [50] and Section 8.2 of [55] for a similar approach in continuous time. Given

$$\omega = (\omega_1, \ldots, \omega_N) \in \Omega = \{-1, 1\}^N,$$

and $k \in \{1, 2, \ldots, N\}$, let

$$\omega_+^k = (\omega_1, \ldots, \omega_{k-1}, +1, \omega_{k+1}, \ldots, \omega_N)$$

and

$$\omega_-^k = (\omega_1, \ldots, \omega_{k-1}, -1, \omega_{k+1}, \ldots, \omega_N).$$

We also assume that the return $R_t(\omega)$ is constructed as

$$R_t(\omega_+^t) = b \quad \text{and} \quad R_t(\omega_-^t) = a, \qquad t = 1, \ldots, N, \quad \omega \in \Omega.$$

Definition 3.1 *The operator D_t is defined on any random variable F and $t \geq 1$ by*

$$D_t F(\omega) = F(\omega_+^t) - F(\omega_-^t), \qquad t = 1, \ldots, N. \tag{3.16}$$

Recall the following predictable representation formula for the functionals of the binomial process.

Definition 3.2 *Let the centered and normalized return Y_t be defined by*

$$Y_t := \frac{R_t - r}{b-a} = \begin{cases} \dfrac{b-r}{b-a} = q, & \omega_t = +1, \\ \\ \dfrac{a-r}{b-a} = -p, & \omega_t = -1, \end{cases} \qquad t = 1, \ldots, N.$$

Note that under the risk-neutral measure $\mathbb{P}^*$ we have

$$\begin{aligned} \mathbb{E}^*[Y_t] &= \mathbb{E}^*\left[\frac{R_t - r}{b-a}\right] \\ &= \frac{a-r}{b-a}\mathbb{P}^*(R_t = a) + \frac{b-r}{b-a}\mathbb{P}^*(R_t = b) \\ &= \frac{a-r}{b-a}\frac{b-r}{b-a} + \frac{b-r}{b-a}\frac{r-a}{b-a} \\ &= 0, \end{aligned}$$

and

$$\mathrm{Var}\,[Y_t] = pq^2 + qp^2 = pq, \qquad t = 1, \ldots, N.$$

In addition the discounted asset price increment reads

$$\begin{aligned} X_t - X_{t-1} &= X_{t-1}\frac{1+R_t}{1+r} - X_{t-1} \\ &= \frac{1}{1+r}X_{t-1}(R_t - r) \\ &= \frac{b-a}{1+r}Y_t X_{t-1}, \qquad t = 1, \ldots, N. \end{aligned}$$

We also have

$$D_t Y_t = \frac{b-r}{b-a} + \frac{r-a}{b-a} = 1, \qquad t = 1, \ldots, N,$$

and

$$\begin{aligned} D_k S_N &= S_0(1+b)\prod_{\substack{t=1 \\ t \neq k}}^N (1+R_t) - S_0(1+a)\prod_{\substack{t=1 \\ t \neq k}}^N (1+R_t) \\ &= S_0(b-a)\prod_{\substack{t=1 \\ t \neq k}}^N (1+R_t) \\ &= S_0\frac{b-a}{1+R_k}\prod_{t=1}^N (1+R_t) \\ &= \frac{b-a}{1+R_k}S_N, \qquad k = 1, \ldots, N. \end{aligned}$$

The next proposition is the Clark–Ocone predictable representation formula in discrete time, cf. e.g., [55], Proposition 1.7.1.

Proposition 3.3 *For any square-integrable random variables F on Ω we have*

$$F = \mathbb{E}^*[F] + \sum_{k=1}^{\infty} \mathbb{E}^*[D_k F | \mathcal{F}_{k-1}] Y_k. \tag{3.17}$$

The Clark–Ocone formula has the following consequence.

Corollary 3.1 *Assume that $(M_k)_{k\in\mathbb{N}}$ is a square-integrable $\mathcal{F}_t$-martingale. Then we have*

$$M_N = \mathbb{E}^*[M_N] + \sum_{k=1}^{N} Y_k D_k M_k, \qquad N \geq 0.$$

Proof. We have

$$\begin{aligned} M_N &= \mathbb{E}^*[M_N] + \sum_{k=1}^{\infty} \mathbb{E}^*[D_k M_N | \mathcal{F}_{k-1}] Y_k \\ &= \mathbb{E}^*[M_N] + \sum_{k=1}^{\infty} D_k \, \mathbb{E}^*[M_N | \mathcal{F}_k] Y_k \\ &= \mathbb{E}^*[M_N] + \sum_{k=1}^{\infty} Y_k D_k M_k \\ &= \mathbb{E}^*[M_N] + \sum_{k=1}^{N} Y_k D_k M_k. \end{aligned}$$

□

In addition to the Clark–Ocone formula we also state a discrete-time analog of Itô's change of variable formula, which can be useful for option hedging. The next result extends Proposition 1.13.1 of [55] by removing the unnecessary martingale requirement on $(M_t)_{n\in\mathbb{N}}$.

Proposition 3.4 *Let $(Z_n)_{n\in\mathbb{N}}$ be an $\mathcal{F}_n$-adapted process and let $f : \mathbb{R}\times\mathbb{N} \to \mathbb{R}$ be a given function. We have*

$$\begin{aligned} f(Z_t, t) &= f(Z_0, 0) + \sum_{k=1}^{t} D_k f(Z_k, k) Y_k \\ &\quad + \sum_{k=1}^{t} (\mathbb{E}^*[f(Z_k, k) | \mathcal{F}_{k-1}] - f(Z_{k-1}, k-1)). \end{aligned} \tag{3.18}$$

Proof. First, we note that the process

$$t \longmapsto f(Z_t, t) - \sum_{k=1}^{t} (\mathbb{E}^*[f(Z_k, k) | \mathcal{F}_{k-1}] - f(Z_{k-1}, k-1))$$

is a martingale under $\mathbb{P}^*$. Indeed we have

$$
\begin{aligned}
& \mathbb{E}^*\left[f(Z_t,t) - \sum_{k=1}^t (\mathbb{E}^*[f(Z_k,k)|\mathcal{F}_{k-1}] - f(Z_{k-1},k-1)) \Big| \mathcal{F}_{t-1} \right] \\
&= \mathbb{E}^*[f(Z_t,t)|\mathcal{F}_{t-1}] \\
&\quad - \sum_{k=1}^t (\mathbb{E}^*[\mathbb{E}^*[f(Z_k,k)|\mathcal{F}_{k-1}]|\mathcal{F}_{t-1}] - \mathbb{E}^*[\mathbb{E}^*[f(Z_{k-1},k-1)|\mathcal{F}_{k-1}]|\mathcal{F}_{t-1}]) \\
&= \mathbb{E}^*[f(Z_t,t)|\mathcal{F}_{t-1}] - \sum_{k=1}^t (\mathbb{E}^*[f(Z_k,k)|\mathcal{F}_{k-1}] - f(Z_{k-1},k-1)) \\
&= f(Z_{t-1},t-1) - \sum_{k=1}^{t-1} (\mathbb{E}^*[f(Z_k,k)|\mathcal{F}_{k-1}] - f(Z_{k-1},k-1)), \qquad t \geq 1.
\end{aligned}
$$

□

Note that if $(Z_t)_{t\in\mathbb{N}}$ is a martingale in $L^2(\Omega)$ with respect to $(\mathcal{F}_t)_{t\in\mathbb{N}}$ and written as

$$
Z_t = Z_0 + \sum_{k=1}^t u_k Y_k, \qquad t \in \mathbb{N},
$$

where $(u_t)_{t\in\mathbb{N}}$ is a predictable process locally in $L^2(\Omega\times\mathbb{N})$, (i.e., $u(\cdot)\mathbf{1}_{[0,N]}(\cdot) \in L^2(\Omega \times \mathbb{N})$ for all $N > 0$), then we have

$$
D_t f(Z_t,t) = f(Z_{t-1} + qu_t, t) - f(Z_{t-1} - pu_t, t), \tag{3.19}
$$

$t = 1, \ldots, N$. On the other hand, the term

$$
\mathbb{E}[f(Z_t,t) - f(Z_{t-1},t-1)|\mathcal{F}_{t-1}]
$$

is analog to the finite variation part in the continuous time Itô formula, and can be written as

$$
pf(Z_{t-1} + qu_t, t) + qf(Z_{t-1} - pu_t, t) - f(Z_{t-1}, t-1).
$$

Naturally, if $(f(Z_t,t))_{t\in\mathbb{N}}$ is a martingale we recover the decomposition

$$
\begin{aligned}
f(Z_t,t) &= f(Z_0,0) \\
&\quad + \sum_{k=1}^t (f(Z_{k-1} + qu_k, k) - f(Z_{k-1} - pu_k, k)) Y_k \\
&= f(Z_0,0) + \sum_{k=1}^t Y_k D_k f(Z_k,k).
\end{aligned} \tag{3.20}
$$

This identity follows from Corollary 3.1 as well as from Proposition 3.3. In this case the Clark–Ocone formula (3.17) and the change of variable formula (3.20) both coincide and we have in particular

$$
D_k f(Z_k,k) = \mathbb{E}[D_k f(Z_N,N)|\mathcal{F}_{k-1}],
$$

$k = 1, \ldots, N$. For example this recovers the martingale representation

$$\begin{aligned} X_t &= S_0 + \sum_{k=1}^{t} Y_k D_k X_k \\ &= S_0 + \frac{b-a}{1+r} \sum_{k=1}^{t} X_{k-1} Y_k \\ &= S_0 + \sum_{k=1}^{t} X_{k-1} \frac{R_k - r}{1+r} \\ &= S_0 + \sum_{k=1}^{t} (X_k - X_{k-1}), \end{aligned}$$

of the discounted asset price.

Our goal is to hedge an arbitrary claim C on Ω, i.e., given an $\mathcal{F}_N$-measurable random variable C we search for a portfolio $(\xi_t, \eta_t)_{t=1,\ldots,N}$ such that the equality

$$C = V_N = \eta_N A_N + \xi_N S_N \tag{3.21}$$

holds, where $A_N = A_0(1+r)^N$ denotes the value of the riskless asset at time $N \in \mathbb{N}$.

The next proposition is the main result of this section, and provides a solution to the hedging problem under the constraint (3.21).

Proposition 3.5 *Given C a contingent claim, let*

$$\xi_t = (1+r)^{-(N-t)} \frac{1}{S_{t-1}(b-a)} \mathbb{E}^*[D_t C | \mathcal{F}_{t-1}], \tag{3.22}$$

$t = 1 \ldots, N$, and

$$\eta_t = \frac{1}{A_t} \left((1+r)^{-(N-t)} \mathbb{E}^*[C|\mathcal{F}_t] - \xi_t S_t \right), \tag{3.23}$$

$t = 1 \ldots, N$. Then the portfolio $(\xi_t, \eta_t)_{t=1\ldots,N}$ is self financing and satisfies

$$V_t = \eta_t A_t + \xi_t S_t = (1+r)^{-(N-t)} \mathbb{E}^*[C|\mathcal{F}_t], \qquad t = 1 \ldots, N,$$

in particular we have $V_N = C$, hence $(\xi_t, \eta_t)_{t=1\ldots,N}$ is a hedging strategy leading to C.

Proof. Let $(\xi_t)_{t=1\ldots,N}$ be defined by (3.22), and consider the process $(\eta_t)_{t=0,1\ldots,N}$ defined by

$$\eta_0 = (1+r)^{-N} \frac{\mathbb{E}^*[C]}{S_0} \quad \text{and} \quad \eta_{t+1} = \eta_t - \frac{(\xi_{t+1} - \xi_t) S_t}{A_t},$$

$t = 0, \ldots, N-1$. Then $(\xi_t, \eta_t)_{t=1,\ldots,N}$ satisfies the self-financing condition

$$A_t(\eta_{t+1} - \eta_t) + S_t(\xi_{t+1} - \xi_t) = 0, \qquad t = 1, \ldots, N-1.$$

Let now

$$V_0 = \mathbb{E}^*[C](1+r)^{-N}, \quad \text{and} \quad V_t = \eta_t A_t + \xi_t S_t, \qquad t = 1, \ldots, N,$$

and

$$\widetilde{V}_t = V_t(1+r)^{-t} \qquad t = 0, \ldots, N.$$

Since $(\xi_t, \eta_t)_{t=1,\ldots,N}$ is self-financing, by Lemma 3.1 we have

$$\widetilde{V}_t = \widetilde{V}_0 + (b-a)\sum_{k=1}^{t} Y_k \xi_k S_{k-1}(1+r)^{-k}, \tag{3.24}$$

$t = 1, \ldots, N$. On the other hand, from the Clark–Ocone formula (3.17) and the definition of $(\xi_t)_{t=1,\ldots,N}$ we have

$$\begin{aligned}
&(1+r)^{-N}\,\mathbb{E}^*[C|\mathcal{F}_t] \\
&= \mathbb{E}^*\left[\mathbb{E}^*[C](1+r)^{-N} + \sum_{i=0}^{N} \mathbb{E}^*[D_i C|\mathcal{F}_{i-1}](1+r)^{-N}\Big|\mathcal{F}_t\right] Y_i \\
&= \mathbb{E}^*[C](1+r)^{-N} + \sum_{i=0}^{t} \mathbb{E}^*[D_i C|\mathcal{F}_{i-1}](1+r)^{-N} Y_i \\
&= \mathbb{E}^*[C](1+r)^{-N} + (b-a)\sum_{i=0}^{t} \xi_i S_{i-1}(1+r)^{-i} Y_i \\
&= \widetilde{V}_t
\end{aligned}$$

from (3.24). Hence

$$\widetilde{V}_t = (1+r)^{-N}\,\mathbb{E}^*[C|\mathcal{F}_t], \qquad t = 0, 1, \ldots, N,$$

and

$$V_t = (1+r)^{-(N-t)}\,\mathbb{E}^*[C|\mathcal{F}_t], \qquad t = 0, 1, \ldots, N. \tag{3.25}$$

In particular, (3.25) shows that we have $V_N = C$. To conclude the proof we note that from the relation $V_t = \eta_t A_t + \xi_t S_t$, $t = 1, \ldots, N$, the process $(\eta_t)_{t=1,\ldots,N}$ coincides with $(\eta_t)_{t=1,\ldots,N}$ defined by (3.23). □

From Proposition 3.1, when $C = f(S_N)$, the price $\pi_t(C)$ of the contingent claim $C = f(S_N)$ is given by

$$\pi_t(C) = v(t, S_t),$$

where the function $v(t, x)$ is given by

$$v(t, S_t) = \frac{1}{(1+r)^{N-t}}\,\mathbb{E}^*[C|\mathcal{F}_t] = \frac{1}{(1+r)^{N-t}}\,\mathbb{E}^*\left[f\left(x\prod_{j=t+1}^{N}(1+R_j)\right)\right]_{x=S_t}.$$

Note that in this case we have $C = v(N, S_N)$, $\mathbb{E}[C] = v(0, M_0)$, and the discounted claim payoff $\widetilde{C} = (1+r)^{-N}C = \widetilde{v}(N, S_N)$ satisfies

$$
\begin{aligned}
\widetilde{C} &= \mathbb{E}\left[\widetilde{C}\right] + \sum_{t=1}^{n} Y_t \, \mathbb{E}\left[D_t \widetilde{v}(N, S_N) | \mathcal{F}_{t-1}\right] \\
&= \mathbb{E}\left[\widetilde{C}\right] + \sum_{t=1}^{n} Y_t D_t \widetilde{v}(t, S_t) \\
&= \mathbb{E}\left[\widetilde{C}\right] + \sum_{t=1}^{n} (1+r)^{-t} Y_t D_t v(t, S_t) \\
&= \mathbb{E}\left[\widetilde{C}\right] + \sum_{t=1}^{n} Y_t D_t \, \mathbb{E}\left[\widetilde{v}(N, S_N) | \mathcal{F}_t\right] \\
&= \mathbb{E}\left[\widetilde{C}\right] + (1+r)^{-N} \sum_{t=1}^{n} Y_t D_t \, \mathbb{E}[C | \mathcal{F}_t],
\end{aligned}
$$

hence we have

$$
\mathbb{E}[D_t v(N, S_N) | \mathcal{F}_{t-1}] = (1+r)^{N-t} D_t v(t, S_t), \qquad t = 1, \ldots, N,
$$

and by Proposition 3.5 the hedging strategy for $C = f(S_N)$ is given by

$$
\begin{aligned}
\xi_t &= \frac{(1+r)^{-(N-t)}}{S_{t-1}(b-a)} \mathbb{E}[D_t v(N, S_N) | \mathcal{F}_{t-1}] \\
&= \frac{1}{S_{t-1}(b-a)} D_t v(t, S_t) \\
&= \frac{1}{S_{t-1}(b-a)} \left(v\left(t, S_{t-1}(1+b)\right) - v\left(t, S_{t-1}(1+a)\right)\right) \\
&= \frac{1}{X_{t-1}(b-a)/(1+r)} \left(\widetilde{v}\left(t, S_{t-1}(1+b)\right) - \widetilde{v}\left(t, S_{t-1}(1+a)\right)\right),
\end{aligned}
$$

$t = 1, \ldots, N$, which recovers Proposition 3.2 as a particular case. Note that ξ_t is non-negative (i.e., there is no short-selling) when f is a non decreasing function, because $a < b$. This is in particular true in the case of a European call option for which we have $f(x) = (x - K)^+$.

3.6 Convergence of the CRR Model

In this section we consider the convergence of the discrete time model to the continuous-time Black–Scholes model.

Continuous compounding – riskless asset

Consider the subdivision

$$\left[0, \frac{T}{N}, \frac{2T}{N}, \ldots, \frac{(N-1)T}{N}, T\right]$$

of the time interval $[0, T]$ into N time steps.

Note that

$$\lim_{N\to\infty} (1+r)^N = \infty,$$

thus we need to renormalize r so that the interest rate on each time interval becomes r_N, with $\lim_{N\to\infty} r_N = 0$.

It turns out that the correct renormalization is

$$r_N = r\frac{T}{N},$$

so that

$$\lim_{N\to\infty} (1+r_N)^N = \lim_{N\to\infty} \left(1 + r\frac{T}{N}\right)^N = e^{rT}, \qquad T \in \mathbb{R}_+. \tag{3.26}$$

Hence the price of the riskless asset satifies

$$A_t = A_0 e^{rt},$$

with the differential equation

$$\frac{dA_t}{dt} = rA_t,$$

also written as

$$dA_t = rA_t dt,$$

or

$$\frac{dA_t}{A_t} = rdt,$$

which means that the *return* of the riskless asset is rdt on the small time interval $[t, t+dt]$. Equivalently, one says that r is the instantaneous interest rate per unit of time.

The same equation rewrites in *integral form* as

$$A_T - A_0 = \int_0^T dA_t = r\int_0^T A_t dt.$$

Continuous compounding – risky asset

We need to apply a similar renormalization to the coefficients a and b of the CRR model. Let $\sigma > 0$ denote a positive parameter called the volatility and let a_N, b_N be defined from

$$\frac{1+a_N}{1+r_N} = e^{-\sigma\sqrt{\frac{T}{N}}} \qquad \text{and} \qquad \frac{1+b_N}{1+r_N} = e^{\sigma\sqrt{\frac{T}{N}}},$$

i.e.,

$$a_N = (1+r_N)e^{-\sigma\sqrt{\frac{T}{N}}} - 1 \qquad \text{and} \qquad b_N = (1+r_N)e^{\sigma\sqrt{\frac{T}{N}}} - 1.$$

Consider the random return $R_k^{(N)} \in \{a_N, b_N\}$ and the price process defined as

$$S_t^{(N)} = S_0 \prod_{k=1}^{t} (1 + R_k^{(t)}), \qquad t = 1, \ldots, N.$$

Note that the risk-neutral probabilities are given by

$$\mathbb{P}^*(R_t = a_N) = \frac{b_N - r_N}{b_N - a_N} = \frac{e^{\sigma\sqrt{\frac{T}{N}}} - 1}{2\sinh\sqrt{\frac{\sigma^2 T}{N}}}, \qquad t = 1, \ldots, N,$$

and

$$\mathbb{P}^*(R_t = b_N) = \frac{r_N - a_N}{b_N - a_N} = \frac{1 - e^{-\sigma\sqrt{\frac{T}{N}}}}{2\sinh\sqrt{\frac{\sigma^2 T}{N}}}, \qquad t = 1, \ldots, N,$$

which both converge to $1/2$ as N goes to infinity.

Continuous-time limit

We have the following convergence result.

Proposition 3.6 *Let f be a continuous and bounded function on $\mathbb{R}$. The price at time $t = 0$ of a contingent claim with payoff $C = f\left(S_N^{(N)}\right)$ converges as follows:*

$$\lim_{N\to\infty} \frac{1}{(1+rT/N)^N} \mathbb{E}^*\left[f(S_N^{(N)})\right] = e^{-rT}\,\mathbb{E}\left[f(S_0 e^{\sigma\sqrt{T}X + rT - \sigma^2 T/2})\right] \quad (3.27)$$

where $X \simeq \mathcal{N}(0,1)$ is a standard Gaussian random variable.

Proof. This result is the consequence of the weak convergence of the sequence $(S_N^{(N)})_{N\geq 1}$ to a lognormal distribution, cf. Theorem 5.53 of [24]. The convergence of the discount factor follows directly from (3.26). □

Note that the expectation (3.27) can be written as a Gaussian integral:

$$e^{-rT}\,\mathbb{E}\left[f(S_0 e^{\sigma\sqrt{T}X+rT-\sigma^2T/2})\right] = e^{-rT}\int_{-\infty}^{\infty} f(S_0 e^{\sigma\sqrt{T}x+rT-\sigma^2T/2})\frac{e^{-x^2/2}}{\sqrt{2\pi}}dx,$$

hence we have

$$\lim_{N\to\infty}\frac{1}{(1+rT/N)^N}\,\mathbb{E}^*\left[f(S_N^{(N)})\right] = e^{-rT}\int_{-\infty}^{\infty} f(S_0 e^{\sigma\sqrt{T}x+rT-\sigma^2T/2})\frac{e^{-x^2/2}}{\sqrt{2\pi}}dx.$$

It is a remarkable fact that in case $f(x) = (x-K)^+$, i.e., when $C = (S_T-K)^+$ is the payoff of a European call option with strike K, the above integral can be computed according to the *Black–Scholes formula*:

$$e^{-rT}\,\mathbb{E}\left[(S_0 e^{\sigma\sqrt{T}X+rT-\sigma^2T/2} - K)^+\right] = S_0\Phi(d_+) - Ke^{-rT}\Phi(d_-),$$

where

$$d_- = \frac{(r-\frac{1}{2}\sigma^2)T + \log\frac{S_0}{K}}{\sigma\sqrt{T}}, \qquad d_+ = d_- + \sigma\sqrt{T},$$

and

$$\Phi(x) = \frac{1}{2\pi}\int_{-\infty}^{x} e^{-y^2/2}dy, \qquad x \in \mathbb{R},$$

is the Gaussian cumulative distribution function.

The Black–Scholes formula will be derived explicitly in the subsequent chapters using both the PDE and probabilistic method, cf. Propositions 1.8nd 6.4. It can be considered as a building block for the pricing of financial derivatives, and its importance is not restricted to the pricing of options on stocks. Indeed, the complexity of the interest rate models makes it in general difficult to obtain closed form expressions, and in many situations one has to rely on the Black–Scholes framework in order to find pricing formulas, for example in the case of interest rate derivatives as in the Black caplet formula of the BGM model, cf. Proposition 12.3 in Section 12.3.

Our aim later on will be to price and hedge options directly in continuous time using stochastic calculus, instead of applying the limit procedure described in the previous section. In addition to the construction of the riskless asset price $(A_t)_{t\in\mathbb{R}_+}$ via

$$\frac{dA_t}{A_t} = r_t dt, \qquad A_0 = 1, \qquad t \in \mathbb{R}_+,$$

i.e.,

$$A_t = A_0 e^{rt}, \qquad t \in \mathbb{R}_+,$$

we now need to construct a mathematical model for the price of the risky asset in continuous time.

The return of the risky asset S_t over the time period $[t, d + dt]$ will be defined as

$$\frac{dS_t}{S_t} = \mu dt + \sigma dB_t,$$

where σdB_t is a "small" Gaussian random component, also called Brownian increment, parametrized by the volatility parameter $\sigma > 0$. In Chapter 4 we will turn to the formal definition of the stochastic process $(B_t)_{t\in\mathbb{R}_+}$ which will be used for the modeling of risky assets in continuous time.

Exercises

Exercise 3.1 (Exercise 2.1 continued)

1. We consider a forward contract on S_N with strike K and payoff

$$C := S_N - K.$$

   Find a portfolio allocation $(\eta_N, \xi_N,)$ with price $V_N = \eta_N \pi_N + \xi_N S_N$ at time N, such that

$$V_N = C, \tag{3.28}$$

   by writing Condition (3.28) as a 2 system of equations.

2. Find a portfolio allocation (η_{N-1}, ξ_{N-1}) with price $V_{N-1} = \eta_{N-1}\pi_{N-1} + \xi_{N-1}S_{N-1}$ at time $N - 1$, and verifying the self-financing condition

$$V_{N-1} = \eta_N \pi_{N-1} + \xi_N S_{N-1}.$$

   Next, at all times $t = 1, \ldots, N-1$, find a portfolio allocation (η_t, ξ_t) with price $V_t = \eta_t \pi_t + \xi_t S_t$ verifying (3.28) and the self-financing condition

$$V_t = \eta_{t+1}\pi_t + \xi_{t+1}S_t,$$

   where η_t, resp. ξ_t, represents the quantity of the riskless, resp. risky, asset in the portfolio over the time period $[t - 1, t]$, $t = 1, \ldots, N$.

3. Forward contract C, at time $t = 0, 1, \ldots, N$.

4. Check that the arbitrage price $\pi_t(C)$ satisfies the relation

$$\pi_t(C) = \frac{1}{(1+r)^{N-t}} \mathbb{E}^*[C \mid \mathcal{F}_t], \qquad t = 0, 1, \ldots, N.$$

Exercise 3.2 Consider the discrete-time Cox–Ross–Rubinstein model with $N+1$ time instants $t = 0, 1, \ldots, N$. The price S_t^0 of the riskless asset evolves as $S_t^0 = \pi^0(1+r)^t$, $t = 0, 1, \ldots, N$. The *return* of the risky asset, defined as

$$R_t := \frac{S_t - S_{t-1}}{S_{t-1}}, \qquad t = 1, \ldots, N,$$

is random and allowed to take only two values a and b, with $-1 < a < r < b$.

The discounted asset price is $X_t = S_t/(1+r)^t$, $t = 0, 1, \ldots, N$.

1. Show that this model admits a unique risk-neutral measure $\mathbb{P}^*$ and explicitly compute $P^*(R_t = a)$ and $P(R_t = b)$ for all $t = 1, \ldots, N$.
2. Do there exist arbitrage opportunities in this model ? Explain why.
3. Is this market model complete ? Explain why.
4. Consider a contingent claim with payoff[1]

$$C = (S_N)^2.$$

   Compute the discounted arbitrage price $\tilde{V}_t$, $t = 0, \ldots, N$, of a self-financing portfolio hedging the claim C, i.e., such that

$$\tilde{V}_N = \tilde{C} = \frac{(S_N)^2}{(1+r)^N}.$$

5. Compute the portfolio strategy

$$(\bar{\xi}_t)_{t=1,\ldots,N} = (\xi_t^0, \xi_t^1)_{t=1,\ldots,N}$$

   associated to $\tilde{V}_t$, i.e., such that

$$\tilde{V}_t = \bar{\xi}_t \cdot \bar{X}_t = \xi_t^0 X_t^0 + \xi_t^1 X_t^1, \qquad t = 1, \ldots, N.$$

6. Check that the above portfolio strategy is self-financing, i.e.,

$$\bar{\xi}_{t+1} \cdot \bar{S}_t = \bar{\xi}_t \cdot \bar{S}_t, \qquad t = 1, \ldots, N-1.$$

Exercise 3.3 We consider the discrete-time Cox–Ross–Rubinstein model with $N+1$ time instants $t = 0, 1, \ldots, N$.

[1]This is the payoff of a power call option with strike $K = 0$.

The price π_t of the riskless asset evolves as $\pi_t = \pi_0(1+r)^t$, $t = 0, 1, \ldots, N$. The evolution of S_{t-1} to S_t is given by

$$S_t = \begin{cases} (1+b)S_{t-1} \\ \\ (1+a)S_{t-1} \end{cases}$$

with $-1 < a < r < b$. The *return* of the risky asset is defined as

$$R_t := \frac{S_t - S_{t-1}}{S_{t-1}}, \qquad t = 1, \ldots, N.$$

Let ξ_t, resp. η_t, denote the (possibly fractional) quantities of the risky, resp. riskless, asset held over the time period $[t-1, t]$ in the portfolio with value

$$V_t = \xi_t S_t + \eta_t \pi_t, \qquad t = 0, \ldots, N. \tag{3.29}$$

1. Show that

$$V_t = (1 + R_t)\xi_t S_{t-1} + (1+r)\eta_t \pi_{t-1}, \qquad t = 1, \ldots, N. \tag{3.30}$$

2. Show that under the probability P^* defined by

$$P^*(R_t = a \mid \mathcal{F}_{t-1}) = \frac{b-r}{b-a}, \qquad P^*(R_t = b \mid \mathcal{F}_{t-1}) = \frac{r-a}{b-a},$$

where $\mathcal{F}_{t-1}$ represents the information generated by $\{R_1, \ldots, R_{t-1}\}$, we have

$$E^*[R_t \mid \mathcal{F}_{t-1}] = r.$$

3. Under the self-financing condition

$$V_{t-1} = \xi_t S_{t-1} + \eta_t \pi_{t-1} \qquad t = 1, \ldots, N, \tag{3.31}$$

show that

$$V_{t-1} = \frac{1}{1+r} E^*[V_t \mid \mathcal{F}_{t-1}],$$

using the result of Question 1.

4. Let $a = 5\%$, $b = 25\%$ and $r = 15\%$. Assume that the price V_t at time t of the portfolio is \$3 if $R_t = a$ and \$8 if $R_t = b$, and compute the price V_{t-1} of the portfolio at time $t-1$.

Chapter 4

Brownian Motion and Stochastic Calculus

The modeling of random assets in finance is based on stochastic processes, which are families $(X_t)_{t\in I}$ of random variables indexed by a time interval I. In this chapter we present a description of Brownian motion and a construction of the associated Itô stochastic integral.

4.1 Brownian Motion

We start by recalling the definition of Brownian motion, which is a fundamental example of a stochastic process. The underlying probability space $(\Omega, \mathcal{F}, \mathbb{P})$ of Brownian motion can be constructed on the space $\Omega = \mathcal{C}_0(\mathbb{R}_+)$ of continuous real-valued functions on $\mathbb{R}_+$ started at 0.

Definition 4.1 *The standard Brownian motion is a stochastic process $(B_t)_{t\in\mathbb{R}_+}$ such that*

(i) $B_0 = 0$ almost surely,

(ii) The sample trajectories $t \mapsto B_t$ are continuous, with probability 1.

(iii) For any finite sequence of times $t_0 < t_1 < \cdots < t_n$, the increments

$$B_{t_1} - B_{t_0}, B_{t_2} - B_{t_1}, \ldots, B_{t_n} - B_{t_{n-1}}$$

are independent.

(iv) For any given times $0 \le s < t$, $B_t - B_s$ has the Gaussian distribution $\mathcal{N}(0, t - s)$ with mean zero and variance $t - s$.

We refer to Theorem 10.28 of [23] and to Chapter 1 of [62] for the proof of the existence of Brownian motion as a stochastic process $(B_t)_{t\in\mathbb{R}_+}$ satisfying the above properties (i)–(iv).

In particular, Condition (iv) above implies

$$\mathbb{E}[B_t - B_s] = 0 \quad \text{and} \quad \mathrm{Var}[B_t - B_s] = t - s, \qquad 0 \leq s \leq t.$$

In the sequel the filtration $(\mathcal{F}_t)_{t\in\mathbb{R}_+}$ will be generated by the Brownian paths up to time t, in other words we write

$$\mathcal{F}_t = \sigma(B_s \;:\; 0 \leq s \leq t), \qquad t \geq 0. \tag{4.1}$$

A random variable F is said to be $\mathcal{F}_t$-*measurable* if the knowledge of F depends only on the information known up to time t. As an example, if $t =$ today,

- the date of the past course exam is $\mathcal{F}_t$-measurable, because it belongs to the past;

- the date of the next Chinese new year, although it refers to a future event, is also $\mathcal{F}_t$-measurable because it is known at time t;

- the date of the next typhoon is not $\mathcal{F}_t$-measurable since it is not known at time t;

- the maturity date T of a European option is $\mathcal{F}_t$-measurable for all $t \in \mathbb{R}_+$, because it has been determined at time 0;

- the exercise date τ of an American option after time t (see Section 9.4) is not $\mathcal{F}_t$-measurable because it refers to a future random event.

Property (iii) above shows that $B_t - B_s$ is independent of all Brownian increments taken before time s, i.e.,

$$(B_t - B_s) \perp\!\!\!\perp (B_{t_1} - B_{t_0}, B_{t_2} - B_{t_1}, \ldots, B_{t_n} - B_{t_{n-1}}),$$

$0 \leq t_0 \leq t_1 \leq \cdots \leq t_n \leq s \leq t$, hence $B_t - B_s$ is also independent of the whole Brownian history up to time s, hence $B_t - B_s$ is in fact independent of $\mathcal{F}_s$, $s \geq 0$.

For convenience we will informally regard Brownian motion as a random walk over infinitesimal time intervals of length Δt, with increments

$$\Delta B_t := B_{t+\Delta t} - B_t$$

over the time interval $[t, t + \Delta t]$ given by

$$\Delta B_t = \pm\sqrt{\Delta t} \tag{4.2}$$

with equal probabilities $(1/2, 1/2)$.

The choice of the square root in (4.2) is in fact not fortuitous. Indeed, any choice of $\pm(\Delta t)^\alpha$ with a power $\alpha > 1/2$ would lead to explosion of the process as dt tends to zero, whereas a power $\alpha \in (0, 1/2)$ would lead to a vanishing process.

Note that we have

$$\mathbb{E}[\Delta B_t] = \frac{1}{2}\sqrt{\Delta t} - \frac{1}{2}\sqrt{\Delta t} = 0,$$

and

$$\text{Var}[\Delta B_t] = \mathbb{E}[(\Delta B_t)^2] = \frac{1}{2}\Delta t + \frac{1}{2}\Delta t = \Delta t.$$

According to this representation, the paths of Brownian motion are not differentiable, although they are continuous by Property (ii), as we have

$$\frac{dB_t}{dt} \simeq \frac{\pm\sqrt{dt}}{dt} = \pm\frac{1}{\sqrt{dt}} \simeq \pm\infty. \tag{4.3}$$

After splitting the interval $[0, T]$ into N intervals

$$\left(\frac{k-1}{N}T, \frac{k}{N}T\right], \qquad k = 1, \ldots, N,$$

of length $\Delta t = T/N$ with N "large," and letting

$$X_k = \pm\sqrt{T} = \pm\sqrt{N}\sqrt{\Delta t} = \sqrt{N}\Delta B_t$$

with probabilities $(1/2, 1/2)$ we have $Var(X_k) = T$ and

$$\Delta B_t = \frac{X_k}{\sqrt{N}} = \pm\sqrt{\Delta t}$$

is the increment of B_t over $((k-1)\Delta t, k\Delta t]$, and we get

$$B_T \simeq \sum_{0<t<T} \Delta B_t \simeq \frac{X_1 + \cdots + X_N}{\sqrt{N}}.$$

Hence by the central limit theorem we recover the fact that B_T has a centered Gaussian distribution with variance T, cf. point *(iv)* of the above definition of Brownian motion. Indeed, the central limit theorem states that given any sequence $(X_k)_{k\geq 1}$ of independent identically distributed centered random variables with variance $\sigma^2 = Var(X_k) = T$, the normalized sum

$$\frac{X_1 + \cdots + X_n}{\sqrt{n}}$$

converges (in distribution) to a centered Gaussian random variable $\mathcal{N}(0, \sigma^2)$ with variance σ^2 as n goes to infinity. As a consequence, ΔB_t could in fact be replaced by any centered random variable with variance Δt in the above description.

Note that there is no point in "computing" the value of B_t as it is a *random variable* for all $t > 0$; however, we can generate samples of B_t, which are distributed according to the centered Gaussian distribution with variance t. Below we draw three sample paths of a standard Brownian motion obtained by computer simulation using (4.2). The n-dimensional Brownian motion can

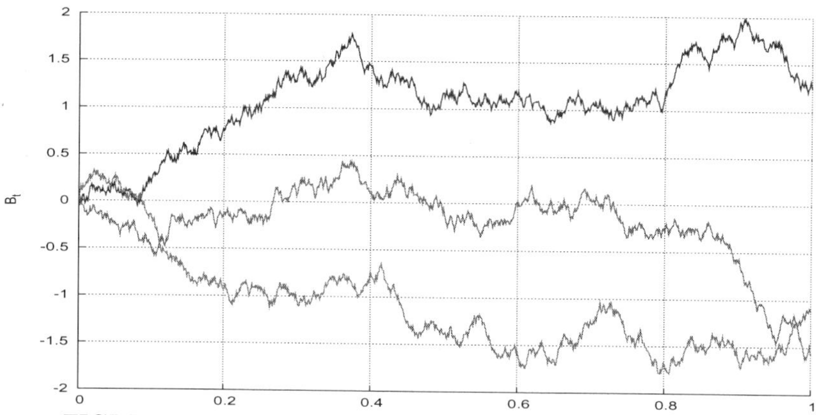

FIGURE 4.1: Sample paths of one-dimensional Brownian motion.

be constructed as $(B_t^1, \ldots, B_t^n)_{t\in\mathbb{R}_+}$ where $(B_t^1)_{t\in\mathbb{R}_+}, \ldots, (B_t^n)_{t\in\mathbb{R}_+}$ are independent copies of $(B_t)_{t\in\mathbb{R}_+}$.

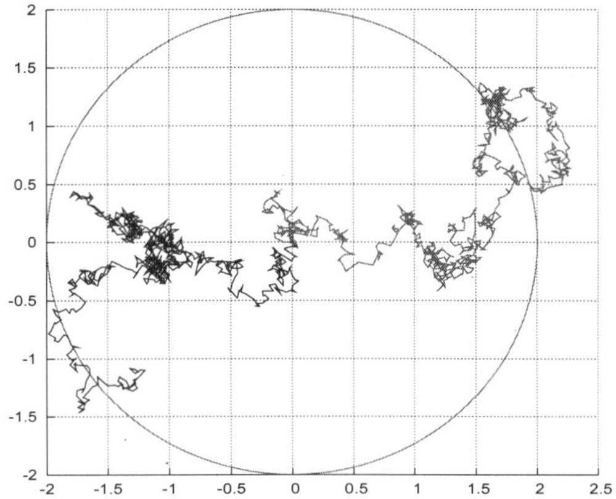

FIGURE 4.2: Two sample paths of a two-dimensional Brownian motion.

Next we turn to simulations of 2 dimensional and 3 dimensional Brownian motions. Recall that the movement of pollen particles originally observed by R. Brown in 1827 was indeed 2-dimensional.

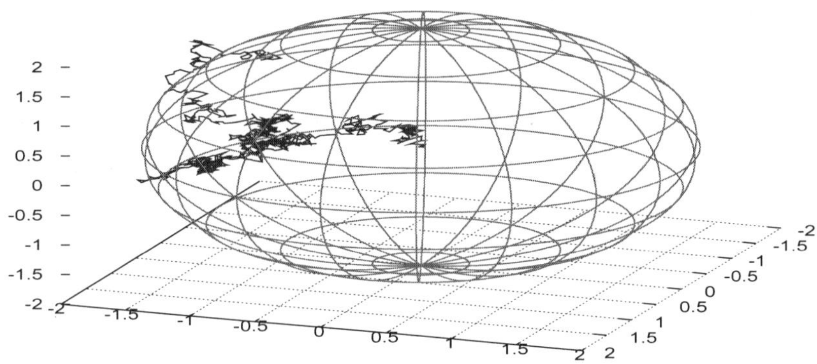

FIGURE 4.3: Sample paths of a three-dimensional Brownian motion.

4.2 Wiener Stochastic Integral

In this section we construct the Itô stochastic integral of square-integrable deterministic function with respect to Brownian motion.

Recall that Bachelier originally modeled the price S_t of a risky asset by

$S_t = \sigma B_t$ where σ is a volatility parameter. The stochastic integral

$$\int_0^T f(t)dS_t = \sigma \int_0^T f(t)dB_t$$

can be used to represent the value of a portfolio as a sum of profits and losses $f(t)dS_t$ where dS_t represents the stock price variation and $f(t)$ is the quantity invested in the asset S_t over the short time interval $[t, t + dt]$.

A naive definition of the stochastic integral with respect to Brownian motion would consist in writing

$$\int_0^\infty f(t)dB_t = \int_0^\infty f(t)\frac{dB_t}{dt}dt,$$

and evaluating the above integral with respect to dt. However, this definition fails because the paths of Brownian motion are not differentiable, cf. (4.3). Next we present Itô's construction of the stochastic integral with respect to Brownian motion. Stochastic integrals will be first constructed as integrals of simple step functions of the form

$$f(t) = \sum_{i=1}^n a_i \mathbf{1}_{(t_{i-1}, t_i]}(t), \qquad t \in \mathbb{R}_+, \tag{4.4}$$

i.e., the function f takes the value a_i on the interval $(t_{i-1}, t_i]$, $i = 1, \ldots, n$, with $0 \leq t_0 < \cdots < t_n$, as illustrated in Figure 4.4.

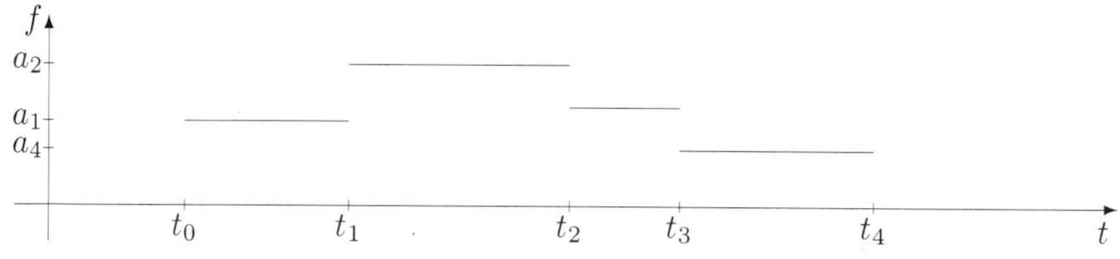

FIGURE 4.4: Step function.

Note that the set of simple step functions f of the form (4.4) is a linear space which is dense in $L^2(\mathbb{R}_+)$ for the norm

$$\|f\|_{L^2(\mathbb{R}_+)} := \sqrt{\int_0^\infty |f(t)|^2 dt}.$$

Recall also that the classical integral of f given in (4.4) is interpreted as the area under the curve f and computed as

$$\int_0^\infty f(t)dt = \sum_{i=1}^n a_i(t_i - t_{i-1}).$$

In the next definition we adapt this construction to the setting of integration with respect to Brownian motion.

Definition 4.2 *The stochastic integral with respect to Brownian motion $(B_t)_{t\in\mathbb{R}_+}$ of the simple step functions f of the form* (4.4) *is defined by*

$$\int_0^\infty f(t)dB_t := \sum_{i=1}^n a_i(B_{t_i} - B_{t_{i-1}}). \tag{4.5}$$

In the next Proposition 4.1 we determine the probability distribution of $\int_0^\infty f(t)dB_t$ and we show that it is independent of the particular representation (4.4) chosen for $f(t)$. In the sequel we will make a repeated use of the space $L^2(\mathbb{R}_+)$ of functions $f : \mathbb{R}_+ \to \mathbb{R}$ such that

$$\|f\|^2_{L^2(\mathbb{R}_+)} := \int_0^\infty |f(t)|^2 dt < \infty,$$

called square-integrable functions.

Proposition 4.1 *The definition of the stochastic integral $\int_0^\infty f(t)dB_t$ can be extended to any measurable function $f \in L^2(\mathbb{R}_+)$,* i.e., *to f such that*

$$\int_0^\infty |f(t)|^2 dt < \infty. \tag{4.6}$$

In this case, $\int_0^\infty f(t)dB_t$ has a centered Gaussian distribution

$$\int_0^\infty f(t)dB_t \simeq \mathcal{N}\left(0, \int_0^\infty |f(t)|^2 dt\right)$$

with variance $\int_0^\infty |f(t)|^2 dt$ and we have the Itô isometry

$$\mathbb{E}\left[\left(\int_0^\infty f(t)dB_t\right)^2\right] = \int_0^\infty |f(t)|^2 dt. \tag{4.7}$$

Proof. Recall that if $X_1, \ldots, X_n$ are independent Gaussian random variables with probability laws $\mathcal{N}(m_1, \sigma_1^2), \ldots, \mathcal{N}(m_n, \sigma_n^2)$ then sum $X_1 + \cdots + X_n$ is a Gaussian random variable with probability law $\mathcal{N}(m_1 + \cdots + m_n, \sigma_1^2 + \cdots + \sigma_n^2)$.

As a consequence, when f is the simple function

$$f(t) = \sum_{i=1}^n a_i \mathbf{1}_{(t_{i-1}, t_i]}(t), \qquad t \in \mathbb{R}_+,$$

the sum

$$\int_0^\infty f(t)dB_t = \sum_{k=1}^n a_k(B_{t_k} - B_{t_{k-1}})$$

has a centered Gaussian distribution with variance

$$\sum_{k=1}^n |a_k|^2 (t_k - t_{k-1}),$$

since

$$\mathrm{Var}\,[a_k(B_{t_k} - B_{t_{k-1}})] = a_k^2 \mathrm{Var}\,[B_{t_k} - B_{t_{k-1}}] = a_k^2(t_k - t_{k-1}),$$

hence the stochastic integral

$$\int_0^\infty f(t)dB_t = \sum_{k=1}^n a_k(B_{t_k} - B_{t_{k-1}})$$

of the step function

$$f(t) = \sum_{k=1}^n a_k \mathbf{1}_{(t_{k-1},t_k]}(t)$$

has a centered Gaussian distribution with variance

$$\begin{aligned}
\mathrm{Var}\left[\int_0^\infty f(t)dB_t\right] &= \sum_{k=1}^n |a_k|^2(t_k - t_{k-1}) \\
&= \sum_{k=1}^n |a_k|^2 \int_{t_{k-1}}^{t_k} dt \\
&= \int_0^\infty \sum_{k=1}^n |a_k|^2 1_{(t_{k-1},t_k]}(t)dt \\
&= \int_0^\infty |f(t)|^2 dt.
\end{aligned}$$

Finally we note that

$$\begin{aligned}
\mathrm{Var}\left[\int_0^\infty f(t)dB_t\right] &= \mathbb{E}\left[\left(\int_0^\infty f(t)dB_t\right)^2\right] - \left(\mathbb{E}\left[\int_0^\infty f(t)dB_t\right]\right)^2 \\
&= \mathbb{E}\left[\left(\int_0^\infty f(t)dB_t\right)^2\right].
\end{aligned}$$

The extension of the stochastic integral to all functions satisfying (4.6) is obtained by density and a Cauchy sequence argument, based on the isometry relation (4.7). Namely, given f a function satisfying (4.6) and $(f_n)_{n\in\mathbb{N}}$ a sequence of simple functions converging to f for the norm

$$\|f - f_n\|_{L^2(\mathbb{R}_+)} := \left(\int_0^\infty |f(t) - f_n(t)|^2 dt\right)^{1/2}$$

i.e., in $L^2(\mathbb{R}_+)$, the isometry (4.7) shows that $\left(\int_0^\infty f_n(t)dB_t\right)_{n\in\mathbb{N}}$ is a Cauchy sequence in the space $L^2(\Omega)$ of square-integrable random variables $F : \Omega \to \mathbb{R}$ such that

$$\|F\|_{L^2(\Omega\times\mathbb{R}_+)}^2 := \mathbb{E}[F^2] < \infty.$$

Indeed, we have

$$\left\|\int_0^\infty f_k(t)dB_t - \int_0^\infty f_n(t)dB_t\right\|_{L^2(\Omega)}$$

$$
\begin{aligned}
&= \left(\mathbb{E}\left[\left(\int_0^\infty f_k(t)dB_t - \int_0^\infty f_n(t)dB_t\right)^2\right]\right)^{1/2} \\
&= \|f_k - f_n\|_{L^2(\mathbb{R}_+)} \\
&\leq \|f - f_k\|_{L^2(\mathbb{R}_+)} + \|f - f_n\|_{L^2(\mathbb{R}_+)},
\end{aligned}
$$

which tends to 0 as k, n tend to infinity, hence $\left(\int_0^\infty f_n(t)dB_t\right)_{n\in\mathbb{N}}$ as it converges for the L^2-norm as $L^2(\Omega)$ is a complete space, cf. e.g., Chapter 4 of [18]. In this case we let

$$
\int_0^\infty f(t)dB_t := \lim_{n\to\infty} \int_0^\infty f_n(t)dB_t
$$

and the limit is unique from (4.7). □

For example, $\displaystyle\int_0^\infty e^{-t}dB_t$ has a centered Gaussian distribution with variance

$$
\int_0^\infty e^{-2t}dt = \left[-\frac{1}{2}e^{-2t}\right]_0^\infty = \frac{1}{2}.
$$

Again, the Wiener stochastic integral $\displaystyle\int_0^\infty f(s)dB_s$ is nothing but a Gaussian random variable and it cannot be "computed" in the way standard integrals are computed via the use of primitives. However, when $f \in L^2(\mathbb{R}_+)$ is $\mathcal{C}^1$ on $\mathbb{R}_+$, we have the following formula

$$
\int_0^\infty f(t)dB_t = -\int_0^\infty f'(t)B_t dt,
$$

provided $\lim_{t\to\infty} t|f(t)|^2 = 0$ and $f \in L^2(\mathbb{R}_+)$, cf. e.g., Remark 2.5.9 in [55].

4.3 Itô Stochastic Integral

In this section we extend the Wiener stochastic integral to square-integrable *adapted* processes. Recall that a process $(X_t)_{t\in\mathbb{R}_+}$ is said to be $\mathcal{F}_t$-adapted if X_t is $\mathcal{F}_t$-measurable for all $t \in \mathbb{R}_+$, where the information flow $(\mathcal{F}_t)_{t\in\mathbb{R}_+}$ has been defined in (4.1).

In other words, a process $(X_t)_{t\in\mathbb{R}_+}$ is $\mathcal{F}_t$-adapted if the value of X_t at time t depends only on information known up to time t. Note that the value of X_t may still depend on "known" future data, for example a fixed future date in the calendar, such as a maturity time $T > t$, as long as its value is known at time t.

The extension of the stochastic integral to adapted random processes is actually necessary in order to compute a portfolio value when the portfolio process is no longer deterministic. This happens in particular when one needs to update the portfolio allocation based on random events occurring on the market.

Stochastic integrals of adapted processes will be first constructed as integrals of simple predictable processes $(u_t)_{t\in\mathbb{R}_+}$ of the form

$$u_t = \sum_{i=1}^{n} F_i \mathbf{1}_{(t_{i-1},t_i]}(t), \qquad t \in \mathbb{R}_+, \tag{4.8}$$

where F_i is an $\mathcal{F}_{t_{i-1}}$-measurable random variable for $i = 1, \ldots, n$. The notion of simple predictable process is natural in the context of portfolio investment, in which F_i will represent an investment allocation decided at time t_{i-1} and to remain unchanged over the time period $(t_{i-1}, t_i]$.

By convention, $u : \Omega \times \mathbb{R}_+ \to \mathbb{R}$ is denoted by $u_t(\omega)$, $t \in \mathbb{R}_+$, $\omega \in \Omega$, and the random outcome ω is often dropped for convenience of notation.

Definition 4.3 *The stochastic integral with respect to Brownian motion $(B_t)_{t\in\mathbb{R}_+}$ of any simple predictable process $(u_t)_{t\in\mathbb{R}_+}$ of the form (4.8) is defined by*

$$\int_0^\infty u_t dB_t := \sum_{i=1}^{n} F_i (B_{t_i} - B_{t_{i-1}}). \tag{4.9}$$

The next proposition gives the extension of the stochastic integral from simple predictable processes to square-integrable $\mathcal{F}_t$-adapted processes $(X_t)_{t\in\mathbb{R}_+}$ for which the value of X_t at time t only depends on information contained in the Brownian path up to time t. This also means that knowing the future is not permitted in the definition of the Itô integral, for example a portfolio strategy that would allow the trader to "buy at the lowest" and "sell at the highest" is not possible as it would require knowledge of future market data.

Note that the difference between Relation (4.10) below and Relation (4.7) is the expectation on the right-hand side.

Proposition 4.2 *The stochastic integral with respect to Brownian motion $(B_t)_{t\in\mathbb{R}_+}$ extends to all adapted processes $(u_t)_{t\in\mathbb{R}_+}$ such that*

$$\mathbb{E}\left[\int_0^\infty |u_t|^2 dt\right] < \infty,$$

with the Itô isometry

$$\mathbb{E}\left[\left(\int_0^\infty u_t dB_t\right)^2\right] = \mathbb{E}\left[\int_0^\infty |u_t|^2 dt\right]. \tag{4.10}$$

Proof. We start by showing that the Itô isometry (4.10) holds for the simple predictable process u of the form (4.8). We have

$$
\begin{aligned}
\mathbb{E}\left[\left(\int_0^\infty u_t dB_t\right)^2\right] &= \mathbb{E}\left[\left(\sum_{i=1}^n F_i(B_{t_i} - B_{t_{i-1}})\right)^2\right] \\
&= \mathbb{E}\left[\sum_{i,j=1}^n F_i F_j (B_{t_i} - B_{t_{i-1}})(B_{t_j} - B_{t_{j-1}})\right] \\
&= \mathbb{E}\left[\sum_{i=1}^n |F_i|^2 (B_{t_i} - B_{t_{i-1}})^2\right] \\
&\quad +2\,\mathbb{E}\left[\sum_{1\leq i<j\leq n} F_i F_j (B_{t_i} - B_{t_{i-1}})(B_{t_j} - B_{t_{j-1}})\right] \\
&= \sum_{i=1}^n \mathbb{E}[\mathbb{E}[|F_i|^2 (B_{t_i} - B_{t_{i-1}})^2 | \mathcal{F}_{t_{i-1}}]] \\
&\quad +2 \sum_{1\leq i<j\leq n} \mathbb{E}[\mathbb{E}[F_i F_j (B_{t_i} - B_{t_{i-1}})(B_{t_j} - B_{t_{j-1}}) | \mathcal{F}_{t_{j-1}}]] \\
&= \sum_{i=1}^n \mathbb{E}[|F_i|^2\, \mathbb{E}[(B_{t_i} - B_{t_{i-1}})^2 | \mathcal{F}_{t_i}]] \\
&\quad +2 \sum_{1\leq i<j\leq n} \mathbb{E}[F_i F_j (B_{t_i} - B_{t_{i-1}})\, \mathbb{E}[(B_{t_j} - B_{t_{j-1}}) | \mathcal{F}_{t_{j-1}}]] \\
&= \sum_{i=1}^n \mathbb{E}[|F_i|^2\, \mathbb{E}[(B_{t_i} - B_{t_{i-1}})^2]] \\
&\quad +2 \sum_{1\leq i<j\leq n} \mathbb{E}[F_i F_j (B_{t_i} - B_{t_{i-1}})\, \mathbb{E}[(B_{t_j} - B_{t_{j-1}})]] \\
&= \sum_{i=1}^n \mathbb{E}[|F_i|^2 (t_i - t_{i-1})] \\
&= \mathbb{E}\left[\sum_{i=1}^n |F_i|^2 (t_i - t_{i-1})\right] \\
&= \mathbb{E}\left[\int_0^\infty |u_t|^2 dt\right],
\end{aligned}
$$

where we used the "tower property" (A.20) of conditional expectations and the facts that $B_{t_i} - B_{t_{i-1}}$ is independent of $\mathcal{F}_{t_{i-1}}$ and

$$
\mathbb{E}[B_{t_i} - B_{t_{i-1}}] = 0, \qquad \mathbb{E}\left[(B_{t_i} - B_{t_{i-1}})^2\right] = t_i - t_{i-1}, \qquad i = 1, \ldots, n.
$$

The extension of the stochastic integral to square-integrable adapted processes $(u_t)_{t\in\mathbb{R}_+}$ is obtained as in Proposition 4.1 by density and a Cauchy sequence argument using the isometry (4.10), in the same way as in the proof of Proposition 4.1. Let $L^2(\Omega \times \mathbb{R}_+)$ denote the space of square-integrable stochastic processes $u : \Omega \times \mathbb{R}_+ \to \mathbb{R}$ such that

$$\|u\|^2_{L^2(\Omega\times\mathbb{R}_+)} := \mathbb{E}\left[\int_0^\infty |u_t|^2 dt\right] < \infty.$$

By Lemma 1.1 of [34], p. 22 and p. 46, or Proposition 2.5.3 of [55], the set of simple predictable processes forms a linear space which is dense in the subspace $L^2_{ad}(\Omega \times \mathbb{R}_+)$ made of square-integrable adapted processes in $L^2(\Omega\times\mathbb{R}_+)$. In other words, given u a square-integrable adapted process there exists a sequence $(u^n)_{n\in\mathbb{N}}$ of simple predictable processes converging to u in $L^2(\Omega \times \mathbb{R}_+)$, and the isometry (4.10) shows that $\left(\int u^n_t dB_t\right)_{n\in\mathbb{N}}$ is a Cauchy sequence in $L^2(\Omega)$, hence it converges in the complete space $L^2(\Omega)$. In this case we let

$$\int_0^\infty u_t dB_t := \lim_{n\to\infty} \int_0^\infty u^n_t dB_t$$

and the limit is unique from (4.10). □

In addition, the Itô integral of an adapted process $(u_t)_{t\in\mathbb{R}_+}$ is always a centered random variable:

$$\mathbb{E}\left[\int_0^\infty u_s dB_s\right] = 0. \tag{4.11}$$

Note also that the Itô isometry (4.10) can also be written as

$$\mathbb{E}\left[\int_0^\infty u_t dB_t \int_0^\infty v_t dB_t\right] = \mathbb{E}\left[\int_0^\infty u_t v_t dt\right],$$

for all square-integrable adapted processes u, v.

In addition, when the integrand $(u_t)_{t\in\mathbb{R}_+}$ is not a deterministic function, the random variable $\int_0^\infty u_s dB_s$ no longer has a Gaussian distribution, except in some exceptional cases.

The stochastic integral of u over the interval $[a, b]$ is defined as

$$\int_a^b u_t dB_t := \int_0^\infty \mathbf{1}_{[a,b]}(t) u_t dB_t.$$

In particular we have

$$\int_0^\infty \mathbf{1}_{[a,b]}(t) dB_t = B_b - B_a, \qquad 0 \le a \le b,$$

and

$$\int_a^b dB_t = B_b - B_a, \qquad 0 \le a \le b.$$

We also have the Chasles relation

$$\int_a^c u_t dB_t = \int_a^b u_t dB_t + \int_b^c u_t dB_t, \qquad 0 \le a \le b \le c,$$

and the stochastic integral has the following linearity property:

$$\int_0^\infty (u_t + v_t)dB_t = \int_0^\infty u_t dB_t + \int_0^\infty v_t dB_t, \qquad u, v \in L^2(\mathbb{R}_+).$$

In the sequel we will define the return at time $t \in \mathbb{R}_+$ of the risky asset $(S_t)_{t\in\mathbb{R}_+}$ as

$$\frac{dS_t}{S_t} = \mu dt + \sigma dB_t,$$

with $\mu \in \mathbb{R}$ and $\sigma > 0$. This equation can be formally rewritten in integral form as

$$S_T = S_0 + \mu \int_0^T S_t dt + \sigma \int_0^T S_t dB_t,$$

hence the need to define an integral with respect to dB_t, in addition to the usual integral with respect to dt.

In Proposition 4.2 we have defined the stochastic integral of square-integrable processes with respect to Brownian motion, thus we have made sense of the equation

$$S_T = S_0 + \mu \int_0^T S_t dt + \sigma \int_0^T S_t dB_t,$$

for $(S_t)_{t\in\mathbb{R}_+}$ an $\mathcal{F}_t$-adapted process, which can be rewritten in differential notation as

$$dS_t = \mu S_t dt + \sigma S_t dB_t,$$

or

$$\frac{dS_t}{S_t} = \mu dt + \sigma dB_t. \tag{4.12}$$

This model will be used to represent the random price S_t of a risky asset at time t. Here the return dS_t/S_t of the asset is made of two components: a constant return μdt and a random return σdB_t parametrized by the coefficient σ, called the volatility.

Our goal is now to solve Equation (4.12) and for this we will need to introduce Itô's calculus in Section 4.5 after reviewing classical deterministic calculus in Section 4.4.

4.4 Deterministic Calculus

The *fundamental theorem of calculus* states that for any continuously differentiable (deterministic) function f we have

$$f(x) = f(0) + \int_0^x f'(y)dy.$$

In differential notation this relation is written as the first order expansion

$$df(x) = f'(x)dx, \tag{4.13}$$

where dx is "small." Higher order expansions can be obtained from *Taylor's formula*, which, letting

$$df(x) = f(x + dx) - f(x),$$

states that

$$df(x) = f'(x)dx + \frac{1}{2}f''(x)(dx)^2 + \frac{1}{3!}f'''(x)(dx)^3 + \frac{1}{4!}f^{(4)}(x)(dx)^4 + \cdots.$$

Note that Relation (4.13) can be obtained by neglecting the terms of order larger than one in Taylor's formula, since $(dx)^n << dx$ when $n \geq 2$ and dx is "small."

4.5 Stochastic Calculus

Let us now apply Taylor's formula to Brownian motion, taking

$$dB_t = B_{t+dt} - B_t,$$

and letting

$$df(B_t) = f(B_{t+dt}) - f(B_t),$$

we have

$$df(B_t) = f'(B_t)dB_t + \frac{1}{2}f''(B_t)(dB_t)^2 + \frac{1}{3!}f'''(B_t)(dB_t)^3 + \frac{1}{4!}f^{(4)}(B_t)(dB_t)^4 + \cdots.$$

From the construction of Brownian motion by its small increments $dB_t = \pm\sqrt{dt}$, it turns out that the terms in $(dt)^2$ and $dtdB_t = \pm(dt)^{3/2}$ can be neglected in Taylor's formula at the first order of approximation in dt. However, the term of order two

$$(dB_t)^2 = (\pm\sqrt{dt})^2 = dt$$

can no longer be neglected in front of dt.

Hence Taylor's formula written at the second-order for Brownian motion reads

$$df(B_t) = f'(B_t)dB_t + \frac{1}{2}f''(B_t)dt, \tag{4.14}$$

for "small" dt. Note that writing this formula as

$$\frac{df(B_t)}{dt} = f'(B_t)\frac{dB_t}{dt} + \frac{1}{2}f''(B_t)$$

does not make sense because the derivative

$$\frac{dB_t}{dt} \simeq \pm\frac{\sqrt{dt}}{dt} \simeq \pm\frac{1}{\sqrt{dt}} \simeq \pm\infty$$

does not exist.

Integrating (4.14) on both sides and using the relation

$$f(B_t) - f(B_0) = \int_0^t df(B_s)$$

we get the integral form of Itô's formula for Brownian motion, i.e.,

$$f(B_t) = f(B_0) + \int_0^t f'(B_s)dB_s + \frac{1}{2}\int_0^t f''(B_s)ds.$$

We now turn to the general expression of Itô's formula which applies to Itô processes of the form

$$X_t = X_0 + \int_0^t v_s ds + \int_0^t u_s dB_s, \qquad t \in \mathbb{R}_+, \tag{4.15}$$

or in differential notation

$$dX_t = v_t dt + u_t dB_t,$$

where $(u_t)_{t\in\mathbb{R}_+}$ and $(v_t)_{t\in\mathbb{R}_+}$ are square-integrable adapted processes.

Given $f(t,x)$ a smooth function of two variables, from now on we let $\frac{\partial f}{\partial x}$ denote partial differentiation with respect to the *second* variable in $f(t,x)$, while $\frac{\partial f}{\partial s}$ denote partial differentiation with respect to the *first* (time) variable in $f(t,x)$.

Theorem 4.1 *(Itô formula for Itô processes). For any Itô process $(X_t)_{t\in\mathbb{R}_+}$ of the form* (4.15) *and any $f \in \mathcal{C}^{1,2}(\mathbb{R}_+ \times \mathbb{R})$ we have*

$$\begin{aligned} f(t,X_t) &= f(0,X_0) + \int_0^t v_s \frac{\partial f}{\partial x}(s,X_s)ds + \int_0^t u_s \frac{\partial f}{\partial x}(s,X_s)dB_s \\ &\quad + \int_0^t \frac{\partial f}{\partial s}(s,X_s)ds + \frac{1}{2}\int_0^t |u_s|^2 \frac{\partial^2 f}{\partial x^2}(s,X_s)ds. \end{aligned} \tag{4.16}$$

Proof. cf. [61]. □

Using the relation

$$\int_0^t df(s,X_s) = f(t,X_t) - f(0,X_0),$$

we get

$$\int_0^t df(s,X_s) = \int_0^t v_s \frac{\partial f}{\partial x}(s,X_s)ds + \int_0^t u_s \frac{\partial f}{\partial x}(s,X_s)dB_s + \int_0^t \frac{\partial f}{\partial s}(s,X_s)ds + \frac{1}{2}\int_0^t |u_s|^2 \frac{\partial^2 f}{\partial x^2}(s,X_s)ds,$$

which allows us to rewrite Itô's formula (4.16) in differential notation as

$$df(t,X_t) = \frac{\partial f}{\partial t}(t,X_t)dt + u_t \frac{\partial f}{\partial x}(t,X_t)dB_t + v_t \frac{\partial f}{\partial x}(t,X_t)dt + \frac{1}{2}|u_t|^2 \frac{\partial^2 f}{\partial x^2}(t,X_t)dt, \tag{4.17}$$

or

$$df(t,X_t) = \frac{\partial f}{\partial t}(t,X_t)dt + \frac{\partial f}{\partial x}(t,X_t)dX_t + \frac{1}{2}|u_t|^2 \frac{\partial^2 f}{\partial x^2}(t,X_t)dt.$$

Next, given two processes $(X_t)_{t\in\mathbb{R}_+}$ and $(Y_t)_{t\in\mathbb{R}_+}$ written as

$$X_t = X_0 + \int_0^t v_s ds + \int_0^t u_s dB_s, \qquad t \in \mathbb{R}_+,$$

and

$$Y_t = Y_0 + \int_0^t b_s ds + \int_0^t a_s dB_s, \qquad t \in \mathbb{R}_+,$$

the Itô formula also shows that

$$d(X_tY_t) = X_t dY_t + Y_t dX_t + dX_t dY_t$$

where the product $dX_t dY_t$ is computed according to the *Itô rule*

$$(dt)^2 = 0, \qquad dt dB_t = 0, \qquad (dB_t)^2 = dt, \tag{4.18}$$

i.e.,

$$\begin{aligned} dX_t dY_t &= (v_t dt + u_t dB_t)(b_t dt + a_t dB_t) \\ &= b_t v_t (dt)^2 + u_t b_t dt dB_t + v_t a_t dt dB_t + u_t a_t (dB_t)^2 \\ &= u_t a_t dt. \end{aligned}$$

Hence we have

$$\begin{aligned} (dX_t)^2 &= (v_t dt + u_t dB_t)^2 \\ &= (v_t)^2 (dt)^2 + (u_t)^2 (dB_t)^2 + 2u_t v_t dt \cdot dB_t \\ &= (u_t)^2 dt, \end{aligned}$$

according to the Itô multiplication table

$\cdot$	dt	dB_t
dt	0	0
dB_t	0	dt

and (4.17) can also be rewriten as

$$df(t, X_t) = \frac{\partial f}{\partial t}(t, X_t)dt + \frac{\partial f}{\partial x}(t, X_t)dX_t + \frac{1}{2}\frac{\partial^2 f}{\partial x^2}(t, X_t)(dX_t)^2.$$

Taking $u_t = 1$ and $v_t = 0$ in (4.15) yields $X_t = B_t$, in which case the Itô formula (4.16) reads

$$f(t, B_t) = f(0, B_0) + \int_0^t \frac{\partial f}{\partial s}(s, B_s)ds + \int_0^t \frac{\partial f}{\partial x}(s, B_s)dB_s + \frac{1}{2}\int_0^t \frac{\partial^2 f}{\partial x^2}(s, B_s)ds,$$

i.e., in differential notation:

$$df(t, B_t) = \frac{\partial f}{\partial t}(t, B_t)dt + \frac{\partial f}{\partial x}(t, B_t)dB_t + \frac{1}{2}\frac{\partial^2 f}{\partial x^2}(t, B_t)dt. \tag{4.19}$$

As another example, applying Itô's formula (4.19) to B_t^2 with

$$B_t^2 = f(t, B_t) \quad \text{and} \quad f(t, x) = x^2,$$

we get

$$\begin{aligned} dB_t^2 &= df(B_t) \\ &= \frac{\partial f}{\partial t}(t, B_t)dt + \frac{\partial f}{\partial x}(t, B_t)dB_t + \frac{1}{2}\frac{\partial^2 f}{\partial x^2}(t, B_t)dt \\ &= 2B_t dB_t + dt, \end{aligned}$$

since

$$\frac{\partial f}{\partial t}(t, x) = 0, \qquad \frac{\partial f}{\partial x}(t, x) = 2x, \qquad \text{and} \qquad \frac{1}{2}\frac{\partial^2 f}{\partial x^2}(t, x) = 1,$$

hence by integration we find

$$\begin{aligned} B_T^2 &= B_0 + 2\int_0^T B_s dB_s + \int_0^T dt \\ &= 2\int_0^T B_s dB_s + T, \end{aligned}$$

and

$$\int_0^T B_s dB_s = \frac{B_T^2}{2} - \frac{T}{2}.$$

We close this section with some comments on the practice of Itô's calculus. In some finance textbooks, Itô's formula for e.g., geometric Brownian motion can be found written in the notation

$$\begin{aligned} f(T, S_T) &= f(0, X_0) + \sigma \int_0^T S_t \frac{\partial f}{\partial S_t}(t, S_t)dB_t + \mu \int_0^T S_t \frac{\partial f}{\partial S_t}(t, S_t)dt \\ &\quad + \int_0^T \frac{\partial f}{\partial t}(t, S_t)dt + \frac{1}{2}\sigma^2 \int_0^T S_t^2 \frac{\partial^2 f}{\partial S_t^2}(t, S_t)dt, \end{aligned}$$

or

$$df(S_t) = \sigma S_t \frac{\partial f}{\partial S_t}(S_t)dB_t + \mu S_t \frac{\partial f}{\partial S_t}(S_t)dt + \frac{1}{2}\sigma^2 S_t^2 \frac{\partial^2 f}{\partial S_t^2}(S_t)dt.$$

The notation $\frac{\partial f}{\partial S_t}(S_t)$ can in fact be easily misused in combination with the fundamental theorem of classical calculus, and lead to the <u>*wrong*</u> identity

$$df(S_t) = \frac{\partial f}{\partial S_t}(S_t)dS_t.$$

Similarly, writing

$$df(B_t) = \frac{df}{dx}(B_t)dB_t + \frac{1}{2}\frac{d^2 f}{dx^2}(B_t)dt$$

is consistent, while writing

$$df(B_t) = \frac{df(B_t)}{dB_t}dB_t + \frac{1}{2}\frac{d^2 f(B_t)}{dB_t^2}dt$$

is potentially a source of confusion.

4.6 Geometric Brownian Motion

Our aim in this section is to solve the stochastic differential equation

$$dS_t = \mu S_t dt + \sigma S_t dB_t \tag{4.20}$$

that will define the price S_t of a risky asset at time t, where $\mu \in \mathbb{R}$ and $\sigma > 0$. This equation is rewritten in *integral form* as

$$S_t = S_0 + \mu \int_0^t S_s ds + \sigma \int_0^t S_s dB_s, \qquad t \in \mathbb{R}_+. \tag{4.21}$$

It can be solved by applying Itô's formula to $f(S_t) = \log S_t$ with $f(x) = \log x$, which shows that

$$\begin{aligned} d\log S_t &= \mu S_t f'(S_t)dt + \sigma S_t f'(S_t)dB_t + \frac{1}{2}\sigma^2 S_t^2 f''(S_t)dt \\ &= \mu dt + \sigma dB_t - \frac{1}{2}\sigma^2 dt, \end{aligned}$$

hence

$$\log S_t - \log S_0 = \int_0^t d\log S_r$$

$$= \int_0^t \left(\mu - \frac{1}{2}\sigma^2\right) dr + \int_0^t \sigma dB_r$$

$$= \left(\mu - \frac{1}{2}\sigma^2\right) t + \sigma B_t, \qquad t \in \mathbb{R}_+,$$

and

$$S_t = S_0 \exp\left(\left(\mu - \frac{1}{2}\sigma^2\right) t + \sigma B_t\right), \qquad t \in \mathbb{R}_+.$$

The above provides a proof of the next proposition.

Proposition 4.3 *The solution of* (4.20) *is given by*

$$S_t = S_0 e^{\mu t + \sigma B_t - \sigma^2 t/2}, \qquad t \in \mathbb{R}_+.$$

Proof. Let us provide an alternative proof by searching for a solution of the form

$$S_t = f(t, B_t)$$

where $f(t, x)$ is a function to be determined. By Itô's formula (4.19) we have

$$dS_t = df(t, B_t) = \frac{\partial f}{\partial t}(t, B_t)dt + \frac{\partial f}{\partial x}(t, B_t)dB_t + \frac{1}{2}\frac{\partial^2 f}{\partial x^2}(t, B_t)dt.$$

Comparing this expression to (4.20) and identifying the terms in dB_t we get

$$\frac{\partial f}{\partial x}(t, B_t) = \sigma S_t,$$

and

$$\frac{\partial f}{\partial t}(t, B_t) + \frac{1}{2}\frac{\partial^2 f}{\partial x^2}(t, B_t) = \mu S_t.$$

Using the relation $S_t = f(t, B_t)$ these two equations rewrite as

$$\frac{\partial f}{\partial x}(t, B_t) = \sigma f(t, B_t),$$

and

$$\frac{\partial f}{\partial t}(t, B_t) + \frac{1}{2}\frac{\partial^2 f}{\partial x^2}(t, B_t) = \mu f(t, B_t).$$

Since B_t is a Gaussian random variable taking all possible values in $\mathbb{R}$, the equations should hold for all $x \in \mathbb{R}$, as follows:

$$\frac{\partial f}{\partial x}(t, x) = \sigma f(t, x), \tag{4.22}$$

and

$$\frac{\partial f}{\partial t}(t, x) + \frac{1}{2}\frac{\partial^2 f}{\partial x^2}(t, x) = \mu f(t, x). \tag{4.23}$$

Letting $g(t,x) = \log f(t,x)$, the first equation (4.22) shows that

$$\frac{\partial g}{\partial x}(t,x) = \frac{\partial \log f}{\partial x}(t,x) = \frac{1}{f(t,x)}\frac{\partial f}{\partial x}(t,x) = \sigma,$$

i.e.,

$$\frac{\partial g}{\partial x}(t,x) = \sigma,$$

which is solved as

$$g(t,x) = g(t,0) + \sigma x,$$

hence

$$f(t,x) = e^{g(t,0)}e^{\sigma x} = f(t,0)e^{\sigma x}.$$

Plugging back this expression into the second equation (4.23) yields

$$e^{\sigma x}\frac{\partial f}{\partial t}(t,0) + \frac{1}{2}\sigma^2 e^{\sigma x} f(t,0) = \mu f(t,0)e^{\sigma x},$$

i.e., after division by $e^{\sigma x}$:

$$\frac{\partial f}{\partial t}(t,0) = \left(\mu - \sigma^2/2\right) f(t,0),$$

or

$$\frac{\partial g}{\partial t}(t,0) = \mu - \sigma^2/2,$$

i.e.,

$$g(t,0) = g(0,0) + \left(\mu - \sigma^2/2\right) t,$$

and

$$\begin{aligned} f(t,x) &= e^{g(t,x)} \\ &= e^{g(t,0)+\sigma x} \\ &= e^{g(0,0)+\sigma x+\left(\mu-\sigma^2/2\right)t} \\ &= f(0,0)e^{\sigma x+\left(\mu-\sigma^2/2\right)t}, \end{aligned}$$

hence

$$S_t = f(t,B_t) = f(0,0)e^{\sigma B_t+\left(\mu-\sigma^2/2\right)t},$$

and the solution to (4.20) is given by

$$S_t = S_0 e^{\sigma B_t+\left(\mu-\sigma^2/2\right)t} \qquad t \in \mathbb{R}_+.$$

□

Conversely, taking $S_t = f(t, B_t)$ with $f(t,x) = S_0 e^{\sigma x - \sigma^2 t/2 + \mu t}$ we may apply Itô's formula to check that

$$\begin{aligned}
dS_t &= df(t, B_t) \\
&= \frac{\partial f}{\partial t}(t, B_t)dt + \frac{\partial f}{\partial x}(t, B_t)dB_t + \frac{1}{2}\frac{\partial^2 f}{\partial x^2}(t, B_t)dt \\
&= \left(\mu - \sigma^2/2\right) S_0 e^{\sigma B_t + \left(\mu - \sigma^2/2\right)t} dt + \sigma S_0 e^{\sigma B_t + \left(\mu - \sigma^2/2\right)t} dB_t \\
&\quad + \frac{1}{2}\sigma^2 S_0 e^{\sigma B_t + \left(\mu - \sigma^2/2\right)t} dt \\
&= \mu S_0 e^{\sigma B_t + \left(\mu - \sigma^2/2\right)t} dt + \sigma S_0 e^{\sigma B_t + \left(\mu - \sigma^2/2\right)t} dB_t \\
&= \mu S_t dt + \sigma S_t dB_t.
\end{aligned}$$

4.7 Stochastic Differential Equations

In addition to geometric Brownian motion there exists a large family of stochastic differential equations that can be studied, although most of the time they cannot be explicitly solved. Now let

$$\sigma : \mathbb{R}_+ \times \mathbb{R}^n \to \mathbb{R}^d \otimes \mathbb{R}^n$$

where $\mathbb{R}^d \otimes \mathbb{R}^n$ denotes the space of $d \times n$ matrices, and

$$b : \mathbb{R}_+ \times \mathbb{R}^n \to \mathbb{R}$$

satisfies the global Lipschitz condition

$$\|\sigma(t,x) - \sigma(t,y)\|^2 + \|b(t,x) - b(t,y)\|^2 \leq K^2 \|x - y\|^2,$$

$t \in \mathbb{R}_+$, $x, y \in \mathbb{R}^n$. Then there exists a unique strong solution to the stochastic differential equation

$$X_t = X_0 + \int_0^t \sigma(s, X_s) dB_s + \int_0^t b(s, X_s) ds, \qquad t \in \mathbb{R}_+,$$

where $(B_t)_{t \in \mathbb{R}_+}$ is a d-dimensional Brownian motion, cf. [61].

Next we consider a few examples of stochastic differential equations that can be solved explicitly using Itô calculus.

Examples

1. Consider the stochastic differential equation

$$dX_t = -\alpha X_t dt + \sigma dB_t, \quad X_0 = x_0,$$

with $\alpha > 0$ and $\sigma > 0$.

Looking for a solution of the form

$$X_t = a(t)\left(x_0 + \int_0^t b(s)dB_s\right)$$

where $a(\cdot)$ and $b(\cdot)$ are deterministic functions, yields

$$X_t = x_0 e^{-\alpha t} + \sigma \int_0^t e^{-\alpha(t-s)} dB_s, \qquad t > 0,$$

after applying Theorem 4.1 to the Itô process $x_0 + \int_0^t b(s)dB_s$ of the form (4.15) with $u_t = b(t)$ and $v(t) = 0$, and to the function $f(t, x) = a(t)x$.

Remark: The solution of this equation *cannot* be written as a function $f(t, B_t)$ of t and B_t as in the proof of Proposition 4.3.

2. Consider the stochastic differential equation

$$dX_t = tX_t dt + e^{t^2/2} dB_t, \qquad X_0 = x_0.$$

Looking for a solution of the form $X_t = a(t)\left(X_0 + \int_0^t b(s)dB_s\right)$, where $a(\cdot)$ and $b(\cdot)$ are deterministic functions we get $a'(t)/a(t) = t$ and $a(t)b(t) = e^{t^2/2}$, hence $a(t) = e^{t^2/2}$ and $b(t) = 1$, which yields $X_t = e^{t^2/2}(X_0 + B_t)$, $t \in \mathbb{R}_+$.

3. Consider the stochastic differential equation

$$dY_t = (2\mu Y_t + \sigma^2)dt + 2\sigma\sqrt{Y_t} dB_t,$$

where $\mu, \sigma > 0$.

Letting $X_t = \sqrt{Y_t}$ we have $dX_t = \mu X_t dt + \sigma dB_t$, hence

$$Y_t = \left(e^{\mu t}\sqrt{Y_0} + \sigma \int_0^t e^{\mu(t-s)} dB_s\right)^2.$$

Exercises

Exercise 4.1 Let $(B_t)_{t\in\mathbb{R}_+}$ denote a standard Brownian motion.

1. Let $c > 0$. Among the following processes, tell which is a standard Brownian motion and which is not. Justify your answer.

   (a) $(B_{c+t} - B_c)_{t\in\mathbb{R}_+}$.

   (b) $(cB_{t/c^2})_{t\in\mathbb{R}_+}$.

   (c) $(B_{ct^2})_{t\in\mathbb{R}_+}$.

2. Compute the stochastic integrals

$$\int_0^T 2dB_t \qquad \text{and} \qquad \int_0^T (2 \times \mathbf{1}_{[0,T/2]}(t) + \mathbf{1}_{(T/2,T]}(t))dB_t$$

   and determine their probability laws (including mean and variance).

3. Determine the probability law (including mean and variance) of the stochastic integral

$$\int_0^{2\pi} \sin(t)\, dB_t.$$

4. Compute $\mathbb{E}[B_t B_s]$ in terms of $s, t \geq 0$.

5. Let $T > 0$. Show that if f is a differentiable function with $f(0) = f(T) = 0$ we have

$$\int_0^T f(t)dB_t = -\int_0^T f'(t)B_t dt.$$

   Hint: Apply Itô's calculus to $t \mapsto f(t)B_t$.

Exercise 4.2 Let $f \in L^2([0,T])$. Compute the conditional expectation

$$E\left[e^{\int_0^T f(s)dB_s} \Big| \mathcal{F}_t\right], \qquad 0 \leq t \leq T,$$

where $(\mathcal{F}_t)_{t\in[0,T]}$ denotes the filtration generated by $(B_t)_{t\in[0,T]}$.

Exercise 4.3 Compute the expectation

$$E\left[\exp\left(\beta \int_0^T B_t dB_t\right)\right]$$

for all $\beta < 1/T$. Hint: expand $(B_T)^2$ using Itô's formula.

Exercise 4.4 Solve the ordinary differential equation $df(t) = cf(t)dt$ and the stochastic differential equation $dS_t = rS_t dt + \sigma S_t dB_t$, $t \in \mathbb{R}_+$, where $r, \sigma \in \mathbb{R}$ are constants and $(B_t)_{t\in\mathbb{R}_+}$ is a standard Brownian motion.

Exercise 4.5 Given $T > 0$, let $(X_t^T)_{t\in[0,T]}$ denote the solution of the stochastic differential equation

$$dX_t^T = \sigma dB_t - \frac{X_t^T}{T-t}dt, \qquad t \in [0,T], \tag{4.24}$$

under the initial condition $X_0^T = 0$ and $\sigma > 0$.

1. Show that

$$X_t^T = \sigma(T-t)\int_0^t \frac{1}{T-s}dB_s, \qquad t \in [0,T].$$

Hint: start by computing $d(X_t^T/(T-t))$ using Itô's calculus.

2. Show that $\mathbb{E}[X_t^T] = 0$ for all $t \in [0,T]$.

3. Show that $\mathrm{Var}[X_t^T] = \sigma^2 t(T-t)/T$ for all $t \in [0,T]$.

4. Show that $X_T^T = 0$. The process $(X_t^T)_{t\in[0,T]}$ is called a *Brownian bridge.*

Exercise 4.6 Exponential Vasicek model. Consider a short term rate interest rate proces $(r_t)_{t\in\mathbb{R}_+}$ in the exponential Vasicek model:

$$dr_t = r_t(\eta - a\log r_t)dt + \sigma r_t dB_t, \tag{4.25}$$

where η, a, σ are positive parameters.

1. Find the solution $(z_t)_{t\in\mathbb{R}_+}$ of the stochastic differential equation

$$dz_t = -az_t dt + \sigma dB_t$$

as a function of the initial condition z_0, where a and σ are positive parameters.

2. Find the solution $(y_t)_{t\in\mathbb{R}_+}$ of the stochastic differential equation

$$dy_t = (\theta - ay_t)dt + \sigma dB_t \tag{4.26}$$

as a function of the initial condition y_0. Hint: let $z_t = y_t - \theta/a$.

3. Let $x_t = e^{y_t}$, $t \in \mathbb{R}_+$. Determine the stochastic differential equation satisfied by $(x_t)_{t\in\mathbb{R}_+}$.

4. Find the solution $(r_t)_{t\in\mathbb{R}_+}$ of (4.25) in terms of the initial condition r_0.

5. Compute the mean[1] $E[r_t]$ of r_t, $t \geq 0$.

6. Compute the asymptotic mean $\lim_{t\to\infty} E[r_t]$.

Exercise 4.7 Cox–Ingerson–Ross model. Consider the equation

$$dr_t = (\alpha - \beta r_t)dt + \sigma\sqrt{r_t}dB_t \tag{4.27}$$

modeling the variations of a short term interest rate process r_t, where α, β, σ and r_0 are positive parameters.

---

[1]You will need to use the generating function $E[e^X] = e^{\alpha^2/2}$ for $X \simeq \mathcal{N}(0,\alpha^2)$.

1. Write down the equation (4.27) in integral form.

2. Let $u(t) = E[r_t]$. Show, using the integral form of (4.27), that $u(t)$ satisfies the differential equation

$$u'(t) = \alpha - \beta u(t).$$

3. By an application of Itô's formula to r_t^2, show that

$$dr_t^2 = r_t(2\alpha + \sigma^2 - 2\beta r_t)dt + 2\sigma r_t^{3/2} dB_t. \tag{4.28}$$

4. Using the integral form of (4.28), find a differential equation satisfied by $v(t) = E[r_t^2]$.

Exercise 4.8 Let $(B_t)_{t\in\mathbb{R}_+}$ denote a standard Brownian motion generating the filtration $(\mathcal{F}_t)_{t\in\mathbb{R}_+}$.

1. Consider the Itô formula

$$f(X_t) = f(X_0)+\int_0^t u_s \frac{\partial f}{\partial x}(X_s)dB_s+\int_0^t v_s \frac{\partial f}{\partial x}(X_s)ds+\frac{1}{2}\int_0^t u_s^2 \frac{\partial^2 f}{\partial x^2}(X_s)ds, \tag{4.29}$$

where $X_t = X_0 + \int_0^t u_s dB_s + \int_0^t v_s ds$.

Compute $S_t := e^{X_t}$ by the Itô formula (4.29) applied to $f(x) = e^x$ and $X_t = \sigma B_t + \nu t$, $\sigma > 0$, $\nu \in \mathbb{R}$.

2. Let $r > 0$. For which value of ν does $(S_t)_{t\in\mathbb{R}_+}$ satisfy the stochastic differential equation

$$dS_t = rS_t dt + \sigma S_t dB_t \quad ?$$

3. Let the process $(S_t)_{t\in\mathbb{R}_+}$ be defined by $S_t = S_0 e^{\sigma B_t + \nu t}$, $t \in \mathbb{R}_+$. Using the result of Exercise A.2, show that the conditional probability $P(S_T > K \mid S_t = x)$ is given by

$$P(S_T > K \mid S_t = x) = \Phi\left(\frac{\log(x/K) + \nu\tau}{\sigma\sqrt{\tau}}\right),$$

where $\tau = T - t$. Hint: use the decomposition $S_T = S_t e^{\sigma(B_T - B_t) + \nu\tau}$.

4. Given $0 \le t \le T$ and $\sigma > 0$, let

$$X = \sigma(B_T - B_t) \qquad \text{and} \qquad \eta^2 = \mathrm{Var}[X], \qquad \eta > 0.$$

What is η equal to ?

Exercise 4.9 Let $(B_t)_{t\in\mathbb{R}_+}$ be a standard Brownian motion generating the information flow $(\mathcal{F}_t)_{t\in\mathbb{R}_+}$.

1. Let $0 \leq t \leq T$. What is the probability law of $B_T - B_t$?

2. From the answer to Exercise A.5, show that

$$\mathbb{E}[(B_T)^+ \mid \mathcal{F}_t] = \sqrt{\frac{\tau}{2\pi}} e^{-\frac{B_t^2}{2\tau}} + B_t \Phi\left(\frac{B_t}{\sqrt{\tau}}\right),$$

$0 \leq t \leq T$, where $\tau = T - t$. Hint: write $B_T = B_T - B_t + B_t$.

3. Let $\sigma > 0$, $\nu \in \mathbb{R}$, and $X_t = \sigma B_t + \nu t$. Compute e^{X_t} using the Itô formula

$$f(X_t) = f(X_0) + \int_0^t u_s \frac{\partial f}{\partial x}(X_s) dB_s + \int_0^t v_s \frac{\partial f}{\partial x}(X_s) ds + \frac{1}{2} \int_0^t u_s^2 \frac{\partial^2 f}{\partial x^2}(X_s) ds$$

stated here for a process $X_t = X_0 + \int_0^t u_s dB_s + \int_0^t v_s ds$, $t \in \mathbb{R}_+$, and applied to $f(x) = e^x$.

4. Let $S_t = e^{X_t}$, $t \in \mathbb{R}_+$, and $r > 0$. For which value of ν does $(S_t)_{t\in\mathbb{R}_+}$ satisfy the stochastic differential equation

$$dS_t = r S_t dt + \sigma S_t dB_t \quad ?$$

Exercise 4.10 From the answer to Exercise A.4-(2), show that

$$\mathbb{E}[(\beta - B_T)^+ \mid \mathcal{F}_t] = \sqrt{\frac{\tau}{2\pi}} e^{-\frac{(\beta - B_t)^2}{2\tau}} + (\beta - B_t)\Phi\left(\frac{\beta - B_t}{\sqrt{\tau}}\right), \qquad 0 \leq t \leq T,$$

where $\tau = T - t$. Hint: write $B_T = B_T - B_t + B_t$.

Chapter 5

The Black–Scholes PDE

In this section we review the notions of assets, self-financing portfolios, risk-neutral measures, and arbitrage in continuous time. We also derive the Black–Scholes PDE for self-financing portfolios, and we solve this equation using the heat kernel method.

5.1 Continuous-Time Market Model

Let $(A_t)_{t\in\mathbb{R}_+}$ be the riskless asset given by

$$\frac{dA_t}{A_t} = rdt, \qquad t \in \mathbb{R}_+,$$

i.e.,

$$A_t = A_0 e^{rt}, \qquad t \in \mathbb{R}_+.$$

For $t > 0$, let $(S_t)_{t\in\mathbb{R}_+}$ be the price process defined as

$$dS_t = \mu S_t dt + \sigma S_t dB_t, \quad t \in \mathbb{R}_+.$$

By Proposition 4.3 we have

$$S_t = S_0 \exp\left(\sigma B_t + \left(\mu - \frac{1}{2}\sigma^2\right)t\right), \qquad t \in \mathbb{R}_+.$$

5.2 Self-Financing Portfolio Strategies

Let ξ_t and η_t denote the (possibly fractional) quantities invested at time t, respectively in the assets S_t and A_t, and let

$$\bar{\xi}_t = (\eta_t, \xi_t), \qquad \bar{S}_t = (A_t, S_t), \qquad t \in \mathbb{R}_+,$$

denote the associated portfolio and asset price processes. The value of the portfolio V_t at time t is given by

$$V_t = \bar{\xi}_t \cdot \bar{S}_t = \eta_t A_t + \xi_t S_t, \qquad t \in \mathbb{R}_+. \tag{5.1}$$

The portfolio strategy $(\eta_t, \xi_t)_{t\in\mathbb{R}_+}$ is self-financing if the portfolio value remains constant after updating the portfolio from (η_t, ξ_t) to $(\eta_{t+dt}, \xi_{t+dt})$, i.e.,

$$\bar{\xi}_t \cdot \bar{S}_{t+dt} = A_{t+dt}\eta_t + S_{t+dt}\xi_t = A_{t+dt}\eta_{t+dt} + S_{t+dt}\xi_{t+dt} = \bar{\xi}_{t+dt} \cdot \bar{S}_{t+dt}, \quad (5.2)$$

which is the continuous-time equivalent of the self-financing condition already encountered in the discrete setting of Chapter 2, see Definition 2.1. A major difference with the discrete-time case, however, is that the continuous-time differentials dS_t and $d\xi_t$ do not make pathwise sense as the stochastic integral is defined by an L^2 limit, cf. Proposition 4.2, or by convergence in probability.

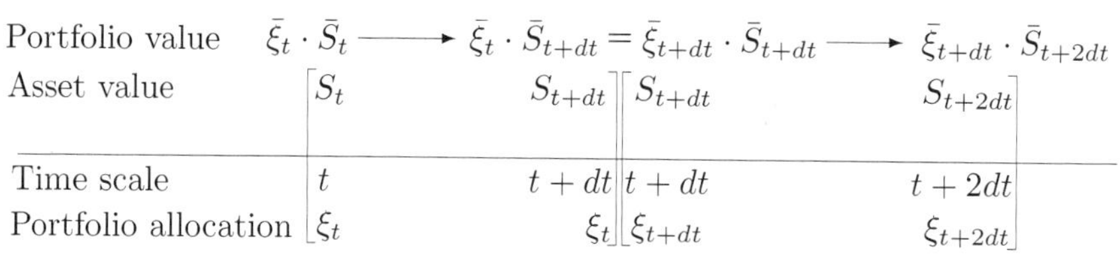

FIGURE 5.1: Illustration of the self-financing condition (5.2).

Equivalently, Condition (5.2) can be rewritten as

$$A_{t+dt}d\eta_t + S_{t+dt}d\xi_t = 0, \quad (5.3)$$

or

$$A_{t+dt}(\eta_{t+dt} - \eta_t) = -S_{t+dt}(\xi_{t+dt} - \xi_t),$$

i.e., when one sells a quantity $-d\xi_t > 0$ of the risky asset S_{t+dt} between the time periods $[t, t+dt]$ and $[t+dt, t+2dt]$ for a total amount $-S_{t+dt}d\xi_t$, one should entirely use this income to buy a quantity $d\eta_t > 0$ of the riskless asset for an amount $A_{t+dt}d\eta_t > 0$.

Similarly, if one sells a (possibly fractional) quantity $-d\eta_t > 0$ of the riskless asset A_{t+dt} between the time periods $[t, t+dt]$ and $[t+dt, t+2dt]$ for a total amount $-A_{t+dt}d\eta_t$, one should entirely use this income to buy a quantity $d\xi_t > 0$ of the risky asset for an amount $S_{t+dt}d\xi_t > 0$, i.e.,

$$S_{t+dt}d\xi_t = -A_{t+dt}d\eta_t,$$

or

$$A_{t+dt}(\eta_{t+dt} - \eta_t) + S_t(\xi_{t+dt} - \xi_t) + (S_{t+dt} - S_t)(\xi_{t+dt} - \xi_t) = 0,$$

which rewrites as

$$A_t d\eta_t + S_t d\xi_t + dS_t \cdot d\xi_t = 0$$

in differential notation, since $dA_t \cdot d\eta_t = (A_{t+dt} - A_t)(\eta_{t+dt} - \eta_t) = r\eta_t dt \cdot dA_t \simeq 0$ in the sense of the Itô calculus. This leads to the following definition.

Definition 5.1 *The portfolio V_t is said to be self-financing if*

$$dV_t = \eta_t dA_t + \xi_t dS_t. \tag{5.4}$$

Again we check that by Itô's calculus we have

$$\begin{aligned} dV_t &= \eta_t dA_t + \xi_t dS_t + A_t d\eta_t + S_t d\xi_t + d\eta_t \cdot dA_t + d\xi_t \cdot dS_t \\ &= \eta_t dA_t + \xi_t dS_t + A_t d\eta_t + S_t d\xi_t + d\xi_t \cdot dS_t, \end{aligned}$$

since $d\eta_t \cdot dA_t = rA_t dt \cdot d\eta_t = 0$, hence Condition (5.4) rewrites as

$$A_t d\eta_t + S_t d\xi_t + d\xi_t \cdot dS_t = 0,$$

which is equivalent to (5.2) and (5.3).

Let

$$\tilde{V}_t = e^{-rt} V_t \qquad \text{and} \qquad X_t = e^{-rt} S_t$$

respectively denote the discounted portfolio value and discounted risky asset prices at time $t \geq 0$. We have

$$\begin{aligned} dX_t &= d(e^{-rt} S_t) \\ &= -re^{-rt} S_t dt + e^{-rt} dS_t \\ &= -re^{-rt} S_t dt + \mu e^{-rt} S_t dt + \sigma e^{-rt} S_t dB_t \\ &= X_t((\mu - r)dt + \sigma dB_t). \end{aligned}$$

In the next lemma we show that when a portfolio is self-financing, its discounted value is a gain process given by the sum of discounted profits and losses (number of risky assets ξ_t times discounted price variation dX_t) over time.

Lemma 5.1 *Let $(\eta_t, \xi_t)_{t \in \mathbb{R}_+}$ be a portfolio strategy with value*

$$V_t = \eta_t A_t + \xi_t S_t, \qquad t \in \mathbb{R}_+.$$

The following statements are equivalent:

i) the portfolio strategy $(\eta_t, \xi_t)_{t \in \mathbb{R}_+}$ is self-financing,

ii) we have

$$\tilde{V}_t = \tilde{V}_0 + \int_0^t \xi_u dX_u, \qquad t \in \mathbb{R}_+. \tag{5.5}$$

Proof. Assuming that (i) holds, the self-financing condition shows that

$$\begin{aligned} dV_t &= \eta_t dA_t + \xi_t dS_t \\ &= r\eta_t A_t dt + \mu \xi_t S_t dt + \sigma \xi_t S_t dB_t \\ &= rV_t dt + (\mu - r)\xi_t S_t dt + \sigma \xi_t S_t dB_t \qquad t \in \mathbb{R}_+, \end{aligned}$$

hence

$$\begin{aligned} d\tilde{V}_t &= d\left(e^{-rt}V_t\right) \\ &= -re^{-rt}V_t dt + e^{-rt}dV_t \\ &= (\mu - r)\xi_t X_t dt + \sigma\xi_t X_t dB_t \\ &= \xi_t dX_t, \qquad t \in \mathbb{R}_+, \end{aligned}$$

i.e., (5.5) holds by integrating on both sides as

$$\tilde{V}_t - \tilde{V}_0 = \int_0^t d\tilde{V}_u = \int_0^t \xi_u dX_u, \qquad t \in \mathbb{R}_+.$$

Conversely, if (5.5) is satisfied we have

$$\begin{aligned} dV_t &= d(e^{rt}\tilde{V}_t) \\ &= re^{rt}\tilde{V}_t dt + e^{rt}d\tilde{V}_t \\ &= re^{rt}\tilde{V}_t dt + e^{rt}\xi_t dX_t \\ &= rV_t dt + e^{rt}\xi_t dX_t \\ &= rV_t dt + e^{rt}\xi_t X_t((\mu - r)dt + \sigma dB_t) \\ &= rV_t dt + \xi_t S_t((\mu - r)dt + \sigma dB_t) \\ &= r\eta_t A_t dt + \mu\xi_t S_t dt + \sigma\xi_t S_t dB_t \\ &= \eta_t dA_t + \xi_t dS_t, \end{aligned}$$

hence the portfolio is self-financing according to Definition 5.1. □

As a consequence of (5.5), the hedging problem of a claim C with maturiry T is reduced to that of finding the representation of the discounted claim $\tilde{C} = e^{-rT}C$ as a stochastic integral:

$$\tilde{C} = \tilde{V}_0 + \int_0^T \xi_u dX_u.$$

Note also that (5.5) shows that the value of a self-financing portfolio can also be written as

$$V_t = e^{rt}V_0 + (\mu - r)\int_0^t e^{r(t-u)}\xi_u S_u dt + \sigma\int_0^t e^{r(t-u)}\xi_u S_u dB_u, \qquad t \in \mathbb{R}_+. \tag{5.6}$$

5.3 Arbitrage and Risk-Neutral Measures

In continuous-time, the definition of arbitrage follows the lines of its analogs in the discrete and two-step models. In the sequel we will only consider *admissible* portfolio strategies whose total value V_t remains non-negative for all times $t \in [0, T]$.

Definition 5.2 *A portfolio strategy* $(\xi_t, \eta_t)_{t\in[0,T]}$ *constitutes an arbitrage opportunity if all* three *following conditions are satisfied:*

i) $V_0 \le 0$,

ii) $V_T \ge 0$,

iii) $\mathbb{P}(V_T > 0) > 0$.

Roughly speaking, (ii) means that the investor wants no loss, (iii) means that he wishes to sometimes make a strictly positive gain, and (i) means that he starts with zero capital or even with a debt.

Next we turn to the definition of risk-neutral measures in continuous time. Recall that the filtration $(\mathcal{F}_t)_{t\in\mathbb{R}_+}$ is generated by Brownian motion $(B_t)_{t\in\mathbb{R}_+}$, i.e.,

$$\mathcal{F}_t = \sigma(B_u \ : \ 0 \le u \le t), \qquad t \in \mathbb{R}_+.$$

Definition 5.3 *A probability measure* $\mathbb{P}^*$ *on* Ω *is called a risk-neutral measure if it satisfies*

$$\mathbb{E}^*[S_t|\mathcal{F}_u] = e^{r(t-u)} S_u, \qquad 0 \le u \le t, \tag{5.7}$$

where $\mathbb{E}^*$ *denotes the expectation under* $\mathbb{P}^*$.

From the relation

$$A_t = e^{r(t-u)} A_u, \qquad 0 \le u \le t,$$

we interpret (5.7) by saying that the expected return of the risky asset S_t under $\mathbb{P}^*$ equals the return of the riskless asset A_t. The discounted price X_t of the risky asset is defined by

$$X_t = e^{-rt} S_t = \frac{S_t}{A_t/A_0}, \qquad t \in \mathbb{R}_+,$$

i.e., A_t/A_0 plays the role of a *numéraire* in the sense of Chapter 10.

Definition 5.4 *A continuous time process* $(Z_t)_{t\in\mathbb{R}_+}$ *of integrable random variables is a martingale with respect to the filtration* $(\mathcal{F}_t)_{t\in\mathbb{R}_+}$ *if*

$$\mathbb{E}[Z_t|\mathcal{F}_s] = Z_s, \qquad 0 \le s \le t.$$

Note that when $(Z_t)_{t\in\mathbb{R}_+}$ is a martingale, Z_t is in particular $\mathcal{F}_t$-measurable for all $t \in \mathbb{R}_+$.

As in the discrete case, the notion of martingale can be used to characterize risk-neutral measures.

Proposition 5.1 *The measure* $\mathbb{P}^*$ *is risk-neutral if and only if the discounted price process* $(X_t)_{t\in\mathbb{R}_+}$ *is a martingale* under $\mathbb{P}^*$.

Proof. This follows from the equalities

$$\begin{aligned}\mathbb{E}^*[X_t|\mathcal{F}_u] &= \mathbb{E}^*[e^{-rt}S_t|\mathcal{F}_u]\\ &= e^{-rt}\,\mathbb{E}^*[S_t|\mathcal{F}_u]\\ &= e^{-rt}e^{r(t-u)}S_u\\ &= e^{-ru}S_u\\ &= X_u, \qquad 0 \le u \le t.\end{aligned}$$

□

As in the discrete time case, $\mathbb{P}^*$ would be called a risk-premium measure if it satisfied

$$\mathbb{E}^*[S_t|\mathcal{F}_u] > e^{r(t-u)}S_u, \qquad 0 \le u \le t,$$

meaning that by taking risks in buying S_t, one could make an expected return higher than that of

$$A_t = e^{r(t-u)}A_u, \qquad 0 \le u \le t.$$

Next we note that the first fundamental theorem of mathematical finance also holds in continuous time, and can be used to check for the existence of arbitrage opportunities.

Theorem 5.1 *A market is* without *an arbitrage opportunity if and only if it admits at least one risk-neutral measure.*

Proof. cf. Chapter VII-4a of [67]. □

5.4 Market Completeness

Definition 5.5 *A contingent claim with payoff C is said to be attainable if there exists a portfolio strategy $(\eta_t, \xi_t)_{t\in[0,T]}$ such that*

$$C = V_T.$$

In this case the price of the claim at time t will be equal to the value V_t of any self-financing portfolio hedging C.

Definition 5.6 *A market model is said to be* complete *if every contingent claim C is attainable.*

The next result is a continuous-time restatement of the second fundamental theorem of mathematical finance.

Theorem 5.2 *A market model without arbitrage is complete if and only if it admits only one risk-neutral measure.*

Proof. cf. Chapter VII-4a of [67]. □

In the Black–Scholes model one can show the existence of a unique risk-neutral measure, hence the model is without arbitrage and complete.

5.5 Black–Scholes PDE

We start by deriving the Black–Scholes partial differential equation (PDE) for the price of a self-financing portfolio.

Proposition 5.2 *Let $(\eta_t, \xi_t)_{t\in\mathbb{R}_+}$ be a portfolio strategy such that*

(i) $(\eta_t, \xi_t)_{t\in\mathbb{R}_+}$ is self-financing,

(ii) the value $V_t := \eta_t A_t + \xi_t S_t$, $t \in \mathbb{R}_+$, takes the form

$$V_t = g(t, S_t), \qquad t \in \mathbb{R}_+,$$

for some $g \in \mathcal{C}^{1,2}((0,\infty)\times(0,\infty))$.

Then the function $g(t,x)$ satisfies the Black–Scholes PDE

$$rg(t,x) = \frac{\partial g}{\partial t}(t,x) + rx\frac{\partial g}{\partial x}(t,x) + \frac{1}{2}x^2\sigma^2\frac{\partial^2 g}{\partial x^2}(t,x), \qquad t, x > 0,$$

and ξ_t is given by

$$\xi_t = \frac{\partial g}{\partial x}(t, S_t), \qquad t \in \mathbb{R}_+. \tag{5.8}$$

Proof. First, note that the self-financing condition implies

$$\begin{aligned} dV_t &= \eta_t dA_t + \xi_t dS_t \\ &= r\eta_t A_t dt + \mu\xi_t S_t dt + \sigma\xi_t S_t dB_t \\ &= rV_t dt + (\mu - r)\xi_t S_t dt + \sigma\xi_t S_t dB_t, \end{aligned} \tag{5.9}$$

$t \in \mathbb{R}_+$.

We now rewrite (4.21) under the form of an Itô process

$$S_t = S_0 + \int_0^t v_s ds + \int_0^t u_s dB_s, \qquad t \in \mathbb{R}_+,$$

as in (4.15), by taking

$$u_t = \sigma S_t, \quad \text{and} \quad v_t = \mu S_t, \qquad t \in \mathbb{R}_+.$$

The application of Itô's formula Theorem 4.1 to $g(t,x)$ leads to

$$\begin{aligned} dg(t,S_t) &= g(0,S_0) + v_t \frac{\partial g}{\partial x}(t,S_t)dt + u_t \frac{\partial g}{\partial x}(t,S_t)dB_t \\ &\quad + \frac{\partial g}{\partial t}(t,S_t)dt + \frac{1}{2}|u_t|^2 \frac{\partial^2 g}{\partial x^2}(t,S_t)dt \\ &= \frac{\partial g}{\partial t}(t,S_t)dt + \mu S_t \frac{\partial g}{\partial x}(t,S_t)dt + \frac{1}{2}S_t^2\sigma^2 \frac{\partial^2 g}{\partial x^2}(t,S_t)dt + \sigma S_t \frac{\partial g}{\partial x}(t,S_t)dB_t. \end{aligned} \tag{5.10}$$

By respective identification of the terms in dB_t and dt in (5.9) and (5.10) we get

$$\begin{cases} r\eta_t A_t dt + \mu \xi_t S_t dt = \dfrac{\partial g}{\partial t}(t,S_t)dt + \mu S_t \dfrac{\partial g}{\partial x}(t,S_t)dt + \dfrac{1}{2}S_t^2\sigma^2 \dfrac{\partial^2 g}{\partial x^2}(t,S_t)dt, \\ \\ \xi_t S_t \sigma dB_t = S_t \sigma \dfrac{\partial g}{\partial x}(t,S_t)dB_t, \end{cases}$$

hence

$$\begin{cases} rV_t - r\xi_t S_t = \dfrac{\partial g}{\partial t}(t,S_t) + \dfrac{1}{2}S_t^2\sigma^2 \dfrac{\partial^2 g}{\partial x^2}(t,S_t), \\ \\ \xi_t = \dfrac{\partial g}{\partial x}(t,S_t), \end{cases}$$

i.e.,

$$\begin{cases} rg(t,S_t) = \dfrac{\partial g}{\partial t}(t,S_t) + rS_t \dfrac{\partial g}{\partial x}(t,S_t) + \dfrac{1}{2}S_t^2\sigma^2 \dfrac{\partial^2 g}{\partial x^2}(t,S_t), \\ \\ \xi_t = \dfrac{\partial g}{\partial x}(t,S_t). \end{cases} \tag{5.11}$$

□

The derivative giving ξ_t in (5.8) is called the *Delta* of the option price.

The amount invested on the riskless asset is

$$\eta_t A_t = V_t - \xi_t S_t = g(t,S_t) - S_t \frac{\partial g}{\partial x}(t,S_t),$$

and η_t is given by

$$\eta_t = \frac{V_t - \xi_t S_t}{A_t}$$

$$= \frac{g(t,S_t) - S_t \frac{\partial g}{\partial x}(t,S_t)}{A_t}$$

$$= \frac{g(t,S_t) - S_t \frac{\partial g}{\partial x}(t,S_t)}{A_0 e^{rt}}.$$

In the next proposition we add a terminal condition $g(T,x) = f(x)$ to the Black–Scholes PDE in order to hedge claim C of the form $C = f(S_T)$.

Proposition 5.3 *The price of any self-financing portfolio of the form $V_t = g(t,S_t)$ hedging an option with payoff $C = f(S_T)$ satisfies the Black–Scholes PDE*

$$\begin{cases} rg(t,x) = \frac{\partial g}{\partial t}(t,x) + rx\frac{\partial g}{\partial x}(t,x) + \frac{1}{2}x^2\sigma^2\frac{\partial^2 g}{\partial x^2}(t,x), \\ g(T,x) = f(x). \end{cases}$$

The Black–Scholes PDE admits an easy solution when $C = S_T - K$ is the (linear) payoff of a forward contract, i.e., $f(x) = x - K$. In this case we find

$$g(t,x) = x - Ke^{-r(T-t)}, \qquad t,x > 0,$$

and the Delta of the option price is given by

$$\xi_t = \frac{\partial g}{\partial x}(t,S_t) = 1, \qquad 0 \le t \le T.$$

Recall that in the case of a European call option with strike K the payoff function is given by $f(x) = (x-K)^+$ and the Black–Scholes PDE reads

$$\begin{cases} rg_c(t,x) = \frac{\partial g_c}{\partial t}(t,x) + rx\frac{\partial g_c}{\partial x}(t,x) + \frac{1}{2}x^2\sigma^2\frac{\partial^2 g_c}{\partial x^2}(t,x) \\ g_c(T,x) = (x-K)^+. \end{cases}$$

In the next sections we will prove that the solution of this PDE is given by the *Black–Scholes* formula

$$g_c(t,x) = \mathrm{BS}(K,x,\sigma,r,T-t) = x\Phi(d_+) - Ke^{-r(T-t)}\Phi(d_-), \tag{5.12}$$

cf. Proposition 5.7 below, where

$$\Phi(x) = \frac{1}{\sqrt{2\pi}}\int_{-\infty}^{x} e^{-y^2/2}dy, \qquad x \in \mathbb{R},$$

denotes the standard Gaussian distribution function and

$$d_+ = \frac{\log(x/K) + (r + \sigma^2/2)(T-t)}{\sigma\sqrt{T-t}}, \qquad d_- = \frac{\log(x/K) + (r - \sigma^2/2)(T-t)}{\sigma\sqrt{T-t}},$$

with

$$d_+ = d_- + \sigma\sqrt{T-t}.$$

One checks easily that when $t = T$,

$$d_+ = d_- = \begin{cases} +\infty, & x > K, \\ -\infty, & x < K, \end{cases}$$

which allows one to recover the boundary condition

$$g_c(T,x) = \left\{ \begin{array}{ll} x\Phi(+\infty) - K\Phi(+\infty) = x - K, & x > K \\ x\Phi(-\infty) - K\Phi(-\infty) = 0, & x < K \end{array} \right\} = (x-K)^+$$

at $t = T$.

Figure 5.2 presents the graph of the solution

$$(t,x) \longmapsto g_c(t,x) = x\Phi(d_+) - Ke^{-r(T-t)}\Phi(d_-)$$

of the Black–Scholes PDE for a call option.

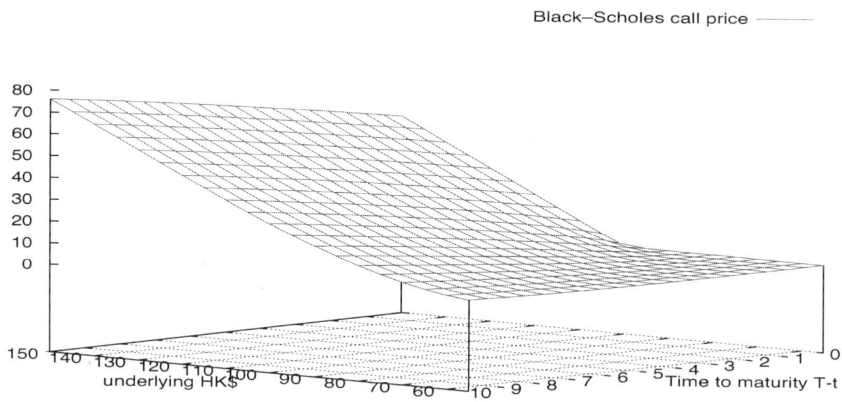

FIGURE 5.2: Graph of the Black–Scholes call price function with strike $K = 100$.

In Figure 5.3 we consider the stock price of HSBC Holdings (0005.HK) over one year:
Consider a call option issued by Societe Generale on 31 December 2008 with strike K=\$63.704, maturity T = October 05, 2009, and an entitlement ratio of

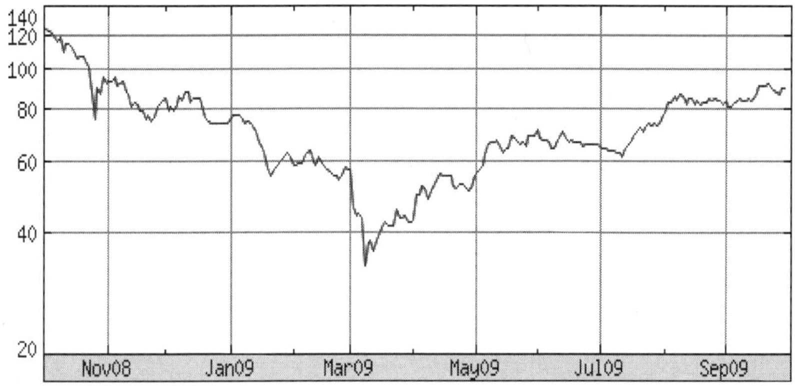

FIGURE 5.3: Graph of the stock price of HSBC Holdings.

100, meaning that one option contract is divided into 100 *warrants*, cf. page 6.
The next graph gives the time evolution of the Black–Scholes portfolio price

$$t \longmapsto g_c(t, S_t)$$

driven by the market price $t \longmapsto S_t$ of the underlying risky asset as given in Figure 5.3, in which the number of days is counted from the origin and not from maturity.

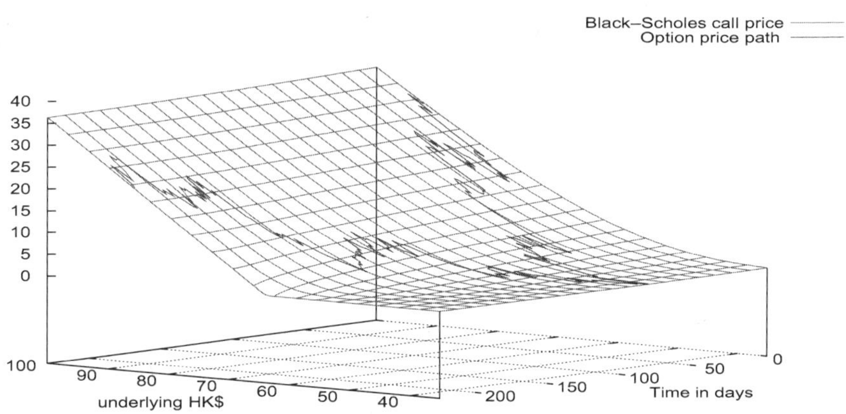

FIGURE 5.4: Path of the Black–Scholes price for a call option on HSBC.

The next proposition is proved by direct differentiation of the Black–Scholes function, and will be recovered later using a probabilistic argument in Proposition 6.7 below.

Proposition 5.4 *The Black–Scholes Delta of a European call option is given by*

$$\xi_t = \Phi(d_+).$$

Proof. We have

$$\begin{aligned}
&\frac{\partial g}{\partial x}(x,t) = \qquad (5.13)\\
&\frac{\partial}{\partial x}\left(x\Phi\left(\frac{\log(x/K)+(r+\sigma^2/2)(T-t)}{\sigma\sqrt{T-t}}\right)-K\Phi\left(\frac{\log(x/K)+(r-\sigma^2/2)(T-t)}{\sigma\sqrt{T-t}}\right)\right)\\
&= x\frac{\partial}{\partial x}\Phi\left(\frac{\log(x/K)+(r+\sigma^2/2)(T-t)}{\sigma\sqrt{T-t}}\right)\\
&\quad -K\frac{\partial}{\partial x}\Phi\left(\frac{\log(x/K)+(r-\sigma^2/2)(T-t)}{\sigma\sqrt{T-t}}\right)\\
&\quad +\Phi\left(\frac{\log(x/K)+(r-\sigma^2/2)(T-t)}{\sigma\sqrt{T-t}}\right)\\
&= \frac{1}{\sqrt{2\pi}\sigma\sqrt{T-t}}\exp\left(-\frac{1}{2}\left(\frac{\log(x/K)+(r+\sigma^2/2)(T-t)}{\sigma\sqrt{T-t}}\right)^2\right)\\
&\quad -\frac{K}{\sqrt{2\pi}\sigma x\sqrt{T-t}}\exp\left(-\frac{1}{2}\left(\frac{\log(x/K)+(r-\sigma^2/2)(T-t)}{\sigma\sqrt{T-t}}\right)^2\right)\\
&\quad +\Phi\left(\frac{\log(x/K)+(r-\sigma^2/2)(T-t)}{\sigma\sqrt{T-t}}\right)\\
&= \Phi\left(\frac{\log(x/K)+(r-\sigma^2/2)(T-t)}{\sigma\sqrt{T-t}}\right)
\end{aligned}$$

□

As a consequence of Proposition 5.4, in the Black–Scholes model the amount invested on the risky asset is

$$S_t\xi_t = S_t\Phi(d_+) = S_t\Phi\left(\frac{\log(S_t/K)+(r+\sigma^2/2)(T-t)}{\sigma\sqrt{T-t}}\right) \geq 0,$$

which is always positive, i.e., there is no short-selling, and the amount invested on the riskless asset is

$$\eta_t A_t = -KA_0 e^{-r(T-t)}\Phi\left(\frac{\log(S_t/K)+(r-\sigma^2/2)(T-t)}{\sigma\sqrt{T-t}}\right) \leq 0,$$

which is always negative, i.e., we are constantly borrowing money. Similarly, in the case of a European put option with strike K the payoff function is given by $f(x) = (K-x)^+$ and the Black–Scholes PDE reads

$$\begin{cases} rg_p(t,x) = \dfrac{\partial g_p}{\partial t}(t,x) + rx\dfrac{\partial g_p}{\partial x}(t,x) + \dfrac{1}{2}x^2\sigma^2\dfrac{\partial^2 g_p}{\partial x^2}(t,x), \\ \\ g_p(T,x) = (K-x)^+, \end{cases}$$

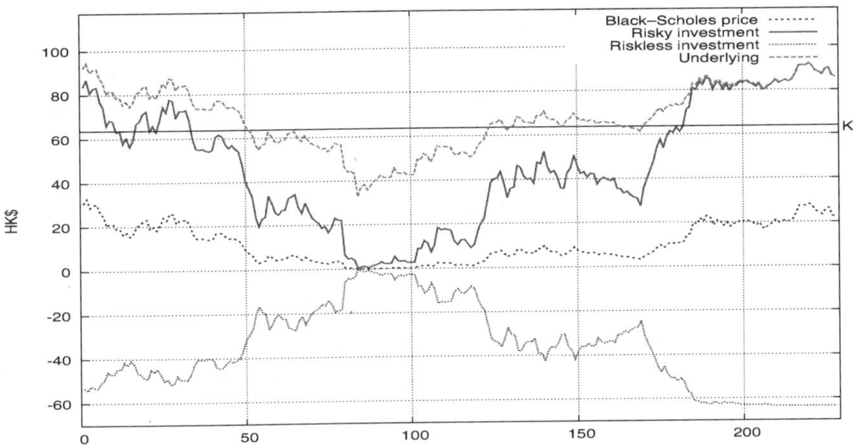

FIGURE 5.5: Time evolution of the hedging portfolio for a call option on HSBC.

with explicit solution

$$g_p(t,x) = Ke^{-r(T-t)}\Phi(-d_-) - x\Phi(-d_+),$$

as illustrated in the next graph.

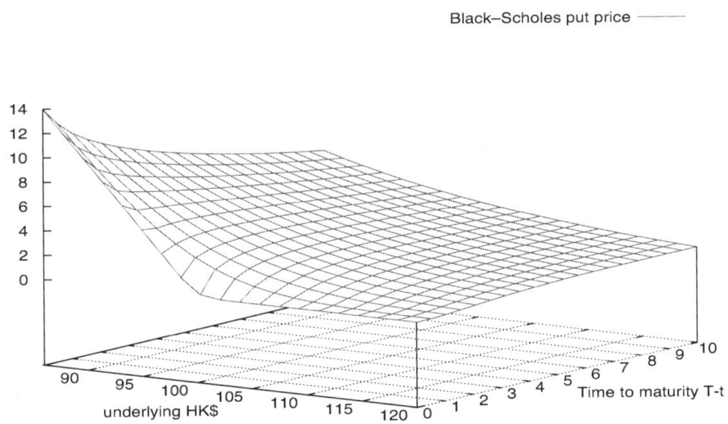

FIGURE 5.6: Graph of the Black–Scholes put price function with strike $K = 100$.

Note that the call-put parity relation

$$g(t, S_t) = g_c(t, S_t) - g_p(t, S_t), \qquad 0 \le t \le T,$$

is satisfied here.

For one more example we consider a put option issued by BNP Paribas on 04 November 2008 with strike K=\$77.667, maturity T = October 05, 2009, and entitlement ratio 92.593, cf. page 6. In Figure 5.7 the number of days is counted from the origin and not from maturity.

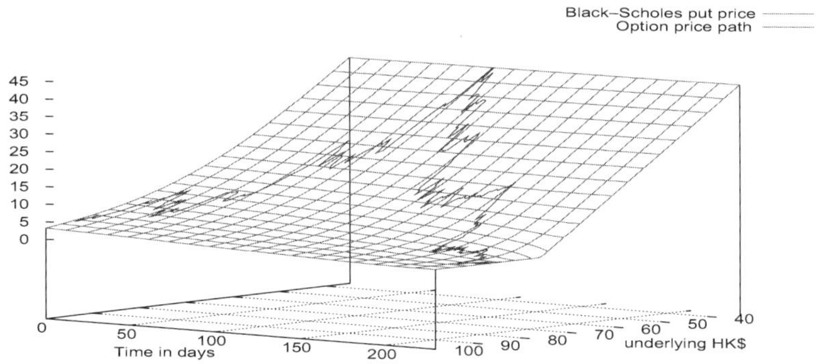

FIGURE 5.7: Path of the Black–Scholes price for a put option on HSBC.

In the case of a Black–Scholes put option the Delta is given by

$$\xi_t = -\Phi(-d_+),$$

and the amount invested on the risky asset is

$$S_t\Phi(d_+) = -S_t\Phi\left(-\frac{\log(S_t/K) + (r+\sigma^2/2)(T-t)}{\sigma\sqrt{T-t}}\right) \leq 0,$$

i.e., there is always short-selling, and the amount invested on the riskless asset is

$$Ke^{-r(T-t)}\Phi\left(-\frac{\log(S_t/K) + (r-\sigma^2/2)(T-t)}{\sigma\sqrt{T-t}}\right) \geq 0,$$

which is always positive, i.e., we are constantly investing on the riskless asset.

5.6 The Heat Equation

In this section we study the *heat equation* which is used to model the diffusion of heat in solids. We refer the reader to [72] for a complete treatment of this topic.

In Section 5.7 this equation will be shown to be equivalent to the Black–Scholes PDE after a change of variables. In particular this will lead to the explicit solution of the Black–Scholes PDE.

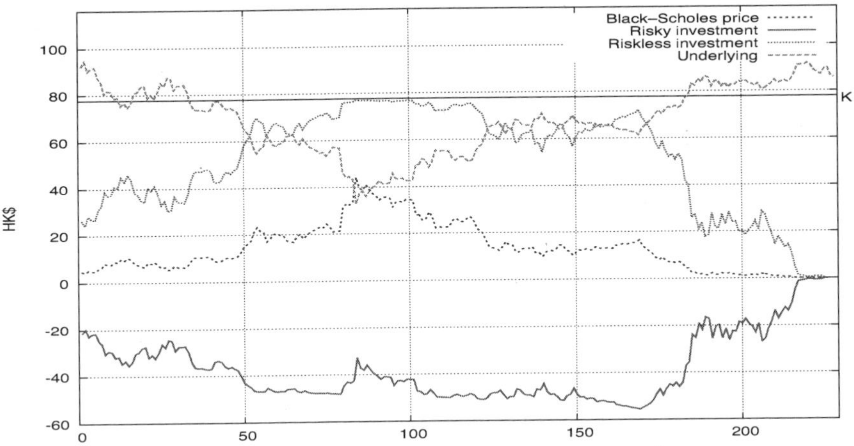

FIGURE 5.8: Time evolution of the hedging portfolio for a put option on HSBC.

Proposition 5.5 *The heat equation*

$$\begin{cases} \dfrac{\partial g}{\partial t}(t,y) = \dfrac{1}{2}\dfrac{\partial^2 g}{\partial y^2}(t,y) \\ \\ g(0,y) = \psi(y) \end{cases} \tag{5.14}$$

with initial condition $\psi(y)$ *has solution*

$$g(t,y) = \int_{-\infty}^{\infty} \psi(z) e^{-\frac{(y-z)^2}{2t}} \frac{dz}{\sqrt{2\pi t}}. \tag{5.15}$$

Proof. We have

$$\begin{aligned} \frac{\partial g}{\partial t}(t,y) &= \frac{\partial}{\partial t}\int_{-\infty}^{\infty} \psi(z) e^{-\frac{(y-z)^2}{2t}} \frac{dz}{\sqrt{2\pi t}} \\ &= \int_{-\infty}^{\infty} \psi(z) \frac{\partial}{\partial t}\left(\frac{e^{-\frac{(y-z)^2}{2t}}}{\sqrt{2\pi t}}\right) dz \\ &= \frac{1}{2}\int_{-\infty}^{\infty} \psi(z) \left(\frac{(y-z)^2}{t^2} - \frac{1}{t}\right) e^{-\frac{(y-z)^2}{2t}} \frac{dz}{\sqrt{2\pi t}} \\ &= \frac{1}{2}\int_{-\infty}^{\infty} \psi(z) \frac{\partial^2}{\partial z^2} e^{-\frac{(y-z)^2}{2t}} \frac{dz}{\sqrt{2\pi t}} \\ &= \frac{1}{2}\int_{-\infty}^{\infty} \psi(z) \frac{\partial^2}{\partial y^2} e^{-\frac{(y-z)^2}{2t}} \frac{dz}{\sqrt{2\pi t}} \\ &= \frac{1}{2}\frac{\partial^2}{\partial y^2}\int_{-\infty}^{\infty} \psi(z) e^{-\frac{(y-z)^2}{2t}} \frac{dz}{\sqrt{2\pi t}} \\ &= \frac{1}{2}\frac{\partial^2 g}{\partial y^2}(t,y). \end{aligned}$$

On the other hand it can be checked that at time $t = 0$,

$$\lim_{t\to 0}\int_{-\infty}^{\infty}\psi(z)e^{-\frac{(y-z)^2}{2t}}\frac{dz}{\sqrt{2\pi t}} = \lim_{t\to 0}\int_{-\infty}^{\infty}\psi(y+z)e^{-\frac{z^2}{2t}}\frac{dz}{\sqrt{2\pi t}} = \psi(y),$$

$y \in \mathbb{R}$. □

Let us provide a second proof of Proposition 5.5 using stochastic calculus and Brownian motion. Note that under the change of variable $x = z - y$ we have

$$\begin{aligned} g(t,y) &= \int_{-\infty}^{\infty}\psi(z)e^{-\frac{(y-z)^2}{2t}}\frac{dz}{\sqrt{2\pi t}} \\ &= \int_{-\infty}^{\infty}\psi(y+x)e^{-\frac{x^2}{2t}}\frac{dx}{\sqrt{2\pi t}} \\ &= \mathbb{E}[\psi(y+B_t)], \end{aligned}$$

where $(B_t)_{t\in\mathbb{R}_+}$ is a standard Brownian motion. Applying Itô's formula we have

$$\begin{aligned} \mathbb{E}[\psi(y+B_t)] &= \psi(y) + \mathbb{E}\left[\int_0^t \psi'(y+B_s)dB_s\right] + \frac{1}{2}\mathbb{E}\left[\int_0^t \psi''(y+B_s)ds\right] \\ &= \psi(y) + \frac{1}{2}\int_0^t \mathbb{E}\left[\psi''(y+B_s)\right]ds \\ &= \psi(y) + \frac{1}{2}\int_0^t \frac{\partial^2}{\partial y^2}\mathbb{E}\left[\psi(y+B_s)\right]ds, \end{aligned}$$

since the expectation of the stochastic integral is zero. Hence

$$\begin{aligned} \frac{\partial g}{\partial t}(t,y) &= \frac{\partial}{\partial t}\mathbb{E}[\psi(y+B_t)] \\ &= \frac{1}{2}\frac{\partial^2}{\partial y^2}\mathbb{E}\left[\psi(y+B_t)\right] \\ &= \frac{1}{2}\frac{\partial^2 g}{\partial y^2}(t,y). \end{aligned}$$

Concerning the initial condition we check that

$$\begin{aligned} g(0,y) &= \mathbb{E}[\psi(y+B_0)] \\ &= \mathbb{E}[\psi(y)] \\ &= \psi(y). \end{aligned}$$

5.7 Solution of the Black–Scholes PDE

In this section we will solve the Black–Scholes PDE by the kernel method of Section 5.6 and a change of variables.

Proposition 5.6 *Assume that $f(t,x)$ solves the Black–Scholes PDE*

$$\begin{cases} rf(t,x) = \dfrac{\partial f}{\partial t}(t,x) + rx\dfrac{\partial f}{\partial x}(t,x) + \dfrac{1}{2}x^2\sigma^2\dfrac{\partial^2 f}{\partial x^2}(t,x), \\ \\ f(T,x) = (x-K)^+, \end{cases} \tag{5.16}$$

with terminal condition $h(x) = (x-K)^+$. Then the function $g(t,y)$ defined by

$$g(t,y) = e^{rt} f\left(T-t, e^{\sigma y + (\frac{\sigma^2}{2}-r)t}\right) \tag{5.17}$$

solves the heat equation (5.14), i.e.,

$$\begin{cases} \dfrac{\partial g}{\partial t}(t,y) = \dfrac{1}{2}\dfrac{\partial^2 g}{\partial y^2}(t,y) \\ \\ g(0,y) = h(e^{\sigma y}), \end{cases}$$

with initial condition

$$g(0,y) = h(e^{\sigma y}). \tag{5.18}$$

Proof. Letting $s = T-t$ and $x = e^{\sigma y + (\frac{\sigma^2}{2}-r)t}$ we have

$$\begin{aligned} \frac{\partial g}{\partial t}(t,y) &= re^{rt} f(T-t, e^{\sigma y + (\frac{\sigma^2}{2}-r)t}) - e^{rt}\frac{\partial f}{\partial s}(T-t, e^{\sigma y + (\frac{\sigma^2}{2}-r)t}) \\ &\quad + \left(\frac{\sigma^2}{2}-r\right) e^{rt} e^{\sigma y + (\frac{\sigma^2}{2}-r)t}\frac{\partial f}{\partial x}(T-t, e^{\sigma y + (\frac{\sigma^2}{2}-r)t}) \\ &= re^{rt} f(T-t,x) - e^{rt}\frac{\partial f}{\partial s}(T-t,x) + \left(\frac{\sigma^2}{2}-r\right) e^{rt} x \frac{\partial f}{\partial x}(T-t,x) \\ &= \frac{1}{2}e^{rt}x^2\sigma^2\frac{\partial^2 f}{\partial x^2}(T-t,x) + \frac{\sigma^2}{2}e^{rt}x\frac{\partial f}{\partial x}(T-t,x), \end{aligned} \tag{5.19}$$

where on the last step we used the Black–Scholes PDE.

On the other hand we have

$$\frac{\partial g}{\partial y}(t,y) = \sigma e^{rt} e^{\sigma y + (\frac{\sigma^2}{2}-r)t}\frac{\partial f}{\partial x}(T-t, e^{\sigma y + (\frac{\sigma^2}{2}-r)t})$$

and

$$\begin{aligned} \frac{1}{2}\frac{\partial g^2}{\partial y^2}(t,y) &= \frac{\sigma^2}{2}e^{rt}e^{\sigma y + \frac{1}{2}(\frac{\sigma^2}{2}-r)t}\frac{\partial f}{\partial x}(T-t, e^{\sigma y + (\frac{\sigma^2}{2}-r)t}) \\ &\quad + \frac{\sigma^2}{2}e^{rt}e^{2\sigma y + 2(\frac{\sigma^2}{2}-r)t}\frac{\partial^2 f}{\partial x^2}(T-t, e^{\sigma y + (\frac{\sigma^2}{2}-r)t}) \\ &= \frac{\sigma^2}{2}e^{rt}x\frac{\partial f}{\partial x}(T-t,x) + \frac{\sigma^2}{2}e^{rt}x^2\frac{\partial^2 f}{\partial x^2}(T-t,x), \end{aligned}$$

which, in view of (5.19), means that $g(t,x)$ satisfies the heat equation (5.14) with initial condition

$$g(0,y) = f(T, e^{\sigma y}) = h(e^{\sigma y}).$$

□

In the next proposition we recover the Black–Scholes formula by solving the PDE (5.16). The Black–Scholes will also be recovered by probabilistic arguments and the computation of an expectation in Proposition 6.4.

Proposition 5.7 *When $h(x) = (x-K)^+$, the solution of the Black–Scholes PDE* (5.16) *is given by*

$$f(t,x) = x\Phi(d_+) - Ke^{-r(T-t)}\Phi(d_-),$$

where

$$\Phi(x) = \frac{1}{\sqrt{2\pi}}\int_{-\infty}^{x} e^{-y^2/2}dy, \qquad x \in \mathbb{R},$$

and

$$d_+ = \frac{\log(x/K) + (r+\sigma^2/2)(T-t)}{\sigma\sqrt{T-t}}, \qquad d_- = \frac{\log(x/K) + (r-\sigma^2/2)(T-t)}{\sigma\sqrt{T-t}}.$$

Proof. By inversion of (5.17) with $s = T-t$ and $x = e^{\sigma y + (\frac{\sigma^2}{2}-r)t}$ we get

$$f(s,x) = e^{-r(T-s)}g\left(T-s, \frac{-(\frac{\sigma^2}{2}-r)(T-s)+\log x}{\sigma}\right).$$

Hence using the solution (5.15) and Relation (5.18) we get

$$\begin{aligned}
f(t,x) &= e^{-r(T-t)}g\left(T-t, \frac{-(\frac{\sigma^2}{2}-r)(T-t)+\log x}{\sigma}\right) \\
&= e^{-r(T-t)}\int_{-\infty}^{\infty}\psi\left(\frac{-(\frac{\sigma^2}{2}-r)(T-t)+\log x}{\sigma}+z\right)e^{-\frac{z^2}{2(T-t)}}\frac{dz}{\sqrt{2\pi(T-t)}} \\
&= e^{-r(T-t)}\int_{-\infty}^{\infty}h\left(xe^{\sigma z-(\frac{\sigma^2}{2}-r)(T-t)}\right)e^{-\frac{z^2}{2(T-t)}}\frac{dz}{\sqrt{2\pi(T-t)}} \\
&= e^{-r(T-t)}\int_{-\infty}^{\infty}\left(xe^{\sigma z-(\frac{\sigma^2}{2}-r)(T-t)}-K\right)^+e^{-\frac{z^2}{2(T-t)}}\frac{dz}{\sqrt{2\pi(T-t)}} \\
&= e^{-r(T-t)}\int_{\frac{(-r+\sigma^2/2)(T-t)+\log(K/x)}{\sigma}}^{\infty}\left(xe^{\sigma z-(\frac{\sigma^2}{2}-r)(T-t)}-K\right)e^{-\frac{z^2}{2(T-t)}} \\
&\quad \frac{dz}{\sqrt{2\pi(T-t)}} = xe^{-r(T-t)}\int_{-d_-\sqrt{T-t}}^{\infty}e^{\sigma z-(\frac{\sigma^2}{2}-r)(T-t)}e^{-\frac{z^2}{2(T-t)}} \\
&\quad \frac{dz}{\sqrt{2\pi(T-t)}} - Ke^{-r(T-t)}\int_{-d_-\sqrt{T-t}}^{\infty}e^{-\frac{z^2}{2(T-t)}}\frac{dz}{\sqrt{2\pi(T-t)}}
\end{aligned}$$

$$
\begin{aligned}
&= x\int_{-d_-\sqrt{T-t}}^{\infty} e^{\sigma z-\frac{\sigma^2}{2}(T-t)-\frac{z^2}{2(T-t)}}\frac{dz}{\sqrt{2\pi(T-t)}}\\
&\quad -Ke^{-r(T-t)}\int_{-d_-\sqrt{T-t}}^{\infty} e^{-\frac{z^2}{2(T-t)}}\frac{dz}{\sqrt{2\pi(T-t)}}\\
&= x\int_{-d_-\sqrt{T-t}}^{\infty} e^{-\frac{1}{2(T-t)}(z-\sigma(T-t))^2}\frac{dz}{\sqrt{2\pi(T-t)}}\\
&\quad -Ke^{-r(T-t)}\int_{-d_-\sqrt{T-t}}^{\infty} e^{-\frac{z^2}{2(T-t)}}\frac{dz}{\sqrt{2\pi(T-t)}}\\
&= x\int_{-d_-\sqrt{T-t}-\sigma(T-t)}^{\infty} e^{-\frac{z^2}{2(T-t)}}\frac{dz}{\sqrt{2\pi(T-t)}}\\
&\quad -Ke^{-r(T-t)}\int_{-d_-\sqrt{T-t}}^{\infty} e^{-\frac{z^2}{2(T-t)}}\frac{dz}{\sqrt{2\pi(T-t)}}\\
&= x\int_{-d_--\sigma\sqrt{T-t}}^{\infty} e^{-\frac{z^2}{2}}\frac{dz}{\sqrt{2\pi}} - Ke^{-r(T-t)}\int_{d_-}^{\infty} e^{-\frac{z^2}{2}}\frac{dz}{\sqrt{2\pi}}\\
&= x\left(1-\Phi\left(-d_+\right)\right) - Ke^{-r(T-t)}\left(1-\Phi\left(-d_-\right)\right)\\
&= x\Phi\left(d_+\right) - Ke^{-r(T-t)}\Phi\left(d_-\right),
\end{aligned}
$$

where we used the relation

$$
1-\Phi(a)=\Phi(-a), \qquad a\in\mathbb{R}.
$$

□

Exercises

Exercise 5.1

1. Solve the stochastic differential equation

$$
dS_t = \alpha S_t dt + \sigma dB_t
$$

in terms of $\alpha, \sigma > 0$, and the initial condition S_0.

2. For which values α_M of α is the discounted price process $\tilde{S}_t = e^{-rt}S_t$, $t\in[0,T]$, a martingale under P ?

3. Compute the arbitrage price $C(t,S_t) = e^{-r(T-t)}\,\mathbb{E}[\exp(S_T)\mid\mathcal{F}_t]$ at time $t\in[0,T]$ of the contingent claim of $\exp(S_T)$, with $\alpha=\alpha_M$.

4. Explicitly compute the strategy $(\zeta_t,\eta_t)_{t\in[0,T]}$ that hedges the contingent claim $\exp(S_T)$.

Exercise 5.2 In the Black–Scholes model, the price at time t of a European claim on the underlying asset S_t, with strike price K, maturity T, interest rate r and volatility σ is given by the Black–Scholes formula as

$$f(t, S_t) = S_t\Phi(d_+) - Ke^{-r(T-t)}\Phi(d_-),$$

where

$$d_- = \frac{(r - \frac{1}{2}\sigma^2)(T-t) + \log(S_t/K)}{\sigma\sqrt{T-t}} \quad \text{and} \quad d_+ = d_- + \sigma\sqrt{T-t}.$$

Recall that

$$\frac{\partial f}{\partial x}(t, S_t) = \Phi(d_+).$$

On December 18, 2007, a call warrant has been issued by Fortis Bank on the stock price S of the MTR Corporation with maturity $T = 23/12/2008$, Strike $K =$ HK\$ 36.08 and Entitlement ratio=10.

1. Using the values of the Gaussian cumulative distribution function, compute the Black–Scholes price of the corresponding call option at time $t =$November 07, 2008 with $S_t =$ HK\$ 17.200, assuming a volatility $\sigma = 90\% = 0.90$ and an *annual* risk-free interest rate $r = 4.377\% = 0.04377$,

2. Still using the values of the Gaussian cumulative distribution function, compute the quantity of the risky asset required in your portfolio at time $t =$November 07, 2008 in order to hedge one such option at maturity $T = 23/12/2008$.

3. Figure 1 represents the Black–Scholes price of the call option as a function of $\sigma \in [0.5, 1.5] = [50\%, 150\%]$.

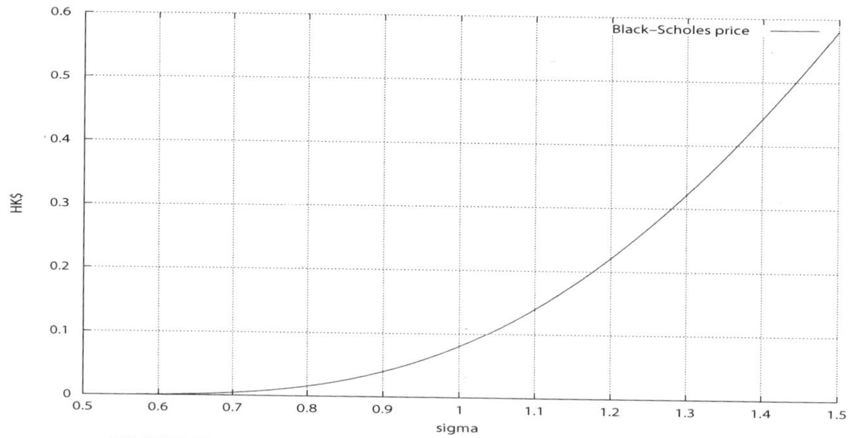

FIGURE 5.9: Option price as a function of the volatility σ.

Knowing that the closing price of the warrant on November 07, 2008 was HK\$ 0.023, which value can you infer for the implied volatility σ at this date?

Exercise 5.3 (Forward contracts). Recall that the price $\pi_t(C)$ of a claim $C = h(S_T)$ of maturity T can be written as $\pi_t(C) = g(t, S_t)$, where the function $g(t, x)$ satisfies the *Black–Scholes PDE*

$$\begin{cases} rg(t,x) = \dfrac{\partial g}{\partial t}(t,x) + rx\dfrac{\partial g}{\partial x}(t,x) + \dfrac{1}{2}x^2\sigma^2\dfrac{\partial^2 g}{\partial x^2}(t,x), \\ \\ g(T,x) = h(x), \end{cases} \tag{1}$$

with terminal condition $g(T, x) = h(x)$.

1. Assume that C is a forward contract with payoff

   $$C = S_T - K,$$

   at time T. Find the function $h(x)$ in (1).

2. Find the solution $g(t, x)$ of the above PDE and compute the price $\pi_t(C)$ at time $t \in [0, T]$.
   Hint: search for a solution of the form $g(t, x) = x - \alpha(t)$ where $\alpha(t)$ is a function of t to be determined.

3. Compute the quantity

   $$\xi_t = \frac{\partial g}{\partial x}(t, S_t)$$

   of risky assets in a self-financing portfolio hedging C.

Exercise 5.4 (Forward contracts revisited). Consider a risky asset whose price S_t is given by $S_t = S_0 e^{\sigma B_t + rt - \sigma^2 t/2}$, $t \in \mathbb{R}_+$, where $(B_t)_{t\in\mathbb{R}_+}$ is a standard Brownian motion. Consider a forward contract with maturity T and payoff $S_T - \kappa$.

1. Compute the price C_t of this claim at any time $t \in [0, T]$.

2. Compute a hedging strategy for the option with payoff $S_T - \kappa$.

Chapter 6

Martingale Approach to Pricing and Hedging

In this chapter we present the probabilistic *martingale approach* method to the pricing and hedging of options. In particular, this allows one to compute option prices as the expectations of the discounted option payoffs, and to determine the associated hedging portfolios.

6.1 Martingale Property of the Itô Integral

Recall (Definition 5.4) that an integrable process $(X_t)_{t\in\mathbb{R}_+}$ is said to be a *martingale* with respect to the filtration $(\mathcal{F}_t)_{t\in\mathbb{R}_+}$ if

$$\mathbb{E}[X_t \mid \mathcal{F}_s] = X_s, \qquad 0 \leq s \leq t.$$

The following result shows that the indefinite Itô integral is a martingale with respect to the Brownian filtration $(\mathcal{F}_t)_{t\in\mathbb{R}_+}$. It is the continuous-time analog of the discrete-time Proposition 2.1.

Proposition 6.1 *The indefinite stochastic integral* $\left(\int_0^t u_s dB_s\right)_{t\in\mathbb{R}_+}$ *of a square-integrable adapted process* $u \in L^2_{ad}(\Omega \times \mathbb{R}_+)$ *is a martingale,* i.e.,*:*

$$\mathbb{E}\left[\int_0^t u_\tau dB_\tau \middle| \mathcal{F}_s\right] = \int_0^s u_\tau dB_\tau, \quad 0 \leq s \leq t.$$

Proposition 6.1 is a consequence of Proposition 6.2 below.

Proposition 6.2 *For any* $u \in L^2_{ad}(\Omega \times \mathbb{R}_+)$ *we have*

$$\mathbb{E}\left[\int_0^\infty u_s dB_s \middle| \mathcal{F}_t\right] = \int_0^t u_s dB_s, \qquad t \in \mathbb{R}_+.$$

In particular, $\int_0^t u_s dB_s$ *is* $\mathcal{F}_t$*-measurable,* $t \in \mathbb{R}_+$.

Proof. The statement is first proved in case u is a simple predictable process, and then extended to the general case, cf. e.g., Proposition 2.5.7 in [55]. □

In particular, since $\mathcal{F}_0 = \{\emptyset, \Omega\}$, this recovers the fact that the Itô integral is a centered random variable:

$$\mathbb{E}\left[\int_0^\infty u_s dB_s\right] = \mathbb{E}\left[\int_0^\infty u_s dB_s \Big| \mathcal{F}_0\right] = \int_0^0 u_s dB_s = 0.$$

Examples

1. Given any square-integrable random variable $F \in L^2(\Omega)$, the process $(X_t)_{t\in\mathbb{R}_+}$ defined by $X_t := \mathbb{E}[F \mid \mathcal{F}_t]$, $t \in \mathbb{R}_+$, is a martingale under $\mathbb{P}$, as follows from the "tower property" (A.20):

$$\mathbb{E}[X_t \mid \mathcal{F}_s] = \mathbb{E}[\mathbb{E}[F \mid \mathcal{F}_t] \mid \mathcal{F}_s] = \mathbb{E}[F \mid \mathcal{F}_s] = X_s, \qquad 0 \le s \le t. \tag{6.1}$$

2. Any integrable stochastic process $(X_t)_{t\in\mathbb{R}_+}$ with centered and independent is a martingale:

$$\begin{aligned} \mathbb{E}[X_t|\mathcal{F}_s] &= \mathbb{E}[X_t - X_s + X_s|\mathcal{F}_s] \\ &= \mathbb{E}[X_t - X_s|\mathcal{F}_s] + \mathbb{E}[X_s|\mathcal{F}_s] \\ &= \mathbb{E}[X_t - X_s] + X_s \\ &= X_s, \qquad 0 \le s \le t. \end{aligned} \tag{6.2}$$

In particular, the standard Brownian motion $(B_t)_{t\in\mathbb{R}_+}$ is a martingale because it has centered, independent increments. This fact can also be recovered from Proposition 6.1 since B_t can be written as

$$B_t = \int_0^t dB_s, \qquad t \in \mathbb{R}_+.$$

3. The discounted asset price

$$X_t = X_0 e^{(\mu-r)t+\sigma B_t - \sigma^2 t/2}$$

is a martingale when $\mu = r$. Indeed we have

$$\begin{aligned} \mathbb{E}[X_t|\mathcal{F}_s] &= \mathbb{E}[X_0 e^{\sigma B_t - \sigma^2 t/2}|\mathcal{F}_s] \\ &= X_0 e^{-\sigma^2 t/2}\,\mathbb{E}[e^{\sigma B_t}|\mathcal{F}_s] \\ &= X_0 e^{-\sigma^2 t/2}\,\mathbb{E}[e^{\sigma(B_t-B_s)+\sigma B_s}|\mathcal{F}_s] \\ &= X_0 e^{-\sigma^2 t/2+\sigma B_s}\,\mathbb{E}[e^{\sigma(B_t-B_s)}|\mathcal{F}_s] \\ &= X_0 e^{-\sigma^2 t/2+\sigma B_s}\,\mathbb{E}[e^{\sigma(B_t-B_s)}] \\ &= X_0 e^{-\sigma^2 t/2+\sigma B_s} e^{\sigma^2(t-s)/2} \\ &= X_0 e^{\sigma B_s - \sigma^2 s/2} \\ &= X_s, \qquad 0 \le s \le t. \end{aligned}$$

This fact can also be recovered from Proposition 6.1 since X_t satisfies the equation

$$dX_t = \sigma X_t dB_t,$$

i.e., it can be written as the Brownian stochastic integral

$$X_t = X_0 + \sigma \int_0^t X_u dB_u, \qquad t \in \mathbb{R}_+.$$

4. The discounted value

$$\widetilde{V}_t = e^{-rt} V_t$$

of a self-financing portfolio given by

$$\widetilde{V}_t = \widetilde{V}_0 + \int_0^t \xi_u dX_u, \qquad t \in \mathbb{R}_+,$$

is a martingale when $\mu = r$ because

$$\widetilde{V}_t = \widetilde{V}_0 + \sigma \int_0^t \xi_u X_u dB_u, \qquad t \in \mathbb{R}_+,$$

since

$$dX_t = X_t((\mu - r)dt + \sigma dB_t).$$

Since the Black–Scholes theory is in fact valid for any value of the parameter μ we will look forward to including the case $\mu \neq r$ in the sequel.

6.2 Risk-Neutral Measures

Recall that by definition, a risk-neutral measure is a probability measure $\mathbb{P}^*$ under which the discounted asset price $(X_t)_{t\in\mathbb{R}_+} = (e^{-rt}S_t)_{t\in\mathbb{R}_+}$ is a martingale. From the analysis of Section 6.1 it appears that when $\mu = r$, $(X_t)_{t\in\mathbb{R}_+}$ is a martingale and $\mathbb{P}^* = \mathbb{P}$ is risk-neutral.

In this section we address the construction of a risk-neutral measure in the general case $\mu \neq r$ and for this we will use the Girsanov theorem.

Note that the relation

$$dX_t = X_t((\mu - r)dt + \sigma dB_t)$$

can be written as

$$dX_t = \sigma X_t d\tilde{B}_t,$$

where

$$\tilde{B}_t := \frac{\mu - r}{\sigma} t + B_t, \qquad t \in \mathbb{R}_+.$$

Therefore the search for a risk-neutral measure can be replaced by the search for a probability measure $\mathbb{P}^*$ under which $(\tilde{B}_t)_{t\in\mathbb{R}_+}$ is a standard Brownian motion.

Let us come back to the informal interpretation of Brownian motion via its infinitesimal increments:

$$\Delta B_t = \pm\sqrt{dt},$$

with

$$\mathbb{P}(\Delta B_t = +\sqrt{dt}) = \mathbb{P}(\Delta B_t = -\sqrt{dt}) = \frac{1}{2}.$$

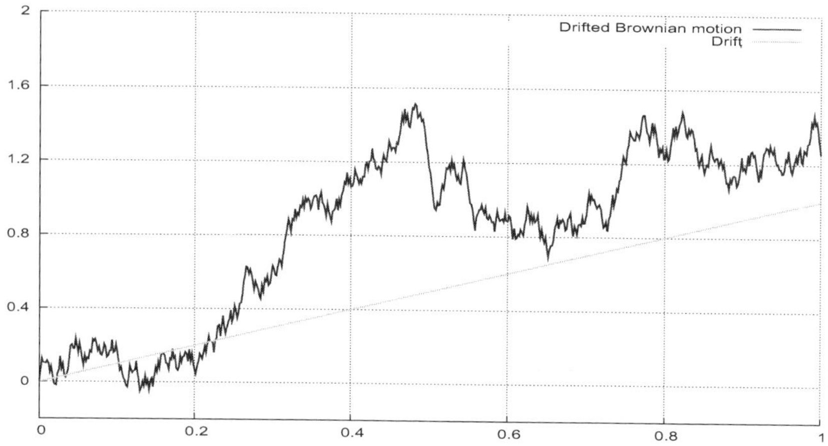

FIGURE 6.1: Drifted Brownian path.

Clearly, given $\nu \in \mathbb{R}$, the drifted process $\nu t + B_t$ is no longer a standard Brownian motion because it is not centered:

$$\mathbb{E}[\nu t + B_t] = \nu t + \mathbb{E}[B_t] = \nu t \neq 0,$$

cf. Figure 6.1. This identity can be formulated in terms of infinitesimal increments as

$$\mathbb{E}[\nu dt + dB_t] = \frac{1}{2}(\nu dt + \sqrt{dt}) + \frac{1}{2}(\nu dt - \sqrt{dt}) = \nu dt \neq 0.$$

In order to make $\nu t+B_t$ a centered process (i.e., a standard Brownian motion), since $\nu t + B_t$ conserves all the other properties (i)–(iii) in the definition of Brownian motion, one may change the probabilities of ups and downs, which have been fixed so far equal to $1/2$.

That is, the problem is now to find two numbers $p, q \in [0,1]$ such that

$$\begin{cases} p(\nu dt + \sqrt{dt}) + q(\nu dt - \sqrt{dt}) = 0 \\ \\ p + q = 1. \end{cases}$$

The solution to this problem is given by

$$p = \frac{1}{2}(1 - \nu\sqrt{dt}) \quad \text{and } q = \frac{1}{2}(1 + \nu\sqrt{dt}).$$

Still considering Brownian motion as a discrete random walk with independent increments $\pm\sqrt{dt}$, the corresponding probability density will be obtained by taking the product of the above probabilities divided by $1/2^N$, that is:

$$2^N \prod_{0<t<T} \left(\frac{1}{2} \mp \frac{1}{2}\nu\sqrt{dt}\right)$$

where 2^N is a normalization factor and $N = T/dt$ is the (infinitely large) number of discrete time steps. Using elementary calculus, this density can be informally shown to converge as follows:

$$\begin{aligned} 2^N \prod_{0<t<T} \left(\frac{1}{2} \mp \frac{1}{2}\nu\sqrt{dt}\right) &= \prod_{0<t<T} \left(1 \mp \nu\sqrt{dt}\right) \\ &= \exp\left(\log \prod_{0<t<T} \left(1 \mp \nu\sqrt{dt}\right)\right) \\ &= \exp\left(\sum_{0<t<T} \log\left(1 \mp \nu\sqrt{dt}\right)\right) \\ &\simeq \exp\left(\nu \sum_{0<t<T} \mp\sqrt{dt} - \frac{1}{2} \sum_{0<t<T} (\mp\nu\sqrt{dt})^2\right) \\ &= \exp\left(\nu \sum_{0<t<T} \mp\sqrt{dt} - \frac{1}{2}\nu^2 \sum_{0<t<T} dt\right) \\ &= \exp\left(-\nu B_T - \frac{1}{2}\nu^2 T\right). \end{aligned}$$

6.3 Girsanov Theorem and Change of Measure

In this section we restate the Girsanov theorem in a more rigorous way, using changes of probability measures. Given $\mathbb{Q}$ a probability measure on Ω, the notation

$$\frac{d\mathbb{Q}}{d\mathbb{P}} = F$$

means that the probability measure $\mathbb{Q}$ has a density F with respect to $\mathbb{P}$, where F is a non-negative random variable such that $\mathbb{E}[F] = 1$. We also write

$$d\mathbb{Q} = F d\mathbb{P},$$

which is equivalent to stating that

$$\int_\Omega \xi(\omega) d\mathbb{Q}(\omega) = \int_\Omega F(\omega)\xi(\omega) d\mathbb{P}(\omega),$$

or, under a different notation,

$$\hat{\mathbb{E}}^{\mathbb{Q}}[\xi] = \mathbb{E}\left[F\xi\right].$$

In addition we say that $\mathbb{Q}$ is *equivalent* to $\mathbb{P}$ when $F > 0$ with $\mathbb{P}$-probability one.

Recall that here, $\Omega = \mathcal{C}_0([0,T])$ is the Wiener space and $\omega \in \Omega$ is a continuous function on $[0,T]$ starting at 0 in $t = 0$. Consider the probability $\mathbb{Q}$ defined by

$$d\mathbb{Q}(\omega) = \exp\left(-\nu B_T - \frac{1}{2}\nu^2 T\right) d\mathbb{P}(\omega).$$

Then the process $\nu t + B_t$ is a standard (centered) Brownian motion under $\mathbb{Q}$.

For example, the fact that $\nu T + B_T$ has a standard (centered) Gaussian law under $\mathbb{Q}$ can be recovered as follows:

$$\begin{aligned}
\mathbb{E}_{\mathbb{Q}}[f(\nu T + B_T)] &= \int_\Omega f(\nu T + B_T) d\mathbb{Q} \\
&= \int_\Omega f(\nu T + B_T) \exp\left(-\nu B_T - \frac{1}{2}\nu^2 T\right) d\mathbb{P} \\
&= \int_{-\infty}^{\infty} f(\nu T + x) \exp\left(-\nu x - \frac{1}{2}\nu^2 T\right) e^{-\frac{x^2}{2T}} \frac{dx}{\sqrt{2\pi T}} \\
&= \int_{-\infty}^{\infty} f(y) e^{-\frac{y^2}{2T}} \frac{dy}{\sqrt{2\pi T}} \\
&= \int_\Omega f(B_T) d\mathbb{P} \\
&= \mathbb{E}_{\mathbb{P}}[f(B_T)].
\end{aligned}$$

The Girsanov theorem can actually be extended to shifts by adapted processes as follows, cf. e.g., [61], Theorem III-42. Section 14.6 will cover the extension of the Girsanov theorem to jump processes.

Theorem 6.1 *Let $(\psi_t)_{t\in[0,T]}$ be an adapted process satisfying the Novikov integrability condition*

$$\mathbb{E}\left[\exp\left(\frac{1}{2}\int_0^T |\psi_t|^2 dt\right)\right] < \infty, \tag{6.3}$$

and let $\mathbb{Q}$ *denote the probability measure defined by*

$$\frac{d\mathbb{Q}}{d\mathbb{P}} = \exp\left(-\int_0^T \psi_s dB_s - \frac{1}{2}\int_0^T \psi_s^2 ds\right).$$

Then

$$\hat{B}_t := B_t + \int_0^t \psi_s ds, \qquad t \in [0, T],$$

is a standard Brownian motion under $\mathbb{Q}$.

When applied to

$$\psi_t := \frac{\mu - r}{\sigma},$$

the Girsanov theorem shows that

$$\tilde{B}_t := \frac{\mu - r}{\sigma} t + B_t, \qquad t \in \mathbb{R}_+, \tag{6.4}$$

is a standard Brownian motion under the probability measure $\mathbb{P}^*$ defined by

$$\frac{d\mathbb{P}^*}{d\mathbb{P}} = \exp\left(-\frac{\mu - r}{\sigma} B_t - \frac{(\mu - r)^2}{2\sigma^2} t\right). \tag{6.5}$$

Hence the discounted price process given by

$$\frac{dX_t}{X_t} = (\mu - r)dt + \sigma dB_t = \sigma d\tilde{B}_t, \qquad t \in \mathbb{R}_+,$$

is a martingale under $\mathbb{P}^*$, hence $\mathbb{P}^*$ is a risk-neutral measure. We easily find that $\mathbb{P} = \mathbb{P}^*$ when $\mu = r$.

6.4 Pricing by the Martingale Method

In this section we give the expression of the Black–Scholes price using expectations of discounted payoffs.

Recall that from the first fundamental theorem of mathematical finance, a continuous market is without arbitrage opportunities if there exists (at least) a risk-neutral probability measure $\mathbb{P}^*$ under which the discounted price process

$$X_t := e^{-rt} S_t, \qquad t \in \mathbb{R}_+,$$

is a martingale under $\mathbb{P}^*$. In addition, when the risk-neutral measure is unique, the market is said to be *complete*.

In case the price process $(S_t)_{s\in[t,\infty)}$ satisfies the equation

$$\frac{dS_t}{S_t} = \mu dt + \sigma dB_t, \qquad t \in \mathbb{R}_+, \qquad S_0 > 0$$

we have

$$S_t = S_0 e^{\sigma B_t - \sigma^2 t/2 + \mu t}, \quad \text{and} \quad X_t = S_0 e^{(\mu - r)t + \sigma B_t - \sigma^2 t/2}, \qquad t \in \mathbb{R}_+,$$

hence from Section 6.2 the discounted price process is a martingale under the probability measure $\mathbb{P}^*$ defined by (6.5), and $\mathbb{P}^*$ is a martingale measure.

We have

$$dX_t = (\mu - r)X_t dt + \sigma X_t dB_t = \sigma X_t d\tilde{B}_t, \qquad t \in \mathbb{R}_+, \tag{6.6}$$

hence the discounted value $\widetilde{V}_t$ of a self-financing portfolio is written as

$$\begin{aligned} \widetilde{V}_t &= \widetilde{V}_0 + \int_0^t \xi_u dX_u \\ &= \widetilde{V}_0 + \sigma \int_0^t \xi_u X_u d\widetilde{B}_u, \qquad t \in \mathbb{R}_+, \end{aligned}$$

and becomes a martingale under $\mathbb{P}^*$.

As in Chapter 3, the value V_t at time t of a self-financing portfolio strategy $(\xi_t)_{t\in[0,T]}$ hedging an attainable claim C will be called an *arbitrage price* of the claim C at time t and denoted by $\pi_t(C)$, $t \in [0, T]$.

Proposition 6.3 *Let $(\xi_t, \eta_t)_{t\in[0,T]}$ be a portfolio strategy with price*

$$V_t = \eta_t A_t + \xi_t S_t, \qquad t \in [0, T],$$

and let C be a contingent claim, such that

(i) $(\xi_t, \eta_t)_{t\in[0,T]}$ is a self-financing portfolio, and

(ii) $(\xi_t, \eta_t)_{t\in[0,T]}$ hedges the claim C, i.e., *we have $V_T = C$.*

Then the arbitrage price of the claim C is given by

$$V_t = e^{-r(T-t)} \, \mathbb{E}^*[C|\mathcal{F}_t], \qquad 0 \le t \le T, \tag{6.7}$$

where $\mathbb{E}^*$ denotes expectation under the risk-neutral measure $\mathbb{P}^*$.

Proof. Since the portfolio strategy $(\xi_t, \eta_t)_{t\in\mathbb{R}_+}$ is self-financing, by Lemma 5.1 and (6.6) we have

$$\widetilde{V}_t = \widetilde{V}_0 + \sigma \int_0^t \xi_u X_u d\tilde{B}_u, \qquad t \in \mathbb{R}_+,$$

which is a martingale under $\mathbb{P}^*$ from Proposition 6.1, hence

$$\begin{aligned}\widetilde{V}_t &= \mathbb{E}^*\left[\widetilde{V}_T \mid \mathcal{F}_t\right] \\ &= e^{-rT}\,\mathbb{E}^*[V_T \mid \mathcal{F}_t] \\ &= e^{-rT}\,\mathbb{E}^*[C \mid \mathcal{F}_t],\end{aligned}$$

which implies

$$V_t = e^{rt}\widetilde{V}_t = e^{-r(T-t)}\,\mathbb{E}^*[C \mid \mathcal{F}_t].$$

□

In case the value V_t of the portfolio at time $t \in [0,T]$ is a function $C(t,S_t)$ of t and S_t, it can be computed from (6.7) as

$$V_t = C(t,S_t) = e^{-r(T-t)}\,\mathbb{E}^*[\phi(S_T)|\mathcal{F}_t], \qquad 0 \le t \le T,$$

and by Proposition 5.2 the function $C(t,x)$ solves the Black–Scholes PDE:

$$\begin{cases} rC(t,x) = \dfrac{\partial C}{\partial t}(t,x) + \dfrac{1}{2}x^2\sigma^2\dfrac{\partial^2 C}{\partial x^2}(t,x) + rx\dfrac{\partial C}{\partial x}(t,x) \\ \\ C(T,x) = \phi(x). \end{cases}$$

In the case of European options with payoff function $\phi(x) = (x-K)^+$ we recover the Black–Scholes formula (5.12), cf. Proposition 5.7, by a probabilistic argument.

Proposition 6.4 *The price at time t of a European call option with strike K and maturity T is given by*

$$C(t,S_t) = S_t\Phi(d_+) - Ke^{-r(T-t)}\Phi(d_-), \qquad t \in [0,T].$$

Proof. The proof of Proposition 6.4 is a consequence of (6.7) and the Lemma 6.1 below. Using the relation

$$S_T = S_t e^{r(T-t)+\sigma(\tilde{B}_T-\tilde{B}_t)-\sigma^2(T-t)/2}, \qquad t \in [0,T].$$

By Proposition 6.3 the price of the portfolio hedging C is given by

$$\begin{aligned} V_t &= e^{-r(T-t)}\,\mathbb{E}^*[C|\mathcal{F}_t] \\ &= e^{-r(T-t)}\,\mathbb{E}^*[(S_T-K)^+|\mathcal{F}_t] \\ &= e^{-r(T-t)}\,\mathbb{E}^*[(S_t e^{r(T-t)+\sigma(\tilde{B}_T-\tilde{B}_t)-\sigma^2(T-t)/2}-K)^+|\mathcal{F}_t] \\ &= e^{-r(T-t)}\,\mathbb{E}^*[(xe^{r(T-t)+\sigma(\tilde{B}_T-\tilde{B}_t)-\sigma^2(T-t)/2}-K)^+]_{x=S_t} \\ &= e^{-r(T-t)}\,\mathbb{E}^*[(e^{m(x)+X}-K)^+]_{x=S_t}, \qquad 0 \le t \le T, \end{aligned}$$

where

$$m(x) = r(T-t) - \sigma^2(T-t)/2 + \log x$$

and $X = \sigma(\tilde{B}_T - \tilde{B}_t)$ is a centered Gaussian random variable with variance

$$\mathrm{Var}\,[X] = \mathrm{Var}\,[\sigma(\tilde{B}_T - \tilde{B}_t)] = \sigma^2 \mathrm{Var}\,[\tilde{B}_T - \tilde{B}_t] = \sigma^2 (T-t)$$

under $\mathbb{P}^*$. Hence by Lemma 6.1 below we have

$$\begin{aligned} V_t &= e^{-r(T-t)}\, \mathbb{E}^*[(e^{m(x)+X} - K)^+]_{x=S_t} \\ &= e^{-r(T-t)} e^{m(S_t)+\sigma^2(T-t)/2} \Phi(v + (m(S_t) - \log K)/v) \\ &\quad - K e^{-r(T-t)} \Phi((m(S_t) - \log K)/v) \\ &= S_t \Phi(v + (m(S_t) - \log K)/v) - K e^{-r(T-t)} \Phi((m(S_t) - \log K)/v) \\ &= S_t \Phi(d_+) - K e^{-r(T-t)} \Phi(d_-), \end{aligned}$$

$0 \le t \le T$. □

Lemma 6.1 *Let X be a centered Gaussian random variable with variance v^2. We have*

$$\mathbb{E}[(e^{m+X} - K)^+] = e^{m+\frac{v^2}{2}} \Phi(v + (m - \log K)/v) - K\Phi((m - \log K)/v).$$

Proof. We have

$$\begin{aligned} \mathbb{E}[(e^{m+X} - K)^+] &= \frac{1}{\sqrt{2\pi v^2}} \int_{-\infty}^{\infty} (e^{m+x} - K)^+ e^{-\frac{x^2}{2v^2}} dx \\ &= \frac{1}{\sqrt{2\pi v^2}} \int_{-m+\log K}^{\infty} (e^{m+x} - K) e^{-\frac{x^2}{2v^2}} dx \\ &= \frac{e^m}{\sqrt{2\pi v^2}} \int_{-m+\log K}^{\infty} e^{x-\frac{x^2}{2v^2}} dx - \frac{K}{\sqrt{2\pi v^2}} \int_{-m+\log K}^{\infty} e^{-\frac{x^2}{2v^2}} dx \\ &= \frac{e^{m+\frac{v^2}{2}}}{\sqrt{2\pi v^2}} \int_{-m+\log K}^{\infty} e^{-\frac{(v^2-x)^2}{2v^2}} dx - \frac{K}{\sqrt{2\pi}} \int_{(-m+\log K)/v}^{\infty} e^{-x^2/2} dx \\ &= \frac{e^{m+\frac{v^2}{2}}}{\sqrt{2\pi v^2}} \int_{-v^2-m+\log K}^{\infty} e^{-\frac{x^2}{2v^2}} dx - K\Phi((m - \log K)/v) \\ &= e^{m+\frac{v^2}{2}} \Phi(v + (m - \log K)/v) - K\Phi((m - \log K)/v). \end{aligned}$$

□

Denoting by

$$P(t, S_t) = e^{-r(T-t)}\, \mathbb{E}^*[(K - S_T)^+ | \mathcal{F}_t]$$

the price of the put option with strike K and maturity T, we check from Proposition 6.3 that

$$\begin{aligned} &C(t, S_t) - P(t, S_t) \\ &= e^{-r(T-t)}\, \mathbb{E}^*[(S_T - K)^+ | \mathcal{F}_t] - e^{-r(T-t)}\, \mathbb{E}^*[(K - S_T)^+ | \mathcal{F}_t] \\ &= e^{-r(T-t)}\, \mathbb{E}^*[(S_T - K)^+ - (K - S_T)^+ | \mathcal{F}_t] \end{aligned}$$

$$\begin{aligned}&= e^{-r(T-t)}\mathbb{E}^*[S_T - K|\mathcal{F}_t]\\ &= S_t - e^{-r(T-t)}K.\end{aligned}$$

This relation is called the *put-call parity*, and it shows that

$$\begin{aligned}P(t, S_t) &= C(t, S_t) - S_t + e^{-r(T-t)}K\\ &= S_t\Phi(d_+) + e^{-r(T-t)}K - S_t - e^{-r(T-t)}K\Phi(d_-)\\ &= -S_t(1 - \Phi(d_+)) + e^{-r(T-t)}K(1 - \Phi(d_-))\\ &= -S_t\Phi(-d_+) + e^{-r(T-t)}K\Phi(-d_-).\end{aligned}$$

6.5 Hedging Strategies

In the next proposition we compute a self-financing hedging strategy leading to an arbitrary square-integrable random variable C admitting a stochastic integral representation formula of the form

$$C = \mathbb{E}[C] + \int_0^T \zeta_t dB_t, \tag{6.8}$$

where $(\zeta_t)_{t\in[0,t]}$ is a square-integrable adapted process. Consequently, the mathematical problem of finding the predictable representation (6.8) of a given random variable has important applications in finance. For example we have

$$B_T^2 = T + 2\int_0^T B_t dB_t.$$

Recall that the risky asset follows the equation

$$\frac{dS_t}{S_t} = \mu dt + \sigma dB_t, \qquad t \in \mathbb{R}_+, \qquad S_0 > 0,$$

and the discounted asset price satisfies

$$dX_t = \sigma X_t d\tilde{B}_t, \qquad t \in \mathbb{R}_+, \qquad X_0 = S_0 > 0,$$

where $(\tilde{B}_t)_{t\in\mathbb{R}_+}$ is a standard Brownian motion under the risk-neutral probability measure $\mathbb{P}^*$.

The following proposition applies to arbitrary square-integrable payoff functions, i.e., it covers exotic and path-dependent options.

Proposition 6.5 *Consider a random payoff $C \in L^2(\Omega)$ such that (6.8) holds, and let*

$$\xi_t = \frac{e^{-r(T-t)}}{\sigma S_t}\zeta_t, \tag{6.9}$$

$$\eta_t = \frac{e^{-r(T-t)}\,\mathbb{E}^*[C|\mathcal{F}_t] - \xi_t S_t}{A_t}, \qquad t \in [0,T]. \tag{6.10}$$

Then the portfolio $(\xi_t, \eta_t)_{t\in[0,T]}$ *is self-financing, and letting*

$$V_t = \eta_t A_t + \xi_t S_t, \qquad t \in [0,T], \tag{6.11}$$

we have

$$V_t = e^{-r(T-t)}\,\mathbb{E}^*[C|\mathcal{F}_t], \qquad t \in [0,T]. \tag{6.12}$$

In particular we have

$$V_T = C, \tag{6.13}$$

i.e., *the portfolio* $(\xi_t, \eta_t)_{t\in[0,T]}$ *yields a hedging strategy leading to* C, *starting from the initial value*

$$V_0 = e^{-rT}\,\mathbb{E}^*[C].$$

Proof. Relation (6.12) follows from (6.10) and (6.11), and it implies

$$V_0 = e^{-rT}\,\mathbb{E}^*[C] = \eta_0 A_0 + \xi_0 S_0$$

at $t = 0$, and (6.13) at $t = T$. It remains to show that the portfolio strategy $(\xi_t, \eta_t)_{t\in[0,T]}$ is self-financing. By (6.8) we have

$$\begin{aligned}
V_t = \eta_t A_t + \xi_t S_t &= e^{-r(T-t)}\,\mathbb{E}^*[C|\mathcal{F}_t] \\
&= e^{-r(T-t)}\,\mathbb{E}^*\left[\mathbb{E}^*[C] + \int_0^T \zeta_u d\tilde{B}_u \Big| \mathcal{F}_t\right] \\
&= e^{-r(T-t)}\left(\mathbb{E}^*[C] + \int_0^t \zeta_u d\tilde{B}_u\right) \\
&= e^{rt} V_0 + e^{-r(T-t)} \int_0^t \zeta_u d\tilde{B}_u \\
&= e^{rt} V_0 + \sigma \int_0^t \xi_u S_u e^{r(t-u)} d\tilde{B}_u \\
&= e^{rt} V_0 + \sigma \int_0^t \xi_u X_u e^{rt} d\tilde{B}_u \\
&= e^{rt} V_0 + e^{rt} \int_0^t \xi_u dX_u, \qquad t \in [0,T],
\end{aligned}$$

which shows that the discounted portfolio value $\widetilde{V}_t = e^{-rt} V_t$ satisfies

$$\widetilde{V}_t = V_0 + \int_0^t \xi_u dX_u, \qquad t \in [0,T],$$

and this implies that $(\xi_t, \eta_t)_{t\in[0,T]}$ is self-financing by Lemma 5.1. □

The above proposition shows that there always exists a hedging strategy starting from

$$V_0 = \mathbb{E}^*[C] e^{-rT}.$$

In addition, since there exists a hedging strategy leading to

$$\widetilde{V}_T = e^{-rT} C,$$

then $(\widetilde{V}_t)_{t\in[0,T]}$ is necessarily a martingale with

$$\widetilde{V}_t = \mathbb{E}^* \left[\widetilde{V}_T | \mathcal{F}_t \right] = e^{-rT} \, \mathbb{E}^*[C|\mathcal{F}_t], \qquad t \in [0,T],$$

and initial value

$$\widetilde{V}_0 = \mathbb{E}^* \left[\widetilde{V}_T \right] = e^{-rT} \, \mathbb{E}^*[C].$$

In practice, the hedging problem can now be reduced to the computation of the process $(\zeta_t)_{t\in[0,T]}$ appearing in (6.8). This computation, called the Delta hedging, can be performed by application of the Itô formula and the Markov property, see e.g., [60]. Consider the (non-homogeneous) semi-group $(P_{s,t})_{0\leq s\leq t\leq T}$ associated to $(S_t)_{t\in[0,T]}$ and defined by

$$P_{s,t} f(S_s) = \mathbb{E}[f(S_t) \mid \mathcal{F}_s] = \mathbb{E}[f(S_t) \mid S_s], \quad 0 \leq s \leq t,$$

which acts on functions $f \in \mathcal{C}_b^2(\mathbb{R}^n)$, with

$$P_{s,t} P_{t,u} = P_{s,u}, \qquad 0 \leq s \leq t \leq u.$$

Note that $(P_{t,T} f(S_t))_{t\in[0,T]}$ is an $\mathcal{F}_t$-martingale, i.e.,:

$$\begin{aligned} \mathbb{E}[P_{t,T} f(S_t) \mid \mathcal{F}_s] &= \mathbb{E}[\mathbb{E}[f(S_T) \mid \mathcal{F}_t] \mid \mathcal{F}_s] \\ &= \mathbb{E}[f(S_T) \mid \mathcal{F}_s] \\ &= P_{s,T} f(S_s), \end{aligned} \tag{6.14}$$

$0 \leq s \leq t \leq T$, and we have

$$P_{s,t} f(x) = \mathbb{E}[f(S_t) \mid S_s = x] = \mathbb{E}[f(x S_t / S_s)], \qquad 0 \leq s \leq t. \tag{6.15}$$

The next lemma allows us to compute the process $(\zeta_t)_{t\in[0,T]}$ in case the payoff C is of the form $C = \phi(S_T)$ for some function ϕ. In case $C \in L^2(\Omega)$ is the payoff of an exotic option, the process $(\zeta_t)_{t\in[0,T]}$ can be computed using the Malliavin gradient on the Wiener space, cf. [50], [55].

Lemma 6.2 *Let $\phi \in \mathcal{C}_b^2(\mathbb{R}^n)$. The predictable representation*

$$\phi(S_T) = \mathbb{E}[\phi(S_T)] + \int_0^T \zeta_t dB_t \tag{6.16}$$

is given by

$$\zeta_t = \sigma S_t \frac{\partial}{\partial x} (P_{t,T} \phi)(S_t), \qquad t \in [0,T]. \tag{6.17}$$

Proof. Since $P_{t,T}\phi$ is in $\mathcal{C}^2(\mathbb{R})$, we can apply the Itô formula to the process

$$t \mapsto P_{t,T}\phi(S_t) = \mathbb{E}[\phi(S_T) \mid \mathcal{F}_t],$$

which is a martingale from the "tower property" of conditional expectations as in (6.14), cf. also Relation (6.1). From the fact that the finite variation term in the Itô formula vanishes when $(P_{t,T}\phi(S_t))_{t\in[0,T]}$ is a martingale, (see e.g., Corollary 1, p. 72 of [61]), we obtain:

$$P_{t,T}\phi(S_t) = P_{0,T}\phi(S_0) + \sigma \int_0^t S_s \frac{\partial}{\partial x}(P_{s,T}\phi)(S_s)dB_s, \quad t \in [0,T], \tag{6.18}$$

with $P_{0,T}\phi(S_0) = \mathbb{E}[\phi(S_T)]$. Letting $t = T$, we obtain (6.17) by uniqueness of the predictable representation (6.16) of $C = \phi(S_T)$. □

By (6.15) we also have

$$\begin{aligned}\zeta_t &= \sigma S_t \frac{\partial}{\partial x} \mathbb{E}[\phi(S_T) \mid S_t = x]_{x=S_t} \\ &= \sigma S_t \frac{\partial}{\partial x} \mathbb{E}[\phi(xS_T/S_t)]_{x=S_t}, \qquad t \in [0,T],\end{aligned}$$

hence

$$\begin{aligned}\xi_t &= \frac{1}{\sigma S_t} e^{-r(T-t)} \zeta_t \\ &= e^{-r(T-t)} \frac{\partial}{\partial x} \mathbb{E}[\phi(xS_T/S_t)]_{x=S_t}, \qquad t \in [0,T],\end{aligned} \tag{6.19}$$

which recovers the formula (5.8) for the Delta of a vanilla option. As a consequence we have $\xi_t \geq 0$ and there is no short selling when the payoff function ϕ is non-decreasing.

In the case of European options, the process ζ can be computed via the next proposition.

Proposition 6.6 *Assume that $C = (S_T - K)^+$. Then for $0 \leq t \leq T$ we have*

$$\zeta_t = \sigma S_t \, \mathbb{E}\left[\frac{S_T}{S_t} \mathbf{1}_{[K,\infty)}\left(x \frac{S_T}{S_t}\right)\right]_{x=S_t}.$$

Proof. This result follows from Lemma 6.2 and the relation $P_{t,T}f(x) = \mathbb{E}[f(S_{t,T}^x)]$, after approximation of $x \mapsto \phi(x) = (x-K)^+$ with $\mathcal{C}^2$ functions. □

From the above proposition we recover the formula for the Delta of a European call option in the Black–Scholes model. Proposition 6.7 shows that the Black–Scholes self-financing hedging strategy is to hold a (possibly fractional) quantity

$$\xi_t = \Phi(d_+) = \Phi\left(\frac{\log(S_t/K) + (r + \sigma^2/2)(T-t)}{\sigma\sqrt{T-t}}\right) \geq 0 \tag{6.20}$$

of the risky asset, and to borrow a quantity

$$-\eta_t = Ke^{-rT}\Phi\left(\frac{\log(S_t/K)+(r-\sigma_t^2/2)(T-t)}{\sigma\sqrt{T-t}}\right) \leq 0 \qquad (6.21)$$

of the riskless (savings) account, cf. also Corollary 10.2 in Chapter 10. In the next proposition we prove another proof of the result of Proposition 5.4.

Proposition 6.7 *The Delta of a European call option with payoff function $f(x) = (x-K)^+$ is given by*

$$\xi_t = \Phi(d_+) = \Phi\left(\frac{\log(S_t/K)+(r+\sigma^2/2)(T-t)}{\sigma\sqrt{T-t}}\right), \qquad 0 \leq t \leq T.$$

Proof. By Propositions 6.5 and 6.6 we have

$$\begin{aligned}
\xi_t &= \frac{1}{\sigma S_t}e^{-r(T-t)}\zeta_t \\
&= e^{-r(T-t)}\,\mathbb{E}\left[\frac{S_T}{S_t}\mathbf{1}_{[K,\infty)}\left(x\frac{S_T}{S_t}\right)\right]_{x=S_t} \\
&= e^{-r(T-t)} \\
&\times \mathbb{E}\left[e^{\sigma(B_T-B_t)-\sigma^2(T-t)/2+r(T-t)}\mathbf{1}_{[K,\infty)}(xe^{\sigma(B_T-B_t)-\sigma^2(T-t)/2+r(T-t)})\right]_{x=S_t} \\
&= \frac{1}{\sqrt{2\pi(T-t)}}\int_{\sigma(T-t)/2-r(T-t)/\sigma+\frac{1}{\sigma}\log\frac{K}{S_t}}^{\infty} e^{\sigma y-\sigma^2(T-t)/2-y^2/(2(T-t))}dy \\
&= \frac{1}{\sqrt{2\pi(T-t)}}\int_{-d_-/\sqrt{T-t}}^{\infty} e^{-\frac{1}{2(T-t)}(y-\sigma(T-t))^2}dy \\
&= \frac{1}{\sqrt{2\pi}}\int_{-d_-}^{\infty} e^{-\frac{1}{2}(y-\sigma(T-t))^2}dy \\
&= \frac{1}{\sqrt{2\pi}}\int_{-d_+}^{\infty} e^{-\frac{1}{2}y^2}dy \\
&= \frac{1}{\sqrt{2\pi}}\int_{-\infty}^{d_+} e^{-\frac{1}{2}y^2}dy \\
&= \Phi(d_+).
\end{aligned}$$

□

The result of Proposition 6.7 can also be recovered by (5.8) or (6.19) and direct differentiation of the Black–Scholes function, cf. (5.13).

Exercises

Exercise 6.1 In this problem, $(\eta_t, \xi_t)_{t\in[0,T]}$ denotes a portfolio strategy with value

$$V_t = \eta_t A_t + \xi_t S_t, \qquad 0 \le t \le T,$$

where S_t, resp. A_t, denotes the price at time t of a risky, resp. riskless, asset. Consider the price process $(S_t)_{t\in[0,T]}$ given by

$$\frac{dS_t}{S_t} = rdt + \sigma dB_t$$

and a riskless asset of value $A_t = A_0 e^{rt}$, $t \in [0, T]$, with $r > 0$.

1. Compute the arbitrage price

$$C(t, S_t) = e^{-r(T-t)} \mathbb{E}_Q[|S_T|^2 \mid \mathcal{F}_t],$$

at time $t \in [0, T]$, of the contingent claim of payoff $|S_T|^2$.

2. Compute the portfolio strategy $(\eta_t, \xi_t)_{t\in[0,T]}$ hedging the claim $|S_T|^2$.
3. Check that this strategy is self-financing.

Exercise 6.2 Again, $(\eta_t, \xi_t)_{t\in[0,T]}$ denotes a portfolio strategy with value

$$V_t = \eta_t A_t + \xi_t S_t, \qquad 0 \le t \le T,$$

where S_t, resp. A_t, denotes the price at time t of a risky, resp. riskless, asset.

1. Recall (without proof) the solution of the stochastic differential equation

$$dS_t = rS_t dt + \sigma dB_t.$$

2. Show that the discounted price process $\tilde{S}_t = e^{-rt} S_t$, $t \in [0, T]$, is a martingale under $\mathbb{P}$.
3. Compute the arbitrage price

$$C(t, S_t) = e^{-r(T-t)} \mathbb{E}[\exp(S_T) \mid \mathcal{F}_t]$$

at time $t \in [0, T]$ of the contingent claim of $\exp(S_T)$.

4. Explicitly compute the portfolio strategy $(\eta_t, \xi_t)_{t\in[0,T]}$ that hedges the contingent claim $\exp(S_T)$.
5. Check that this strategy is self-financing.

Exercise 6.3 Let $(B_t)_{t\in\mathbb{R}_+}$ be a standard Brownian motion generating a filtration $(\mathcal{F}_t)_{t\in\mathbb{R}_+}$. Recall that for $f \in \mathcal{C}^2(\mathbb{R}_+ \times \mathbb{R})$, Itô's formula for Brownian motion reads

$$\begin{aligned} f(t, B_t) &= f(0, B_0) + \int_0^t \frac{\partial f}{\partial s}(s, B_s)ds \\ &\quad + \int_0^t \frac{\partial f}{\partial x}(s, B_s)dB_s + \frac{1}{2}\int_0^t \frac{\partial^2 f}{\partial x^2}(s, B_s)ds. \end{aligned}$$

1. Let $r \in \mathbb{R}$, $\sigma > 0$, $f(x,t) = e^{rt+\sigma x-\sigma^2 t/2}$, and $S_t = f(t, B_t)$. Compute $df(t, B_t)$ by Itô's formula, and show that S_t solves the stochastic differential equation

$$dS_t = rS_t dt + \sigma S_t dB_t,$$

where $r > 0$ and $\sigma > 0$.

2. Show that

$$E[e^{\sigma B_T} \mid \mathcal{F}_t] = e^{\sigma B_t + \sigma^2(T-t)/2}, \qquad 0 \le t \le T.$$

Hint: use the independence of the increments of $B_T = (B_T - B_t) + (B_t - B_0)$ and the Laplace transform $E[e^{\alpha X}] = e^{\alpha^2\eta^2/2}$ when $X \simeq \mathcal{N}(0, \eta^2)$.

3. Show that the process $(S_t)_{t\in\mathbb{R}_+}$ satisfies

$$E[S_T \mid \mathcal{F}_t] = e^{r(T-t)} S_t, \qquad 0 \le t \le T.$$

4. Let $C = S_T - K$ denote the payoff of a forward contract with exercise price K and maturity T. Compute the discounted expected payoff

$$V_t := e^{-r(T-t)} E[C \mid \mathcal{F}_t].$$

5. Find a self-financing portfolio strategy $(\xi_t, \eta_t)_{t\in\mathbb{R}_+}$ such that

$$V_t = \xi_t S_t + \eta_t A_t, \qquad 0 \le t \le T,$$

where $A_t = A_0 e^{rt}$ is the price of a riskless asset with interest rate $r > 0$. Show that it recovers the result of Exercise 5.3-(3).

6. Show that the portfolio $(\xi_t, \eta_t)_{t\in[0,T]}$ found in Question 5 *hedges* the payoff $C = S_T - K$ at time T, i.e., show that $V_T = C$.

Exercise 6.4 Digital options. Consider a price process $(S_t)_{t\in\mathbb{R}_+}$ given by

$$\frac{dS_t}{S_t} = rdt + \sigma dB_t, \qquad S_0 = 1,$$

under the risk-neutral measure $\mathbb{P}$.

The digital *call*, resp. *put*, option is a contract with maturity T, strike K, and payoff

$$C_d := \begin{cases} \$1 & \text{if } S_T \geq K, \\ 0 & \text{if } S_T < K, \end{cases} \qquad \text{resp.} \qquad P_d := \begin{cases} \$1 & \text{if } S_T \leq K, \\ 0 & \text{if } S_T > K. \end{cases}$$

Recall that the prices $\pi_t(C_d)$ and $\pi_t(P_d)$ at time t of the digital call and put options are given by the discounted expected payoffs

$$\pi_t(C_d) = e^{-r(T-t)} \mathbb{E}[C_d \mid \mathcal{F}_t] \qquad \text{and} \qquad \pi_t(P_d) = e^{-r(T-t)} \mathbb{E}[P_d \mid \mathcal{F}_t]. \tag{6.22}$$

1. Show that the payoffs C_d and P_d can be rewritten as

$$C_d = \mathbf{1}_{[K,\infty)}(S_T) \qquad \text{and} \qquad P_d = \mathbf{1}_{[0,K]}(S_T).$$

2. Using Relation (6.22) and Question 1, prove the call-put parity relation

$$\pi_t(C_d) + \pi_t(P_d) = e^{-r(T-t)}, \qquad 0 \leq t \leq T. \tag{6.23}$$

   If needed, you may use the fact that $\mathbb{P}(S_T = K) = 0$.

3. Using Relation (6.22), Question 1, and the relation

$$\mathbb{E}[\mathbf{1}_{[K,\infty)}(S_T) \mid S_t = x] = P(S_T \geq K \mid S_t = x),$$

   show that the price $\pi_t(C_d)$ is given by

$$\pi_t(C_d) = C_d(t, S_t),$$

   where $C_d(t,x)$ is the function defined by

$$C_d(t,x) = e^{-r(T-t)} P(S_T \geq K \mid S_t = x).$$

4. Using the results of Exercise 4.8-(3) and of Question 3, show that

$$C_d(t,x) = e^{-r(T-t)} \Phi\left(\frac{(r - \sigma^2/2)\tau + \log(x/K)}{\sigma\sqrt{\tau}}\right),$$

   where $\tau = T - t$.

5. Show that the price $\pi_t(C_d)$ of the digital call option is given by

$$\pi_t(C_d) = e^{-r(T-t)} \Phi(d_-),$$

   where

$$d_- = \frac{(r - \sigma^2/2)\tau + \log(S_t/K)}{\sigma\sqrt{\tau}}.$$

6. Using the results of Questions 2 and 5, show that the price $\pi_t(P_d)$ of the digital put is given by

$$\pi_t(P_d) = e^{-r(T-t)}\Phi(-d_-).$$

7. Using the result of Question 5, compute the Delta

$$\xi_t := \frac{\partial C_d}{\partial x}(t, S_t)$$

of the digital call option. Does the Black–Scholes hedging strategy of such a call option involve short-selling? Why?

8. Using the result of Question 6, compute the Delta

$$\xi_t := \frac{\partial P_d}{\partial x}(t, S_t)$$

of the digital put option. Does the Black–Scholes hedging strategy of such a put option involve short-selling? Why?

Exercise 6.5

1. Consider a market model made of a risky asset with price $(S_t)_{t\in\mathbb{R}_+}$ as in Exercise 4.9-(4) and a riskless asset with price $A_t = \$1 \times e^{rt}$ and riskless interest rate $r = \sigma^2/2$. From the answer to Exercise 4.9-(2), show that the arbitrage price

$$V_t = e^{-r(T-t)}\,\mathbb{E}[(\log S_T)^+ \mid \mathcal{F}_t]$$

at time $t \in [0, T]$ of a log call option with payoff $(\log S_T)^+$ is equal to

$$V_t = \sigma e^{-r\tau}\sqrt{\frac{\tau}{2\pi}}e^{-\frac{B_t^2}{2\tau}} + \sigma e^{-r\tau}B_t\Phi\left(\frac{B_t}{\sqrt{\tau}}\right),$$

where $\tau = T - t$ denote the time to maturity.

2. Show that V_t can be written as

$$V_t = g(\tau, S_t),$$

where $g(\tau, x) = e^{-r\tau}f(\tau, \log x)$, and

$$f(\tau, y) = \sigma\sqrt{\frac{\tau}{2\pi}}e^{-\frac{y^2}{2\sigma^2\tau}} + y\Phi\left(\frac{y}{\sigma\sqrt{\tau}}\right).$$

3. Figure 6.2 represents the graph of $(\tau, x) \mapsto g(\tau, x)$, with $r = 0.05 = 5\%$ *per year* and $\sigma = 0.1$. Assume that the current underlying price is \$1 and there remain 700 days to maturity. What is the price of the option?

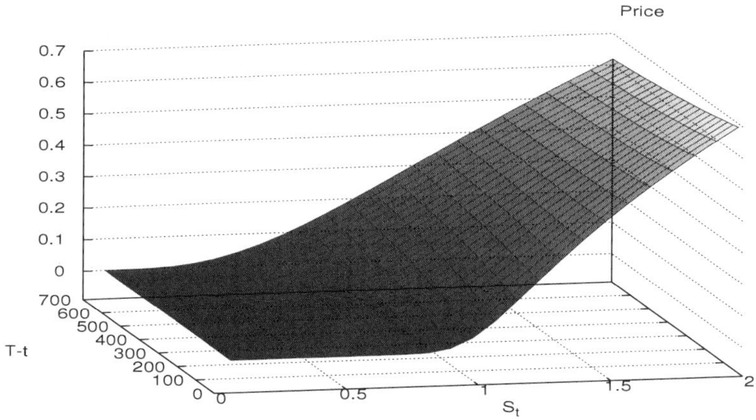

FIGURE 6.2: Option price as a function of the underlying asset price and of time to maturity.

4. Show[1] that the (possibly fractional) quantity $\Delta_t = \frac{\partial g}{\partial x}(\tau, S_t)$ of S_t at time t in a portfolio hedging the payoff $(\log S_T)^+$ is equal to

$$\Delta_t = e^{-r\tau} \frac{1}{S_t} \Phi\left(\frac{\log S_t}{\sigma\sqrt{\tau}}\right), \qquad 0 \leq t \leq T.$$

5. Figure 6.3 represents the graph of $(\tau, x) \mapsto \frac{\partial g}{\partial x}(\tau, x)$. Assuming that the current underlying price is \$1 and that there remain 700 days to maturity, how much of the risky asset should you hold in your portfolio in order to hedge one log option?

[1]Recall the chain rule of derivation $\frac{\partial}{\partial x} f(\tau, \log x) = \frac{1}{x} \frac{\partial f}{\partial y}(\tau, y)_{|y=\log x}$.

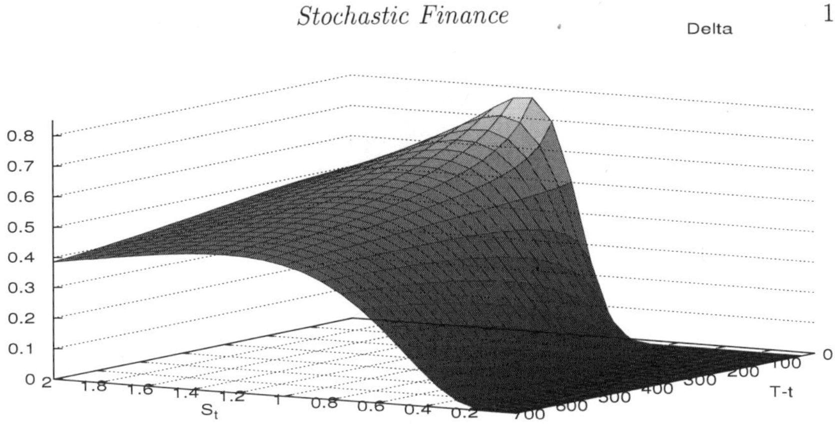

FIGURE 6.3: Delta as a function of the underlying asset price and of time to maturity.

6. Based on the framework and answers of Questions 3 and 5, should you borrow or lend the riskless asset $A_t = \$1 \times e^{rt}$, and for what amount?

7. Show that the Gamma of the portfolio, defined as $\Gamma_t = \dfrac{\partial^2 g}{\partial x^2}(\tau, S_t)$, equals

$$\Gamma_t = e^{-r\tau}\frac{1}{S_t^2}\left(\frac{1}{\sigma\sqrt{2\pi\tau}}e^{\frac{-(\log S_t)^2}{2\sigma^2\tau}} - \Phi\left(\frac{\log S_t}{\sigma\sqrt{\tau}}\right)\right), \qquad 0 \le t \le T.$$

8. Figure 6.4 represents the graph of Gamma. Assume that there remain 60 days to maturity and that S_t, currently at \$1, is expected to increase. Should you buy or (short) sell the underlying asset in order to hedge the option?

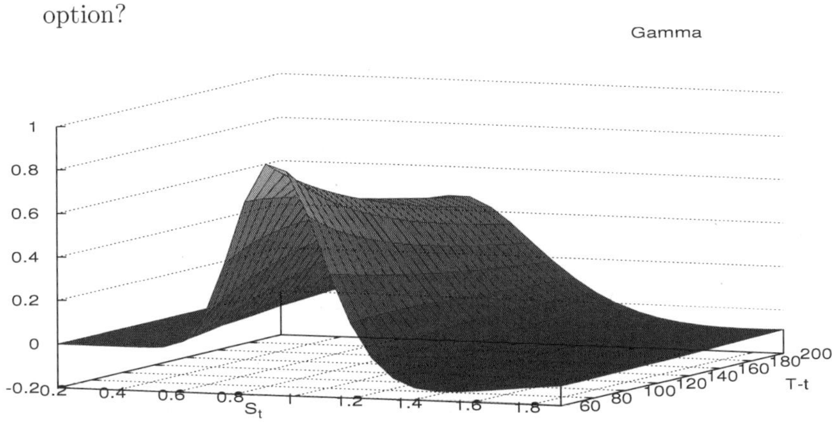

FIGURE 6.4: Gamma as a function of the underlying asset price and of time to maturity.

9. Let now $\sigma = 1$. Show that the function $f(\tau, y)$ of Question 2 solves the *heat equation*
$$\begin{cases} \dfrac{\partial f}{\partial \tau}(\tau, y) = \dfrac{1}{2}\dfrac{\partial^2 f}{\partial y^2}(\tau, y) \\ \\ f(0, y) = (y)^+. \end{cases}$$

Exercise 6.6

1. Consider a market model made of a risky asset with price $(S_t)_{t\in\mathbb{R}_+}$ as in Exercise 4.8 and a riskless asset with price $A_t = \$1 \times e^{rt}$ and riskless interest rate $r = \sigma^2/2$. From the answer to Exercise A.4-(2), show that the arbitrage price
$$V_t = e^{-r(T-t)}\,\mathbb{E}[(K - \log S_T)^+ \mid \mathcal{F}_t]$$
at time $t \in [0, T]$ of a log call option with strike K and payoff $(K - \log S_T)^+$ is equal to
$$V_t = \sigma e^{-r\tau}\sqrt{\frac{\tau}{2\pi}}e^{-\frac{(B_t - K/\sigma)^2}{2\tau}} + e^{-r\tau}(K - \sigma B_t)\Phi\left(\frac{K/\sigma - B_t}{\sqrt{\tau}}\right),$$
where $\tau = T - t$ denote the time to maturity.

2. Show that V_t can be written as
$$V_t = g(\tau, S_t),$$
where $g(\tau, x) = e^{-r\tau} f(\tau, \log x)$, and
$$f(\tau, y) = \sigma\sqrt{\frac{\tau}{2\pi}}e^{-\frac{(K-y)^2}{2\sigma^2\tau}} + (K - y)\Phi\left(\frac{K - y}{\sigma\sqrt{\tau}}\right).$$

3. Figure 6.5 represents the graph of $(\tau, x) \mapsto g(\tau, x)$, with $r = 0.125$ *per year* and $\sigma = 0.5$. Assume that the current underlying price is \$3 and there remains 700 days to maturity. What is the price of the option?

4. Show[2] that the quantity $\Delta_t = \dfrac{\partial g}{\partial x}(\tau, S_t)$ of S_t at time t in a portfolio hedging the payoff $(K - \log S_T)^+$ is equal to
$$\Delta_t = -e^{-r\tau}\frac{1}{S_t}\Phi\left(\frac{K - \log S_t}{\sigma\sqrt{\tau}}\right), \qquad 0 \le t \le T.$$

[2] Recall the chain rule of derivation $\dfrac{\partial}{\partial x} f(\tau, \log x) = \dfrac{1}{x}\dfrac{\partial f}{\partial y}(\tau, y)_{|y=\log x}$.

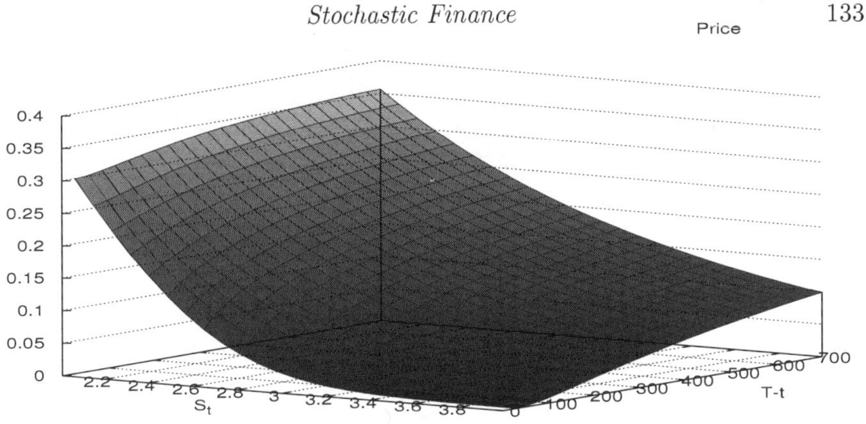

FIGURE 6.5: Option price as a function of the underlying asset price and of time to maturity.

5. Figure 6.6 represents the graph of $(\tau, x) \mapsto \frac{\partial g}{\partial x}(\tau, x)$. Assuming that the current underlying price is \$3 and that there remain 700 days to maturity, how much of the risky asset should you hold in your portfolio in order to hedge one log option?

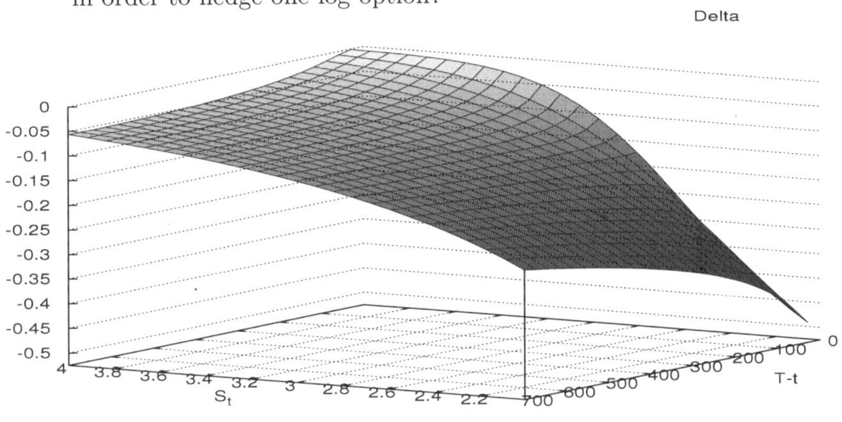

FIGURE 6.6: Delta as a function of the underlying asset price and of time to maturity.

6. Based on the framework and answers of Questions 3 and 5, should you borrow or lend the riskless asset $A_t = \$1 \times e^{rt}$, and for what amount?

7. Show that the Gamma of the portfolio, defined as $\Gamma_t = \dfrac{\partial^2 g}{\partial x^2}(\tau, S_t)$, equals

$$\Gamma_t = e^{-r\tau} \frac{1}{S_t^2} \left(\frac{1}{\sigma\sqrt{2\pi\tau}} e^{-\frac{(K - \log S_t)^2}{2\sigma^2\tau}} + \Phi\left(\frac{K - \log S_t}{\sigma\sqrt{\tau}} \right) \right), \qquad 0 \leq t \leq T.$$

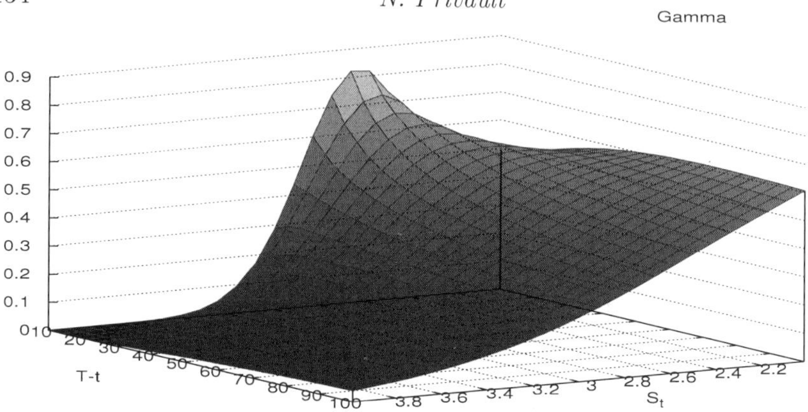

FIGURE 6.7: Gamma as a function of the underlying asset price and of time to maturity.

8. Figure 6.7 represents the graph of Gamma. Assume that there remain 10 days to maturity and that S_t, currently at \$3, is expected to increase. Should you buy or (short) sell the underlying asset in order to hedge the option?

9. Show that the function $f(\tau, y)$ of Question 2 solves the *heat equation*

$$\begin{cases} \dfrac{\partial f}{\partial \tau}(\tau, y) = \dfrac{1}{2}\sigma^2 \dfrac{\partial^2 f}{\partial y^2}(\tau, y) \\ \\ f(0, y) = (K - y)^+. \end{cases}$$

Chapter 7

Estimation of Volatility

The values of the parameters r, t, S_t, T, and K used to price a call option via the Black–Scholes formula can be easily obtained from market data. Estimating the volatility coefficient σ can be a more difficult task, and several estimation methods are considered in this section with some examples of how the Black–Scholes formula can be fitted to market data. We cover the historical, implied, and local volatility models, and refer to [25] for stochastic volatility models.

7.1 Historical Volatility

We consider the problem of estimating the parameters μ and σ from market data in the stock price model

$$\frac{dS_t}{S_t} = \mu dt + \sigma dB_t.$$

A natural estimator for the trend parameter μ can be written as

$$\hat{\mu}_N := \frac{1}{N} \sum_{k=0}^{N-1} \frac{1}{t_{k+1} - t_k} \frac{S_{t_k+1} - S_{t_k}}{S_{t_k}},$$

where $(S_{t_k+1} - S_{t_k})/S_{t_k}$, $k = 0, \ldots, N-1$ is a family of log-returns observed at different times $t_0, \ldots, t_N$ on the market.

Similarly the parameter σ can be estimated as by the estimator $\hat{\sigma}_N$ built as

$$\hat{\sigma}_N^2 := \frac{1}{N-1} \sum_{k=0}^{N-1} \frac{1}{t_{k+1} - t_k} \left(\frac{S_{t_k+1} - S_{t_k}}{S_{t_k}} - \hat{\mu}_N (t_{k+1} - t_k) \right)^2.$$

Parameter estimation based on historical data requires a lot of samples and it can only be valid on a given time interval, or as a moving average.

7.2 Implied Volatility

In contrast with the historical volatility, the computation of the implied volatility can be done at a fixed time and requires much less data.

Recall that when $h(x) = (x - K)^+$, the solution of the Black–Scholes PDE is given by

$$f(t, x, K, \sigma, r, T) = x\Phi(d_+) - Ke^{-(T-t)r}\Phi(d_-),$$

where

$$\Phi(x) = \frac{1}{\sqrt{2\pi}} \int_{-\infty}^{x} e^{-y^2/2} dy, \qquad x \in \mathbb{R},$$

and

$$d_+ = \frac{\log(x/K) + (r + \sigma^2/2)(T-t)}{\sigma\sqrt{T-t}}, \qquad d_- = \frac{\log(x/K) + (r - \sigma^2/2)(T-t)}{\sigma\sqrt{T-t}}.$$

Equating

$$f(t, S_t, K, \sigma, r, T) = M$$

to the observed value M of a given market price, when t, S_t, r, T are known, allows one to infer a value for σ. This value is called the implied volatility and denoted here by $\sigma^{\mathrm{imp}}(K, T)$. The implied volatility value can be used as an alternative way to quote the option price, based on the knowledge of the remaining parameters (such as underlying asset price, time to maturity, interest rate, and strike price).

Given two European call options with strikes K_1, resp. K_2 and maturities T_1, resp. T_2, on the same stock S, this procedure should yield two estimates $\sigma^{\mathrm{imp}}(K_1, T_1)$ and $\sigma^{\mathrm{imp}}(K_2, T_2)$ of implied volatilites.

Clearly, there is no reason a priori for the implied volatilites $\sigma^{\mathrm{imp}}(K_1, T_1)$ and $\sigma^{\mathrm{imp}}(K_2, T_2)$ to coincide. However, in the standard Black–Scholes model the value of the parameter σ should be unique for a given stock S. This contradiction between a model and market data is a reason for the development of more sophisticated stochastic volatility models.

Plotting the different values of the implied volatility σ as a function of K and T will yield a planar curve called the volatility surface.

Figure 7.1 presents an estimation of implied volatility for Asian options whose underlying asset is the price of light sweet crude oil futures traded on the New York Mercantile Exchange (NYMEX), based on contract specifications and market data obtained from the Chicago Mercantile Exchange.

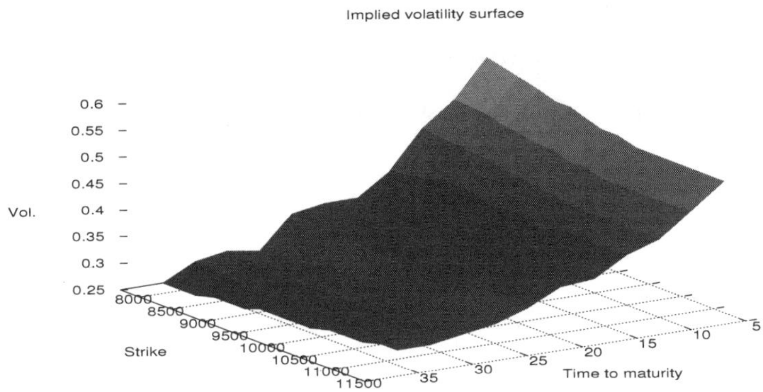

FIGURE 7.1: Implied volatility of Asian options on light sweet crude oil futures.[1]

As observed in Figure 7.1, the volatility surface can exhibit a *smile* phenomenon, in which implied volatility is higher at a given end (or at both ends) of the range of strike values.

7.3 Black–Scholes Formula vs. Market Data

On July 28, 2009 a call warrant was issued by Merrill Lynch on the stock price S of Cheung Kong Holdings (0001.HK) with Strike K=\$109.99 and Maturity T = December 13, 2010.

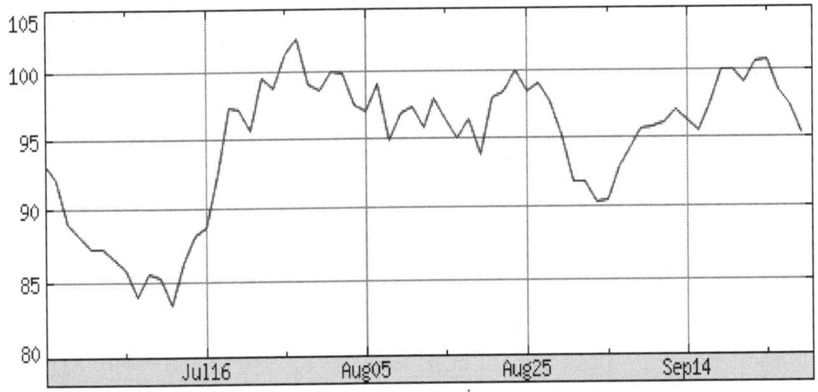

FIGURE 7.2: Graph of the (market) stock price of Cheung Kong Holdings.

The market price of the option (17838.HK) on September 28 was \$12.30, as

[1]This graph is courtesy of Tan Yu Jia.

obtained from `http://www.hkex.com.hk/dwrc/search/listsearch.asp`

The next graph shows the evolution of the market price of the option over time. One sees that the option price is much more volatile than the underlying stock price.

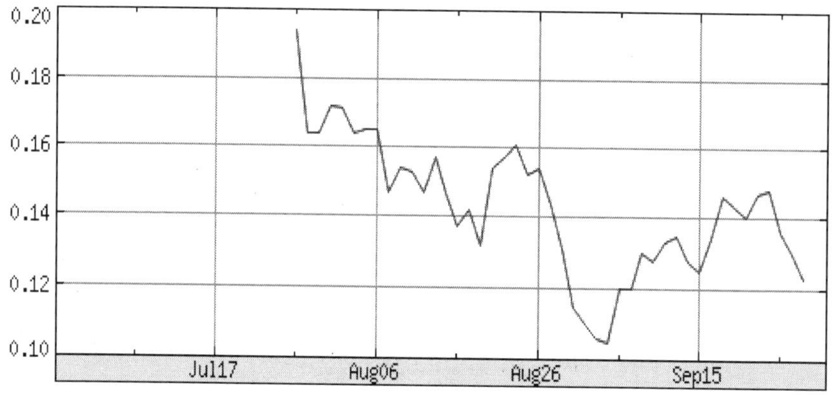

FIGURE 7.3: Graph of the (market) call option price on Cheung Kong Holdings.

In Figure 7.4 we have fitted the path

$$t \longmapsto g_c(t, S_t)$$

of the Black–Scholes price to the data of Figure 7.3 using the stock price data of Figure 7.2, by varying the values of the volatility σ.

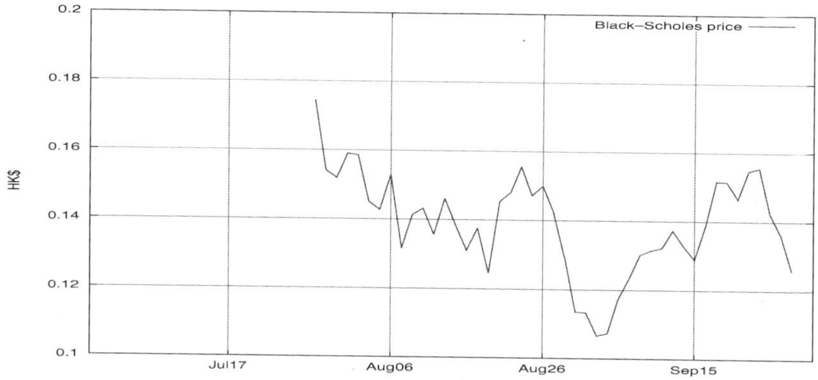

FIGURE 7.4: Graph of the Black–Scholes call option price on Cheung Kong Holdings.

Another example

Let us consider the stock price of HSBC Holdings (0005.HK) over one year:

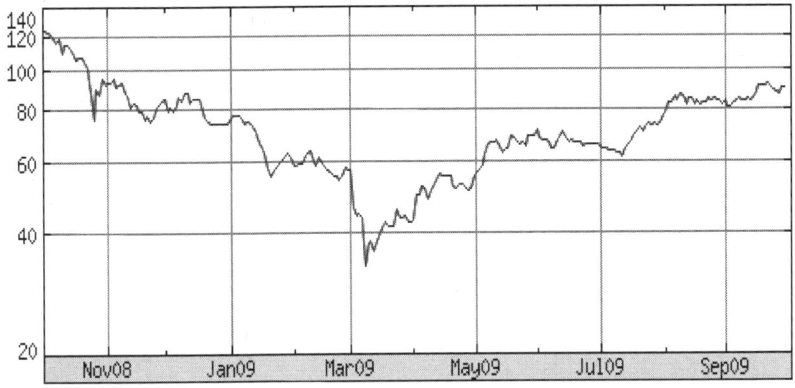

FIGURE 7.5: Graph of the (market) stock price of HSBC Holdings.

Next we consider the graph of the price of a call option issued by Societe Generale on 31 December 2008 with strike K=\$63.704, maturity T = October 05, 2009, and entitlement ratio 100, cf. page 6.

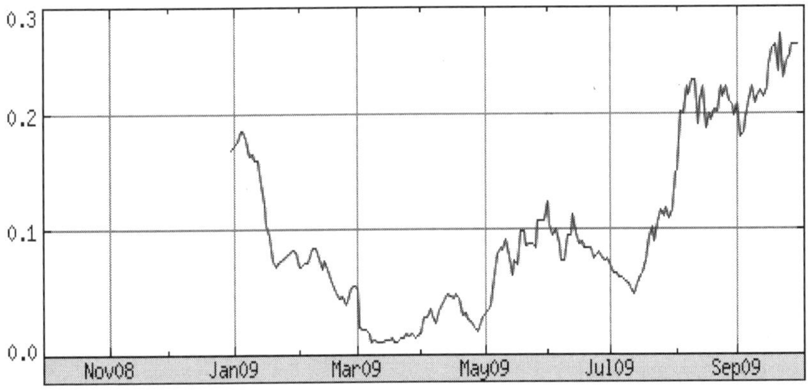

FIGURE 7.6: Graph of the (market) call option price on HSBC Holdings.

As above, in Figure 7.7 we have fitted the path $t \longmapsto g_c(t, S_t)$ of the Black–Scholes price to the data of Figure 7.6 using the stock price data of Figure 7.5. In this case we are in the money at maturity, and we also check that the option is worth $100 \times 0.2650 = \$26.650$ at that time which, by absence of arbitrage, is very close to the value $\$90 - \$63.703 = \$26.296$ of its payoff.

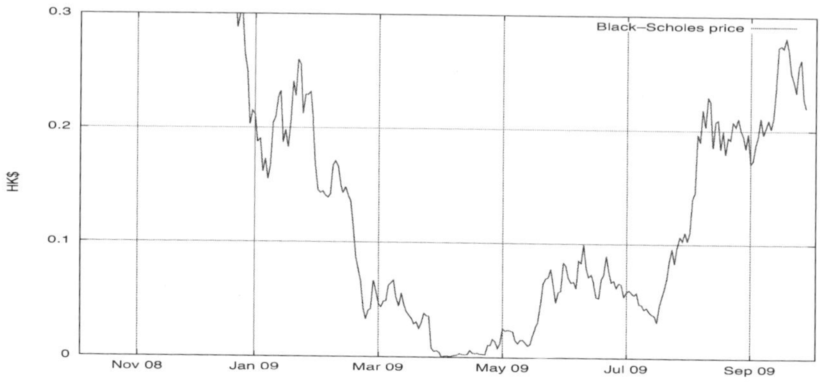

FIGURE 7.7: Graph of the Black–Scholes call option price on HSBC Holdings.

For one more example, consider the graph of the price of a put option issued by BNP Paribas on 04 November 2008 with strike K=\$77.667, maturity T = October 05, 2009, and entitlement ratio 92.593.

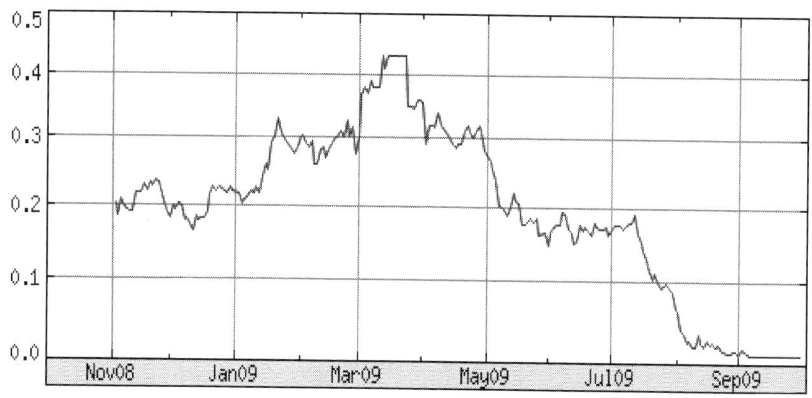

FIGURE 7.8: Graph of the (market) put option price on HSBC Holdings.

One checks easily that at maturity, the price of the put option is worth \$0.01 (a market price cannot be lower), which almost equals the option payoff \$0, by absence of arbitrage opportunities. Figure 7.9 is a fit of the Black–Scholes put price graph

$$t \longmapsto g_p(t, S_t)$$

to Figure 7.8 as a function of the stock price data of Figure 7.7. Note that the Black–Scholes price at maturity is strictly equal to 0 while the corresponding market price cannot be lower than one cent.

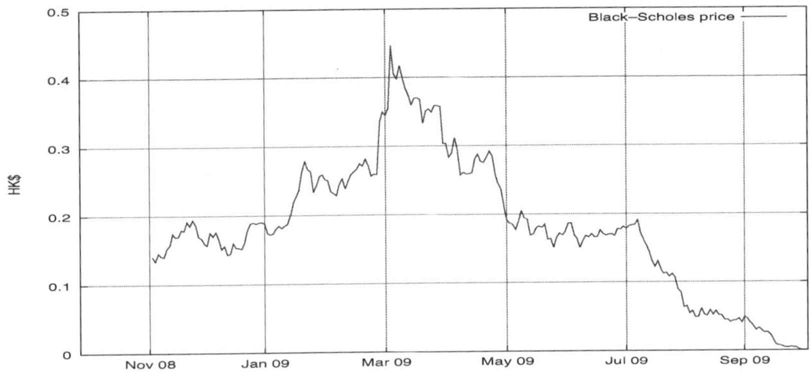

FIGURE 7.9: Graph of the Black–Scholes put option price on HSBC Holdings.

7.4 Local Volatility

Since the constant volatility assumption in the Black–Scholes model appears to be not satisfying due to the existence of volatility smiles, it makes sense to consider models of the form

$$\frac{dS_t}{S_t} = rdt + \sigma_t dB_t$$

where σ_t is a random process. Such models are called stochastic volatility models.

A particular class of stochastic volatility models can be written as

$$\frac{dS_t}{S_t} = rdt + \sigma(t, S_t)dB_t \tag{7.1}$$

where $\sigma(t, x)$ is a deterministic function of time and the stock price. Such models are called local volatility models. The corresponding Black–Scholes PDE can be written as

$$\begin{cases} rg(t,x,K) = \dfrac{\partial g}{\partial t}(t,x,K) + rx\dfrac{\partial g}{\partial x}(t,x,K) + \dfrac{1}{2}x^2\sigma^2(t,x)\dfrac{\partial^2 g}{\partial x^2}(t,x,K), \\ \\ g(T,x,K) = (x-K)^+, \end{cases} \tag{7.2}$$

with terminal condition $g(T, x, K) = (x - K)^+$, i.e., we consider European call options.

Note that the Black–Scholes PDE would allow one to recover the value of $\sigma(t,x)$ as a function of the option price $g(t,x,K)$, as

$$\sigma(t,x) = \sqrt{\frac{2rg(t,x,K) - 2\frac{\partial g}{\partial t}(t,x,K) - 2rx\frac{\partial g}{\partial x}(t,x,K)}{x^2\frac{\partial^2 g}{\partial x^2}(t,x,K)}}, \qquad x,t > 0,$$

however, this formula requires the knowledge of the option price for different values of the underlying x.

The Dupire formula provides a solution to the local volatility calibration problem by providing as estimator of $\sigma(t,x)$ based on different values of the strike price K.

Proposition 7.1 *Assume that a family $(C(T,K))_{T,K>0}$ of market call option prices with maturities T and strikes K is given at time t with $S_t = x$, while the values of r and x are fixed.*

The Dupire formula states that defining the volatility function $\sigma(t,y)$ by

$$\sigma(t,y) = \sqrt{\frac{2\frac{\partial C}{\partial t}(t,y) + 2ry\frac{\partial C}{\partial y}(t,y)}{y^2\frac{\partial^2 C}{\partial y^2}(t,y)}} \tag{7.3}$$

the prices $g(t,x,K)$ computed from the Black–Scholes PDE (7.2) will match the option prices $C(T,K)$ in the sense that

$$g(t,x,K) = C(T,K), \qquad T,K > 0. \tag{7.4}$$

Proof. We use the probabilistic approach that allows us to write $g(t,x,K)$ as

$$g(t,x,K) = e^{-r(T-t)}\,\mathbb{E}[(S_T - K)^+ \mid S_t = x], \tag{7.5}$$

where $(S_t)_{t\in\mathbb{R}_+}$ is defined by (7.1). Hence the condition (7.4) can be written at $t = 0$ as

$$C(T,K) = e^{-rT}\int_{-\infty}^{\infty}(y-K)^+\varphi_T(y)dy,$$

where $\varphi_T(y)$ is the probability density of S_T. After differentiating both sides twice with respect to K one gets

$$\frac{\partial^2 C}{\partial K^2}(T,K) = e^{-rT}\varphi_T(K). \tag{7.6}$$

On the other hand, for any sufficiently smooth function f, using the Itô formula we have

$$\int_{-\infty}^{\infty}\varphi_T(y)f(y)dy = \mathbb{E}[f(S_T)]$$

$$= \mathbb{E}\left[f(S_0) + \int_0^T f'(S_t)dS_t + \frac{1}{2}\int_0^T f''(S_t)\sigma^2(t, S_t)dt\right]$$

$$= \mathbb{E}\left[f(S_0) + r\int_0^T f'(S_t)S_t dt + \sigma\int_0^T f'(S_t)S_t dB_t + \frac{1}{2}\int_0^T f''(S_t)\sigma^2(t, S_t)dt\right]$$

$$= f(S_0) + \mathbb{E}\left[r\int_0^T f'(S_t)S_t dt + \frac{1}{2}\int_0^T f''(S_t)S_t^2\sigma^2(t, S_t)dt\right]$$

$$= f(S_0) + r\int_{-\infty}^{\infty}\int_0^T yf'(y)\varphi_t(y)dtdy + \frac{1}{2}\int_{-\infty}^{\infty}\int_0^T y^2 f''(y)\sigma^2(t, y)\varphi_t(y)dtdy,$$

hence after differentiating both sides of the equality with respect to T,

$$\int_{-\infty}^{\infty}\frac{\partial\varphi_T}{\partial T}(y)f(y)dy = r\int_{-\infty}^{\infty} yf'(y)\varphi_T(y)dy + \frac{1}{2}\int_{-\infty}^{\infty} y^2 f''(y)\sigma^2(T, y)\varphi_T(y)dy.$$

Integrating by parts in the above relation yields

$$\begin{aligned}&\int_{-\infty}^{\infty}\frac{\partial\varphi_T}{\partial T}(y)f(y)dy\\ &= -r\int_{-\infty}^{\infty} f(y)\frac{\partial}{\partial y}(y\varphi_T(y))dy + \frac{1}{2}\int_{-\infty}^{\infty} f(y)\frac{\partial^2}{\partial y^2}(y^2\sigma^2(T, y)\varphi_T(y))dy,\end{aligned}$$

for all smooth functions $f(y)$ with compact support, hence

$$\frac{\partial\varphi_T}{\partial T}(y) = -r\frac{\partial}{\partial y}(y\varphi_T(y)) + \frac{1}{2}\frac{\partial^2}{\partial y^2}(y^2\sigma^2(T, y)\varphi_T(y)), \qquad y \in \mathbb{R}.$$

Making use of (7.6) we get

$$\begin{aligned}&-r\frac{\partial^2 C}{\partial y^2}(T, y) - \frac{\partial}{\partial T}\frac{\partial^2 C}{\partial y^2}(T, y)\\ &= r\frac{\partial}{\partial y}\left(y\frac{\partial^2 C}{\partial y^2}(T, y)\right) - \frac{1}{2}\frac{\partial^2}{\partial y^2}\left(y^2\sigma^2(T, y)\frac{\partial^2 C}{\partial y^2}(T, y)\right), \qquad y \in \mathbb{R}.\end{aligned}$$

After a first integration with respect to y under the limiting condition $\lim_{K\to+\infty} C(T, K) = 0$, we obtain

$$-r\frac{\partial C}{\partial y}(T, y) - \frac{\partial}{\partial T}\frac{\partial C}{\partial y}(T, y) = ry\frac{\partial^2 C}{\partial y^2}(T, y) - \frac{1}{2}\frac{\partial}{\partial y}\left(y^2\sigma^2(T, y)\frac{\partial^2 C}{\partial y^2}(T, y)\right),$$

i.e.,

$$\begin{aligned}&-r\frac{\partial C}{\partial y}(T, y) - \frac{\partial}{\partial T}\frac{\partial C}{\partial y}(T, y)\\ &= r\frac{\partial}{\partial y}\left(y\frac{\partial C}{\partial y}(T, y)\right) - r\frac{\partial C}{\partial y}(T, y) - \frac{1}{2}\frac{\partial}{\partial y}\left(y^2\sigma^2(T, y)\frac{\partial^2 C}{\partial y^2}(T, y)\right),\end{aligned}$$

or

$$-\frac{\partial}{\partial y}\frac{\partial C}{\partial T}(T, y) = r\frac{\partial}{\partial y}\left(y\frac{\partial C}{\partial y}(T, y)\right) - \frac{1}{2}\frac{\partial}{\partial y}\left(y^2\sigma^2(T, y)\frac{\partial^2 C}{\partial y^2}(T, y)\right).$$

Integrating one more time with respect to y yields

$$-\frac{\partial C}{\partial T}(T, y) = ry\frac{\partial C}{\partial y}(T, y) - \frac{1}{2}y^2\sigma^2(T, y)\frac{\partial^2 C}{\partial y^2}(T, y), \qquad y \in \mathbb{R},$$

which conducts to (7.3) and is called the Dupire [21] PDE. □

From (7.3) the local volatility $\sigma(t, y)$ can be estimated by computing $C(T, y)$ by the Black–Scholes formula, based on a value of the implied volatility σ. See [1] for numerical methods applied to volatility estimation in this framework.

Chapter 8

Exotic Options

Throughout this chapter we work in a continuous geometric Brownian model in which the asset price $(S_t)_{t\in[0,T]}$ has the dynamics

$$dS_t = rS_t dt + \sigma S_t dB_t, \qquad t \in [0,T],$$

and we assume that probability measure $\mathbb{P}$ is risk-neutral. In particular the value V_t of a self-financing portfolio satisfies

$$V_T e^{-rT} = V_0 + \sigma \int_0^T \xi_t S_t e^{-rt} dB_t, \quad t \in [0,T].$$

8.1 Generalities

An exotic option is an option whose payoff may depend on the whole path $\{S_t : t \in [0,T]\}$ of the price process via a "complex" operation such as averaging or computing a maximum. They are opposed to vanilla options whose payoff

$$C = \phi(S_T),$$

where ϕ is called a payoff function, depends only on the terminal value S_T of the price process.

An option with payoff $C = \phi(S_T)$ can be priced as

$$e^{-rT}\,\mathbb{E}[\phi(S_T)] = e^{-rT}\int_{-\infty}^{\infty} \phi(y) f_{S_T}(y) dy$$

where $f_{S_T}(y)$ is the (one parameter) *probability density* function of S_T, which satisfies

$$\mathbb{P}(S_T \leq y) = \int_{-\infty}^{y} f_{S_T}(v) dv, \qquad y \in \mathbb{R}.$$

Recall that typically we have

$$\phi(x) = (x-K)^+ = \begin{cases} x - K & \text{if } x \geq K, \\ 0 & \text{if } x < K, \end{cases}$$

for a European call option with strike K, and

$$\phi(x) = \mathbf{1}_{[K,\infty)}(x) = \begin{cases} \$1 & \text{if } x \geq K, \\ 0 & \text{if } x < K, \end{cases}$$

for a binary call option with strike K.

Exotic Options

Exotic options, also called path-dependent options, are options whose payoff C may depend on the whole path

$$\{S_t \ : \ 0 \leq t \leq T\}$$

of the underlying price process instead of its terminal value S_T. Next we review some examples of exotic options.

Options on Extrema

We take

$$C := \phi(M_T),$$

where

$$M_T = \max_{t \in [0,T]} S_t$$

is the maximum of $(S_t)_{t \in \mathbb{R}_+}$ over the time interval $[0, T]$.

Figure 8.1 represents the running maximum process $(M_t)_{t \in \mathbb{R}_+}$ of Brownian motion $(B_t)_{t \in \mathbb{R}_+}$.

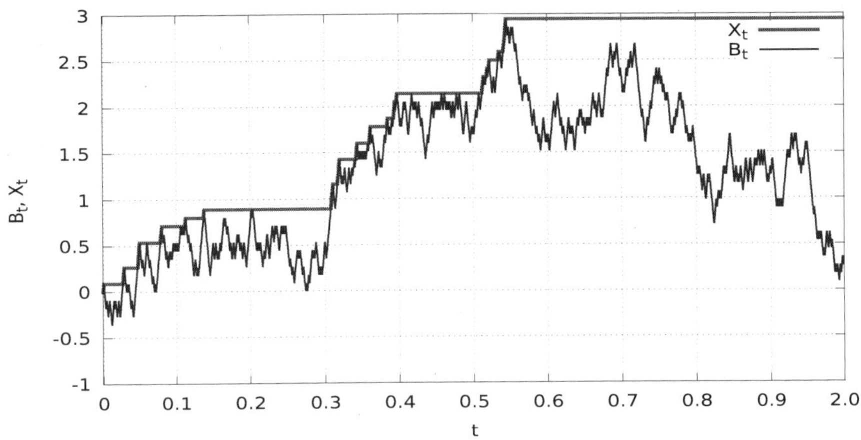

FIGURE 8.1: Brownian motion B_t and its supremum X_t.

Barrier Options

The payoff of an up-and-out barrier put option on the underlying asset S_t with exercise date T, strike K and barrier B is

$$C = (K - S_T)^+ \mathbf{1}_{\left\{\max\limits_{0\le t\le T} S_t < B\right\}} = \begin{cases} K - S_T & \text{if } \max\limits_{0\le t\le T} S_t < B, \\ 0 & \text{if } \max\limits_{0\le t\le T} S_t \ge B. \end{cases}$$

This option is also called a Callable Bear Contract with no residual value, in which the call price B usually satisfies $B \le K$.

The payoff of a down-and-out barrier call option on the underlying asset S_t with exercise date T, strike K and barrier B is

$$C = (S_T - K)^+ \mathbf{1}_{\left\{\min\limits_{0\le t\le T} S_t > B\right\}} = \begin{cases} S_T - K & \text{if } \min\limits_{0\le t\le T} S_t > B, \\ 0 & \text{if } \min\limits_{0\le t\le T} S_t \le B. \end{cases}$$

This option is also called a Callable Bull Contract with no residual value, in which B denotes the call price $B \ge K$. It is also called a turbo warrant with no rebate.

Lookback Options

The payoff of a floating strike lookback call option on the underlying asset S_t with exercise date T is

$$C = S_T - \min_{0\le t\le T} S_t.$$

The payoff of a floating strike lookback put option on the underlying asset S_t with exercise date T is

$$C = \left(\max_{0 \le t \le T} S_t \right) - S_T.$$

Options on Average

In this case we can take

$$C = \phi \left(\frac{1}{T} \int_0^T S_t dt \right)$$

where

$$\frac{1}{T} \int_0^T S_t dt$$

represents the average of $(S_t)_{t \in \mathbb{R}_+}$ over the time interval $[0, T]$ and $\phi : \mathbb{R} \longrightarrow \mathbb{R}$ is a payoff function.

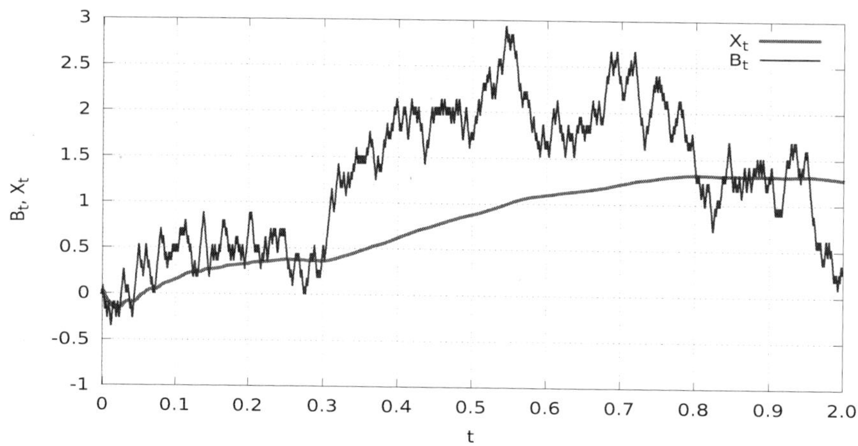

FIGURE 8.2: Brownian motion B_t and its moving average.

Figure 8.2 shows a graph of Brownian motion and its moving average process X_t.

Asian Options

Asian options are particular cases of options on average, and they were first traded in Tokyo in 1987. The payoff of the Asian call option on the underlying asset S_t with exercise date T and strike K is given by

$$C = \left(\frac{1}{T} \int_0^T S_t dt - K \right)^+.$$

Similarly, the payoff of the Asian put option on the underlying asset S_t with exercise date T and strike K is

$$C = \left(K - \frac{1}{T}\int_0^T S_t dt\right)^+ .$$

Because of their dependence on averaged asset prices, Asian options are less volatile than plain vanilla options whose payoffs depend only on the terminal value of the underlying asset. Asian options have become particularly popular in commodities trading.

8.2 Reflexion Principle

In order to price barrier options we will have to derive the probability density of the maximum

$$M_T = \max_{t\in[0,T]} S_t$$

of geometric Brownian motion $(S_t)_{t\in\mathbb{R}_+}$ over a given time interval $[0,T]$.

In such situations the option price at time $t = 0$ can be expressed as

$$e^{-rT}\,\mathbb{E}[\phi(M_T, S_T)] = e^{-rT}\int_{-\infty}^{\infty} \phi(x,y) f_{(M_T,S_T)}(x,y)dxdy$$

where $f_{(M_T,S_T)}$ is the *joint probability density* function of (M_T, S_T), which satisfies

$$\mathbb{P}(M_T \le x,\ S_T \le y) = \int_{-\infty}^{x}\int_{-\infty}^{y} f_{(M_T,S_T)}(u,v)dudv, \qquad x,y\in\mathbb{R}.$$

In order to price such options by the above probabilistic method, we will compute $f_{(M_T,S_T)}(u,v)$ by the *reflection principle.*

Maximum of Standard Brownian Motion

Let $(B_t)_{t\in\mathbb{R}_+}$ denote a standard Brownian motion started at $B_0 = 0$. While it is well-known that $B_T \simeq \mathcal{N}(0,T)$, computing the law of the maximum

$$X_T = \max_{t\in[0,T]} B_t$$

might seem a difficult problem. However this is not the case, due to the *reflection principle.* Note that since $B_0 = 0$ we have

$$X_T \ge 0,$$

almost surely.

Given $a > B_0 = 0$, let

$$\tau_a = \inf\{t \in \mathbb{R}_+ \ : \ B_t = a\}$$

denote the first time $(B_t)_{t\in\mathbb{R}_+}$ hits the level $a > 0$.

Due to the space symmetry of Brownian motion we have the identity

$$\mathbb{P}(B_T > a \mid \tau_a < T) = \frac{1}{2} = \mathbb{P}(B_T < a \mid \tau_a < T).$$

This identity is clearly equivalent to

$$2\mathbb{P}(B_T > a \ \& \ \tau_a < T) = \mathbb{P}(\tau_a < T) = 2\mathbb{P}(B_T < a \ \& \ \tau_a < T),$$

and to

$$2\mathbb{P}(B_T > a \ \& \ X_T \geq a) \ = \ \mathbb{P}(\tau_a < T) = 2\mathbb{P}(B_T < a \ \& \ X_T \geq a),$$

due to the equivalence

$$\{X_T \geq a\} = \{\tau_a < T\}. \tag{8.1}$$

In other words, we have

$$\begin{aligned}
\mathbb{P}(X_T \geq a) &= \mathbb{P}(B_T > a \ \& \ X_T \geq a) + \mathbb{P}(B_T < a \ \& \ X_T \geq a) \\
&= 2\mathbb{P}(B_T > a \ \& \ X_T \geq a) \\
&= 2\mathbb{P}(B_T > a) \\
&= \mathbb{P}(B_T > a) + \mathbb{P}(B_T < -a) \\
&= \mathbb{P}(|B_T| > a),
\end{aligned}$$

where we used the fact that

$$\{B_T > a\} \subset \{B_T > a \ \& \ X_T \geq a\} \subset \{B_T > a\}.$$

Figure 8.3 shows a graph of Brownian motion and its reflected path.

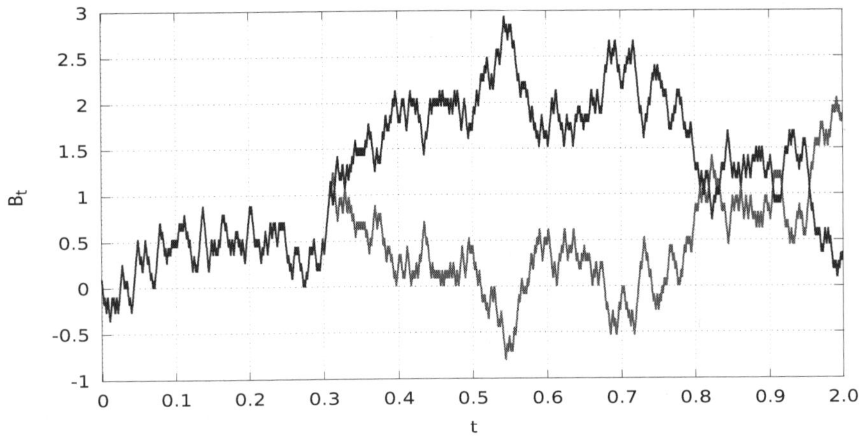

FIGURE 8.3: Reflected Brownian motion with $a = 1$.

Consequently, the maximum X_T of Brownian motion has *same distribution* as the absolute value $|B_T|$ of B_T. In other words, X_T is a non-negative random variable with distribution function

$$\begin{aligned} \mathbb{P}(X_T \leq a) &= \mathbb{P}(|B_T| \leq a) \\ &= \frac{1}{\sqrt{2\pi T}} \int_{-a}^{a} e^{-x^2/(2T)} dx \\ &= \frac{2}{\sqrt{2\pi T}} \int_{0}^{a} e^{-x^2/(2T)} dx, \qquad a \in \mathbb{R}_+, \end{aligned}$$

and probability density

$$f_{X_T}(a) = \frac{d\mathbb{P}(X_T \leq a)}{da} = \sqrt{\frac{2}{\pi T}} e^{-a^2/(2T)} \mathbf{1}_{[0,\infty)}(a), \qquad a \in \mathbb{R}. \tag{8.2}$$

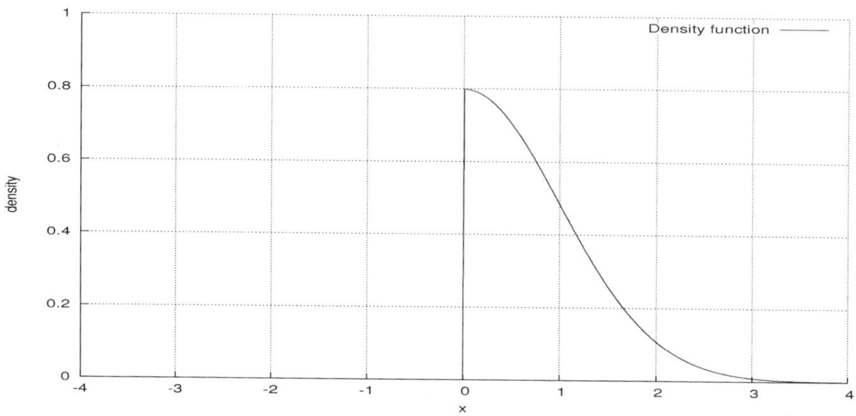

FIGURE 8.4: Probability density of the maximum of Brownian motion over $[0, 1]$.

Using the density of X_T we can price an option with payoff $\phi(X_T)$, as

$$\begin{aligned} e^{-rT}\,\mathbb{E}[\phi(X_T)] &= e^{-rT}\int_{-\infty}^{\infty}\phi(x)d\mathbb{P}(X_T = x) \\ &= e^{-rT}\sqrt{\frac{2}{\pi T}}\int_0^{\infty}\phi(x)e^{-|x|^2/(2T)}dx. \end{aligned}$$

Next we consider

$$\begin{aligned} M_T &= \max_{t\in[0,T]} S_t \\ &= S_0 \max_{t\in[0,T]} e^{\sigma B_t} \\ &= S_0 e^{\sigma \max_{t\in[0,T]} B_t} \\ &= S_0 e^{\sigma X_T}, \end{aligned}$$

since $\sigma > 0$. When the payoff takes the form

$$C = \phi(M_T),$$

where

$$S_T = S_0 e^{\sigma B_T},$$

we have

$$C = \phi(M_T) = \phi(S_0 e^{\sigma X_T}),$$

hence

$$\begin{aligned} & e^{-rT}\,\mathbb{E}[C] = e^{-rT}\,\mathbb{E}\left[\phi(S_0 e^{\sigma X_T})\right] \\ &= e^{-rT}\int_{-\infty}^{\infty}\phi(S_0 e^{\sigma x})d\mathbb{P}(X_T = x) \\ &= \sqrt{\frac{2}{\pi T}}e^{-rT}\int_0^{\infty}\phi(S_0 e^{\sigma x})e^{-x^2/(2T)}dx. \end{aligned}$$

This, however, is not sufficient since it imposes the condition $r = \sigma^2/2$. In order to do away with this condition we need to consider the maximum of *drifted* Brownian motion, and for this we have to compute the *joint density* of X_T and B_T.

Joint Density

The reflection principle also allows us to compute the *joint density* of Brownian motion B_T and its maximum X_T. Indeed, for $b \in [0, a]$ we also have

$$\mathbb{P}(B_T > a + (a - b) \mid \tau_a < T) = \mathbb{P}(B_T < b \mid \tau_a < T),$$

i.e.,

$$\mathbb{P}(B_T > 2a - b \ \& \ \tau_a < T) = \mathbb{P}(B_T < b \ \& \ \tau_a < T),$$

or, by (8.1),

$$\mathbb{P}(B_T > 2a - b \ \& \ X_T \geq a) = \mathbb{P}(B_T < b \ \& \ X_T \geq a),$$

hence, since $2a - b \geq a$,

$$\mathbb{P}(B_T \geq 2a - b) = \mathbb{P}(B_T > 2a - b \ \& \ X_T \geq a) = \mathbb{P}(B_T < b \ \& \ X_T \geq a), \quad (8.3)$$

where we used the fact that

$$\begin{aligned} \{B_T \geq 2a - b\} &\subset \{B_T > 2a - b \ \& \ X_T \geq 2a - b\} \\ &\subset \{B_T > 2a - b \ \& \ X_T \geq a\} \subset \{B_T > a\}, \end{aligned}$$

which shows that $\{B_T \geq 2a - b\} = \{B_T > 2a - b \ \& \ X_T \geq a\}$.

Hence by (8.3) we have

$$\mathbb{P}(B_T < b \ \& \ X_T \geq a) = \mathbb{P}(B_T \geq 2a - b) = \frac{1}{\sqrt{2\pi T}} \int_{2a-b}^{\infty} e^{-x^2/(2T)} dx,$$

$0 \leq b \leq a$, which yields the joint probability density

$$f_{X_T, B_T}(a, b) = -\frac{d\mathbb{P}(X_T \geq a \ \& \ B_T \leq b)}{dadb} = \frac{d\mathbb{P}(X_T \leq a \ \& \ B_T \leq b)}{dadb},$$

$a, b \in \mathbb{R}$, by (A.11), i.e., letting $a \vee b := \max(a, b)$,

$$f_{X_T,B_T}(a,b) = \sqrt{\frac{2}{\pi T}}\frac{(2a-b)}{T}e^{-(2a-b)^2/(2T)}\mathbf{1}_{\{a\geq b\vee 0\}} \tag{8.4}$$

$$= \begin{cases} \sqrt{\frac{2}{\pi T}}\frac{(2a-b)}{T}e^{-(2a-b)^2/(2T)}, & a > b\vee 0, \\ 0, & a < b\vee 0. \end{cases}$$

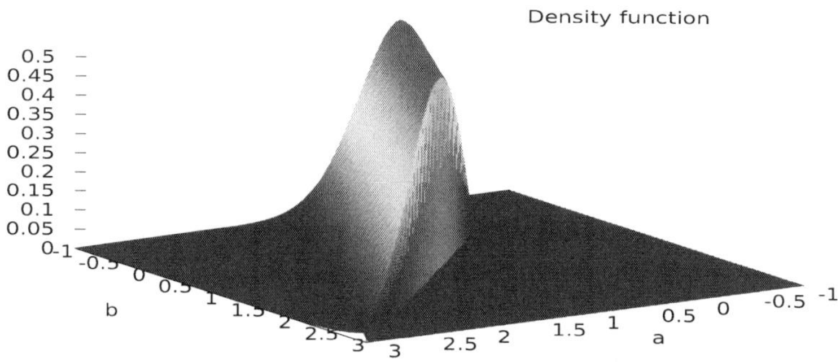

FIGURE 8.5: Joint probability density of B_1 and its maximum over [0,1].

Maximum of Drifted Brownian Motion

Using the Girsanov theorem, it is even possible to compute the probability density function of the maximum

$$\tilde{X}_T = \max_{t\in[0,T]} \tilde{B}_t = \max_{t\in[0,T]} (B_t + \mu t)$$

of the drifted Brownian motion $\tilde{B}_t = B_t + \mu t$, $\mu \in \mathbb{R}$. The arguments previously applied to B_t cannot be directly applied to $\tilde{B}_t$ because drifted Brownian motion is no longer symmetric in space when $\mu \neq 0$.

On the other hand, $\tilde{B}_t$ is a standard Brownian motion under the probability measure $\tilde{\mathbb{P}}$ defined from

$$\frac{d\tilde{\mathbb{P}}}{d\mathbb{P}} = e^{-\mu B_T - \mu^2 T/2}, \tag{8.5}$$

hence the density of $\tilde{X}_T$ under $\tilde{\mathbb{P}}$ is given by (8.4).

Now, using the density (8.5) we get

$$\begin{aligned}
\mathbb{P}(\tilde{X}_T \leq a \ \&\ \tilde{B}_T \leq b) &= \mathbb{E}\left[\mathbf{1}_{\{\tilde{X}_T\leq a \ \&\ \tilde{B}_T\leq b\}}\right] \\
&= \tilde{\mathbb{E}}\left[e^{\mu B_T+\mu^2T/2}\mathbf{1}_{\{\tilde{X}_T\leq a \ \&\ \tilde{B}_T\leq b\}}\right] \\
&= \tilde{\mathbb{E}}\left[e^{\mu \tilde{B}_T-\mu^2T/2}\mathbf{1}_{\{\tilde{X}_T\leq a \ \&\ \tilde{B}_T\leq b\}}\right] \\
&= \sqrt{\frac{2}{\pi T}}\int_0^a\int_{-\infty}^b \mathbf{1}_{(-\infty,x]}(y)e^{\mu y-\mu^2T/2}\frac{(2x-y)}{T}e^{-(2x-y)^2/(2T)}dxdy,
\end{aligned}$$

$0 \leq b \leq a$, which yields the joint probability density

$$f_{\tilde{X}_T,\tilde{B}_T}(a,b) = \frac{d\mathbb{P}(\tilde{X}_T \leq a \ \&\ \tilde{B}_T \leq b)}{dadb},$$

i.e.,

$$\begin{aligned}
f_{\tilde{X}_T,\tilde{B}_T}(a,b) &= \mathbf{1}_{\{a\geq b\vee 0\}}\frac{1}{T}\sqrt{\frac{2}{\pi T}}(2a-b)e^{\mu b-(2a-b)^2/(2T)-\mu^2T/2} \qquad (8.6) \\
&= \begin{cases} \frac{1}{T}\sqrt{\frac{2}{\pi T}}(2a-b)e^{-\mu^2T/2+\mu b-(2a-b)^2/(2T)}, & a > b\vee 0, \\ 0, & a < b\vee 0. \end{cases}
\end{aligned}$$

We also find

$$\begin{aligned}
\mathbb{P}(\tilde{X}_T \leq a) &= \sqrt{\frac{2}{\pi T}}\int_0^a\int_{-\infty}^\infty \mathbf{1}_{(-\infty,x]}(y)e^{\mu y-\mu^2T/2}\frac{(2x-y)}{T}e^{-(2x-y)^2/(2T)}dydx \\
&= \sqrt{\frac{2}{\pi T}}e^{-\mu^2T/2}\int_{-\infty}^a e^{\mu y}\int_{y\vee 0}^a\frac{(2x-y)}{T}e^{-(2x-y)^2/(2T)}dxdy \\
&= \sqrt{\frac{1}{2\pi T}}e^{-\mu^2T/2}\int_{-\infty}^a\left(e^{\mu y-(2(y\vee 0)-y)^2/(2T)} - e^{\mu y-(2a-y)^2/(2T)}\right)dy \\
&= \sqrt{\frac{1}{2\pi T}}\int_{-\infty}^a\left(e^{\mu y-y^2/(2T)-\mu^2T/2} - e^{\mu y-2a^2/T+2ay/T-y^2/(2T)-\mu^2T/2}\right)dy \\
&= \sqrt{\frac{1}{2\pi T}}\int_{-\infty}^a\left(e^{-(y-\mu T)^2/(2T)} - e^{-(y-(\mu T+2a))^2/(2T)+2a\mu}\right)dy
\end{aligned}$$

$$= \sqrt{\frac{1}{2\pi T}} \int_{-\infty}^{a} e^{-(y-\mu T)^2/(2T)} dy - e^{2a\mu} \sqrt{\frac{1}{2\pi T}} \int_{-\infty}^{a} e^{-(y-(\mu T+2a))^2/(2T)} dy$$
$$= \sqrt{\frac{1}{2\pi T}} \int_{-\infty}^{a-\mu T} e^{-y^2/(2T)} dy - e^{2a\mu} \sqrt{\frac{1}{2\pi T}} \int_{-\infty}^{-a-\mu T} e^{-y^2/(2T)} dy$$
$$= \Phi\left(\frac{a-\mu T}{\sqrt{T}}\right) - e^{2\mu a} \Phi\left(\frac{-a-\mu T}{\sqrt{T}}\right), \tag{8.7}$$

cf. Corollary 7.2.2 and pages 297-299 of [68] for another derivation. This yields the density

$$\frac{d\mathbb{P}(\tilde{X}_T \leq a)}{da} = \sqrt{\frac{2}{\pi T}} e^{-(a-\mu T)^2/(2T)} - 2\mu e^{2\mu a} \Phi\left(\frac{-a-\mu T}{\sqrt{T}}\right),$$

of the supremum of drifted Brownian motion, and recovers (8.2) for $\mu = 0$.

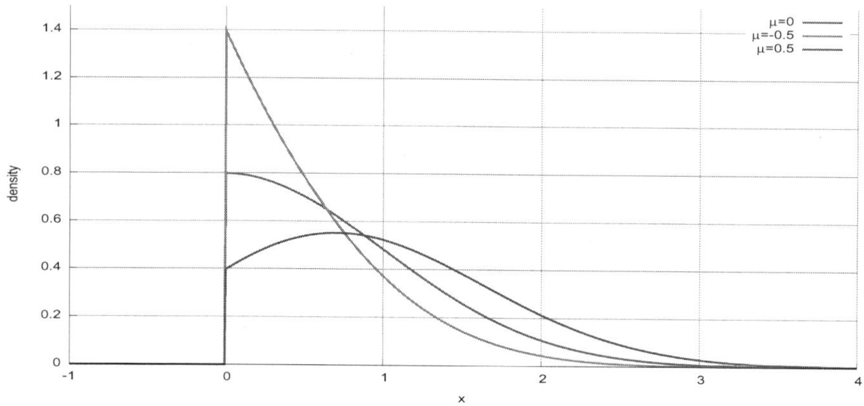

FIGURE 8.6: Probability density of the maximum of drifted Brownian motion.

Note from Figure 8.2 that small values of the maximum are more likely to occur when μ takes large negative values.

The joint density $f_{\tilde{R}_T, \tilde{B}_T}$ of the minimum

$$\tilde{R}_T = \min_{t\in[0,T]} \tilde{B}_t = \min_{t\in[0,T]} (B_t + \mu t)$$

of the drifted Brownian motion $\tilde{B}_t := B_t + \mu t$ and its value $\tilde{B}_T$ at time T can similarly be computed as follows, letting $a \wedge b := \min(a, b)$:

$$f_{\tilde{R}_T,\tilde{B}_T}(a,b) = \mathbf{1}_{\{a\leq b\wedge 0\}}\frac{1}{T}\sqrt{\frac{2}{\pi T}}(b-2a)e^{\mu b-(2a-b)^2/(2T)-\mu^2T/2} \tag{8.8}$$

$$= \begin{cases} \dfrac{1}{T}\sqrt{\dfrac{2}{\pi T}}(b-2a)e^{-\mu^2T/2+\mu b-(2a-b)^2/(2T)}, & a < b\wedge 0, \\ 0, & a > b\wedge 0. \end{cases}$$

8.3 Barrier Options

General Case

Using the joint density of $\tilde{B}_T$ and $\tilde{X}_T$ we are able to price any exotic option with payoff $\phi(\tilde{B}_T, \tilde{X}_T)$, as

$$e^{-r(T-t)}\,\mathbb{E}\left[\phi(\tilde{X}_T,\tilde{B}_T)\middle|\mathcal{F}_t\right],$$

with in particular

$$e^{-rT}\,\mathbb{E}\left[\phi(\tilde{X}_T,\tilde{B}_T)\right] = e^{-rT}\int_{-\infty}^{\infty}\int_{y\vee 0}^{\infty}\phi(x,y)d\mathbb{P}(\tilde{X}_T = x, \tilde{B}_T = y).$$

When the payoff takes the form

$$C = \phi(M_T, S_T),$$

where

$$S_T = S_0e^{\sigma B_T-\sigma^2T/2+rT} = S_0e^{\sigma\tilde{B}_T},$$

with $\mu = -\sigma/2 + r/\sigma$ and $\tilde{B}_T = B_T + \mu T$, and

$$\begin{aligned} M_T &= \max_{t\in[0,T]} S_t \\ &= S_0 \max_{t\in[0,T]} e^{\sigma B_t-\sigma^2t/2+rt} \\ &= S_0 \max_{t\in[0,T]} e^{\sigma\tilde{B}_t} \\ &= S_0e^{\sigma\max_{t\in[0,T]}\tilde{B}_t} \\ &= S_0e^{\sigma\tilde{X}_T}, \end{aligned}$$

we have

$$\begin{aligned} C &= \phi(S_T, M_T) \\ &= \phi(S_0 e^{\sigma B_T - \sigma^2 T/2 + rT}, M_T) \\ &= \phi(S_0 e^{\sigma \tilde{B}_T}, S_0 e^{\sigma \tilde{X}_T}), \end{aligned}$$

hence

$$\begin{aligned} e^{-rT}\,\mathbb{E}\,[C] &= e^{-rT}\,\mathbb{E}\left[\phi(S_0 e^{\sigma \tilde{B}_T}, S_0 e^{\sigma \tilde{X}_T})\right] \\ &= e^{-rT}\int_{-\infty}^{\infty}\int_{y\vee 0}^{\infty} \phi(S_0 e^{\sigma y}, S_0 e^{\sigma x}) d\mathbb{P}(\tilde{X}_T = x, \tilde{B}_T = y) \\ &= \frac{1}{T}\sqrt{\frac{2}{\pi T}} e^{-rT}\int_{-\infty}^{\infty}\int_{y\vee 0}^{\infty} \phi(S_0 e^{\sigma y}, S_0 e^{\sigma x}) \\ &\qquad (2x-y) e^{-\mu^2 T/2 + \mu y - (2x-y)^2/(2T)} dx dy \\ &= \frac{1}{T} e^{-rT}\sqrt{\frac{2}{\pi T}}\int_{0}^{\infty}\int_{y}^{\infty} \phi(S_0 e^{\sigma y}, S_0 e^{\sigma x}) \\ &\qquad (2x-y) e^{-\mu^2 T/2 + \mu y - (2x-y)^2/(2T)} dx dy \\ &\quad + \frac{1}{T} e^{-rT}\sqrt{\frac{2}{\pi T}}\int_{-\infty}^{0}\int_{0}^{\infty} \phi(S_0 e^{\sigma y}, S_0 e^{\sigma x}) \\ &\qquad (2x-y) e^{-\mu^2 T/2 + \mu y - (2x-y)^2/(2T)} dx dy. \end{aligned}$$

We can distinguish 8 different versions of barrier options according to the following table.

We have the following obvious relations between the prices of barrier and vanilla call and put options:

$$C_{\text{up-in}}(t) + C_{\text{up-out}}(t) = C(t) = e^{-r(T-t)}\,\mathbb{E}[(S_T - K)^+],$$

$$C_{\text{down-in}}(t) + C_{\text{down-out}}(t) = C(t) = e^{-r(T-t)}\,\mathbb{E}[(S_T - K)^+],$$

$$P_{\text{up-in}}(t) + P_{\text{up-out}}(t) = P(t) = e^{-r(T-t)}\,\mathbb{E}[(K - S_T)^+],$$

$$P_{\text{down-in}}(t) + P_{\text{down-out}}(t) = P(t) = e^{-r(T-t)}\,\mathbb{E}[(K - S_T)^+],$$

where $C(t)$, resp. $P(t)$ denotes the price of a European call, resp. put option with strike K.

Consequently, in the sequel we will only compute the prices of the up-and-out call and put, and down-and-out barrier call and put options.

option type	behavior	payoff
barrier call option	down-and-out	$(S_T-K)^+\mathbf{1}_{\left\{\min\limits_{0\leq t\leq T} S_t > B\right\}}$
	down-and-in	$(S_T-K)^+\mathbf{1}_{\left\{\min\limits_{0\leq t\leq T} S_t < B\right\}}$
	up-and-out	$(S_T-K)^+\mathbf{1}_{\left\{\max\limits_{0\leq t\leq T} S_t < B\right\}}$
	up-and-in	$(S_T-K)^+\mathbf{1}_{\left\{\max\limits_{0\leq t\leq T} S_t > B\right\}}$
barrier put option	down-and-out	$(K-S_T)^+\mathbf{1}_{\left\{\min\limits_{0\leq t\leq T} S_t > B\right\}}$
	down-and-in	$(K-S_T)^+\mathbf{1}_{\left\{\min\limits_{0\leq t\leq T} S_t < B\right\}}$
	up-and-out	$(K-S_T)^+\mathbf{1}_{\left\{\max\limits_{0\leq t\leq T} S_t < B\right\}}$
	up-and-in	$(K-S_T)^+\mathbf{1}_{\left\{\max\limits_{0\leq t\leq T} S_t > B\right\}}$

Up-and-Out Barrier Call Option

Let us consider an up-and-out call option with maturity T, strike K, barrier (or call price) B, and payoff

$$C = (S_T-K)^+\mathbf{1}_{\left\{\max\limits_{0\leq t\leq T} S_t \leq B\right\}} = \begin{cases} S_T-K & \text{if } \max\limits_{0\leq t\leq T} S_t \leq B, \\ 0 & \text{if } \max\limits_{0\leq t\leq T} S_t > B, \end{cases}$$

with $B > K$. Our goal is to prove the following result.

Proposition 8.1 *When $K < B$, the price*

$$e^{-r(T-t)} \mathbf{1}_{\{M_t \le B\}} \mathbb{E}\left[\left(x \frac{S_{T-t}}{S_0} - K \right)^+ \mathbf{1}_{\left\{ x \max\limits_{0 \le r \le T-t} S_r / S_0 \le B \right\}} \right]_{x = S_t}$$

of the up-and-out call option with maturity T, strike K and barrier B is given by

$$\begin{aligned} & e^{-r(T-t)} \mathbb{E}\left[C \middle| \mathcal{F}_t \right] \qquad (8.9) \\ = \ & S_t \mathbf{1}_{\{M_t \le B\}} \left(\Phi\left(\delta_+^{T-t}\left(\frac{S_t}{K} \right) \right) - \Phi\left(\delta_+^{T-t}\left(\frac{S_t}{B} \right) \right) \right. \\ & - \left. \left(\frac{B}{S_t} \right)^{1+2r/\sigma^2} \left(\Phi\left(\delta_+^{T-t}\left(\frac{B^2}{K S_t} \right) \right) - \Phi\left(\delta_+^{T-t}\left(\frac{B}{S_t} \right) \right) \right) \right) \\ & - e^{-r(T-t)} K \mathbf{1}_{\{M_t \le B\}} \left(\Phi\left(\delta_-^{T-t}\left(\frac{S_t}{K} \right) \right) - \Phi\left(\delta_-^{T-t}\left(\frac{S_t}{B} \right) \right) \right. \\ & - \left. \left(\frac{S_t}{B} \right)^{1-2r/\sigma^2} \left(\Phi\left(\delta_-^{T-t}\left(\frac{B^2}{K S_t} \right) \right) - \Phi\left(\delta_-^{T-t}\left(\frac{B}{S_t} \right) \right) \right) \right), \end{aligned}$$

where

$$\delta_\pm^\tau(s) = \frac{1}{\sigma\sqrt{\tau}} \left(\log s + \left(r \pm \frac{1}{2}\sigma^2 \right) \tau \right), \qquad s > 0. \tag{8.10}$$

Note that taking $B = +\infty$ in the above identity (8.9) recovers the Black–Scholes formula for the price of a European call option, and that the price of the up-and-out barrier call option is 0 when $B < K$.

The following graph represents the up-and-out call option price given the value S_t of the underlying and the time $t \in [0, T]$ with $T = 220$ days.

Proof of Proposition 8.1. We have $C = \phi(S_T, M_T)$ with

$$\phi(x, y) = (x - K)^+ \mathbf{1}_{\{y \le B\}} = \begin{cases} x - K & \text{if } y \le B, \\ 0 & \text{if } y > B, \end{cases}$$

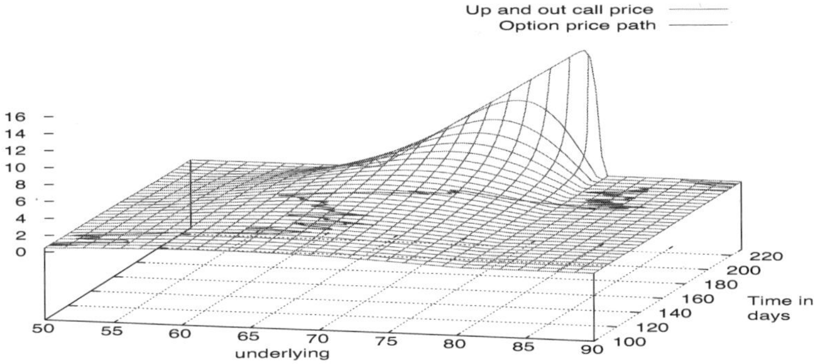

FIGURE 8.7: Graph of the up-and-out call option price.

hence

$$
\begin{aligned}
e^{-r(T-t)} \mathbb{E}\left[C \middle| \mathcal{F}_t\right] &= e^{-r(T-t)} \mathbb{E}\left[(S_T - K)^+ \mathbf{1}_{\{M_T \le B\}} \middle| \mathcal{F}_t\right] \\
&= e^{-r(T-t)} \mathbb{E}\left[(S_T - K)^+ \mathbf{1}_{\{M_T \le B\}} \middle| \mathcal{F}_t\right] \\
&= e^{-r(T-t)} \mathbb{E}\left[(S_T - K)^+ \mathbf{1}_{\{M_t \le B\}} \mathbf{1}_{\left\{\max\limits_{t \le r \le T} S_r \le B\right\}} \middle| \mathcal{F}_t\right] \\
&= e^{-r(T-t)} \mathbf{1}_{\{M_t \le B\}} \mathbb{E}\left[(S_T - K)^+ \mathbf{1}_{\left\{\max\limits_{t \le r \le T} S_r \le B\right\}} \middle| \mathcal{F}_t\right] \\
&= e^{-r(T-t)} \mathbf{1}_{\{M_t \le B\}} \mathbb{E}\left[\left(x \frac{S_T}{S_t} - K\right)^+ \mathbf{1}_{\left\{x \max\limits_{t \le r \le T} S_r / S_t \le B\right\}}\right]_{x = S_t} \\
&= e^{-r(T-t)} \mathbf{1}_{\{M_t \le B\}} \mathbb{E}\left[\left(x \frac{S_{T-t}}{S_0} - K\right)^+ \mathbf{1}_{\left\{x \max\limits_{0 \le r \le T-t} S_r / S_0 \le B\right\}}\right]_{x = S_t}.
\end{aligned}
$$

It suffices to compute

$$
\begin{aligned}
e^{-r\tau} \mathbb{E}[C] &= e^{-r\tau} \mathbb{E}\left[(S_\tau - K)^+ \mathbf{1}_{\{M_\tau \le B\}}\right] \\
&= e^{-r\tau} \mathbb{E}\left[\left(S_0 e^{\sigma \tilde{B}_\tau} - K\right)^+ \mathbf{1}_{\{S_0 e^{\sigma \tilde{X}_\tau} \le B\}}\right] \\
&= e^{-r\tau} \int_{-\infty}^{\infty} \int_{y \vee 0}^{\infty} (S_0 e^{\sigma y} - K)^+ \mathbf{1}_{\{S_0 e^{\sigma x} \le B\}} d\mathbb{P}(\tilde{X}_\tau = x, \tilde{B}_\tau = y) \\
&= \frac{1}{\tau} \sqrt{\frac{2}{\pi \tau}} e^{-r\tau} \int_{-\infty}^{\sigma^{-1} \log(B/S_0)} \\
&\quad \int_{y \vee 0}^{\infty} (S_0 e^{\sigma y} - K)^+ \mathbf{1}_{\{S_0 e^{\sigma x} \le B\}} (2x - y) e^{-\mu^2 \tau/2 + \mu y - (2x-y)^2/(2\tau)} dx dy
\end{aligned}
$$

$$
\begin{aligned}
&= \frac{1}{\tau}e^{-r\tau}\sqrt{\frac{2}{\pi\tau}}\int_0^{\sigma^{-1}\log(B/S_0)} \\
&\quad \int_y^{\infty}(S_0e^{\sigma y}-K)^+\mathbf{1}_{\{S_0e^{\sigma x}\le B\}}(2x-y)e^{-\mu^2\tau/2+\mu y-(2x-y)^2/(2\tau)}dxdy \\
&\quad +\frac{1}{\tau}e^{-r\tau}\sqrt{\frac{2}{\pi\tau}}\int_{-\infty}^{0} \\
&\quad \int_0^{\infty}(S_0e^{\sigma y}-K)^+\mathbf{1}_{\{S_0e^{\sigma x}\le B\}}(2x-y)e^{-\mu^2\tau/2+\mu y-(2x-y)^2/(2\tau)}dxdy \\
&= \frac{1}{\tau}e^{-r\tau}\sqrt{\frac{2}{\pi\tau}}\int_0^{\sigma^{-1}\log(B/S_0)} \\
&\quad \int_y^{\infty}(S_0e^{\sigma y}-K)^+\mathbf{1}_{\{x\le\sigma^{-1}\log(B/S_0)\}}(2x-y)e^{-\mu^2\tau/2+\mu y-(2x-y)^2/(2\tau)}dxdy \\
&\quad +\frac{1}{\tau}e^{-r\tau}\sqrt{\frac{2}{\pi\tau}}\int_{-\infty}^{0} \\
&\quad \int_0^{\infty}(S_0e^{\sigma y}-K)^+\mathbf{1}_{\{x\le\sigma^{-1}\log(B/S_0)\}}(2x-y)e^{-\mu^2\tau/2+\mu y-(2x-y)^2/(2\tau)}dxdy \\
&= \frac{1}{\tau}e^{-r\tau}\sqrt{\frac{2}{\pi\tau}}\int_0^{\sigma^{-1}\log(B/S_0)} \\
&\quad \int_y^{\sigma^{-1}\log(B/S_0)}(S_0e^{\sigma y}-K)^+(2x-y)e^{-\mu^2\tau/2+\mu y-(2x-y)^2/(2\tau)}dxdy \\
&\quad +\frac{1}{\tau}e^{-r\tau}\sqrt{\frac{2}{\pi\tau}}\int_{-\infty}^{0} \\
&\quad \int_0^{\sigma^{-1}\log(B/S_0)}(S_0e^{\sigma y}-K)^+(2x-y)e^{-\mu^2\tau/2+\mu y-(2x-y)^2/(2\tau)}dxdy \\
&= \frac{1}{\tau}e^{-r\tau}\sqrt{\frac{2}{\pi\tau}}\int_{\sigma^{-1}\log(K/S_0)}^{\sigma^{-1}\log(B/S_0)} \\
&\quad \int_{y\vee 0}^{\sigma^{-1}\log(B/S_0)}(S_0e^{\sigma y}-K)(2x-y)e^{-\mu^2\tau/2+\mu y-(2x-y)^2/(2\tau)}dxdy \\
&= \frac{1}{\tau}e^{-r\tau-\mu^2\tau/2}\sqrt{\frac{2}{\pi\tau}}\int_{\sigma^{-1}\log(K/S_0)}^{\sigma^{-1}\log(B/S_0)}(S_0e^{\sigma y}-K)\,e^{\mu y-y^2/(2\tau)} \\
&\quad \int_{y\vee 0}^{\sigma^{-1}\log(B/S_0)}(2x-y)e^{2x(y-x)/\tau}dxdy,
\end{aligned}
$$

if $B \ge S_0$ (otherwise the option price is 0), with $\mu = r/\sigma - \sigma/2$ and $y\vee 0 = \max(y,0)$.

Letting $a = y\vee 0$ and $b = \sigma^{-1}\log(B/S_0)$, we have

$$
\begin{aligned}
\int_a^b (2x-y)e^{2x(y-x)/\tau}dx &= \int_a^b (2x-y)e^{2x(y-x)/\tau}dx \\
&= -\frac{\tau}{2}\left[e^{2x(y-x)/\tau}\right]_{x=a}^{x=b} \\
&= \frac{\tau}{2}(e^{2a(y-a)/\tau}-e^{2b(y-b)/\tau})
\end{aligned}
$$

$$= \frac{\tau}{2}(e^{2(y\vee 0)(y-y\vee 0)/\tau} - e^{2b(y-b)/\tau})$$
$$= \frac{\tau}{2}(1 - e^{2b(y-b)/\tau}),$$

hence, letting $c = \sigma^{-1}\log(K/S_0)$, we have

$$\begin{aligned}
e^{-r\tau}\,\mathbb{E}[C] &= e^{-\tau(r+\mu^2/2)}\frac{1}{\sqrt{2\pi\tau}}\int_c^b (S_0e^{\sigma y} - K)\,e^{\mu y - y^2/(2\tau)}(1 - e^{2b(y-b)/\tau})dy \\
&= S_0e^{-\tau(r+\mu^2/2)}\frac{1}{\sqrt{2\pi\tau}}\int_c^b e^{y(\sigma+\mu)-y^2/(2\tau)}(1 - e^{2b(y-b)/\tau})dy \\
&\quad -Ke^{-\tau(r+\mu^2/2)}\frac{1}{\sqrt{2\pi\tau}}\int_c^b e^{\mu y-y^2/(2\tau)}(1 - e^{2b(y-b)/\tau})dy \\
&= S_0e^{-\tau(r+\mu^2/2)}\frac{1}{\sqrt{2\pi\tau}}\int_c^b e^{y(\sigma+\mu)-y^2/(2\tau)}dy \\
&\quad -S_0e^{-\tau(r+\mu^2/2)-2b^2/\tau}\frac{1}{\sqrt{2\pi\tau}}\int_c^b e^{y(\sigma+\mu+2b/\tau)-y^2/(2\tau)}dy \\
&\quad -Ke^{-\tau(r+\mu^2/2)}\frac{1}{\sqrt{2\pi\tau}}\int_c^b e^{\mu y-y^2/(2\tau)}dy \\
&\quad +Ke^{-\tau(r+\mu^2/2)-2b^2/\tau}\frac{1}{\sqrt{2\pi\tau}}\int_c^b e^{y(\mu+2b/\tau)-y^2/(2\tau)}dy.
\end{aligned}$$

Using the relation

$$\frac{1}{\sqrt{2\pi\tau}}\int_c^b e^{\gamma y-y^2/(2\tau)}dy = e^{\gamma^2\tau/2}\left(\Phi\left(\frac{-c+\gamma\tau}{\sqrt{\tau}}\right) - \Phi\left(\frac{-b+\gamma\tau}{\sqrt{\tau}}\right)\right),$$

we find

$$\begin{aligned}
e^{-r\tau}\,\mathbb{E}[C] &= e^{-rT}\,\mathbb{E}\left[(S_T-K)^+\,\mathbf{1}_{\{M_T\le B\}}\right] \\
&= S_0e^{-\tau(r+\mu^2/2)+(\sigma+\mu)^2\tau/2}\left(\Phi\left(\frac{-c+(\sigma+\mu)\tau}{\sqrt{\tau}}\right) - \Phi\left(\frac{-b+(\sigma+\mu)\tau}{\sqrt{\tau}}\right)\right) \\
&\quad -S_0e^{-\tau(r+\mu^2/2)-2b^2/\tau+(\sigma+\mu+2b/\tau)^2\tau/2} \\
&\quad \times\left(\Phi\left(\frac{-c+(\sigma+\mu+2b/\tau)\tau}{\sqrt{\tau}}\right) - \Phi\left(\frac{-b+(\sigma+\mu+2b/\tau)\tau}{\sqrt{\tau}}\right)\right) \\
&\quad -Ke^{-r\tau}\left(\Phi\left(\frac{-c+\mu\tau}{\sqrt{\tau}}\right) - \Phi\left(\frac{-b+\mu\tau}{\sqrt{\tau}}\right)\right) \\
&\quad +Ke^{-\tau(r+\mu^2/2)-2b^2/\tau+(\mu+2b/\tau)^2\tau/2} \\
&\quad \times\left(\Phi\left(\frac{-c+(\mu+2b/\tau)\tau}{\sqrt{\tau}}\right) - \Phi\left(\frac{-b+(\mu+2b/\tau)\tau}{\sqrt{\tau}}\right)\right) \\
&= S_0\left(\Phi\left(\delta_+^\tau\left(\frac{S_0}{K}\right)\right) - \Phi\left(\delta_+^\tau\left(\frac{S_0}{B}\right)\right)\right) \\
&\quad -S_0e^{-\tau(r+\mu^2/2)-2b^2/\tau+(\sigma+\mu+2b/\tau)^2\tau/2}\left(\Phi\left(\delta_+^\tau\left(\frac{B^2}{KS_0}\right)\right) - \Phi\left(\delta_+^\tau\left(\frac{B}{S_0}\right)\right)\right)
\end{aligned}$$

$$-Ke^{-r\tau}\left(\Phi\left(\delta_-^\tau\left(\frac{S_0}{K}\right)\right)-\Phi\left(\delta_-^\tau\left(\frac{S_0}{B}\right)\right)\right)$$
$$+Ke^{-\tau(r+\mu^2/2)-2b^2/\tau+(\mu+2b/\tau)^2\tau/2}\left(\Phi\left(\delta_-\left(\frac{B^2}{KS_0}\right)\right)-\Phi\left(\delta_-\left(\frac{B}{S_0}\right)\right)\right),$$

$0 \leq x \leq B$, where $\delta_\pm^\tau(s)$ is defined in (8.10). Given the relations

$$-\tau(r+\mu^2/2)-2b^2/\tau+(\sigma+\mu+2b/\tau)^2\tau/2 = 2b(r/\sigma+\sigma/2) = (1+2r/\sigma^2)\log(B/S_0),$$

and

$$-\tau(r+\mu^2/2)-2b^2/\tau+(\mu+2b/\tau)^2\tau/2 = -r\tau+2\mu b = -r\tau+(-1+2r/\sigma^2)\log(B/S_0),$$

this yields

$$\begin{aligned}
e^{-r\tau}\,\mathbb{E}[C] &= e^{-r\tau}\,\mathbb{E}\left[(S_\tau-K)^+\mathbf{1}_{\{M_\tau\leq B\}}\right] \qquad (8.11)\\
&= S_0\left(\Phi\left(\delta_+^\tau\left(\frac{S_0}{K}\right)\right)-\Phi\left(\delta_+^\tau\left(\frac{S_0}{B}\right)\right)\right)\\
&\quad -e^{-r\tau}K\left(\Phi\left(\delta_-^\tau\left(\frac{S_0}{K}\right)\right)-\Phi\left(\delta_-^\tau\left(\frac{S_0}{B}\right)\right)\right)\\
&\quad -B\left(\frac{B}{S_0}\right)^{2r/\sigma^2}\left(\Phi\left(\delta_+^\tau\left(\frac{B^2}{KS_0}\right)\right)-\Phi\left(\delta_+^\tau\left(\frac{B}{S_0}\right)\right)\right)\\
&\quad +e^{-r\tau}K\left(\frac{S_0}{B}\right)^{1-2r/\sigma^2}\left(\Phi\left(\delta_-^\tau\left(\frac{B^2}{KS_0}\right)\right)-\Phi\left(\delta_-^\tau\left(\frac{B}{S_0}\right)\right)\right)\\
&= S_0\left(\Phi\left(\delta_+^\tau\left(\frac{S_0}{K}\right)\right)-\Phi\left(\delta_+^\tau\left(\frac{S_0}{B}\right)\right)\right)\\
&\quad -S_0\left(\frac{B}{S_0}\right)^{1+2r/\sigma^2}\left(\Phi\left(\delta_+^\tau\left(\frac{B^2}{KS_0}\right)\right)-\Phi\left(\delta_+^\tau\left(\frac{B}{S_0}\right)\right)\right)\\
&\quad -e^{-r\tau}K\left(\Phi\left(\delta_-^\tau\left(\frac{S_0}{K}\right)\right)-\Phi\left(\delta_-^\tau\left(\frac{S_0}{B}\right)\right)\right)\\
&\quad -e^{-r\tau}K\left(\frac{S_0}{B}\right)^{1-2r/\sigma^2}\left(\Phi\left(\delta_-^\tau\left(\frac{B^2}{KS_0}\right)\right)-\Phi\left(\delta_-^\tau\left(\frac{B}{S_0}\right)\right)\right),
\end{aligned}$$

and this yields the result of Proposition 8.1, cf. § 7.3.3 pages 304–307 of [68] for a different calculation.

This concludes the proof of Proposition 8.1. □

Up-and-Out Barrier Put Option

The price

$$e^{-r(T-t)}\mathbf{1}_{\{M_t\leq B\}}\,\mathbb{E}\left[\left(K-x\frac{S_{T-t}}{S_0}\right)^+\mathbf{1}_{\left\{x\max\limits_{0\leq r\leq T-t}S_r/S_0\leq B\right\}}\right]_{x=S_t}$$

of the up-and-out put option with maturity T, strike K and barrier B is given by

$$
\begin{aligned}
& e^{-r(T-t)}\,\mathbb{E}\left[P\middle|\mathcal{F}_t\right] \\
&= S_t\mathbf{1}_{\{M_t \le B\}}\left(\Phi\left(\delta_+^{T-t}\left(\frac{S_t}{K}\right)\right)-1\right. \\
&\quad \left.-\left(\frac{B}{S_t}\right)^{1+2r/\sigma^2}\left(\Phi\left(\delta_+^{T-t}\left(\frac{B^2}{KS_t}\right)\right)-1\right)\right) \\
&\quad -e^{-r(T-t)}K\mathbf{1}_{\{M_t \le B\}}\left(\Phi\left(\delta_-^{T-t}\left(\frac{S_t}{K}\right)\right)-1\right. \\
&\quad \left.-\left(\frac{S_t}{B}\right)^{1-2r/\sigma^2}\left(\Phi\left(\delta_-^{T-t}\left(\frac{B^2}{KS_t}\right)\right)-1\right)\right) \\
&= S_t\mathbf{1}_{\{M_t \le B\}}\left(-\Phi\left(-\delta_+^{T-t}\left(\frac{S_t}{K}\right)\right)+\left(\frac{B}{S_t}\right)^{1+2r/\sigma^2}\right. \\
&\quad \left.\Phi\left(-\delta_+^{T-t}\left(\frac{B^2}{KS_t}\right)\right)\right)-Ke^{-r(T-t)} \\
&\quad \times\mathbf{1}_{\{M_t \le B\}}\left(-\Phi\left(-\delta_-^{T-t}\left(\frac{S_t}{K}\right)\right)+\left(\frac{S_t}{B}\right)^{1-2r/\sigma^2}\right. \\
&\quad \left.\Phi\left(-\delta_-^{T-t}\left(\frac{B^2}{KS_t}\right)\right)\right),
\end{aligned}
$$

if $B > K$, and

$$
\begin{aligned}
& e^{-r(T-t)}\,\mathbb{E}\left[P\middle|\mathcal{F}_t\right] \\
&= S_t\mathbf{1}_{\{M_t \le B\}}\left(\Phi\left(\delta_+^{T-t}\left(\frac{S_t}{B}\right)\right)-1\right. \\
&\quad \left.-\left(\frac{B}{S_t}\right)^{1+2r/\sigma^2}\left(\Phi\left(\delta_+^{T-t}\left(\frac{B}{S_t}\right)\right)-1\right)\right) \\
&\quad -e^{-r(T-t)}K\mathbf{1}_{\{M_t \le B\}}\left(\Phi\left(\delta_-^{T-t}\left(\frac{S_t}{B}\right)\right)-1\right. \\
&\quad \left.-\left(\frac{S_t}{B}\right)^{1-2r/\sigma^2}\left(\Phi\left(\delta_-^{T-t}\left(\frac{B}{S_t}\right)\right)-1\right)\right) \\
&= S_t\mathbf{1}_{\{M_t \le B\}}\left(-\Phi\left(-\delta_+^{T-t}\left(\frac{S_t}{B}\right)\right)+\left(\frac{B}{S_t}\right)^{1+2r/\sigma^2}\Phi\left(-\delta_+^{T-t}\left(\frac{B}{S_t}\right)\right)\right) \\
&\quad -Ke^{-r(T-t)} \\
&\quad \times\mathbf{1}_{\{M_t \le B\}}\left(-\Phi\left(-\delta_-^{T-t}\left(\frac{S_t}{B}\right)\right)+\left(\frac{S_t}{B}\right)^{1-2r/\sigma^2}\Phi\left(-\delta_-^{T-t}\left(\frac{B}{S_t}\right)\right)\right),
\end{aligned}
$$

if $B < K$.

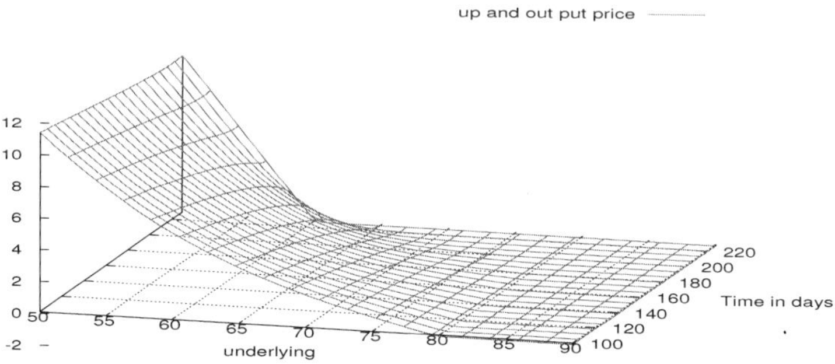

FIGURE 8.8: Graph of the up-and-out put option price with $B > K$.

Down-and-Out Barrier Call Option

Let us now consider a down-and-out barrier call option on the underlying asset S_t with exercise date T, strike K, barrier B, and payoff

$$C = (S_T - K)^+ \mathbf{1}_{\left\{ \min\limits_{0 \leq t \leq T} S_t > B \right\}} = \begin{cases} S_T - K & \text{if } \min\limits_{0 \leq t \leq T} S_t > B, \\ \\ 0 & \text{if } \min\limits_{0 \leq t \leq T} S_t \leq B, \end{cases}$$

with $0 \leq B \leq K$. This option is also called a Callable Bull Contract with no residual value, in which B denotes the call price, or a turbo warrant with no rebate.

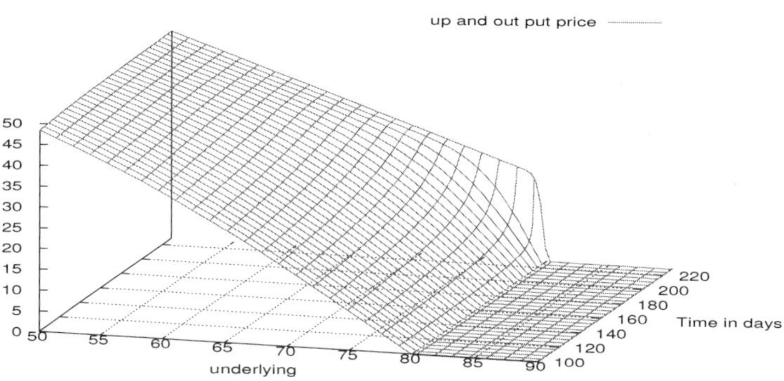

FIGURE 8.9: Graph of the up-and-out put option price with $K > B$.

We have

$$
\begin{aligned}
e^{-r(T-t)}\,\mathbb{E}\left[C\middle|\mathcal{F}_t\right] &= g(t,S_t) = S_t\Phi\left(\delta_+^{T-t}\left(\frac{S_t}{K}\right)\right)\\
&\quad -e^{-r(T-t)}K\Phi\left(\delta_-^{T-t}\left(\frac{S_t}{K}\right)\right) - B\left(\frac{B}{S_t}\right)^{2r/\sigma^2}\Phi\left(\delta_+^{T-t}\left(\frac{B^2}{Kx}\right)\right)\\
&\quad +e^{-r(T-t)}K\left(\frac{S_t}{B}\right)^{1-2r/\sigma^2}\Phi\left(\delta_-^{T-t}\left(\frac{B^2}{KS_t}\right)\right) \qquad (8.12)\\
&= \mathrm{BS}_c(S_t,r,T-t,K)\\
&\quad -B\left(\frac{B}{S_t}\right)^{2r/\sigma^2}\Phi\left(\delta_+^{T-t}\left(\frac{B^2}{KS_t}\right)\right)\\
&\quad +e^{-r(T-t)}K\left(\frac{S_t}{B}\right)^{1-2r/\sigma^2}\Phi\left(\delta_-^{T-t}\left(\frac{B^2}{KS_t}\right)\right)\\
&= \mathrm{BS}_c(S_t,r,T-t,K) - \frac{1}{B}\left(\frac{S_t}{B}\right)^{1-2r/\sigma^2}\mathrm{BS}_c(B/S_t,r,T-t,K/B),
\end{aligned}
$$

$S_t > B$, $0 \le t \le T$, and

$$
\mathbb{E}\left[(S_T-K)^+\mathbf{1}_{\left\{\min\limits_{0\le t\le T} S_t > B\right\}}\middle|\mathcal{F}_t\right] = \mathbf{1}_{\{\min_{t\in[0,T]} S_t > B\}}\,g(t,S_t),
$$

$t \in [0,T]$. When $B > K$ we find

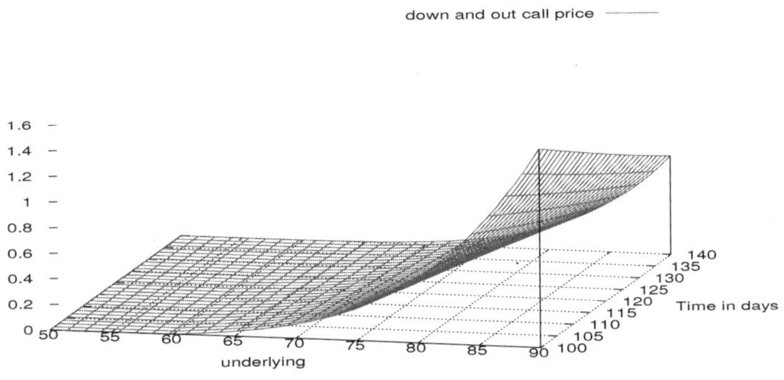

FIGURE 8.10: Graph of the down-and-out call option price with $B < K$.

$$e^{-r(T-t)}\,\mathbb{E}\left[C\Big|\mathcal{F}_t\right] = g(t,S_t) = S_t\Phi\left(\delta_+^{T-t}\left(\frac{S_t}{B}\right)\right) -$$
$$e^{-r(T-t)}K\Phi\left(\delta_-^{T-t}\left(\frac{S_t}{B}\right)\right) - B\left(\frac{B}{S_t}\right)^{2r/\sigma^2}\Phi\left(\delta_+^{T-t}\left(\frac{B}{S_t}\right)\right)$$
$$+e^{-r(T-t)}K\left(\frac{S_t}{B}\right)^{1-2r/\sigma^2}\Phi\left(\delta_-^{T-t}\left(\frac{B}{S_t}\right)\right), \tag{8.13}$$

$S_t > B$, $0 \leq t \leq T$, cf. Exercise 8.2 below.

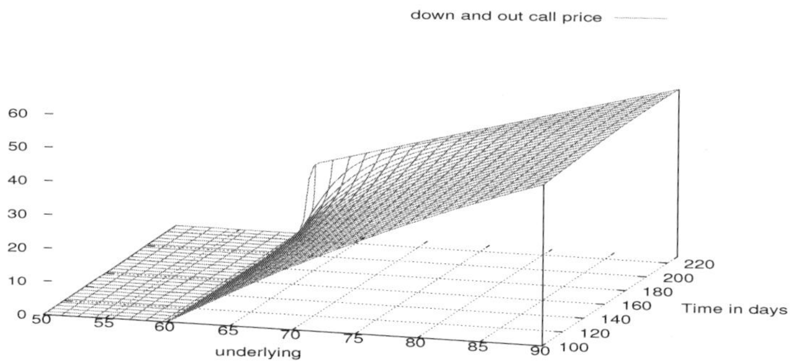

FIGURE 8.11: Graph of the down-and-out call option price with $K > B$.

Down-and-Out Barrier Put Option

When $B > K$, the price

$$e^{-r(T-t)}\mathbf{1}_{\{m_t \geq B\}}\,\mathbb{E}\left[\left(K - x\frac{S_{T-t}}{S_0}\right)^+ \mathbf{1}_{\left\{x \min\limits_{0\leq r\leq T-t} S_r/S_0 \geq B\right\}}\right]_{x=S_t}$$

of the down-and-out put option with maturity T, strike K and barrier B is given by

$$e^{-r(T-t)}\,\mathbb{E}\left[P\Big|\mathcal{F}_t\right] \tag{8.14}$$
$$= S_t\mathbf{1}_{\{m_t \geq B\}}\left(\Phi\left(\delta_+^{T-t}\left(\frac{S_t}{K}\right)\right) - \Phi\left(\delta_+^{T-t}\left(\frac{S_t}{B}\right)\right)\right.$$
$$\left. - \left(\frac{B}{S_t}\right)^{1+2r/\sigma^2}\left(\Phi\left(\delta_+^{T-t}\left(\frac{B^2}{KS_t}\right)\right) - \Phi\left(\delta_+^{T-t}\left(\frac{B}{S_t}\right)\right)\right)\right)$$
$$-e^{-r(T-t)}K\mathbf{1}_{\{m_t \geq B\}}\left(\Phi\left(\delta_-^{T-t}\left(\frac{S_t}{K}\right)\right) - \Phi\left(\delta_-^{T-t}\left(\frac{S_t}{B}\right)\right)\right.$$
$$\left. + \left(\frac{S_t}{B}\right)^{1-2r/\sigma^2}\left(\Phi\left(\delta_-^{T-t}\left(\frac{B^2}{KS_t}\right)\right) - \Phi\left(\delta_-^{T-t}\left(\frac{B}{S_t}\right)\right)\right)\right),$$

while the corresponding price vanishes when $B < K$.

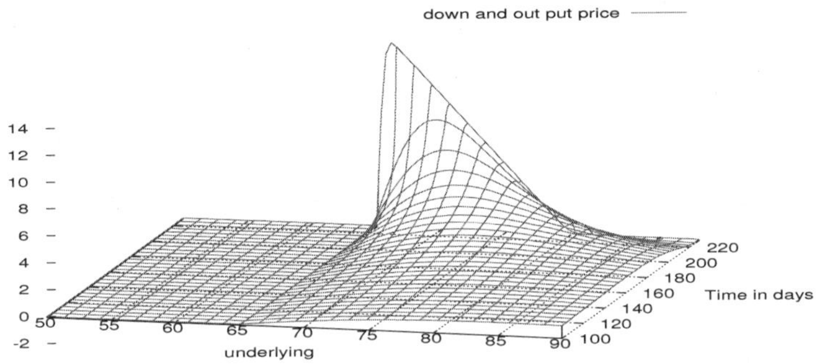

FIGURE 8.12: Graph of the down-and-out put option price with $K > B$.

Note that although Figures 8.8 and 8.10, resp. 8.9 and 8.11, appear to share some symmetry property, the functions themselves are not exactly symmetric. Concerning 8.7 and 8.12 the pricing function is actually the same, but the conditions $B < K$ and $B > K$ play opposite roles.

PDE Method

Having computed the up-and-out call option price by probabilistic arguments, we are now interested in deriving a PDE for this price.

The option price can be written as

$$\begin{aligned}
&e^{-r(T-t)}\,\mathbb{E}\left[(S_T-K)^+\,\mathbf{1}_{\{M_T\le B\}}\Big|\mathcal{F}_t\right]\\
&= e^{-r(T-t)}\mathbf{1}_{\left\{\max\limits_{0\le r\le t} S_r\le B\right\}}\,\mathbb{E}\left[(S_T-K)^+\,\mathbf{1}_{\left\{\max\limits_{t\le r\le T} S_r\le B\right\}}\Bigg|\mathcal{F}_t\right]\\
&= g(t,S_t,M_t),
\end{aligned}$$

where the function $g(t,x)$ of t and S_t is given by

$$g(t,x,y)=\mathbf{1}_{\{y\le B\}}e^{-r(T-t)}\,\mathbb{E}\left[(S_T-K)^+\,\mathbf{1}_{\left\{\max\limits_{t\le r\le T} S_r\le B\right\}}\Bigg|S_t=x\right].$$

Next, by the same argument as in the proof of Proposition 5.2 we derive the Black–Scholes partial differential equation (PDE) satisfied by $g(t,x)$, for the price of a self-financing portfolio.

Proposition 8.2 *Let $(\eta_t,\xi_t)_{t\in\mathbb{R}_+}$ be a portfolio strategy such that* *(i)* $(\eta_t, \xi_t)_{t\in\mathbb{R}_+}$ *is self-financing,*

(ii) the value $V_t := \eta_t A_t + \xi_t S_t$, $t \in \mathbb{R}_+$, *takes the form*

$$V_t = g(t, S_t, M_t), \qquad t \in \mathbb{R}_+.$$

Then the function $g(t, x, y)$ *satisfies the Black–Scholes PDE*

$$rg(t,x,y) = \frac{\partial g}{\partial t}(t,x,y) + ry\frac{\partial g}{\partial x}(t,x,y) + \frac{1}{2}x^2\sigma^2\frac{\partial^2 g}{\partial x^2}(t,x,y), \tag{8.15}$$

$t > 0$, $x > 0$, $0 < y < B$, *and* ξ_t *is given by*

$$\xi_t = \frac{\partial g}{\partial x}(t, S_t, M_t), \qquad t \in [0,T], \tag{8.16}$$

provided $M_t < B$.

In the sequel we will drop the variable y in $g(t, x, y)$ and simply write $g(t, x)$ since

$$\frac{\partial g}{\partial y}(t,x,y) = 0, \qquad 0 < y < B,$$

and the function $g(t, x, y)$ is constant in $y \in (0, B)$.

In the next proposition we add a boundary condition to the Black–Scholes PDE (8.15) in order to hedge the up-and-out call option with maturity T, strike K, barrier (or call price) B, and payoff

$$C = (S_T - K)^+ \mathbf{1}_{\left\{\max\limits_{0\le t\le T} S_t \le B\right\}} = \begin{cases} S_T - K & \text{if } \max\limits_{0\le t\le T} S_t \le B, \\ 0 & \text{if } \max\limits_{0\le t\le T} S_t > B, \end{cases}$$

with $B > K$.

Proposition 8.3 *The price of any self-financing portfolio of the form* $V_t = g(t, S_t)$ *hedging the up-and-out barrier call option satisfies the Black–Scholes PDE*

$$\begin{cases} rg(t,x) = \dfrac{\partial g}{\partial t}(t,x) + rx\dfrac{\partial g}{\partial x}(t,x) + \dfrac{1}{2}x^2\sigma^2\dfrac{\partial^2 g}{\partial x^2}(t,x), \\ g(t,x) = 0, \qquad x \ge B, \quad t \in [0,T], \\ g(T,x) = (x-K)^+\mathbf{1}_{\{x<B\}}, \end{cases}$$

on the time-space domain $[0,T] \times [0,B]$ *with terminal condition*

$$g(T,x) = (x-K)^+\mathbf{1}_{\{x<B\}}$$

and additional boundary condition

$$g(t, B) = 0. \tag{8.17}$$

Condition (8.17) holds since the price of the claim at time t is 0 whenever $S_t = B$, cf. e.g., [22].

The closed-form solution for this PDE is given by (8.11), as

$$\begin{aligned} g(t,x) = {} & x\left(\Phi\left(\delta_+^{T-t}\left(\frac{x}{K}\right)\right) - \Phi\left(\delta_+^{T-t}\left(\frac{x}{B}\right)\right)\right) \\ & -x\left(\frac{x}{B}\right)^{1-2r/\sigma^2}\left(\Phi\left(\delta_+^{T-t}\left(\frac{B^2}{Kx}\right)\right) - \Phi\left(\delta_+^{T-t}\left(\frac{B}{x}\right)\right)\right) \\ & -Ke^{-r(T-t)}\left(\Phi\left(\delta_-^{T-t}\left(\frac{x}{K}\right)\right) - \Phi\left(\delta_-^{T-t}\left(\frac{x}{B}\right)\right)\right) \\ & +Ke^{-r(T-t)}\left(\frac{x}{B}\right)^{1-2r/\sigma^2}\left(\Phi\left(\delta_-^{T-t}\left(\frac{B^2}{Kx}\right)\right) - \Phi\left(\delta_-^{T-t}\left(\frac{B}{x}\right)\right)\right), \end{aligned} \tag{8.18}$$

$0 < x \le B,\ 0 \le t \le T.$

We note that the expression (8.18) can be rewritten using the standard Black–Scholes formula

$$\mathrm{BS}_c(S, K, r, \sigma, \tau) = S\Phi\left(\delta_+^\tau\left(\frac{S}{K}\right)\right) - Ke^{-r\tau}\Phi\left(\delta_-^\tau\left(\frac{S}{K}\right)\right)$$

for the price of a European call option, as

$$\begin{aligned} g(t,x) &= \mathrm{BS}_c(x, K, r, \sigma, T-t) - x\Phi\left(\delta_+^{T-t}\left(\frac{x}{B}\right)\right) + e^{-r(T-t)}K\Phi\left(\delta_-^{T-t}\left(\frac{x}{B}\right)\right) \\ & \quad -B\left(\frac{B}{x}\right)^{2r/\sigma^2}\left(\Phi\left(\delta_+^{T-t}\left(\frac{B^2}{Kx}\right)\right) - \Phi\left(\delta_+^{T-t}\left(\frac{B}{x}\right)\right)\right) \\ & \quad +e^{-r(T-t)}K\left(\frac{x}{B}\right)^{1-2r/\sigma^2}\left(\Phi\left(\delta_-^{T-t}\left(\frac{B^2}{Kx}\right)\right) - \Phi\left(\delta_-^{T-t}\left(\frac{B}{x}\right)\right)\right), \end{aligned}$$

$0 < x \le B,\ 0 \le t \le T.$

Figure 8.13 represents the value of Delta obtained from (8.16) for the up-and-out call option, cf. Exercise 8.2-(1).

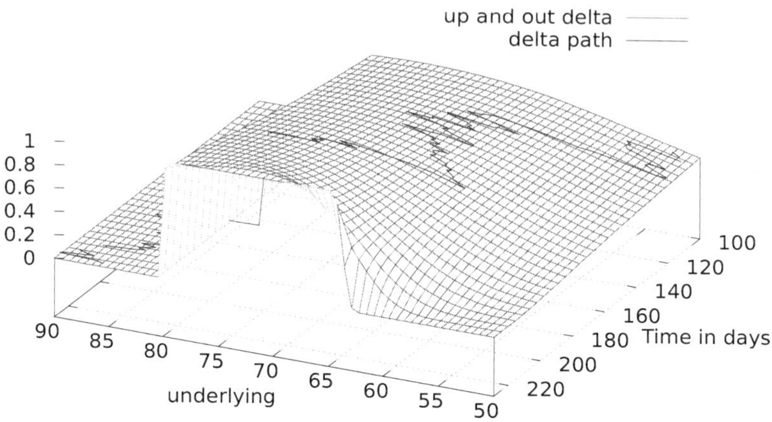

FIGURE 8.13: Delta for the up-and-out option.

Checking the Boundary Conditions

For $x = B$ we check that

$$\begin{aligned} g(t,B) &= B\left(\Phi\left(\delta_+^{T-t}\left(\frac{B}{K}\right)\right) - \Phi\left(\delta_+^{T-t}(1)\right)\right) \\ &\quad - e^{-r(T-t)}K\left(\Phi\left(\delta_-^{T-t}\left(\frac{B}{K}\right)\right) - \Phi\left(\delta_-^{T-t}(1)\right)\right) \\ &\quad - B\left(\Phi\left(\delta_+^{T-t}\left(\frac{B}{K}\right)\right) - \Phi\left(\delta_+^{T-t}(1)\right)\right) \\ &\quad + e^{-r(T-t)}K\left(\Phi\left(\delta_-^{T-t}\left(\frac{B}{K}\right)\right) - \Phi\left(\delta_-^{T-t}(1)\right)\right) \\ &= 0, \end{aligned}$$

and the function $g(t,x)$ is extended to $x > B$ by letting

$$g(t,x) = 0, \qquad x > B.$$

For $x = K$ and $t = T$ we find

$$\delta_\pm^0(s) = -\infty \times \mathbf{1}_{\{s<1\}} + \infty \times \mathbf{1}_{\{s>1\}} = \begin{cases} +\infty & \text{if } s > 1, \\ 0 & \text{if } s = 1, \\ -\infty & \text{if } s < 1, \end{cases}$$

hence when $x < K < B$ we have

$$g(T,K) = x\left(\Phi(-\infty) - \Phi(-\infty)\right)$$

$$
\begin{aligned}
& -K\left(\Phi(-\infty)-\Phi(-\infty)\right) \\
& -B\left(\frac{B}{x}\right)^{2r/\sigma^2}\left(\Phi(+\infty)-\Phi(+\infty)\right) \\
& +K\left(\frac{B}{K}\right)^{2r/\sigma^2}\left(\Phi(+\infty)-\Phi(+\infty)\right) \\
= \; & 0,
\end{aligned}
$$

when $\underline{K < x < B}$ we get

$$
\begin{aligned}
g(T,K) &= x\left(\Phi(+\infty)-\Phi(-\infty)\right) \\
&\quad -K\left(\Phi(+\infty)-\Phi(-\infty)\right) \\
&\quad -B\left(\frac{B}{x}\right)^{2r/\sigma^2}\left(\Phi(+\infty)-\Phi(+\infty)\right) \\
&\quad +K\left(\frac{B}{K}\right)^{2r/\sigma^2}\left(\Phi(+\infty)-\Phi(+\infty)\right) \\
&= x-K,
\end{aligned}
$$

and for $x > B$ we obtain

$$
\begin{aligned}
g(T,K) &= x\left(\Phi(+\infty)-\Phi(+\infty)\right) \\
&\quad -K\left(\Phi(+\infty)-\Phi(+\infty)\right) \\
&\quad -B\left(\frac{B}{x}\right)^{2r/\sigma^2}\left(\Phi(-\infty)-\Phi(-\infty)\right) \\
&\quad +K\left(\frac{B}{K}\right)^{2r/\sigma^2}\left(\Phi(-\infty)-\Phi(-\infty)\right) \\
&= 0.
\end{aligned}
$$

Down-and-Out Barrier Call Option

Similarly the price $g(t, S_t)$ at time t of the down-and-out barrier call option satisfies the Black–Scholes PDE

$$
\begin{cases}
rg(t,x) = \dfrac{\partial g}{\partial t}(t,x) + rx\dfrac{\partial g}{\partial x}(t,x) + \dfrac{1}{2}x^2\sigma^2\dfrac{\partial^2 g}{\partial x^2}(t,x), \\[2ex]
g(t,B) = 0, \quad t \in [0,T], \\[2ex]
g(T,x) = (x-K)^+\mathbf{1}_{\{x\geq B\}},
\end{cases}
$$

on the time-space domain $[0,T] \times [0,B]$ with terminal condition $g(T,x) = (x-K)^+\mathbf{1}_{\{x\geq B\}}$ and the additional boundary condition $g(t,B) = 0$ since the price of the claim at time t is 0 whenever $S_t = B$.

8.4 Lookback Options

Let

$$m_s^t = \inf_{u \in [s,t]} S_u$$

and

$$M_s^t = \sup_{u \in [s,t]} S_u,$$

$0 \leq s \leq t \leq T$, and let $\mathcal{M}_s^t$ be either m_s^t or M_s^t. In the lookback option case the payoff $\phi(S_T, \mathcal{M}_0^T)$ depends not only on the price of the underlying asset at maturity but it also depends on all price values of the underlying asset over the period which starts from the initial time and ends at maturity.

The payoff of such of an option is of the form $\phi(S_T, \mathcal{M}_0^T)$ with $\phi(x,y) = x - y$ in the case of lookback call options, and $\phi(x,y) = y - x$ in the case of lookback put options. We let

$$e^{-r(T-t)} \mathbb{E}[\phi(S_T, \mathcal{M}_0^T) | \mathcal{F}_t]$$

denote the price at time $t \in [0,T]$ of such an option.

The Lookback Put Option

The standard lookback put option gives its holder the right to sell the underlying asset at its historically highest price. In this case the strike is M_0^T and the payoff is

$$C = M_0^T - S_T.$$

Our goal is to prove the following pricing formula for lookback put options.

Proposition 8.4 *The price at time $t \in [0,T]$ of the lookback put option with payoff $M_0^T - S_T$ is given by*

$$\begin{aligned} & e^{-r(T-t)} \mathbb{E}[M_0^T - S_T \mid \mathcal{F}_t] \\ &= M_0^t e^{-r(T-t)} \Phi\left(-\delta_-^{T-t}\left(\frac{S_t}{M_0^t}\right)\right) + S_t\left(1 + \frac{\sigma^2}{2r}\right)\Phi\left(\delta_+^{T-t}\left(\frac{S_t}{M_0^t}\right)\right) \\ & \quad -S_t e^{-r(T-t)} \frac{\sigma^2}{2r}\left(\frac{M_0^t}{S_t}\right)^{2r/\sigma^2} \Phi\left(-\delta_-^{T-t}\left(\frac{M_0^t}{S_t}\right)\right) - S_t. \end{aligned}$$

Figure 8.14 represents the lookback put price as a function of S_t and M_0^t, for different values of the time to maturity $T - t$.

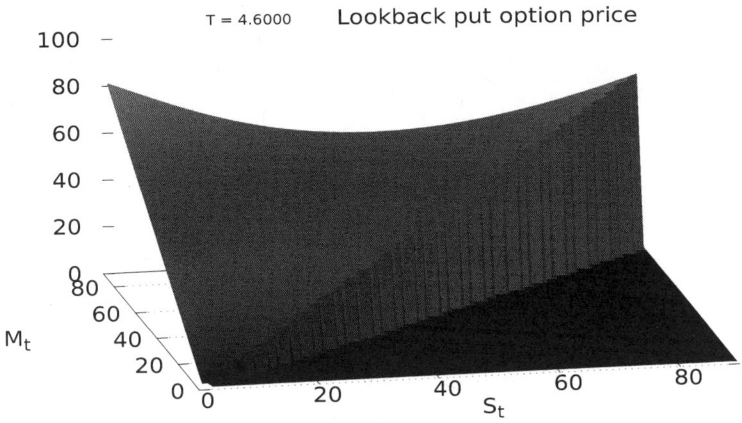

FIGURE 8.14: Graph of the lookback put option price.

Proof of Proposition 8.4. We have

$$\begin{aligned}\mathbb{E}[M_0^T - S_T \mid \mathcal{F}_t] &= \mathbb{E}[M_0^T \mid \mathcal{F}_t] - \mathbb{E}[S_T \mid \mathcal{F}_t] \\ &= \mathbb{E}[M_0^T \mid \mathcal{F}_t] - e^{r(T-t)} S_t,\end{aligned}$$

and

$$\begin{aligned}\mathbb{E}[M_0^T \mid \mathcal{F}_t] &= \mathbb{E}[M_0^t \vee M_t^T \mid \mathcal{F}_t] \\ &= \mathbb{E}[M_0^t \mathbf{1}_{\{M_0^t > M_t^T\}} \mid \mathcal{F}_t] + \mathbb{E}[M_t^T \mathbf{1}_{\{M_t^T > M_0^t\}} \mid \mathcal{F}_t] \\ &= M_0^t\, \mathbb{E}[\mathbf{1}_{\{M_0^t > M_t^T\}} \mid \mathcal{F}_t] + \mathbb{E}[M_t^T \mathbf{1}_{\{M_t^T > M_0^t\}} \mid \mathcal{F}_t] \\ &= M_0^t \mathbb{P}(M_0^t > M_t^T \mid \mathcal{F}_t) + \mathbb{E}[M_t^T \mathbf{1}_{\{M_t^T > M_0^t\}} \mid \mathcal{F}_t].\end{aligned}$$

Next, we have

$$\begin{aligned}\mathbb{P}(M_0^t > M_t^T \mid \mathcal{F}_t) &= \mathbb{P}\left(\frac{M_0^t}{S_t} > \frac{M_t^T}{S_t} \middle| \mathcal{F}_t \right) \\ &= \mathbb{P}\left(x > \frac{M_t^T}{S_t} \middle| \mathcal{F}_t \right)_{x = M_0^t / S_t} \\ &= \mathbb{P}\left(\frac{M_0^{T-t}}{S_0} < x \right)_{x = M_0^t / S_t}.\end{aligned}$$

On the other hand, letting $\mu = r/\sigma - \sigma/2$, from (8.7) we have

$$\begin{aligned}&\mathbb{P}\left(\frac{M_0^\tau}{S_0} < x \right) = \mathbb{P}(\tilde{X}_\tau < \sigma^{-1} \log x) \\ &= \Phi\left(\frac{-\mu\tau + \sigma^{-1} \log x}{\sqrt{\tau}} \right) - e^{2\mu\sigma^{-1} \log x} \Phi\left(\frac{-\mu\tau - \sigma^{-1} \log x}{\sqrt{\tau}} \right)\end{aligned}$$

$$= \Phi\left(-\delta_-^\tau(1/x)\right) - x^{-1+2r/\sigma^2}\Phi\left(-\delta_-^\tau(x)\right).$$

Hence

$$\begin{aligned}\mathbb{P}(M_0^t > M_t^T) &= \mathbb{P}\left(\frac{M_0^{T-t}}{S_0} < x\right)_{x=M_0^t/S_t} \\ &= \Phi\left(-\delta_-^T\left(\frac{S_t}{M_0^t}\right)\right) - \left(\frac{M_0^t}{S_t}\right)^{-1+2r/\sigma^2}\Phi\left(-\delta_-^T\left(\frac{M_0^t}{S_t}\right)\right).\end{aligned}$$

Next, we have

$$\begin{aligned}&\mathbb{E}[M_t^T \mathbf{1}_{\{M_t^T > M_0^t\}} \mid \mathcal{F}_t] = S_t\,\mathbb{E}\left[\frac{M_t^T}{S_t}\mathbf{1}_{\{M_t^T/S_t > M_0^t/S_t\}} \mid \mathcal{F}_t\right] \\ &= S_t\,\mathbb{E}\left[\max_{r\in[t,T]}\frac{S_r}{S_t}\mathbf{1}_{\{\max_{r\in[t,T]} S_r/S_t > x\}}\Big|\mathcal{F}_t\right]_{x=M_0^t/S_t} \\ &= S_t\,\mathbb{E}\left[\max_{r\in[0,T-t]}\frac{S_r}{S_0}\mathbf{1}_{\{\max_{r\in[0,T-t]} S_r/S_0 > x\}}\right]_{x=M_0^t/S_t},\end{aligned}$$

and

$$\begin{aligned}&\mathbb{E}\left[\max_{r\in[0,\tau]}\frac{S_r}{S_0}\mathbf{1}_{\{\max_{r\in[0,\tau]} S_r/S_0 > x\}}\right] \\ &= \mathbb{E}\left[\max_{r\in[0,\tau]} e^{\sigma\tilde{B}_r}\mathbf{1}_{\{\max_{r\in[0,\tau]} e^{\sigma\tilde{B}_r} > x\}}\right] \\ &= \mathbb{E}\left[e^{\sigma\max_{r\in[0,\tau]}\tilde{B}_r}\mathbf{1}_{\{\max_{r\in[0,\tau]}\tilde{B}_r > \sigma^{-1}\log x\}}\right] \\ &= \mathbb{E}\left[e^{\sigma\tilde{X}_\tau}\mathbf{1}_{\{\tilde{X}_\tau > \sigma^{-1}\log x\}}\right] \\ &= \int_{\sigma^{-1}\log x}^\infty e^{\sigma x} f_{\tilde{X}_\tau}(z)dz \\ &= \int_{\sigma^{-1}\log x}^\infty e^{\sigma z}\left(\sqrt{\frac{2}{\pi\tau}}e^{-(z-\mu\tau)^2/(2\tau)} - 2\mu e^{2\mu z}\Phi\left(\frac{-z-\mu\tau}{\sqrt{\tau}}\right)\right)dz \\ &= \sqrt{\frac{2}{\pi\tau}}\int_{\sigma^{-1}\log x}^\infty e^{\sigma z-(z-\mu\tau)^2/(2\tau)}dz - 2\mu\int_{\sigma^{-1}\log x}^\infty e^{z(\sigma+2\mu)}\Phi\left(\frac{-z-\mu\tau}{\sqrt{\tau}}\right)dz.\end{aligned}$$

By standard arguments we have

$$\begin{aligned}&\frac{1}{\sqrt{2\pi\tau}}\int_{\sigma^{-1}\log x}^\infty e^{\sigma z-(z-\mu\tau)^2/(2\tau)}dz \\ &= \frac{1}{\sqrt{2\pi\tau}}\int_{\sigma^{-1}\log x}^\infty e^{-(z^2+\mu^2\tau^2-2(\mu+\sigma)\tau z)/(2\tau)}dz \\ &= \frac{1}{\sqrt{2\pi\tau}}e^{\sigma^2\tau/2+\mu\sigma\tau}\int_{\sigma^{-1}\log x}^\infty e^{-(z-(\mu+\sigma)\tau)^2/(2\tau)}dz\end{aligned}$$

$$
\begin{aligned}
&= \frac{1}{\sqrt{2\pi\tau}} e^{r\tau} \int_{-(\mu+\sigma)\tau+\sigma^{-1}\log x}^{\infty} e^{-z^2/(2\tau)} dz \\
&= e^{r\tau} \Phi\left(\delta_+^\tau\left(\frac{1}{x}\right)\right),
\end{aligned}
$$

since $\mu\sigma + \sigma^2/2 = r$. The second integral

$$
\int_{\sigma^{-1}\log x}^{\infty} e^{z(\sigma+2\mu)} \Phi\left(\frac{-z-\mu\tau}{\sqrt{\tau}}\right) dz
$$

can be computed by integration by parts using the identity

$$
\int_a^\infty v'(z)u(z)dz = u(+\infty)v(+\infty) - u(a)v(a) - \int_a^\infty v(z)u'(z)dz,
$$

with $a = \sigma^{-1}\log x$. We let

$$
u(z) = \Phi\left(\frac{-z-\mu\tau}{\sqrt{\tau}}\right) \quad \text{and} \quad v'(z) = e^{z(\sigma+2\mu)}
$$

which satisfy

$$
u'(z) = -\frac{1}{\sqrt{2\pi\tau}} e^{-(z+\mu\tau)^2/(2\tau)} \quad \text{and} \quad v(z) = \frac{1}{\sigma+2\mu} e^{z(\sigma+2\mu)},
$$

and

$$
\begin{aligned}
&\int_a^\infty e^{z(\sigma+2\mu)} \Phi\left(\frac{-z-\mu\tau}{\sqrt{\tau}}\right) dz = \int_a^\infty v'(z)u(z)dz \\
&= u(+\infty)v(+\infty) - u(a)v(a) - \int_a^\infty v(z)u'(z)dz \\
&= -\frac{1}{\sigma+2\mu} e^{a(\sigma+2\mu)} \Phi\left(\frac{-a-\mu\tau}{\sqrt{\tau}}\right) \\
&\quad + \frac{1}{(\sigma+2\mu)\sqrt{2\pi\tau}} \int_a^\infty e^{z(\sigma+2\mu)} e^{-(z+\mu\tau)^2/(2\tau)} dz \\
&= -\frac{1}{\sigma+2\mu} e^{a(\sigma+2\mu)} \Phi\left(\frac{-a-\mu\tau}{\sqrt{\tau}}\right) \\
&\quad + \frac{1}{(\sigma+\mu)\sqrt{2\pi\tau}} e^{(\tau(\sigma+\mu)^2-\mu^2\tau)/2} \int_a^\infty e^{-(z-\tau(\sigma+\mu))^2/(2\tau)} dz \\
&= -\frac{1}{\sigma+2\mu} e^{a(\sigma+2\mu)} \Phi\left(\frac{-a-\mu\tau}{\sqrt{\tau}}\right) \\
&\quad + \frac{1}{(\sigma+2\mu)\sqrt{2\pi}} e^{(\tau(\sigma+\mu)^2-\mu^2\tau)/2} \int_{(a-\tau(\sigma+\mu))/\sqrt{\tau}}^\infty e^{-z^2/2} dz \\
&= -\frac{1}{\sigma+2\mu} e^{a(\sigma+2\mu)} \Phi\left(\frac{-a-\mu\tau}{\sqrt{\tau}}\right) \\
&\quad + \frac{1}{\sigma+2\mu} e^{(\tau(\sigma+\mu)^2-\mu^2\tau)/2} \Phi\left(\frac{-a+\tau(\sigma+\mu)}{\sqrt{\tau}}\right)
\end{aligned}
$$

$$\begin{aligned}
&= -\frac{2r}{\sigma}(x)^{2r/\sigma^2}\Phi\left(\frac{-(r/\sigma-\sigma/2)\tau-\sigma^{-1}\log x}{\sqrt{\tau}}\right) \\
&\quad +\frac{2r}{\sigma}e^{\sigma\tau(\sigma+2\mu)/2}\Phi\left(\frac{+\tau(r/\sigma+\sigma/2)-\sigma^{-1}\log x}{\sqrt{\tau}}\right) \\
&= \frac{\sigma}{2r}e^{r\tau}\Phi\left(\delta_+^\tau\left(\frac{1}{x}\right)\right)-\frac{\sigma}{2r}x^{2r/\sigma^2}\Phi\left(-\delta_-^\tau(x)\right),
\end{aligned}$$

cf. pages 317–319 of [68] for a different derivation using double integrals.

Hence we have

$$\begin{aligned}
&\mathbb{E}\left[M_t^T\mathbf{1}_{\{M_t^T>M_0^t\}}\mid\mathcal{F}_t\right]=S_t\,\mathbb{E}\left[\max_{r\in[0,T-t]}\frac{S_r}{S_0}\mathbf{1}_{\{\max_{r\in[0,T-t]}S_r/S_0>x\}}\right]_{x=M_0^t/S_t} \\
&= 2S_te^{r(T-t)}\Phi\left(\delta_+^{T-t}\left(\frac{S_t}{M_0^t}\right)\right)-S_t\frac{\mu\sigma}{r}e^{r(T-t)}\Phi\left(\delta_+^{T-t}\left(\frac{S_t}{M_0^t}\right)\right) \\
&\quad +S_t\frac{\mu\sigma}{r}\left(\frac{M_0^t}{S_t}\right)^{2r/\sigma^2}\Phi\left(-\delta_-^{T-t}\left(\frac{M_0^t}{S_t}\right)\right),
\end{aligned}$$

and consequently this yields, since $\mu\sigma/r=1-\sigma^2/(2r)$,

$$\begin{aligned}
&\mathbb{E}[M_0^T\mid\mathcal{F}_t]=\mathbb{E}[M_0^T\mid M_0^t] \\
&= M_0^t\mathbb{P}(M_0^t>M_t^T\mid M_0^t)+\mathbb{E}[M_t^T\mathbf{1}_{\{M_t^T>M_0^t\}}\mid M_0^t] \\
&= M_0^t\Phi\left(-\delta_-^{T-t}\left(\frac{S_t}{M_0^t}\right)\right)-S_t\left(\frac{M_0^t}{S_t}\right)^{2r/\sigma^2}\Phi\left(-\delta_-^{T-t}\left(\frac{M_0^t}{S_t}\right)\right) \\
&\quad +2S_te^{r(T-t)}\Phi\left(\delta_+^{T-t}\left(\frac{S_t}{M_0^t}\right)\right) \\
&\quad -S_t\left(1-\frac{\sigma^2}{2r}\right)e^{r(T-t)}\Phi\left(\delta_+^{T-t}\left(\frac{S_t}{M_0^t}\right)\right) \\
&\quad +S_t\left(1-\frac{\sigma^2}{2r}\right)\left(\frac{M_0^t}{S_t}\right)^{2r/\sigma^2}\Phi\left(-\delta_-^{T-t}\left(\frac{M_0^t}{S_t}\right)\right) \\
&= M_0^t\Phi\left(-\delta_-^{T-t}\left(\frac{S_t}{M_0^t}\right)\right)+S_te^{r(T-t)}\left(1+\frac{\sigma^2}{2r}\right)\Phi\left(\delta_+^{T-t}\left(\frac{S_t}{M_0^t}\right)\right) \\
&\quad -S_t\frac{\sigma^2}{2r}\left(\frac{M_0^t}{S_t}\right)^{2r/\sigma^2}\Phi\left(-\delta_-^{T-t}\left(\frac{M_0^t}{S_t}\right)\right),
\end{aligned}$$

hence

$$\begin{aligned}
&e^{-r(T-t)}\,\mathbb{E}[M_0^T-S_T\mid\mathcal{F}_t]=e^{-r(T-t)}\,\mathbb{E}[M_0^T\mid\mathcal{F}_t]-e^{-r(T-t)}\,\mathbb{E}[S_T\mid\mathcal{F}_t] \\
&= e^{-r(T-t)}\,\mathbb{E}[M_0^T\mid M_0^t]-S_t \\
&= M_0^te^{-r(T-t)}\Phi\left(-\delta_-^{T-t}\left(\frac{S_t}{M_0^t}\right)\right)-S_t\Phi\left(-\delta_+^{T-t}\left(\frac{S_t}{M_0^t}\right)\right)
\end{aligned}$$

$$+S_t\frac{\sigma^2}{2r}\Phi\left(\delta_+^{T-t}\left(\frac{S_t}{M_0^t}\right)\right)-S_t\frac{\sigma^2}{2r}e^{-r(T-t)}\left(\frac{M_0^t}{S_t}\right)^{2r/\sigma^2}\Phi\left(-\delta_-^{T-t}\left(\frac{M_0^t}{S_t}\right)\right).$$

This concludes the proof of Proposition 8.4.

PDE Method

If the couple (S_t, M_t) is Markov, the price can be written as a function

$$f(t,S_t,M_t)=e^{-rT}\,\mathbb{E}[\phi(S_T,M_T)\mid\mathcal{F}_t],$$

and in this case the function $f(t,x,y)$ can solve a PDE.

Next we derive the Black–Scholes partial differential equation (PDE) for the price of a self-financing portfolio.

Black–Scholes PDE for Lookback Put Options

Proposition 8.5 *Let $(\eta_t,\xi_t)_{t\in\mathbb{R}_+}$ be a portfolio strategy such that*

(i) $(\eta_t,\xi_t)_{t\in\mathbb{R}_+}$ is self-financing,

(ii) the portfolio value $V_t:=\eta_tA_t+\xi_tS_t$, $t\in\mathbb{R}_+$, takes the form

$$V_t=f(t,S_t,M_t),\qquad t\in\mathbb{R}_+,$$

for some $f\in\mathcal{C}^2((0,\infty)\times(0,\infty)^2)$.

Then the function $f(t,x,y)$ satisfies the Black–Scholes PDE

$$rf(t,x,y)=\frac{\partial f}{\partial t}(t,x,y)+rx\frac{\partial f}{\partial x}(t,x,y)+\frac{1}{2}x^2\sigma^2\frac{\partial^2 f}{\partial x^2}(t,x,y),\qquad t,x,y>0,\tag{8.19}$$

under the boundary conditions

$$\begin{cases} f(t,0,y)=e^{-r(T-t)}y, & 0\le t\le T,\quad y\in\mathbb{R}_+, & (8.20a)\\ \dfrac{\partial f}{\partial y}(t,x,y)_{x=y}=0, & 0\le t\le T,\quad y>0, & (8.20b)\\ f(T,x,y)=y-x, & 0\le x\le y. & (8.20c)\end{cases}$$

The replicating portfolio of the lookback put option is given by

$$\xi_t=\frac{\partial f}{\partial x}(t,S_t,M_0^t),\qquad t\in[0,T],\tag{8.21}$$

where $f(t,x,y)$ is given by

$$f(t,S_t,M_0^t) = e^{-r(T-t)}\,\mathbb{E}[\phi(S_T,M_0^T) \mid \mathcal{F}_t], \qquad 0 \le t \le T. \tag{8.22}$$

Proof. The existence of $f(t,x,y)$ follows from the Markov property, more precisely the function $f(t,x,y)$ satisfies

$$\begin{aligned} f(t,x,y) &= e^{-r(T-t)}\,\mathbb{E}[\phi(S_T,M_0^T) \mid S_t = x,\ M_0^t = y] \\ &= e^{-r(T-t)}\,\mathbb{E}\left[\phi\left(x\frac{S_T}{S_t}, \frac{y}{x} \wedge \frac{\mathcal{M}_t^T}{S_t}\right)\right] \\ &= e^{-r(T-t)}\,\mathbb{E}\left[\phi\left(x\frac{S_{T-t}}{S_0}, \frac{y}{x} \wedge \frac{\mathcal{M}_0^{T-t}}{x}\right)\right], \qquad t \in [0,T], \end{aligned}$$

from the time homogeneity of the asset price process $(S_t)_{t\in\mathbb{R}_+}$. Applying the change of variable formula to the discounted portfolio value

$$\tilde{f}(t,x,y) = e^{-rt} f(t,x,y) = e^{-rT}\,\mathbb{E}[\phi(S_T,M_0^T) \mid S_t = x,\ M_0^t = y]$$

which is a martingale for $t \in [0,T]$, we have

$$\begin{aligned} d\tilde{f}(t,S_t,M_0^t) &= -re^{-rt} f(t,S_t,M_0^t)dt + e^{-rt} df(t,S_t,M_0^t) \\ &= -re^{-rt} f(t,S_t,M_0^t)dt + e^{-rt}\frac{\partial f}{\partial t}(t,S_t,M_0^t)dt + re^{-rt}S_t\frac{\partial f}{\partial x}(t,S_t,M_0^t)dt \\ &\quad +\frac{1}{2}e^{-rt}\sigma^2 S_t^2\frac{\partial^2 f}{\partial x^2}(t,S_t,M_0^t)dt \\ &\quad +e^{-rt}\frac{\partial f}{\partial y}(t,S_t,M_0^t)dM_0^t + e^{-rt}\sigma S_t\frac{\partial f}{\partial x}(t,S_t,M_0^t)dB_t. \end{aligned}$$

Since $\left(\mathbb{E}[\phi(S_T,M_0^T) \mid \mathcal{F}_t]\right)_{t\in[0,T]}$ is a $\mathbb{P}$-martingale and $(M_0^t)_{t\in[0,T]}$ has finite variation (it is in fact a non-decreasing process), we have:

$$df(t,S_t,M_0^t) = \sigma S_t\frac{\partial f}{\partial x}(t,S_t,M_0^t)dB_t, \qquad t \in [0,T], \tag{8.23}$$

and the function $f(t,x,y)$ satisfies the equation

$$\begin{aligned} &\frac{\partial f}{\partial t}(t,S_t,M_0^t)dt + rS_t\frac{\partial f}{\partial x}f(t,S_t,M_0^t)dt \\ &\qquad +\frac{1}{2}\sigma^2 S_t^2\frac{\partial^2 f}{\partial x^2}(t,S_t,M_0^t)dt + \frac{\partial f}{\partial y}(t,S_t,M_0^t)dM_0^t = rf(t,S_t,M_0^t), \end{aligned}$$

which implies

$$\frac{\partial f}{\partial t}(t,S_t,M_0^t) + rS_t\frac{\partial f}{\partial x}(t,S_t,M_0^t) + \frac{1}{2}\sigma^2 S_t^2\frac{\partial^2 f}{\partial x^2}(t,S_t,M_0^t) = rf(t,S_t,M_0^t),$$

which is (8.19), and

$$\frac{\partial f}{\partial y}(t,S_t,M_0^t)dM_0^t = 0,$$

because M_0^t increases only on a set of zero measure (which has no isolated points). This implies

$$\frac{\partial f}{\partial y}(t, S_t, S_t) = 0,$$

which shows the boundary condition (8.20b), since M_0^t hits S_t when M_0^t increases. On the other hand, (8.23) shows that

$$\phi(S_T, M_0^T) = \mathbb{E}[\phi(S_T, M_0^T)] + \sigma \int_0^T S_t \frac{\partial f}{\partial x}(t, x, M_0^t)_{|x=S_t} dB_t,$$

$0 \leq t \leq T$, which implies (8.21) as in the proof of Proposition 5.2. □

In other words, the price of the lookback put option takes the form

$$f(t, S_t, M_t) = e^{-r(T-t)} \, \mathbb{E}\left[M_0^T - S_T \mid \mathcal{F}_t\right],$$

where the function $f(t, x, y)$ is given by

$$\boxed{\begin{aligned} f(t,x,y) &= ye^{-r(T-t)}\Phi\left(-\delta_-^{T-t}(x/y)\right) + x\left(1+\frac{\sigma^2}{2r}\right)\Phi\left(\delta_+^{T-t}(x/y)\right) \\ &\quad -x\frac{\sigma^2}{2r}e^{-r(T-t)}\left(y/x\right)^{2r/\sigma^2}\Phi\left(-\delta_-^{T-t}(y/x)\right) - x. \end{aligned}}$$

Checking the Boundary Conditions

The boundary condition (8.20a) is explained by the fact that

$$\begin{aligned} f(t,0,y) &= e^{-r(T-t)}\,\mathbb{E}[M_0^T - S_T \mid S_t = 0,\ M_0^t = x] \\ &= e^{-r(T-t)}\,\mathbb{E}[M_0^t - S_T \mid S_t = 0,\ M_0^t = x] \\ &= e^{-r(T-t)}\,\mathbb{E}[M_0^t \mid M_0^t = x] - e^{-r(T-t)}\,\mathbb{E}[S_T \mid S_t = 0] \\ &= xe^{-r(T-t)}, \end{aligned}$$

since $\mathbb{E}[S_T \mid S_t = 0] = 0$ as $S_t = 0$ implies $S_T = 0$. On the other hand, (8.20c) follows from the fact that

$$f(T,x,y) = \mathbb{E}[M_0^T - S_T \mid S_T = x,\ M_0^T = y] = y - x.$$

Note that we have

$$f(t,x,x) = xC(T-t),$$

with

$$C(\tau) = e^{-r\tau}\Phi\left(-\delta_-^{\tau}(1)\right) + \left(1+\frac{\sigma^2}{2r}\right)\Phi\left(\delta_+^{\tau}(1)\right) - \frac{\sigma^2}{2r}e^{-r\tau}\Phi\left(-\delta_-^{\tau}(1)\right) - 1,$$

$\tau > 0$, hence

$$\frac{\partial f}{\partial x}(t,x,x) = C(T-t), \qquad t \in [0,T],$$

while we also have

$$\frac{\partial f}{\partial y}(t,x,y)_{y=x} = 0, \qquad 0 \le x \le y.$$

Scaling Property of Lookback Put Prices

We note the scaling property

$$\begin{aligned}
f(t,x,y) &= e^{-r(T-t)}\,\mathbb{E}\left[M_0^T - S_T \middle| S_t = x,\ M_t = y\right] \\
&= e^{-r(T-t)}\,\mathbb{E}\left[M_0^t \vee M_t^T - S_T \middle| S_t = x,\ M_t = y\right] \\
&= e^{-r(T-t)} x\,\mathbb{E}\left[\frac{M_0^t}{S_t} \vee \frac{M_t^T}{S_t} - 1 \middle| S_t = x,\ M_t = y\right] \\
&= e^{-r(T-t)} x\,\mathbb{E}\left[\frac{y}{x} \vee \frac{M_t^T}{S_t} - 1 \middle| S_t = x,\ M_t = y\right] \\
&= e^{-r(T-t)} x\,\mathbb{E}\left[M_0^t \vee M_t^T - 1 \middle| S_t = 1,\ M_t = \frac{y}{x}\right] \\
&= x f(t,1,y/x),
\end{aligned}$$

hence letting

$$\begin{aligned}
g(\tau,z) &= \frac{1}{z}e^{-r\tau}\Phi\left(-\delta_-^\tau(z)\right) + \left(1+\frac{\sigma^2}{2r}\right)\Phi\left(\delta_+^\tau(z)\right) \\
&\quad -\frac{\sigma^2}{2r}e^{-r\tau}\left(\frac{1}{z}\right)^{2r/\sigma^2}\Phi\left(-\delta_-^\tau\left(\frac{1}{z}\right)\right) - 1,
\end{aligned}$$

we note that we have

$$f(t,x,y) = xg(T-t,x/y)$$

and the boundary condition

$$\left\{\begin{array}{ll}
\dfrac{\partial g}{\partial z}(\tau,1) = 0, & \tau > 0, \qquad (8.24a)\\[2ex]
g(0,z) = \dfrac{1}{z} - 1, & z \in (0,1]. \qquad (8.24b)
\end{array}\right.$$

The next Figure 8.15 shows a graph of the function $g(\tau,z)$.

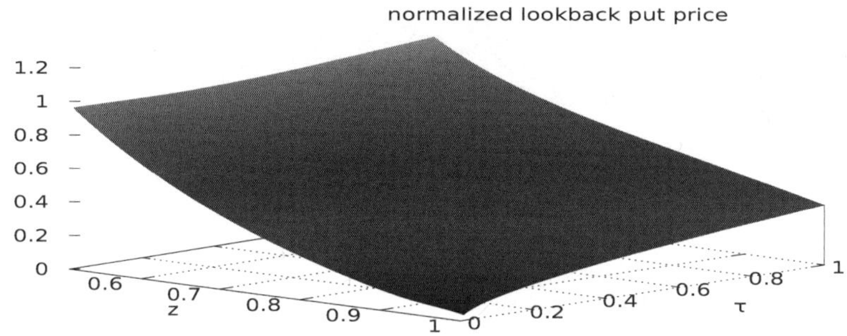

FIGURE 8.15: Graph of the normalized lookback put option price.

Black–Scholes Approximation of Lookback Put Prices

Letting

$$\mathrm{BS}_p(S,K,r,\sigma,\tau) = Ke^{-r\tau}\Phi\left(-\delta_-^\tau\left(\frac{S}{K}\right)\right) - S\Phi\left(-\delta_+^\tau\left(\frac{S}{K}\right)\right)$$

denote the standard Black–Scholes formula for the price of a European put option, we observe that the lookback put option price satisfies

$$\begin{aligned} e^{-r(T-t)}\,\mathbb{E}[M_0^T - S_T \mid \mathcal{F}_t] &= \mathrm{BS}_p(S_t, M_0^t, r, \sigma, T-t) \\ &+ S_t\frac{\sigma^2}{2r}\left(\Phi\left(\delta_+^{T-t}\left(\frac{S_t}{M_0^t}\right)\right) - e^{-r(T-t)}\left(\frac{M_0^t}{S_t}\right)^{2r/\sigma^2}\Phi\left(-\delta_-^{T-t}\left(\frac{M_0^t}{S_t}\right)\right)\right), \end{aligned}$$

i.e.,

$$e^{-r(T-t)}\,\mathbb{E}\left[M_0^T - S_T \mid \mathcal{F}_t\right] = \mathrm{BS}_p(S_t, M_0^t, r, \sigma, T-t) + S_t h_p\left(T-t, \frac{S_t}{M_0^t}\right)$$

where the function

$$h_p(\tau, z) = \frac{\sigma^2}{2r}\left(\Phi\left(\delta_+^\tau(z)\right) - e^{-r\tau}z^{-2r/\sigma^2}\Phi\left(-\delta_-^\tau(1/z)\right)\right), \tag{8.25}$$

depends only on $z = S_t/M_0^t$. In other words, due to the relation

$$\mathrm{BS}_p(x,y,r,\sigma,\tau) \quad = \quad ye^{-r\tau}\Phi\left(-\delta_-^\tau\left(\frac{x}{y}\right)\right) - x\Phi\left(-\delta_+^\tau\left(\frac{x}{y}\right)\right)$$

$$= \quad x\mathrm{BS}_p(1, y/x, r, \sigma, \tau)$$

for the standard Black–Scholes put formula, we observe that $f(t, x, y)$ satisfies

$$f(t, x, y) = x\mathrm{BS}_p(1, y/x, r, \sigma, T - t) + xh(T - t, x/y),$$

i.e.,

$$f(t, x, y) = xg(T - t, x/y),$$

with

$$g(\tau, z) = \mathrm{BS}_p(1, 1/z, r, \sigma, \tau) + h_p(\tau, z), \tag{8.26}$$

where $h_p(\tau, z)$ is the function given by (8.25), and $(x, y) \mapsto xh_p(T - t, x/y)$ also satisfies the Black–Scholes PDE (8.19), i.e., $(\tau, z) \mapsto \mathrm{BS}_p(1, 1/z, r, \sigma, \tau)$ and $h_p(\tau, z)$ both satisfy the PDE

$$\frac{\partial h_p}{\partial \tau}(\tau, z) = z\left(r + \sigma^2\right)\frac{\partial h_p}{\partial z}(\tau, z) + \frac{1}{2}\sigma^2 z^2 \frac{\partial^2 h_p}{\partial z^2}(\tau, z), \tag{8.27}$$

$\tau \in \mathbb{R}_+$, $z \in [0, 1]$, under the boundary condition

$$h_p(0, z) = 0, \qquad 0 \leq z \leq 1.$$

The next Figures 8.16 and 8.17 show the decompositions (8.26) of the normalized lookback put option price $g(\tau, z)$ in Figure 8.15 into the Black–Scholes put function $\mathrm{BS}_p(1, 1/z, r, \sigma, \tau)$ and $h_p(\tau, z)$.

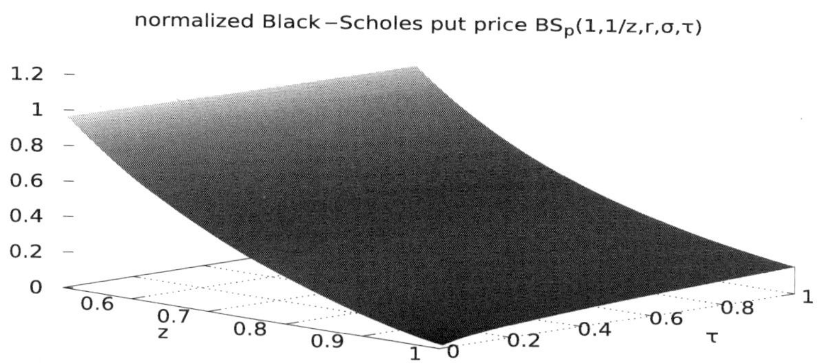

FIGURE 8.16: Black–Scholes put price in the decomposition (8.26).

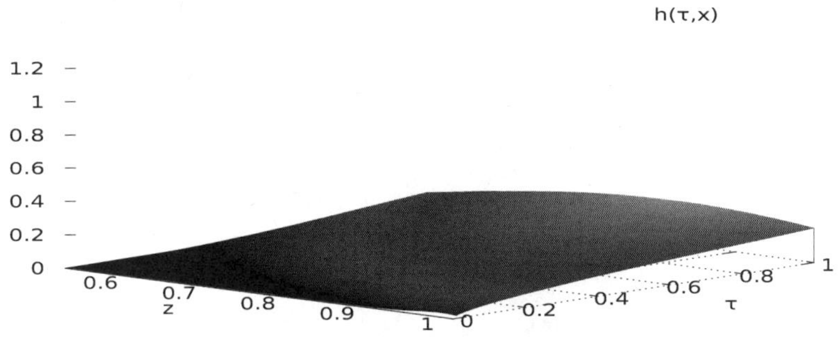

FIGURE 8.17: Function $h_p(\tau, z)$ in the decomposition (8.26).

Note that in Figures 8.16–8.17 the condition $h_p(0, z) = 0$ is not fully respected as $z \to 1$ due to numerical error in the approximation of the function Φ.

The Lookback Call Option

The standard lookback call option gives the right to buy the underlying asset at its historically lowest price. In this case the strike is m_0^T and the payoff is

$$C = S_T - m_0^T.$$

The following result gives the price of the lookback call option, cf. e.g., Proposition 9.5.1, page 270 of [15].

Proposition 8.6 *The price at time $t \in [0, T]$ of the lookback call option with payoff $S_T - m_0^T$ is given by*

$$\begin{aligned} & e^{-r(T-t)} \mathbb{E}[S_T - m_0^T \mid \mathcal{F}_t] \\ = \ & S_t \Phi\left(\delta_+^{T-t}\left(\frac{S_t}{m_0^t}\right)\right) - m_0^t e^{-r(T-t)} \Phi\left(\delta_-^{T-t}\left(\frac{S_t}{m_0^t}\right)\right) \\ & + e^{-r(T-t)} S_t \frac{\sigma^2}{2r}\left(\frac{m_0^t}{S_t}\right)^{2r/\sigma^2} \Phi\left(\delta_-^{T-t}\left(\frac{m_0^t}{S_t}\right)\right) - S_t \frac{\sigma^2}{2r} \Phi\left(-\delta_+^{T-t}\left(\frac{S_t}{m_0^t}\right)\right) \end{aligned}$$

Figure 8.18 represents the price of the lookback call option as a function of m_0^t and S_t for different values of the time to maturity $T - t$.

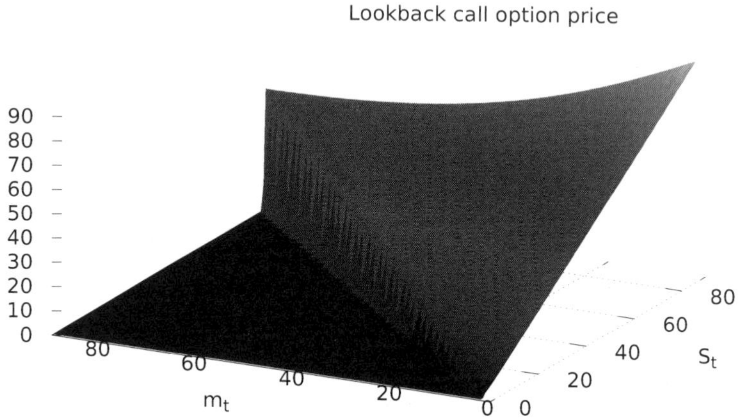

FIGURE 8.18: Graph of the lookback call option price.

Proof of Proposition 8.6. We have

$$e^{-r(T-t)}\,\mathbb{E}[S_T - m_0^T \mid \mathcal{F}_t] = e^{-r(T-t)}\,\mathbb{E}[S_T \mid \mathcal{F}_t] - e^{-r(T-t)}\,\mathbb{E}[m_0^T \mid \mathcal{F}_t],$$

and

$$\begin{aligned}
\mathbb{E}[m_0^T \mid \mathcal{F}_t] &= \mathbb{E}[m_0^t \wedge m_t^T \mid \mathcal{F}_t] \\
&= \mathbb{E}[m_0^t \mathbf{1}_{\{m_0^t < m_t^T\}} \mid \mathcal{F}_t] + \mathbb{E}[m_t^T \mathbf{1}_{\{m_0^t > m_t^T\}} \mid \mathcal{F}_t] \\
&= m_0^t\,\mathbb{E}[\mathbf{1}_{\{m_0^t < m_t^T\}} \mid \mathcal{F}_t] + \mathbb{E}[m_t^T \mathbf{1}_{\{m_0^t > m_t^T\}} \mid \mathcal{F}_t] \\
&= m_0^t \mathbb{P}(m_0^t < m_t^T \mid \mathcal{F}_t) + \mathbb{E}[m_t^T \mathbf{1}_{\{m_0^t > m_t^T\}} \mid \mathcal{F}_t].
\end{aligned}$$

By computations similar to those of the lookback put option case we find

$$\begin{aligned}
\mathbb{P}(m_0^t < m_t^T \mid \mathcal{F}_t) &= \mathbb{P}\left(\frac{m_0^t}{S_t} < \frac{m_t^T}{S_t} \middle| \mathcal{F}_t \right) \\
&= \mathbb{P}\left(x < \frac{m_t^T}{S_t} \middle| \mathcal{F}_t \right)_{x = m_0^t/S_t} \\
&= \mathbb{P}\left(\frac{m_0^{T-t}}{S_0} > x \right)_{x = m_0^t/S_t} \\
&= \Phi\left(\delta_-^{T-t}\left(\frac{S_t}{m_0^t} \right) \right) - \left(\frac{m_0^t}{S_t} \right)^{-1+2r/\sigma^2} \Phi\left(\delta_-^{T-t}\left(\frac{m_0^t}{S_t} \right) \right),
\end{aligned}$$

and

$$\begin{aligned}
\mathbb{E}[m_t^T \mathbf{1}_{\{m_0^t > m_t^T\}} \mid \mathcal{F}_t] &= S_t\,\mathbb{E}\left[\min_{r\in[t,T]} \frac{S_r}{S_t} \mathbf{1}_{\{m_0^t/S_t > m_t^T/S_t\}} \right]_{x = m_0^t, y = S_t} \\
&= S_t\,\mathbb{E}\left[\min_{r\in[t,T]} \frac{S_r}{S_t} \mathbf{1}_{\{\min_{r\in[t,T]} S_r/S_t < x\}} \right]_{x = m_0^t/S_t}
\end{aligned}$$

$$
\begin{aligned}
&= S_t \,\mathbb{E}\left[\min_{r\in[0,T-t]} \frac{S_r}{S_0} \mathbf{1}_{\{\min_{r\in[0,T-t]} S_r/S_0 < x\}}\right]_{x=m_0^t/S_t} \\
&= 2S_t e^{r(T-t)} \Phi\left(-\delta_+^{T-t}\left(\frac{S_t}{m_0^t}\right)\right) - S_t \frac{\mu\sigma}{r} e^{r(T-t)} \Phi\left(-\delta_+^{T-t}\left(\frac{S_t}{m_0^t}\right)\right) \\
&\quad + S_t \frac{\mu\sigma}{r}\left(\frac{m_0^t}{S_t}\right)^{2r/\sigma^2} \Phi\left(\delta_-^{T-t}\left(\frac{m_0^t}{S_t}\right)\right).
\end{aligned}
$$

Given the relation $\mu\sigma/r = 1 - \sigma^2/(2r)$, this yields

$$
\begin{aligned}
& e^{-r(T-t)}\,\mathbb{E}[S_T - m_0^T \mid \mathcal{F}_t] = S_t - m_0^t e^{-r(T-t)} \mathbb{P}\left(\frac{m_0^{T-t}}{S_0} > x\right)_{x=m_0^t/S_t} \\
& - S_t e^{-r(T-t)}\,\mathbb{E}\left[\min_{r\in[0,T-t]} \frac{S_r}{S_0} \mathbf{1}_{\{\min_{r\in[0,T-t]} S_r/S_0 < x\}}\right]_{x=m_0^t/S_t} \\
&= S_t - m_0^t e^{-r(T-t)} \Phi\left(\delta_-^{T-t}\left(\frac{S_t}{m_0^t}\right)\right) + m_0^t e^{-r(T-t)}\left(\frac{m_0^t}{S_t}\right)^{-1+2r/\sigma^2} \\
& \quad \Phi\left(\delta_-^{T-t}\left(\frac{m_0^t}{S_t}\right)\right) - 2S_t \Phi\left(-\delta_+^{T-t}\left(\frac{S_t}{m_0^t}\right)\right) + S_t \frac{\mu\sigma}{r} \Phi\left(-\delta_+^{T-t}\left(\frac{S_t}{m_0^t}\right)\right) \\
& - S_t e^{-r(T-t)} \frac{\mu\sigma}{r}\left(\frac{m_0^t}{S_t}\right)^{2r/\sigma^2} \Phi\left(\delta_-^{T-t}\left(\frac{m_0^t}{S_t}\right)\right) \\
&= S_t - m_0^t e^{-r(T-t)} \Phi\left(\delta_-^{T-t}\left(\frac{S_t}{m_0^t}\right)\right) - S_t\left(1 + \frac{\sigma^2}{2r}\right) \Phi\left(-\delta_+^{T-t}\left(\frac{S_t}{m_0^t}\right)\right) \\
& + S_t e^{-r(T-t)} \frac{\sigma^2}{2r}\left(\frac{m_0^t}{S_t}\right)^{2r/\sigma^2} \Phi\left(\delta_-^{T-t}\left(\frac{m_0^t}{S_t}\right)\right) \\
&= S_t \Phi\left(\delta_+^{T-t}\left(\frac{S_t}{m_0^t}\right)\right) - e^{-r(T-t)} m_0^t \Phi\left(\delta_-^{T-t}\left(\frac{S_t}{m_0^t}\right)\right) \\
& + e^{-r(T-t)} \frac{S_t\sigma^2}{2r}\left(\left(\frac{m_0^t}{S_t}\right)^{2r/\sigma^2} \Phi\left(\delta_-^{T-t}\left(\frac{m_0^t}{S_t}\right)\right) - e^{r(T-t)} \Phi\left(-\delta_+^{T-t}\left(\frac{S_t}{m_0^t}\right)\right)\right).
\end{aligned}
$$

Black–Scholes Approximation of Lookback Call Prices

Letting

$$
\mathrm{BS}_c(S, K, r, \sigma, \tau) = S\Phi\left(\delta_+^\tau\left(\frac{S}{K}\right)\right) - Ke^{-r\tau}\Phi\left(\delta_-^\tau\left(\frac{S}{K}\right)\right)
$$

denote the standard Black–Scholes formula for the price of a European call option, we observe that the lookback call option price satisfies

$$
\begin{aligned}
& e^{-r(T-t)}\,\mathbb{E}[S_T - m_0^T \mid \mathcal{F}_t] = \mathrm{BS}_c(S_t, m_0^t, r, \sigma, T-t) \\
& \quad - S_t \frac{\sigma^2}{2r}\left(\Phi\left(-\delta_+^{T-t}\left(\frac{S_t}{m_0^t}\right)\right) - e^{-r(T-t)}\left(\frac{m_0^t}{S_t}\right)^{2r/\sigma^2} \Phi\left(\delta_-^{T-t}\left(\frac{m_0^t}{S_t}\right)\right)\right),
\end{aligned}
$$

i.e.,

$$e^{-r(T-t)}\,\mathbb{E}[S_T - m_0^T \mid \mathcal{F}_t] = \mathrm{BS}_c(S_t, m_0^t, r, \sigma, T-t) + S_t h_c\left(T-t, \frac{S_t}{m_0^t}\right)$$

where the function

$$h_c(\tau, z) = -\frac{\sigma^2}{2r}\left(\Phi\left(-\delta_+^\tau(z)\right) - e^{-r\tau} z^{-2r/\sigma^2}\,\Phi\left(\delta_-^\tau(1/z)\right)\right), \tag{8.28}$$

depends only on $z = S_t/m_0^t$ and satisfies

$$h_c(\tau, z) = h_p(\tau, z) - \frac{\sigma^2}{2r}\left(1 - e^{-r\tau} z^{-2r/\sigma^2}\right), \qquad \tau \in \mathbb{R}_+, \quad z \in \mathbb{R}_+,$$

where $(z, \tau) \mapsto e^{-r\tau} z^{-2r/\sigma^2}$ also solves the PDE (8.27).

Black–Scholes PDE for Lookback Call Options

By the same argument as in the proof of Proposition 8.5, the function $f(t, x, y)$ satisfies the Black–Scholes PDE

$$rf(t, x, y) = \frac{\partial f}{\partial t}(t, x, y) + rx\frac{\partial f}{\partial x}(t, x, y) + \frac{1}{2}x^2\sigma^2\frac{\partial^2 f}{\partial x^2}(t, x, y), \qquad t, x > 0,$$

under the boundary conditions

$$\begin{cases} \lim\limits_{y \searrow 0} f(t, x, y) = x, \qquad 0 \le t \le T, \quad x > 0, & (8.29\text{a}) \\[2ex] \dfrac{\partial f}{\partial y}(t, x, y)_{x=y} = 0, \qquad 0 \le t \le T, \quad y > 0, & (8.29\text{b}) \\[2ex] f(T, x, y) = x - y, \qquad 0 \le y \le x, & (8.29\text{c}) \end{cases}$$

and the corresponding self-financing hedging strategy is given by

$$\xi_t = \frac{\partial f}{\partial x}(t, S_t, m_0^t), \qquad t \in [0, T], \tag{8.30}$$

which represents the quantity of the risky asset S_t to be held at time t in the hedging portfolio.

In other words, the price of the lookback call option takes the form

$$f(t, S_t, m_t) = e^{-r(T-t)}\,\mathbb{E}[S_T - m_0^T \mid \mathcal{F}_t],$$

where the function $f(t, x, y)$ is given by

$$f(t,x,y) = x\Phi\left(\delta_+^{T-t}\left(\frac{x}{y}\right)\right) - e^{-r(T-t)}y\Phi\left(\delta_-^{T-t}\left(\frac{x}{y}\right)\right)$$
$$+e^{-r(T-t)}x\frac{\sigma^2}{2r}\left(\left(\frac{y}{x}\right)^{2r/\sigma^2}\Phi\left(\delta_-^{T-t}\left(\frac{y}{x}\right)\right) - e^{r(T-t)}\Phi\left(-\delta_+^{T-t}\left(\frac{x}{y}\right)\right)\right)$$
$$= x - ye^{-r(T-t)}\Phi\left(\delta_-^{T-t}\left(\frac{x}{y}\right)\right) - x\left(1+\frac{\sigma^2}{2r}\right)\Phi\left(-\delta_+^{T-t}\left(\frac{x}{y}\right)\right)$$
$$+xe^{-r(T-t)}\frac{\sigma^2}{2r}\left(\frac{y}{x}\right)^{2r/\sigma^2}\Phi\left(\delta_-^{T-t}\left(\frac{y}{x}\right)\right). \tag{8.31}$$

Checking the Boundary Conditions

The boundary condition (8.29a) is explained by the fact that

$$\begin{aligned} f(t,x,0) &= e^{-r(T-t)}\,\mathbb{E}[S_T - m_0^T \mid S_t = x,\ m_0^t = 0] \\ &= e^{-r(T-t)}\,\mathbb{E}[S_T \mid S_t = x,\ m_0^t = 0] \\ &= e^{-r(T-t)}\,\mathbb{E}[S_T \mid S_t = x] \\ &= e^{-r(T-t)}x. \end{aligned}$$

On the other hand, (8.29b) follows from the fact that

$$f(T,x,y) = \mathbb{E}[S_T - m_0^T \mid S_T = x,\ m_0^T = y] = x - y.$$

We have

$$f(t,x,x) = xC(T-t),$$

with

$$C(\tau) = 1 - e^{-r\tau}\Phi\left(\delta_-^\tau(1)\right) - \left(1+\frac{\sigma^2}{2r}\right)\Phi\left(-\delta_+^\tau(1)\right) + e^{-r\tau}\frac{\sigma^2}{2r}\Phi\left(\delta_-^\tau(1)\right),$$

$\tau > 0$, hence

$$\frac{\partial f}{\partial x}(t,x,x) = C(T-t), \qquad t \in [0,T],$$

while we also have

$$\frac{\partial f}{\partial y}(t,x,y)_{y=x} = 0, \qquad 0 \le x \le y.$$

Scaling Property of Lookback Call Prices

We note the scaling property

$$f(t,x,y) \;=\; e^{-r(T-t)}\,\mathbb{E}\left[S_T - m_0^T \mid S_t = x,\ m_t = y\right]$$

$$
\begin{aligned}
&= e^{-r(T-t)}\,\mathbb{E}\left[m_0^t \wedge m_t^T - S_T \mid S_t = x,\ m_t = y\right] \\
&= e^{-r(T-t)} x\,\mathbb{E}\left[\frac{m_0^t}{S_t} \vee \frac{m_t^T}{S_t} - 1 \mid S_t = x,\ m_t = y\right] \\
&= e^{-r(T-t)} x\,\mathbb{E}\left[\frac{y}{x} \vee \frac{m_t^T}{S_t} - 1 \mid S_t = x,\ m_t = y\right] \\
&= e^{-r(T-t)} x\,\mathbb{E}\left[m_0^t \vee m_t^T - 1 \mid S_t = 1,\ m_t = y/x\right] \\
&= x f(t,1,y/x),
\end{aligned}
$$

hence letting

$$
\begin{aligned}
g(\tau,z) &= 1 - \frac{1}{z} e^{-r\tau}\Phi\left(\delta_-^\tau(z)\right) - \left(1+\frac{\sigma^2}{2r}\right)\Phi\left(-\delta_+^\tau(z)\right) \\
&\quad + \frac{\sigma^2}{2r} e^{-r\tau} z^{-2r/\sigma^2}\Phi\left(\delta_-^\tau\left(\frac{1}{z}\right)\right),
\end{aligned}
$$

we have $g(\tau,1) = C(T-t)$, and

$$
f(t,x,y) = x g(T-t, x/y)
$$

and the boundary condition

$$
\begin{cases}
\dfrac{\partial g}{\partial z}(\tau,1) = 0, & \tau > 0, \quad (8.32a) \\[2ex]
g(0,z) = 1 - \dfrac{1}{z}, & z \geq 1. \quad (8.32b)
\end{cases}
$$

Figure 8.19 shows a graph of the function $g(\tau,z)$.

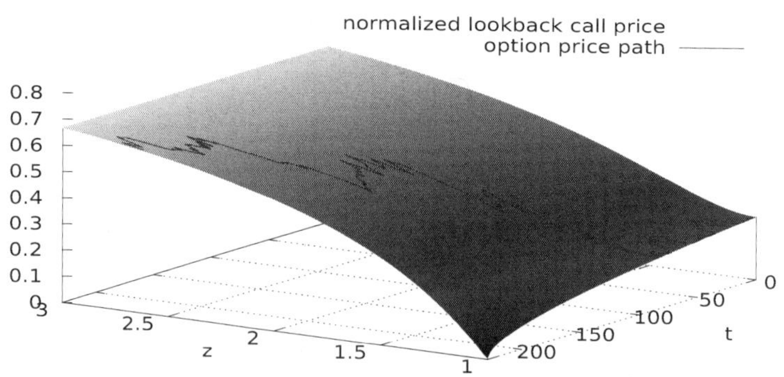

FIGURE 8.19: Normalized lookback call option price.

Figure 8.20 represents the path of the underlying asset price used in Figure 8.19.
Next we represent the option price as a function of time.

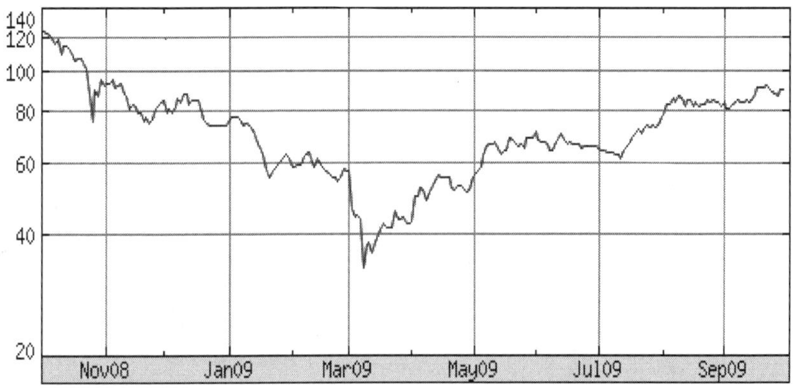

FIGURE 8.20: Graph of the underlying asset price.

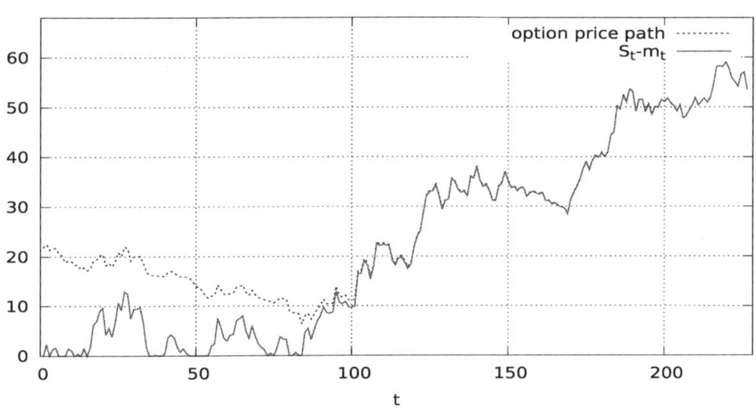

FIGURE 8.21: Graph of the lookback call option price.

Figure 8.22 represents the corresponding underlying asset price and its running minimum.

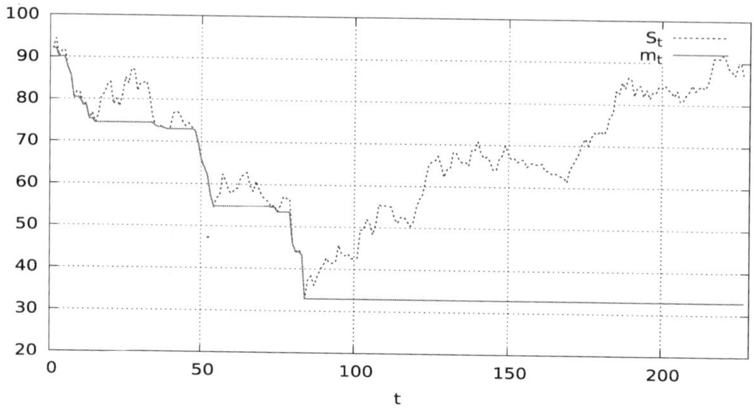

FIGURE 8.22: Running minimum of the underlying asset price.

Due to the relation

$$\begin{aligned} \mathrm{BS}_c(x, y, r, \sigma, \tau) &= x\Phi\left(\delta_+^\tau\left(\frac{x}{y}\right)\right) - ye^{-r\tau}\Phi\left(\delta_-^\tau\left(\frac{x}{y}\right)\right) \\ &= x\mathrm{BS}_c(1, y/x, r, \sigma, \tau) \end{aligned}$$

for the standard Black–Scholes call formula, we observe that $f(t, x, y)$ satisfies

$$f(t, x, y) = x\mathrm{BS}_c(1, y/x, r, \sigma, T - t) + xh_c(T - t, x/y),$$

i.e.,

$$f(t, x, y) = xg(T - t, x/y),$$

with

$$g(\tau, z) = \mathrm{BS}_c(1, 1/z, r, \sigma, \tau) + h_c(\tau, z), \tag{8.33}$$

where $h_c(\tau, z)$ is the function given by (8.28), and $(x, y) \mapsto xh_c(T - t, x/y)$ also satisfies the Black–Scholes PDE (8.19), i.e., $(\tau, z) \mapsto \mathrm{BS}_c(1, 1/z, r, \sigma, \tau)$ and $h_c(\tau, z)$ both satisfy the PDE (8.27) under the boundary condition

$$h_c(0, z) = 0, \qquad z \geq 1.$$

Figures 8.23 and 8.24 show the decomposition of $g(t, z)$ in (8.33) and Figures 8.19–8.20 into the sum of the Black–Scholes call function $\mathrm{BS}_c(1, 1/z, r, \sigma, \tau)$ and $h(t, z)$.

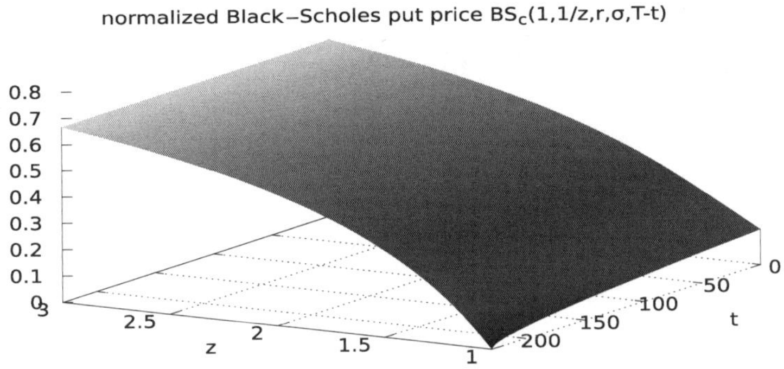

FIGURE 8.23: Black–Scholes call price in the decomposition (8.33) of the normalized lookback call option price $g(\tau, z)$.

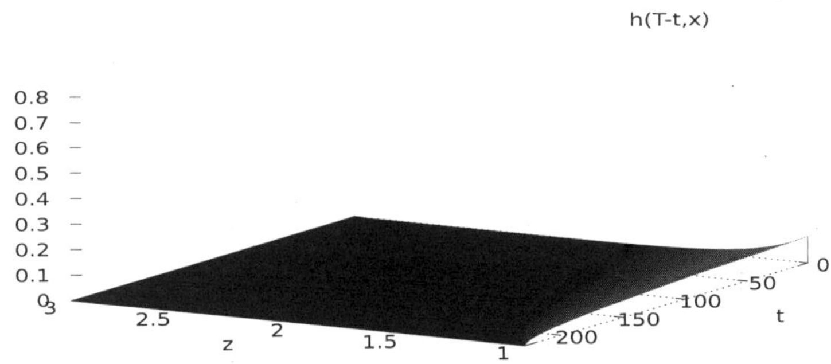

FIGURE 8.24: Function $h_c(\tau, z)$ in the decomposition (8.33) of the normalized lookback call option price $g(\tau, z)$.

We also note that

$$\begin{aligned}\mathbb{E}[M_0^T - m_0^T \mid S_0 = x] &= x - xe^{-r(T-t)}\Phi\left(\delta_-^{T-t}(1)\right) \\ &-x\left(1+\frac{\sigma^2}{2r}\right)\Phi\left(-\delta_+^{T-t}(1)\right) + xe^{-r(T-t)}\frac{\sigma^2}{2r}\Phi\left(\delta_-^{T-t}(1)\right) \\ &+xe^{-r(T-t)}\Phi\left(-\delta_-^{T-t}(1)\right) + x\left(1+\frac{\sigma^2}{2r}\right)\Phi\left(\delta_+^{T-t}(1)\right) \\ &-x\frac{\sigma^2}{2r}e^{-r(T-t)}\Phi\left(-\delta_-^{T-t}(1)\right) - x\end{aligned}$$

$$
\begin{aligned}
&= x\left(1+\frac{\sigma^2}{2r}\right)\left(\Phi\left(\delta_+^{T-t}(1)\right)-\Phi\left(-\delta_+^{T-t}(1)\right)\right)\\
&\quad +xe^{-r(T-t)}\left(\frac{\sigma^2}{2r}-1\right)\left(\Phi\left(\delta_-^{T-t}(1)\right)-\Phi\left(-\delta_-^{T-t}(1)\right)\right).
\end{aligned}
$$

Hedging of Lookback Options

In this section we compute hedging strategies for lookback options by application of the Delta hedging formula (8.30). See [3], § 2.6.1, page 29, for another approach to the following result using the Clark–Ocone formula. Here we use (8.30) instead, cf. Proposition 4.6 of [42].

Proposition 8.7 *The hedging strategy of the lookback call option is given by*

$$
\begin{aligned}
\xi_t &= \Phi\left(\delta_+^{T-t}\left(\frac{S_t}{m_0^t}\right)\right)-\frac{\sigma^2}{2r}\Phi\left(-\delta_+^{T-t}\left(\frac{S_t}{m_0^t}\right)\right) \qquad (8.34)\\
&\quad +e^{-r(T-t)}\left(\frac{m_0^t}{S_t}\right)^{2r/\sigma^2}\left(\frac{\sigma^2}{2r}-1\right)\Phi\left(\delta_-^{T-t}\left(\frac{m_0^t}{S_t}\right)\right), \qquad t\in[0,T].
\end{aligned}
$$

Proof. We need to differentiate

$$
f(t,x,y)=\mathrm{BS}_c(x,y,r,\sigma,T-t)+xh_c\left(T-t,\frac{x}{y}\right)
$$

with respect to the variable x, where

$$
h_c(\tau,z)=-\frac{\sigma^2}{2r}\left(\Phi\left(-\delta_+^\tau(z)\right)-e^{-r\tau}z^{-2r/\sigma^2}\Phi\left(\delta_-^\tau(1/z)\right)\right)
$$

is given by (8.28). First we note that the relation

$$
\frac{\partial}{\partial x}\mathrm{BS}_c(x,y,r,\sigma,\tau)=\Phi\left(\delta_+^\tau\left(\frac{x}{y}\right)\right)
$$

is known. Next, we have

$$
\frac{\partial}{\partial x}\left(xh_c\left(\tau,\frac{x}{y}\right)\right)=h_c\left(\tau,\frac{x}{y}\right)+\frac{x}{y}\frac{\partial h_c}{\partial z}\left(\tau,\frac{x}{y}\right),
$$

and

$$
\begin{aligned}
\frac{\partial h_c}{\partial z}(\tau,z) &= -\frac{\sigma^2}{2r}\left(\frac{\partial}{\partial x}\left(\Phi\left(-\delta_+^\tau(z)\right)\right)-e^{-r\tau}z^{-2r/\sigma^2}\frac{\partial}{\partial z}\left(\Phi\left(\delta_-^\tau\left(\frac{1}{z}\right)\right)\right)\right)\\
&\quad -\frac{\sigma^2}{2r}\left(\frac{2r}{\sigma^2}e^{-r\tau}z^{-1-2r/\sigma^2}\Phi\left(\delta_-^\tau\left(\frac{1}{z}\right)\right)\right)\\
&= \frac{\sigma}{2rz\sqrt{2\pi\tau}}\exp\left(-\frac{1}{2}\left(\delta_+^\tau(z)\right)^2\right)
\end{aligned}
$$

$$-e^{-r\tau}z^{-2r/\sigma^2}\frac{\sigma}{2rz\sqrt{2\pi\tau}}\exp\left(-\frac{1}{2}\left(\delta_-^\tau\left(\frac{1}{z}\right)\right)^2\right.$$
$$\left.+\frac{2r}{\sigma^2}e^{-r\tau}z^{-1-2r/\sigma^2}\Phi\left(\delta_-^\tau\left(\frac{1}{z}\right)\right)\right).$$

Next we note that

$$\begin{aligned}
e^{-\frac{1}{2}\left(\delta_-^\tau\left(\frac{1}{z}\right)\right)^2} &= \exp\left(-\frac{1}{2}\left(\delta_+^\tau(z)\right)^2-\frac{1}{2}\left(\frac{4r^2}{\sigma^2}\tau-\frac{4r}{\sigma}\delta_+^\tau(z)\sqrt{\tau}\right)\right)\\
&= e^{-\frac{1}{2}\left(\delta_+^\tau(z)\right)^2}\exp\left(-\frac{1}{2}\left(\frac{4r^2}{\sigma^2}\tau-\frac{4r}{\sigma^2}\left(\log z+(r+\frac{1}{2}\sigma^2)\tau\right)\right)\right)\\
&= e^{-\frac{1}{2}\left(\delta_+^\tau(z)\right)^2}\exp\left(\frac{-2r^2}{\sigma^2}\tau+\frac{2r}{\sigma^2}\log z+\frac{2r^2}{\sigma^2}\tau+r\tau\right)\\
&= e^{r\tau}z^{2r/\sigma^2}\exp\left(-\frac{1}{2}\left(\delta_+^\tau(z)\right)^2\right),
\end{aligned}$$

hence

$$\frac{\partial h_c}{\partial z}\left(\tau,\frac{x}{y}\right)=-e^{-r\tau}z^{-1-2r/\sigma^2}\Phi\left(\delta_-^\tau\left(\frac{1}{z}\right)\right),$$

and

$$\frac{\partial}{\partial x}\left(xh_c\left(\tau,\frac{x}{y}\right)\right)=h_c\left(\tau,\frac{x}{y}\right)-e^{-r\tau}\left(\frac{y}{x}\right)^{2r/\sigma^2}\Phi\left(\delta_-^\tau\left(\frac{y}{x}\right)\right),$$

which concludes the proof. □

Similar calculations using (8.21) can be carried out for other types of lookback options, such as options on extrema and partial lookback options, cf. [41].

As a consequence of (8.34) we have

$$\begin{aligned}
&e^{-r(T-t)}\,\mathbb{E}[S_T-m_0^T\mid\mathcal{F}_t]\\
&=S_t\Phi\left(\delta_+^{T-t}\left(\frac{S_t}{m_0^t}\right)\right)-m_0^te^{-r(T-t)}\Phi\left(\delta_-^{T-t}\left(\frac{S_t}{m_0^t}\right)\right)\\
&\quad+e^{-r(T-t)}S_t\frac{\sigma^2}{2r}\left(\frac{S_t}{m_0^t}\right)^{-2r/\sigma^2}\Phi\left(\delta_-^{T-t}\left(\frac{m_0^t}{S_t}\right)\right)-S_t\frac{\sigma^2}{2r}\Phi\left(-\delta_+^{T-t}\left(\frac{S_t}{m_0^t}\right)\right)\\
&=\xi_tS_t-m_0^te^{-r(T-t)}\left(\Phi\left(\delta_-^{T-t}\left(\frac{S_t}{m_0^t}\right)\right)+\left(\frac{S_t}{m_0^t}\right)^{1-2r/\sigma^2}\Phi\left(\delta_-^{T-t}\left(\frac{m_0^t}{S_t}\right)\right)\right),
\end{aligned}$$

and the quantity of the riskless asset e^{rt} in the portfolio is given by

$$\begin{aligned}
\eta_t &= -m_0^te^{-rT}\left(\Phi\left(\delta_-^{T-t}\left(\frac{S_t}{m_0^t}\right)\right)+\left(\frac{S_t}{m_0^t}\right)^{1-2r/\sigma^2}\Phi\left(\delta_-^{T-t}\left(\frac{m_0^t}{S_t}\right)\right)\right)\\
&\leq 0,
\end{aligned}$$

so that the portfolio value V_t at time t satisfies

$$V_t = \xi_t S_t + \eta_t e^{rt}, \qquad t \in \mathbb{R}_+,$$

and one has to constantly borrow from the riskless account in order to hedge the lookback option.

8.5 Asian Options

As we will see below there exists no easily tractable closed form solution for the price of an arithmetically averaged Asian option.

General Results

An option on average is an option whose payoff has the form

$$C = \phi(Y_T, S_T),$$

where

$$Y_T = S_0 \int_0^T e^{\sigma B_u + ru - \sigma^2 u/2} du = \int_0^T S_u du, \qquad T \in \mathbb{R}_+.$$

- For example when $\phi(y,x) = (y/T - K)^+$ this yields the Asian call option with payoff

$$\left(\frac{1}{T}\int_0^T S_u du - K\right)^+ = \left(\frac{Y_T}{T} - K\right)^+, \tag{8.35}$$

which is a path-dependent option whose price at time $t \in [0,T]$ is given by

$$e^{-r(T-t)}\,\mathbb{E}^*\left[\left(\frac{1}{T}\int_0^T S_u du - K\right)^+ \Big| \mathcal{F}_t\right]. \tag{8.36}$$

- As another example, when $\phi(y,x) = e^{-y}$ this yields the price

$$P(0,T) = \mathbb{E}^*\left[e^{-\int_0^T S_u du}\right] = \mathbb{E}^*\left[e^{-Y_T}\right]$$

at time 0 of a bond with underlying short term rate process S_t.

The option with payoff $C = \phi(Y_T, S_T)$ can be priced as

$$\begin{aligned} e^{-r(T-t)}\,\mathbb{E}^*[\phi(Y_T,S_T) \mid \mathcal{F}_t] &= e^{-r(T-t)}\,\mathbb{E}^*\left[\phi\left(Y_t + \int_t^T S_u du, S_T\right) \Big| \mathcal{F}_t\right] \\ &= e^{-r(T-t)}\,\mathbb{E}^*\left[\phi\left(y + x\int_t^T \frac{S_u}{S_t} du, x\frac{S_T}{S_t}\right) \Big| \mathcal{F}_t\right]_{y=Y_t, x=S_t} \end{aligned}$$

$$= e^{-r(T-t)} \mathbb{E}^*\left[\phi\left(y + x\int_0^{T-t}\frac{S_u}{S_0}du, x\frac{S_{T-t}}{S_0}\right)\right]_{y=Y_t, x=S_t}.$$

Hence the option can be priced as

$$f(t, S_t, Y_t) = e^{-r(T-t)} \mathbb{E}^*[\phi(Y_T, S_T) \mid \mathcal{F}_t],$$

where the function $f(t, x, y)$ is given by

$$f(t, x, y) = e^{-r(T-t)} \mathbb{E}^*\left[\phi\left(y + x\int_0^{T-t}\frac{S_u}{S_0}du, x\frac{S_{T-t}}{S_0}\right)\right].$$

First we note that the numerical computation of Asian option prices can be done using the joint probability density $\psi_{Y_{T-t},B_{T-t}}$ of (Y_{T-t}, B_{T-t}), as follows:

$$f(t, x, y) =$$
$$e^{-r(T-t)} \int_0^\infty \int_{-\infty}^\infty \phi\left(y + xz, xe^{\sigma u + r(T-t) - \sigma^2(T-t)/2}\right) \psi_{Y_{T-t},B_{T-t}}(z, u) dz du.$$

In [75], Proposition 2, the joint probability density of

$$(Y_\tau, B_\tau) = \left(\int_0^\tau S_0 e^{\sigma B_s - p\sigma^2 s/2} ds, B_\tau\right), \qquad \tau > 0,$$

has been computed in the case $\sigma = 2$, cf. also [48]. In the next proposition we restate this result for an arbitrary variance parameter σ after rescaling. Let $\theta(v, t)$ denote the function defined as

$$\theta(v, t) = \frac{ve^{\pi^2/(2t)}}{\sqrt{2\pi^3 t}} \int_0^\infty e^{-\xi^2/(2t)} e^{-v\cosh\xi} \sinh(\xi) \sin(\pi\xi/t) d\xi, \qquad v, t > 0. \tag{8.37}$$

Proposition 8.8 *For all $\tau > 0$ we have*

$$\mathbb{P}\left(\int_0^\tau e^{\sigma B_s - p\sigma^2 s/2} ds \in du, \ B_\tau \in dy\right)$$
$$= \frac{\sigma}{2} e^{-p\sigma y/2 - p^2\sigma^2\tau/8} \exp\left(-2\frac{1 + e^{\sigma y}}{\sigma^2 u}\right) \theta\left(\frac{4e^{\sigma y/2}}{\sigma^2 u}, \frac{\sigma^2\tau}{4}\right) \frac{du}{u} dy,$$

$u > 0$, $y \in \mathbb{R}$.

This probability density can then be used for the pricing of options on average, as

$$f(t, x, y) = e^{-r(T-t)} \mathbb{E}^*\left[\phi\left(y + x\int_0^{T-t}\frac{S_u}{S_0}du, x\frac{S_{T-t}}{S_0}\right)\right]$$
$$= e^{-r(T-t)}$$
$$\times \int_0^\infty \phi\left(y + xz, xe^{\sigma u + r(T-t) - \sigma^2(T-t)/2}\right) \mathbb{P}\left(\int_0^{T-t}\frac{S_r}{S_0}dr \in dz, B_{T-t} \in du\right)$$

$$= e^{-r(T-t)}\frac{\sigma}{2}e^{-p^2\sigma^2\tau/8}\int_0^\infty\int_{-\infty}^\infty \phi\left(y+xz, xe^{\sigma u+r(T-t)-\sigma^2(T-t)/2}\right)$$

$$\times \exp\left(-2\frac{1+e^{\sigma v}}{\sigma^2 z}-p\sigma\frac{v}{2}\right)\theta\left(\frac{4e^{\sigma v/2}}{\sigma^2 z},\frac{\sigma^2\tau}{4}\right)dv\frac{dz}{z},$$

which is actually a triple integral due to the definition (8.37) of $\theta(v,t)$. Note that here the order of integration between dv and dz cannot be exchanged without particular precautions, at the risk of wrong computations.

We refer to e.g., [2], [9], [20], and references therein for more on Asian option pricing using the probability density of the averaged geometric Brownian motion.

The Asian Call Option

We have

$$\begin{aligned}
& e^{-r(T-t)}\,\mathbb{E}^*\left[\left(\frac{1}{T}\int_0^T S_u du - K\right)^+ \Big| \mathcal{F}_t\right] \\
&= e^{-r(T-t)}\,\mathbb{E}^*\left[\left(\frac{1}{T}\left(Y_t+\int_t^T S_u du\right) - K\right)^+ \Big| \mathcal{F}_t\right] \\
&= e^{-r(T-t)}\,\mathbb{E}^*\left[\left(\frac{1}{T}\left(y+x\int_t^T \frac{S_u}{S_t} du\right) - K\right)^+ \Big| \mathcal{F}_t\right]_{x=S_t,\ y=Y_t} \\
&= e^{-r(T-t)}\,\mathbb{E}^*\left[\left(\frac{1}{T}\left(y+x\int_0^{T-t} \frac{S_u}{S_0} du\right) - K\right)^+ \right]_{x=S_t,\ y=Y_t}.
\end{aligned}$$

Hence the option can be priced as

$$f(t,S_t,Y_t) = e^{-r(T-t)}\,\mathbb{E}^*\left[\left(\frac{1}{T}\int_0^T S_u du - K\right)^+ \Big| \mathcal{F}_t\right],$$

where the function $f(t,x,y)$ is defined by

$$\begin{aligned}
f(t,x,y) &= e^{-r(T-t)}\,\mathbb{E}^*\left[\left(\frac{1}{T}\left(y+x\int_0^{T-t}\frac{S_u}{S_0}du\right)-K\right)^+\right] \\
&= e^{-r(T-t)}\,\mathbb{E}^*\left[\left(\frac{1}{T}\left(y+\frac{x}{S_0}Y_{T-t}\right)-K\right)^+\right].
\end{aligned}$$

Probabilistic Approach

First we note that the numerical computation of Asian option prices can be done using the probability density of

$$Y_T = \int_0^T S_t dt.$$

From Proposition 8.8 we deduce the marginal density of

$$Y_\tau = \int_0^\tau e^{\sigma B_s - p\sigma^2 s/2} ds,$$

as follows:

$$\begin{aligned}
&\mathbb{P}\left(\int_0^\tau e^{\sigma B_s - p\sigma^2 s/2} ds \in du\right) \\
&= \frac{\sigma}{2u} e^{-p^2\sigma^2\tau/8} \int_{-\infty}^{\infty} \exp\left(-2\frac{1+e^{\sigma v}}{\sigma^2 u} - p\sigma\frac{v}{2}\right) \theta\left(\frac{4e^{\sigma v/2}}{\sigma^2 u}, \frac{\sigma^2\tau}{4}\right) dvdu,
\end{aligned}$$

$u > 0$. From this we get

$$\begin{aligned}
&\mathbb{P}(Y_t/S_0 \in du) = \mathbb{P}\left(\int_0^\tau S_t dt \in du\right) \\
&= \frac{\sigma}{2u} e^{-p^2\sigma^2\tau/8} \int_{-\infty}^{\infty} \exp\left(-2\frac{1+e^{\sigma v}}{\sigma^2 u} - p\sigma\frac{v}{2}\right) \theta\left(\frac{4e^{\sigma v/2}}{\sigma^2 u}, \frac{\sigma^2\tau}{4}\right) dvdu,
\end{aligned} \tag{8.38}$$

where $S_t = S_0 e^{\sigma B_t - p\sigma^2 t/2}$ and

$$p = 1 - \frac{2r}{\sigma^2}.$$

This probability density can then be used for the pricing of Asian options, as

$$\begin{aligned}
f(t,x,y) &= e^{-r(T-t)} \mathbb{E}^*\left[\left(\frac{1}{T}\left(y + \frac{x}{S_0} Y_{T-t} du\right) - K\right)^+\right] \\
&= e^{-r(T-t)} \int_0^\infty \left(\frac{y+xz}{T} - K\right)^+ \mathbb{P}(Y_{T-t}/S_0 \in dz) \\
&= e^{-r(T-t)} \frac{\sigma}{2} e^{-p^2\sigma^2\tau/8} \int_0^\infty \int_{-\infty}^\infty \left(\frac{y+xz}{T} - K\right)^+ \\
&\quad \times \exp\left(-2\frac{1+e^{\sigma v}}{\sigma^2 z} - p\sigma\frac{v}{2}\right) \theta\left(\frac{4e^{\sigma v/2}}{\sigma^2 z}, \frac{\sigma^2\tau}{4}\right) dv \frac{dz}{z} \\
&= \frac{\sigma}{2T} e^{-p^2\sigma^2\tau/8 - r(T-t)} \int_{0 \vee (KT-y)/x}^\infty \int_{-\infty}^\infty (xz + y - KT) \\
&\quad \times \exp\left(-2\frac{1+e^{\sigma v}}{\sigma^2 z} - p\sigma\frac{v}{2}\right) \theta\left(\frac{4e^{\sigma v/2}}{\sigma^2 z}, \frac{\sigma^2\tau}{4}\right) dv \frac{dz}{z},
\end{aligned}$$

which is actually a triple integral due to the definition (8.37) of $\theta(v,t)$. Note that here the order of integration between dv and dz cannot be exchanged without particular precautions, at the risk of wrong computations.

The time Laplace transform of Asian option prices has been computed in [28], and this expression can be used for pricing by numerical inversion of the

Laplace transform. The following Figure 8.25 represents Asian option prices computed by the Geman-Yor [28] method.

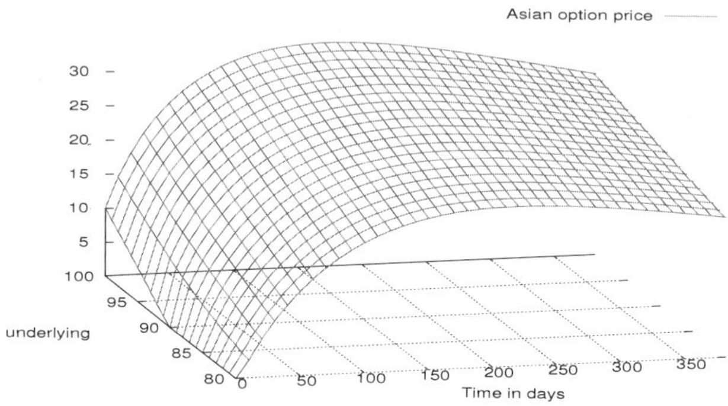

FIGURE 8.25: Graph of the Asian option price with $\sigma = 0.3$, $r = 0.1$ and $K = 90$

Other numerical approaches to the pricing of Asian options include [70] which relies on approximations of the average price probability based on the Lognormal distribution.

We refer to Figure 7.1 for a graph of implied volatility surface for Asian options on light sweet crude oil futures.

PDE Method – Two Variables

The price at time t of the Asian option with payoff (8.35) can be written as

$$f(t, S_t, Y_t) = e^{-r(T-t)}\, \mathbb{E}^*\left[\left(\frac{1}{T}\int_0^T S_u du - K\right)^+ \Big| \mathcal{F}_t\right], \qquad t \in [0, T]. \quad (8.39)$$

Next, we derive the Black–Scholes partial differential equation (PDE) for the price of a self-financing portfolio.

Proposition 8.9 *Let $(\eta_t, \xi_t)_{t\in\mathbb{R}_+}$ be a portfolio strategy such that*

(i) $(\eta_t, \xi_t)_{t\in\mathbb{R}_+}$ is self-financing,

(ii) the value $V_t := \eta_t A_t + \xi_t S_t$, $t \in \mathbb{R}_+$, takes the form

$$V_t = f(t, Y_t, S_t), \qquad t \in \mathbb{R}_+,$$

for some $f \in \mathcal{C}^2((0,\infty) \times (0,\infty)^2)$.

Then the function $f(t, x, y)$ in (8.39) satisfies the PDE

$$rf(t,x,y) = \frac{\partial f}{\partial t}(t,x,y) + x\frac{\partial f}{\partial y}(t,x,y) + rx\frac{\partial f}{\partial x}(t,x,y) + \frac{1}{2}x^2\sigma^2\frac{\partial^2 f}{\partial x^2}(t,x,y),$$

$t, x > 0$, under the boundary conditions

$$\begin{cases} f(t,0,y) = e^{-r(T-t)}\left(\frac{y}{T} - K\right)^+, & 0 \le t \le T, \quad y \in \mathbb{R}_+, \quad (8.40a) \\ \lim_{y\to-\infty} f(t,x,y) = 0, & 0 \le t \le T, \quad x \in \mathbb{R}_+, \quad (8.40b) \\ f(T,x,y) = \left(\frac{y}{T} - K\right)^+, & x, y \in \mathbb{R}_+, \quad (8.40c) \end{cases}$$

and ξ_t is given by

$$\xi_t = \frac{\partial f}{\partial x}(t, S_t, Y_t), \qquad t \in \mathbb{R}_+.$$

Proof. We note that the self-financing condition implies

$$\begin{aligned} dV_t &= \eta_t dA_t + \xi_t dS_t \\ &= r\eta_t A_t dt + \mu\xi_t S_t dt + \sigma\xi_t S_t dB_t \qquad (8.41) \\ &= rV_t dt + (\mu - r)\xi_t S_t dt + \sigma\xi_t S_t dB_t \\ &= r\eta_t A_t dt + \mu\xi_t S_t dt + \sigma\xi_t S_t dB_t, \qquad (8.42) \end{aligned}$$

$t \in \mathbb{R}_+$. Noting that $dY_t = S_t dt$, the application of Itô's formula to $f(t,x,y)$ leads to

$$df(t,S_t,Y_t) = \frac{\partial f}{\partial t}(t,S_t,Y_t)dt + S_t\frac{\partial f}{\partial y}(t,S_t,Y_t)dt$$
$$+\mu S_t\frac{\partial f}{\partial x}(t,S_t,Y_t)dt + \frac{1}{2}S_t^2\sigma^2\frac{\partial^2 f}{\partial x^2}(t,S_t,Y_t)dt + \sigma S_t\frac{\partial f}{\partial x}(t,S_t,Y_t)dB_t. \qquad (8.43)$$

By respective identification of the terms in dB_t and dt in (8.41) and (8.43) we get

$$\begin{cases} r\eta_t A_t dt + \mu\xi_t S_t dt = \frac{\partial f}{\partial t}(t,S_t,Y_t)dt + S_t\frac{\partial f}{\partial y}(t,S_t,Y_t)dt + \mu S_t\frac{\partial f}{\partial x}(t,S_t,Y_t)dt \\ \qquad\qquad + \frac{1}{2}S_t^2\sigma^2\frac{\partial^2 f}{\partial x^2}(t,S_t,Y_t)dt, \\ \xi_t S_t \sigma dB_t = S_t\sigma\frac{\partial f}{\partial x}(t,S_t,Y_t)dB_t, \end{cases}$$

hence

$$\begin{cases} rV_t - r\xi_t S_t = \dfrac{\partial f}{\partial t}(t, S_t, Y_t) + S_t \dfrac{\partial f}{\partial y}(t, S_t, Y_t)dt + \dfrac{1}{2} S_t^2 \sigma^2 \dfrac{\partial^2 f}{\partial x^2}(t, S_t, Y_t), \\ \xi_t = \dfrac{\partial f}{\partial x}(t, S_t, Y_t), \end{cases}$$

i.e.,

$$\begin{cases} rf(t, S_t, Y_t) = & \dfrac{\partial f}{\partial t}(t, S_t, Y_t) + S_t \dfrac{\partial f}{\partial y}(t, S_t, Y_t) + rS_t \dfrac{\partial f}{\partial x}(t, S_t, Y_t) \\ & + \dfrac{1}{2} S_t^2 \sigma^2 \dfrac{\partial^2 f}{\partial x^2}(t, S_t, Y_t), \\ \xi_t = \dfrac{\partial f}{\partial x}(t, S_t, Y_t). \end{cases}$$

□

Next we examine two methods which allow one to reduce the Asian option pricing PDE from two variables to one variable.

PDE Method - One Variable (1) - Time Independent Coefficients

Following [44], page 91, we define the auxiliary process

$$Z_t = \frac{1}{S_t}\left(\frac{1}{T}\int_0^t S_u du - K\right) = \frac{1}{S_t}\left(\frac{Y_t}{T} - K\right), \qquad t \in [0, T].$$

With this notation, the price of the Asian option at time t becomes

$$e^{-r(T-t)} \mathbb{E}^* \left[\left(\frac{1}{T} \int_0^T S_u du - K \right)^+ \Big| \mathcal{F}_t \right] = e^{-r(T-t)} \mathbb{E}^* \left[S_T (Z_T)^+ \mid \mathcal{F}_t \right].$$

Lemma 8.1 *The price* (8.36) *at time t of the Asian option with payoff* (8.35) *can be written as*

$$S_t g(t, Z_t), \qquad t \in [0, T],$$

where

$$\begin{aligned} g(t, z) &= e^{-r(T-t)} \mathbb{E}^* \left[\left(z + \frac{1}{T} \int_0^{T-t} \frac{S_u}{S_0} du \right)^+ \right] \qquad (8.44) \\ &= e^{-r(T-t)} \mathbb{E}^* \left[\left(z + \frac{Y_{T-t}}{S_0 T} \right)^+ \right]. \end{aligned}$$

Proof. For $0 \le s \le t \le T$, we have

$$d\left(S_t Z_t\right) = \frac{1}{T} d\left(\int_0^t S_u du - K\right) = \frac{S_t}{T} dt,$$

hence

$$\frac{S_t Z_t}{S_s} = Z_s + \frac{1}{T}\int_s^t \frac{S_u}{S_s} du, \qquad t \ge s.$$

Since for any $t \in [0, T]$, S_t is positive and $\mathcal{F}_t$-measurable, and S_u/S_t is independent of $\mathcal{F}_t$, $u \ge t$, we have:

$$\begin{aligned}
e^{-r(T-t)}\,\mathbb{E}^*\left[S_T(Z_T)^+ \mid \mathcal{F}_t\right] &= e^{-r(T-t)} S_t\,\mathbb{E}^*\left[\left(\frac{S_T}{S_t} Z_T\right)^+ \Big| \mathcal{F}_t\right] \\
&= e^{-r(T-t)} S_t\,\mathbb{E}^*\left[\left(Z_t + \frac{1}{T}\int_t^T \frac{S_u}{S_t} du\right)^+ \Big| \mathcal{F}_t\right] \\
&= e^{-r(T-t)} S_t\,\mathbb{E}^*\left[\left(z + \frac{1}{T}\int_t^T \frac{S_u}{S_t} du\right)^+ \Big| \mathcal{F}_t\right]_{z=Z_t} \\
&= e^{-r(T-t)} S_t\,\mathbb{E}^*\left[\left(z + \frac{1}{T}\int_0^{T-t} \frac{S_u}{S_0} du\right)^+\right]_{z=Z_t} \\
&= e^{-r(T-t)} S_t\,\mathbb{E}^*\left[\left(z + \frac{Y_{T-t}}{S_0 T}\right)^+\right]_{z=Z_t} \\
&= S_t g(t, Z_t),
\end{aligned}$$

which proves (8.44). □

Note that $g(t, z)$ can be computed from the density of Y_{T-t} as

$$\begin{aligned}
g(t,z) &= \mathbb{E}^*\left[\left(z + \frac{Y_{T-t}}{S_0 T}\right)^+\right] \\
&= \int_0^\infty \left(z + \frac{u}{T}\right)^+ \mathbb{P}(Y_t/S_0 \in du) \\
&= e^{-p^2\sigma^2\tau/8} \\
&\times \int_0^\infty \left(z + \frac{u}{T}\right)^+ \frac{\sigma}{2u}\int_{-\infty}^\infty \exp\left(-2\frac{1+e^{\sigma v}}{\sigma^2 u} - p\sigma\frac{v}{2}\right)\theta\left(\frac{4e^{\sigma v/2}}{\sigma^2 u}, \frac{\sigma^2\tau}{4}\right) dv du \\
&= e^{-p^2\sigma^2\tau/8} \\
&\times \int_{(-zT)\vee 0}^\infty \left(z + \frac{u}{T}\right) \frac{\sigma}{2u}\int_{-\infty}^\infty \exp\left(-2\frac{1+e^{\sigma v}}{\sigma^2 u} - p\sigma\frac{v}{2}\right)\theta\left(\frac{4e^{\sigma v/2}}{\sigma^2 u}, \frac{\sigma^2\tau}{4}\right) dv du \\
&= \frac{z\sigma}{2} e^{-p^2\sigma^2\tau/8}\int_{(-zT)\vee 0}^\infty \int_{-\infty}^\infty \exp\left(-2\frac{1+e^{\sigma v}}{\sigma^2 u} - p\sigma\frac{v}{2}\right)\theta\left(\frac{4e^{\sigma v/2}}{\sigma^2 u}, \frac{\sigma^2\tau}{4}\right) dv\frac{du}{u} \\
&+ \frac{\sigma}{2T} e^{-p^2\sigma^2\tau/8}\int_{(-zT)\vee 0}^\infty \int_{-\infty}^\infty \exp\left(-2\frac{1+e^{\sigma v}}{\sigma^2 u} - p\sigma\frac{v}{2}\right)\theta\left(\frac{4e^{\sigma v/2}}{\sigma^2 u}, \frac{\sigma^2\tau}{4}\right) dv du.
\end{aligned}$$

The next proposition gives a replicating hedging strategy for Asian options.

Proposition 8.10 *Let $(\eta_t, \xi_t)_{t\in\mathbb{R}_+}$ be a portfolio strategy such that*

(i) $(\eta_t, \xi_t)_{t\in\mathbb{R}_+}$ is self-financing,

(ii) the value $V_t := \eta_t A_t + \xi_t S_t$, $t \in \mathbb{R}_+$, takes the form

$$V_t = S_t g(t, Z_t), \qquad t \in \mathbb{R}_+,$$

for some $f \in \mathcal{C}^2((0,\infty) \times (0,\infty)^2)$.

Then the function $g(t,x)$ satisfies the PDE

$$\frac{\partial g}{\partial t}(t,z) + \left(\frac{1}{T} - rz\right)\frac{\partial g}{\partial z}(t,z) + \frac{1}{2}\sigma^2 z^2 \frac{\partial^2 g}{\partial z^2}(t,z) = 0, \tag{8.45}$$

under the terminal condition

$$g(T,z) = z^+,$$

and the corresponding replicating portfolio is given by

$$\xi_t = g(t,Z_t) - Z_t \frac{\partial g}{\partial z} g(t,Z_t), \qquad t \in [0,T].$$

Proof. We proceed as in [63]. From the expression of $1/S_t$ we have

$$d\left(\frac{1}{S_t}\right) = \frac{1}{S_t}\left(\left(-\mu + \sigma^2\right)dt - \sigma dB_t\right),$$

hence

$$\begin{aligned}
dZ_t &= d\left(\frac{1}{S_t}\left(\frac{Y_t}{T} - K\right)\right) \\
&= d\left(\frac{Y_t}{TS_t} - \frac{K}{S_t}\right) \\
&= \frac{1}{T} d\left(\frac{Y_t}{S_t}\right) - K d\left(\frac{1}{S_t}\right) \\
&= \frac{1}{T}\frac{dY_t}{S_t} + \left(\frac{Y_t}{T} - K\right) d\left(\frac{1}{S_t}\right) \\
&= \frac{dt}{T} + S_t Z_t d\left(\frac{1}{S_t}\right) \\
&= \frac{dt}{T} + Z_t\left(-\mu + \sigma^2\right)dt - Z_t \sigma dB_t.
\end{aligned}$$

By self-financing we have

$$\begin{aligned} dV_t &= \eta_t dA_t + \xi_t dS_t \\ &= r\eta_t A_t dt + \mu\xi_t S_t dt + \sigma\xi_t S_t dB_t, \end{aligned} \tag{8.46}$$

$t \in \mathbb{R}_+$. The application of Itô's formula to $f(t,x,y)$ leads to

$$\begin{aligned} &d(S_t g(t,Z_t)) = g(t,Z_t)dS_t + S_t dg(t,Z_t) + dS_t \cdot dg(t,Z_t) \\ &= \frac{\partial g}{\partial t}(t,Z_t)dt + \frac{\partial g}{\partial z}(t,Z_t)dZ_t \\ &\quad + \frac{1}{2}\frac{\partial^2 g}{\partial z^2}(t,Z_t)(dZ_t)^2 + dS_t \cdot dg(t,Z_t) \\ &= S_t\frac{\partial g}{\partial t}(t,Z_t)dt + \mu S_t g(t,Z_t)dt + \sigma S_t g(t,Z_t)dB_t \\ &\quad + S_t Z_t\left(-\mu+\sigma^2\right)\frac{\partial g}{\partial z}(t,Z_t)dt + \frac{1}{T}S_t\frac{\partial g}{\partial z}(t,Z_t)dt - \sigma S_t Z_t\frac{\partial g}{\partial z}(t,Z_t)dB_t \\ &\quad + \frac{1}{2}\sigma^2 Z_t^2 S_t\frac{\partial^2 g}{\partial z^2}(t,Z_t)dt - \sigma^2 S_t Z_t\frac{\partial g}{\partial z}(t,Z_t)dt \\ &= \mu S_t g(t,Z_t)dt + S_t\frac{\partial g}{\partial t}(t,Z_t)dt + S_t Z_t\left(-\mu+\sigma^2\right)\frac{\partial g}{\partial z}(t,Z_t)dt \\ &\quad + \frac{1}{T}S_t\frac{\partial g}{\partial z}(t,Z_t)dt + \frac{1}{2}\sigma^2 Z_t^2 S_t\frac{\partial^2 g}{\partial z^2}(t,Z_t)dt - \sigma^2 S_t Z_t\frac{\partial g}{\partial z}(t,Z_t)dt \\ &\quad + \sigma S_t g(t,Z_t)dB_t - \sigma S_t Z_t\frac{\partial g}{\partial z}(t,Z_t)dB_t. \end{aligned}$$

By respective identification of the terms in dB_t and dt in (8.46) and (8.43) we get

$$\begin{cases} r\eta_t A_t + \mu\xi_t S_t = \mu S_t g(t,Z_t) + S_t\dfrac{\partial g}{\partial t}(t,Z_t) - \mu S_t Z_t\dfrac{\partial g}{\partial z}(t,Z_t) \\ \qquad + \dfrac{1}{T}S_t\dfrac{\partial g}{\partial z}(t,Z_t) + \dfrac{1}{2}\sigma^2 Z_t^2 S_t\dfrac{\partial^2 g}{\partial z^2}(t,Z_t), \\ \\ \xi_t S_t\sigma = \sigma S_t g(t,Z_t) - \sigma S_t Z_t\dfrac{\partial g}{\partial z}(t,Z_t), \end{cases}$$

hence

$$\begin{cases} rV_t - r\xi_t S_t = S_t\dfrac{\partial g}{\partial t}(t,Z_t) + \dfrac{1}{T}S_t\dfrac{\partial g}{\partial z}(t,Z_t) + \dfrac{1}{2}\sigma^2 Z_t^2 S_t\dfrac{\partial^2 g}{\partial z^2}(t,Z_t), \\ \\ \xi_t = g(t,Z_t) - Z_t\dfrac{\partial g}{\partial z}(t,Z_t), \end{cases}$$

i.e.,

$$\begin{cases} \dfrac{\partial g}{\partial t}(t,z) + \left(\dfrac{1}{T} - rz\right)\dfrac{\partial g}{\partial z}(t,z) + \dfrac{1}{2}\sigma^2 z^2\dfrac{\partial^2 g}{\partial z^2}(t,z) = 0, \\ \\ \xi_t = g(t,Z_t) - Z_t\dfrac{\partial g}{\partial z}(t,Z_t), \end{cases}$$

under the terminal condition

$$g(T, z) = z^+.$$

$\square$

We check that

$$\begin{aligned}
\xi_t &= e^{-r(T-t)}\sigma S_t \frac{\partial f}{\partial x} f(t, S_t, Z_t) - \sigma Z_t \frac{\partial f}{\partial z} f(t, S_t, Z_t) \\
&= e^{-r(T-t)}\left(-Z_t \frac{\partial g}{\partial z}(t, Z_t) + g(t, Z_t)\right) \\
&= e^{-r(T-t)}\left(S_t \frac{\partial g}{\partial x}\left(t, \frac{1}{x}\left(\frac{1}{T}\int_0^t S_u du - K\right)\right)_{|x=S_t} + g(t, Z_t)\right) \\
&= \frac{\partial}{\partial x}\left(xe^{-r(T-t)} g\left(t, \frac{1}{x}\left(\frac{1}{T}\int_0^t S_u du - K\right)\right)\right)_{|x=S_t}, \quad t \in [0, T].
\end{aligned}$$

We also find that the amount invested on the riskless asset is given by

$$\eta_t A_t = Z_t S_t \frac{\partial g}{\partial z}(t, Z_t).$$

Next we note that a PDE with no first order derivative term can be obtained using time-dependent coefficients.

PDE Method - One Variable (2) - Time Dependent Coefficients

Define now the auxiliary process

$$\begin{aligned}
U_t &:= \frac{1}{rT}(1 - e^{-r(T-t)}) + e^{-r(T-t)} \frac{1}{S_t}\left(\frac{1}{T}\int_0^t S_u du - K\right) \\
&= \frac{1}{rT}(1 - e^{-r(T-t)}) + e^{-r(T-t)} Z_t, \qquad t \in [0, T],
\end{aligned}$$

i.e.,

$$Z_t = e^{r(T-t)} U_t + \frac{e^{r(T-t)} - 1}{rT}, \qquad t \in [0, T].$$

We have

$$\begin{aligned}
dU_t &= -\frac{1}{T} e^{-r(T-t)} dt + re^{-r(T-t)} Z_t dt + e^{-r(T-t)} dZ_t \\
&= e^{-r(T-t)}\sigma^2 Z_t dt - e^{-r(T-t)}\sigma Z_t dB_t - (\mu - r)e^{-r(T-t)} Z_t dt \\
&= -e^{-r(T-t)}\sigma Z_t d\hat{B}_t, \qquad t \in \mathbb{R}_+,
\end{aligned}$$

where

$$d\hat{B}_t = dB_t - \sigma dt + \frac{\mu - r}{\sigma} dt = d\tilde{B}_t - \sigma dt$$

is a standard Brownian motion under

$$d\hat{\mathbb{P}} = e^{\sigma B_T - \sigma^2 t/2} d\mathbb{P}^* = e^{-rT}\frac{S_T}{S_0} d\mathbb{P}^*.$$

Lemma 8.2 *The Asian option price can be written as*

$$S_t h(t, U_t) = e^{-r(T-t)} \mathbb{E}^* \left[\left(\frac{1}{T} \int_0^T S_u du - K \right)^+ \Big| \mathcal{F}_t \right],$$

where the function $h(t, y)$ is given by

$$h(t, y) = \hat{\mathbb{E}} \left[(U_T)^+ \mid y = U_t \right], \qquad 0 \leq t \leq T.$$

Proof. We have

$$U_T = \frac{1}{S_T} \left(\frac{1}{T} \int_0^T S_u du - K \right) = Z_T,$$

and

$$\frac{d\hat{\mathbb{P}}_{|\mathcal{F}_t}}{d\mathbb{P}^*_{|\mathcal{F}_t}} = e^{\sigma(B_T - B_t) - \sigma^2 (T-t)/2} = \frac{e^{-rT} S_T}{e^{-rt} S_t},$$

hence the price of the Asian option is

$$\begin{aligned} e^{-r(T-t)} \mathbb{E}^* \left[S_T (Z_T)^+ \mid \mathcal{F}_t \right] &= e^{-r(T-t)} \mathbb{E}^* \left[S_T (U_T)^+ \mid \mathcal{F}_t \right] \\ &= S_t \mathbb{E}^* \left[\frac{e^{-rT} S_T}{e^{-rt} S_t} (U_T)^+ \mid \mathcal{F}_t \right] \\ &= S_t \mathbb{E}^* \left[\frac{d\hat{\mathbb{P}}_{|\mathcal{F}_t}}{d\mathbb{P}^*_{|\mathcal{F}_t}} (U_T)^+ \mid \mathcal{F}_t \right] \\ &= S_t \hat{\mathbb{E}} \left[(U_T)^+ \mid \mathcal{F}_t \right]. \end{aligned}$$

□

The next proposition gives a replicating hedging strategy for Asian options. See § 7.5.3 of [68] and references therein for a different derivation of the PDE (8.47).

Proposition 8.11 *Let $(\eta_t, \xi_t)_{t \in \mathbb{R}_+}$ be a portfolio strategy such that*

(i) $(\eta_t, \xi_t)_{t \in \mathbb{R}_+}$ is self-financing,

(ii) the value $V_t := \eta_t A_t + \xi_t S_t$, $t \in \mathbb{R}_+$, takes the form

$$V_t = S_t h(t, U_t) \qquad t \in \mathbb{R}_+,$$

for some $f \in \mathcal{C}^2((0, \infty) \times (0, \infty)^2)$.

Then the function $h(t, z)$ satisfies the PDE

$$\frac{\partial h}{\partial t}(t,y) + \frac{1}{2}\sigma^2 \left(\frac{1-e^{-r(T-t)}}{rT} - y \right)^2 \frac{\partial^2 h}{\partial y^2}(t,y) = 0, \tag{8.47}$$

under the terminal condition

$$h(T,z) = z^+,$$

and the corresponding replicating portfolio is given by

$$\xi_t = h(t,U_t) - Z_t \frac{\partial h}{\partial y}(t,U_t), \qquad t \in [0,T].$$

Proof. By the self-financing condition (8.42) we have

$$dV_t = rV_t dt + (\mu - r)\xi_t S_t dt + \sigma \xi_t S_t dB_t, \tag{8.48}$$

$t \in \mathbb{R}_+$. By Itô's formula we get

$$\begin{aligned}
d(S_t h(t,U_t)) &= h(t,U_t)dS_t + S_t dh(t,U_t) + dS_t \cdot dh(t,U_t) \\
&= \mu S_t h(t,U_t)dt + \sigma S_t h(t,U_t) dB_t \\
&\quad + S_t \left(\frac{\partial h}{\partial t}(t,U_t)dt + \frac{\partial h}{\partial y}(t,U_t)dU_t + \frac{1}{2}\frac{\partial^2 h}{\partial y^2}(t,U_t)(dU_t)^2 \right) \\
&\quad + \frac{\partial h}{\partial y}(t,U_t) dS_t \cdot dU_t \\
&= \mu S_t h(t,U_t)dt + \sigma S_t h(t,U_t) dB_t - S_t(\mu - r)\frac{\partial h}{\partial y}(t,U_t) Z_t dt \\
&\quad + S_t \left(\frac{\partial h}{\partial t}(t,U_t)dt - \sigma \frac{\partial h}{\partial y}(t,U_t) Z_t d\tilde{B}_t + \frac{1}{2}\sigma^2 \frac{\partial^2 h}{\partial y^2}(t,U_t) Z_t^2 dt \right) \\
&\quad - \sigma^2 S_t \frac{\partial h}{\partial y}(t,U_t) Z_t dt \\
&= \mu S_t h(t,U_t)dt + \sigma S_t h(t,U_t) dB_t - S_t(\mu - r)\frac{\partial h}{\partial y}(t,U_t) Z_t dt \\
&\quad + S_t \left(\frac{\partial h}{\partial t}(t,U_t)dt - \sigma \frac{\partial h}{\partial y}(t,U_t) Z_t (dB_t - \sigma dt) + \frac{1}{2}\sigma^2 \frac{\partial^2 h}{\partial y^2}(t,U_t) Z_t^2 dt \right) \\
&\quad - \sigma^2 S_t \frac{\partial h}{\partial y}(t,U_t) Z_t dt.
\end{aligned}$$

By respective identification of the terms in dB_t and dt in (8.48) and (8.43) we get

$$\begin{cases} r\eta_t A_t + \mu\xi_t S_t & = \mu S_t h(t,U_t) - (\mu - r) S_t Z_t \frac{\partial h}{\partial y}(t,U_t)dt + S_t \frac{\partial h}{\partial t}(t,U_t) \\ & + \frac{1}{2} S_t \sigma^2 Z_t^2 \frac{\partial^2 h}{\partial y^2}(t,U_t), \\ \\ \xi_t = h(t,U_t) - Z_t \frac{\partial h}{\partial y}(t,U_t), \end{cases}$$

hence

$$\begin{cases} r\eta_t A_t = -rS_t(\xi_t - h(t,U_t)) + S_t \frac{\partial h}{\partial t}(t,U_t) + \frac{1}{2} S_t \sigma^2 \frac{\partial^2 h}{\partial y^2}(t,U_t) Z_t^2, \\ \\ \xi_t = h(t,U_t) - Z_t \frac{\partial h}{\partial y}(t,U_t), \end{cases}$$

and

$$\begin{cases} \frac{\partial h}{\partial t}(t,y) + \frac{1}{2}\sigma^2 \left(\frac{1 - e^{-r(T-t)}}{rT} - y \right)^2 \frac{\partial^2 h}{\partial y^2}(t,y) = 0, \\ \\ \xi_t = h(t,U_t) + \left(\frac{1 - e^{-r(T-t)}}{rT} - U_t \right) \frac{\partial h}{\partial y}(t,U_t), \end{cases}$$

under the terminal condition

$$h(T,z) = z^+.$$

□

We also find

$$\eta_t A_t = e^{r(T-t)} S_t \left(U_t - \frac{1 - e^{-r(T-t)}}{rT} \right) \frac{\partial h}{\partial y}(t,U_t) = S_t Z_t \frac{\partial h}{\partial y}(t,U_t).$$

Exercises

Exercise 8.1 Recall that the maximum $X_t := \sup_{s\in[0,t]} B_s$ over $[0,t]$ of standard Brownian motion $(B_s)_{s\in[0,t]}$ has the probability density

$$\varphi_{X_t}(x) = \sqrt{\frac{2}{\pi t}} e^{-x^2/(2t)} \mathbf{1}_{[0,\infty)}(x), \qquad x \in \mathbb{R}.$$

1. Let $\tau_a = \inf\{s \in \mathbb{R}_+ : B_s = a\}$ denote the first hitting time of $a > 0$ by $(B_s)_{s\in\mathbb{R}_+}$. Using the relation between $\{\tau_a \le t\}$ and $\{X_t \ge a\}$, write down the probability $P(\tau_a \le t)$ as an integral from a to ∞.

2. Using integration by parts on $[a, \infty)$, compute the probability density of τ_a.

   Hint: the derivative of $e^{-x^2/(2t)}$ with respect to x is $-xe^{-x^2/(2t)}/t$.

3. Compute the mean value $E[(\tau_a)^{-2}]$ of $1/\tau_a^2$.

Exercise 8.2 Barrier options.

1. Compute the hedging strategy of the up-and-out barrier call option on the underlying asset S_t with exercise date T, strike K and barrier B, with $B > K$.

2. Compute the joint probability density

$$f_{Y_T,B_T}(a,b) = \frac{d\mathbb{P}(Y_T \le a \;\&\; B_T \le b)}{dadb}, \qquad a, b \in \mathbb{R},$$

   of standard Brownian motion B_T and its *minimum*

$$Y_T = \min_{t\in[0,T]} B_t.$$

3. Compute the joint probability density

$$f_{\tilde{Y}_T,\tilde{B}_T}(a,b) = \frac{d\mathbb{P}(\tilde{Y}_T \le a \;\&\; \tilde{B}_T \le b)}{dadb}, \qquad a, b \in \mathbb{R},$$

   of *drifted* Brownian motion $\tilde{B}_T = B_T + \mu T$ and its *minimum*

$$\tilde{Y}_T = \min_{t\in[0,T]} \tilde{B}_t = \min_{t\in[0,T]} (B_t + \mu t).$$

4. Compute the price at time $t \in [0, T]$ of the down-and-out barrier call option on the underlying asset S_t with exercise date T, strike K, barrier B, and payoff

$$C = (S_T - K)^+ \mathbf{1}_{\left\{\min\limits_{0\le t\le T} S_t > B\right\}} = \begin{cases} S_T - K & \text{if } \min\limits_{0\le t\le T} S_t > B, \\ 0 & \text{if } \min\limits_{0\le t\le T} S_t \le B, \end{cases}$$

   in cases $0 < B < K$ and $B > K$.

Exercise 8.3 Lookback options. Compute the hedging strategy of the lookback **put** option.

Exercise 8.4 Asian call options with *negative* strike. Consider the asset price process

$$S_t = S_0 e^{rt+\sigma B_t-\sigma^2 t/2}, \qquad t \in \mathbb{R}_+,$$

where $(B_t)_{t\in\mathbb{R}_+}$ is a standard Brownian motion. Assuming that $\kappa \leq 0$, compute the price

$$e^{-r(T-t)} \mathbb{E}\left[\left(\frac{1}{T}\int_0^T S_u du - \kappa \right)^+ \Big| \mathcal{F}_t \right]$$

of the Asian option at time $t \in [0, T]$.

Exercise 8.5 Pricing of Asian options by PDEs. Show that the functions $g(t, z)$ and $h(t, y)$ are linked by the relation

$$g(t, z) = h\left(t, \frac{1 - e^{-r(T-t)}}{rT} + e^{-r(T-t)} z\right), \qquad t \in [0, T], \quad z > 0,$$

and that the PDE (1.35) for $h(t, y)$ can be derived from the PDE (1.33) for $g(t, z)$ and the above relation.

Chapter 9

American Options

In contrast with European options which have fixed maturities, the holder of an American option is allowed to exercise it at any given (random) time. This transforms the valuation problem into an optimization problem in which one has to find the optimal time to exercise in order to maximize the payoff of the option. As will be seen in the first section below, not all random times can be considered in this process, and we restrict ourselves to *stopping times* whose value at time t be can decided based on the historical data available.

9.1 Filtrations and Information Flow

Let $(\mathcal{F}_t)_{t\in\mathbb{R}_+}$ denote the *filtration* generated by a stochastic process $(X_t)_{t\in\mathbb{R}_+}$. In other words, $\mathcal{F}_t$ denotes the collection of all events possibly generated by $\{X_s : 0 \leq s \leq t\}$ up to time t. Examples of such events include the event

$$\{X_{t_0} \leq a_0, \ X_{t_1} \leq a_1, \ \ldots, \ X_{t_n} \leq a_n\}$$

for $a_0, a_1, \ldots, a_n$ a given fixed sequence of real numbers and $0 \leq t_1 < \cdots < t_n < t$, and $\mathcal{F}_t$ is said to represent the *information* generated by $(X_s)_{s\in[0,t]}$ up to time t.

By construction, $(\mathcal{F}_t)_{t\in\mathbb{R}_+}$ is an *increasing* family of σ-algebras in the sense that we have $\mathcal{F}_s \subset \mathcal{F}_t$ (information known at time s is contained in the information known at time t) when $0 < s < t$.

One refers sometimes to $(\mathcal{F}_t)_{t\in\mathbb{R}_+}$ as the increasing flow of information generated by $(X_t)_{t\in\mathbb{R}_+}$.

9.2 Martingales, Submartingales, and Supermartingales

Let us recall the definition of *martingale* (cf. Definition 5.4) and introduce in addition the definitions of *super*martingale and *sub*martingale.

Definition 9.1 *An integrable stochastic process* $(Z_t)_{t\in\mathbb{R}_+}$ *is a martingale (resp. a* super*martingale, resp. a* sub*martingale) with respect to* $(\mathcal{F}_t)_{t\in\mathbb{R}_+}$ *if it satisfies the property*

$$Z_s = \mathbb{E}[Z_t \mid \mathcal{F}_s], \qquad 0 \leq s \leq t,$$

resp.

$$Z_s \geq \mathbb{E}[Z_t \mid \mathcal{F}_s], \qquad 0 \leq s \leq t,$$

resp.

$$Z_s \leq \mathbb{E}[Z_t \mid \mathcal{F}_s], \qquad 0 \leq s \leq t.$$

Clearly, a process $(Z_t)_{t\in\mathbb{R}_+}$ is a martingale if and only if it is both a supermartingale and a submartingale.

A particular property of martingales is that their expectation is constant.

Proposition 9.1 *Let* $(Z_t)_{t\in\mathbb{R}_+}$ *be a martingale. We have*

$$\mathbb{E}[Z_t] = \mathbb{E}[Z_s], \qquad 0 \leq s \leq t.$$

The above proposition follows from the "tower property" (A.20) of conditional expectations, which shows that

$$\mathbb{E}[Z_t] = \mathbb{E}[\mathbb{E}[Z_t \mid \mathcal{F}_s]] = \mathbb{E}[Z_s], \qquad 0 \leq s \leq t. \tag{9.1}$$

Similarly, a *super*martingale has a *decreasing* expectation, while a *sub*martingale has a *increasing* expectation.

Proposition 9.2 *Let* $(Z_t)_{t\in\mathbb{R}_+}$ *be a* super*martingale, resp. a* sub*martingale. Then we have*

$$\mathbb{E}[Z_t] \leq \mathbb{E}[Z_s], \qquad 0 \leq s \leq t,$$

resp.

$$\mathbb{E}[Z_t] \geq \mathbb{E}[Z_s], \qquad 0 \leq s \leq t.$$

Proof. As in (9.1) above we have

$$\mathbb{E}[Z_t] = \mathbb{E}[\mathbb{E}[Z_t \mid \mathcal{F}_s]] \leq \mathbb{E}[Z_s], \qquad 0 \leq s \leq t.$$

The proof is similar in the *sub*martingale case. □

Independent increments processes whose increments have negative expectation give examples of *super*martingales. For example, if $(X_t)_{t\in\mathbb{R}_+}$ is such a process then we have

$$\begin{aligned}\mathbb{E}[X_t \mid \mathcal{F}_s] &= \mathbb{E}[X_s \mid \mathcal{F}_s] + \mathbb{E}[X_t - X_s \mid \mathcal{F}_s] \\ &= \mathbb{E}[X_s \mid \mathcal{F}_s] + \mathbb{E}[X_t - X_s] \\ &\leq \mathbb{E}[X_s \mid \mathcal{F}_s] \\ &= X_s, \qquad 0 \leq s \leq t.\end{aligned}$$

Similarly, a process with independent increments which have positive expectation will be a *sub*martingale. Brownian motion $B_t + \mu t$ with positive drift $\mu > 0$ is such an example, as in Figure 9.1 below.

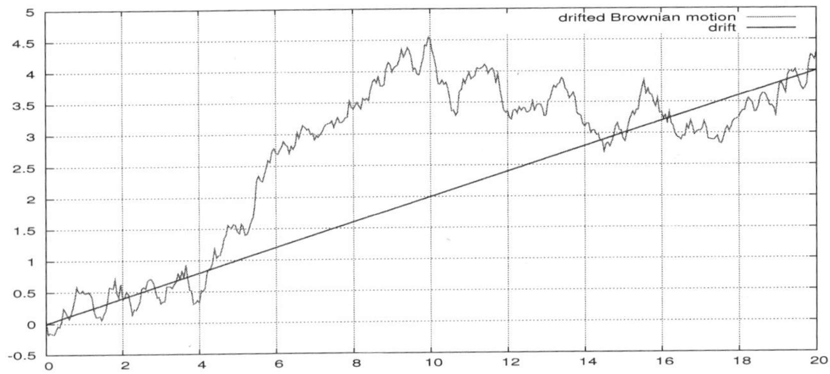

FIGURE 9.1: Drifted Brownian path.

The following example comes from gambling.

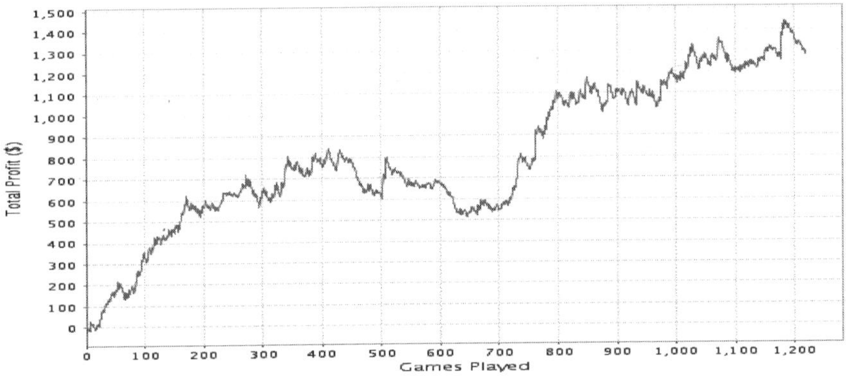

FIGURE 9.2: Evolution of the fortune of a poker player vs. number of games played.

A natural way to construct *sub*martingales is to take convex functions of martingales. Indeed, if $(M_t)_{t\in\mathbb{R}_+}$ is a martingale and ϕ is a convex function,

Jensen's inequality states that

$$\phi(\mathbb{E}[M_t \mid \mathcal{F}_s]) \leq \mathbb{E}[\phi(M_t) \mid \mathcal{F}_s], \qquad 0 \leq s \leq t, \tag{9.2}$$

which shows that

$$\phi(M_s) = \phi(\mathbb{E}[M_t \mid \mathcal{F}_s]) \leq \mathbb{E}[\phi(M_t) \mid \mathcal{F}_s], \qquad 0 \leq s \leq t,$$

i.e., $(\phi(M_t))_{t\in\mathbb{R}_+}$ is a *sub*martingale. More generally, the above shows that $\phi(M_t)_{t\in\mathbb{R}_+}$ remains a *sub*martingale when ϕ is convex non-decreasing and $(M_t)_{\in\mathbb{R}_+}$ is a *sub*martingale. Similarly, $(\phi(M_t))_{t\in\mathbb{R}_+}$ will be *super*martingale when $(M_t)_{\in\mathbb{R}_+}$ is a martingale and the function ϕ is *concave*.

Other examples of (super, sub)-martingales include geometric Brownian motion

$$S_t = S_0 e^{rt+\sigma B_t - \sigma^2 t/2}, \qquad t \in \mathbb{R}_+,$$

which is a martingale for $r = 0$, a *super*martingale for $r \leq 0$, and a *sub*martingale for $r \geq 0$.

9.3 Stopping Times

Next we turn to the definition of *stopping time.*

Definition 9.2 *A stopping time is a random variable $\tau : \Omega \longrightarrow \mathbb{R}_+ \cup \{+\infty\}$ such that*

$$\{\tau > t\} \in \mathcal{F}_t, \qquad t \in \mathbb{R}_+. \tag{9.3}$$

The meaning of Relation (9.3) is that the knowledge of the event $\{\tau > t\}$ depends only on the information present in $\mathcal{F}_t$ up to time t, i.e., on the knowledge of $(X_s)_{0\leq s\leq t}$.

In other words, an event occurs at a *stopping time* τ if at any time t it can be decided whether the event has already occured $(\tau \leq t)$ or not $(\tau > t)$ based on the information generated by $(X_s)_{s\in\mathbb{R}_+}$ up to time t.

For example, the day you bought your first car is a stopping time (one can always answer the question "did I ever buy a car?"), whereas the day you will buy your *last* car may not be a stopping time (one may not be able to answer the question "will I ever buy another car?").

Note that a constant time is always a stopping time, and if τ and θ are stopping times then the smallest $\tau \wedge \theta$ of τ and θ is also a stopping time, since

$$\{\tau \wedge \theta > t\} = \{\tau > t \quad \text{and} \quad \theta > t\} = \{\tau > t\} \cap \{\theta > t\} \in \mathcal{F}_t, \qquad t \in \mathbb{R}_+.$$

Hitting times provide natural examples of stopping times. The hitting time of level x by the process $(X_t)_{t\in\mathbb{R}_+}$, defined as

$$\tau_x = \inf\{t \in \mathbb{R}_+ \ : \ X_t = x\},$$

is a stopping time,[1] as we have (here in discrete time)

$$\begin{aligned}\{\tau_x > t\} &= \{X_s \neq x, \ 0 \leq s \leq t\} \\ &= \{X_0 \neq x\} \cap \{X_1 \neq x\} \cap \cdots \cap \{X_t \neq x\} \in \mathcal{F}_t, \qquad t \in \mathbb{N}.\end{aligned}$$

In gambling, a hitting time can be used as an exit strategy from the game. For example, letting

$$\tau_{x,y} := \inf\{t \in \mathbb{R}_+ \ : \ X_t = x \text{ or } X_t = y\} \tag{9.4}$$

defines a hitting time (hence a stopping time) which allows a gambler to exit the game as soon as losses become equal to $x = -10$, or gains become equal to $y = +100$, whichever comes first.

However, not every $\mathbb{R}_+$-valued random variable is a stopping time. For example the random time

$$\tau = \inf\left\{t \in [0,T] \ : \ X_t = \sup_{s\in[0,T]} X_s\right\},$$

which represents the first time the process $(X_t)_{t\in[0,T]}$ reaches its maximum over $[0,T]$, is not a stopping time with respect to the filtration generated by $(X_t)_{t\in[0,T]}$. Indeed, the information known at time $t \in (0,T)$ is not sufficient to determine whether $\{\tau > t\}$.

Given $(Z_t)_{t\in\mathbb{R}_+}$ a stochastic process and $\tau : \Omega \longrightarrow \mathbb{R}_+ \cup \{+\infty\}$ a stopping time, the *stopped process* $(Z_{t\wedge\tau})_{t\in\mathbb{R}_+}$ is defined as

$$Z_{t\wedge\tau} = \begin{cases} Z_\tau & \text{if } t \geq \tau, \\ Z_t & \text{if } t < \tau, \end{cases}$$

Using indicator functions we may also write

$$Z_{t\wedge\tau} = Z_\tau \mathbf{1}_{\{\tau\leq t\}} + Z_t \mathbf{1}_{\{\tau>t\}}, \qquad t \in \mathbb{R}_+.$$

The following figure is an illustration of the path of a stopped process.

[1]As a convention we let $\tau = +\infty$ in case there exists no $t \in \mathbb{R}_+$ such that $X_t = x$.

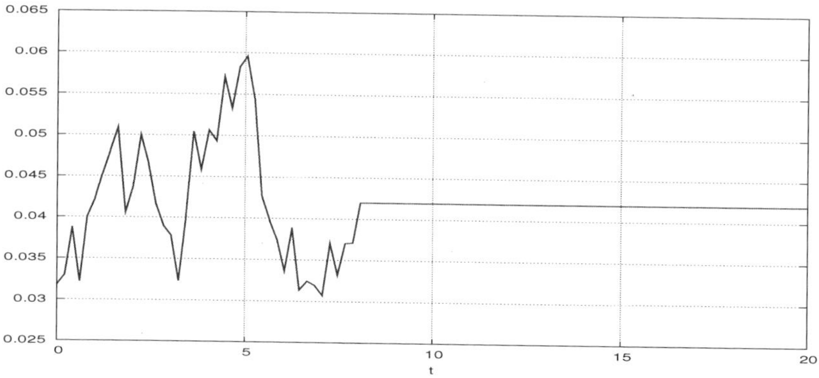

FIGURE 9.3: Stopped process.

Theorem 9.1 below is called the *stopping time* (or *optional sampling*, or *optional stopping*) theorem. It is due to the mathematician J.L. Doob (1910–2004). It is also used in Exercise 9.2 below.

Theorem 9.1 *Assume that* $(M_t)_{t\in\mathbb{R}_+}$ *is a martingale with respect to* $(\mathcal{F}_t)_{t\in\mathbb{R}_+}$. *Then the* stopped process $(M_{t\wedge\tau})_{t\in\mathbb{R}_+}$ *is also a martingale with respect to* $(\mathcal{F}_t)_{t\in\mathbb{R}_+}$.

Proof. We only give the proof in discrete time by applying the martingale transform argument of Proposition 2.1. Writing

$$M_n = M_0 + \sum_{l=1}^{n}(M_l - M_{l-1}) = M_0 + \sum_{l=1}^{\infty} \mathbf{1}_{\{l\leq n\}}(M_l - M_{l-1}),$$

we have

$$M_{\tau\wedge n} = M_0 + \sum_{l=1}^{\tau\wedge n}(M_l - M_{l-1}) = M_0 + \sum_{l=1}^{\infty} \mathbf{1}_{\{l\leq \tau\wedge n\}}(M_l - M_{l-1}),$$

and for $k \leq n$,

$$\begin{aligned}
\mathbb{E}[M_{\tau\wedge n} \mid \mathcal{F}_k] &= M_0 + \sum_{l=1}^{\infty} \mathbb{E}[\mathbf{1}_{\{l\leq \tau\wedge n\}}(M_l - M_{l-1}) \mid \mathcal{F}_k] \\
&= M_0 + \sum_{l=1}^{k} \mathbb{E}[\mathbf{1}_{\{l\leq \tau\wedge n\}}(M_l - M_{l-1}) \mid \mathcal{F}_k] \\
&\quad + \sum_{l=k+1}^{\infty} \mathbb{E}[\mathbf{1}_{\{l\leq \tau\wedge n\}}(M_l - M_{l-1}) \mid \mathcal{F}_k]
\end{aligned}$$

$$
\begin{aligned}
&= M_0 + \sum_{l=1}^{k}(M_l - M_{l-1})\,\mathbb{E}[\mathbf{1}_{\{l\le\tau\wedge n\}} \mid \mathcal{F}_k] \\
&\quad + \sum_{l=k+1}^{\infty} \mathbb{E}[\mathbb{E}[(M_l - M_{l-1})\mathbf{1}_{\{l\le\tau\wedge n\}} \mid \mathcal{F}_{l-1}] \mid \mathcal{F}_k] \\
&= M_0 + \sum_{l=1}^{k}(M_l - M_{l-1})\mathbf{1}_{\{l\le\tau\wedge n\}} \\
&\quad + \sum_{l=k+1}^{\infty} \mathbb{E}[\mathbf{1}_{\{l\le\tau\wedge n\}}\,\mathbb{E}[(M_l - M_{l-1}) \mid \mathcal{F}_{l-1}] \mid \mathcal{F}_k] \\
&= M_0 + \sum_{l=1}^{\tau\wedge n\wedge k}(M_l - M_{l-1})\mathbf{1}_{\{l\le\tau\wedge n\}} \\
&= M_0 + \sum_{l=1}^{\tau\wedge k}(M_l - M_{l-1})\mathbf{1}_{\{l\le\tau\wedge n\}} \\
&= M_{\tau\wedge k}, \qquad k = 0, 1, \ldots, n.
\end{aligned}
$$

□

Since the stopped process $(M_{\tau\wedge t})_{t\in\mathbb{R}_+}$ is a martingale by Theorem 9.1 we find that its expectation is constant by Proposition 9.1. More generally, if $(M_t)_{t\in\mathbb{R}_+}$ is a *super*martingale with respect to $(\mathcal{F}_t)_{t\in\mathbb{R}_+}$, then the *stopped process* $(M_{t\wedge\tau})_{t\in\mathbb{R}_+}$ remains a *super*martingale with respect to $(\mathcal{F}_t)_{t\in\mathbb{R}_+}$.

As a consequence, if τ is a stopping time bounded by $T > 0$, i.e., $\tau \le T$ almost surely, we have

$$
\mathbb{E}[M_\tau] = \mathbb{E}[M_{\tau\wedge T}] = \mathbb{E}[M_{\tau\wedge 0}] = \mathbb{E}[M_0]. \tag{9.5}
$$

In case τ is finite with probability one but not bounded we may also write

$$
\mathbb{E}[M_\tau] = \mathbb{E}\left[\lim_{t\to\infty} M_{\tau\wedge t}\right] = \lim_{t\to\infty}\mathbb{E}[M_{\tau\wedge t}] = \mathbb{E}[M_0], \tag{9.6}
$$

provided

$$
|M_{\tau\wedge t}| \le C, \qquad a.s., \quad t\in\mathbb{R}_+. \tag{9.7}
$$

More generally, (9.6) will hold provided the limit and expectation signs can be exchanged, and this can be done using e.g., the *Dominated Convergence Theorem.*

In case $\mathbb{P}(\tau = +\infty) > 0$, (9.6) will hold under the above conditions, provided

$$
M_\infty := \lim_{t\to\infty} M_t \tag{9.8}
$$

exists with probability one.

In addition, if τ and ν are two *bounded* stopping times such that $\tau \leq \nu$, a.s., we have

$$\mathbb{E}[M_\tau] \geq \mathbb{E}[M_\nu] \tag{9.9}$$

if $(M_t)_{t\in\mathbb{R}_+}$ is a *super*martingale, and

$$\mathbb{E}[M_\tau] \leq \mathbb{E}[M_\nu] \tag{9.10}$$

if $(M_t)_{t\in\mathbb{R}_+}$ is a *sub*martingale, cf. Exercise 9.2 below for a proof in discrete time. As a consequence of (9.9) and (9.10) (or directly from (9.5)), if τ and ν are two *bounded* stopping times such that $\tau \leq \nu$, a.s., we have

$$\mathbb{E}[M_\tau] = \mathbb{E}[M_\nu] \tag{9.11}$$

if $(M_t)_{t\in\mathbb{R}_+}$ is a martingale.

Relations (9.9), (9.10) and (9.11) can be extended to unbounded stopping times along the same lines and conditions as (9.6), such as (9.7) applied to both τ and ν. Dealing with unbounded stopping times can be necessary in the case of hitting times.

In general, for all stopping times τ (bounded or unbounded) it remains true that

$$\mathbb{E}[M_\tau] = \mathbb{E}[\lim_{t\to\infty} M_{\tau\wedge t}] \leq \lim_{t\to\infty} \mathbb{E}[M_{\tau\wedge t}] \leq \lim_{t\to\infty} \mathbb{E}[M_0] = \mathbb{E}[M_0], \tag{9.12}$$

provided $(M_t)_{t\in\mathbb{R}_+}$ is a nonnegative *super*martingale, where we used Fatou's lemma.[2] As in (9.6), the limit (9.8) is required to exist with probability one if $\mathbb{P}(\tau = +\infty) > 0$.

In the case of the exit strategy $\tau_{x,y}$ of (9.4) the stopping time theorem shows that $\mathbb{E}[M_{\tau_{x,y}}] = 0$ if $M_0 = 0$, which shows that on average this exit strategy does not increase the average gain of the player. More precisely we have

$$\begin{aligned} 0 &= M_0 = \mathbb{E}[M_{\tau_{x,y}}] = x\mathbb{P}(M_{\tau_{x,y}} = x) + y\mathbb{P}(M_{\tau_{x,y}} = y) \\ &= -10\mathbb{P}(M_{\tau_{x,y}} = -10) + 100\mathbb{P}(M_{\tau_{x,y}} = 100), \end{aligned}$$

which shows that

$$\mathbb{P}(M_{\tau_{x,y}} = -10) = \frac{10}{11} \quad \text{and} \quad \mathbb{P}(M_{\tau_{x,y}} = 100) = \frac{1}{11},$$

provided the relation $\mathbb{P}(M_{\tau_{x,y}} = x) + \mathbb{P}(M_{\tau_{x,y}} = y) = 1$ is satisfied, see below for further applications to Brownian motion.

[2] $\mathbb{E}[\lim_{n\to\infty} F_n] \leq \lim_{n\to\infty} \mathbb{E}[F_n]$ for any sequence $(F_n)_{n\in\mathbb{N}}$ of *nonnegative* random variables, provided the limits exist.

As a counterexample to (9.11), the random time

$$\tau := \inf \left\{ t \in [0,T] \ : \ M_t = \sup_{s \in [0,T]} M_s \right\},$$

which is not a stopping time, will satisfy

$$\mathbb{E}[M_\tau] > \mathbb{E}[M_T],$$

although $\tau \leq T$ almost surely.

Similarly,

$$\tau := \inf \left\{ t \in [0,T] \ : \ M_t = \inf_{s \in [0,T]} M_s \right\},$$

is not a stopping time and satisfies

$$\mathbb{E}[M_\tau] < \mathbb{E}[M_T].$$

When X_t is a martingale, e.g., a centered random walk with independent increments, the message of the stopping time Theorem 9.1 is that the expected gain of the strategy (9.4) remains zero on average since

$$\mathbb{E}[X_{\tau_{x,y}}] = \mathbb{E}[X_0] = 0.$$

Similar arguments are used in the examples below.

In the table below we summarize some of the results of this section for bounded stopping times.

		bounded stopping times $\tau \leq \nu$
M_t	*super*martingale	$\mathbb{E}[M_\tau] \geq \mathbb{E}[M_\nu]$
	martingale	$\mathbb{E}[M_\tau] = \mathbb{E}[M_\nu]$
	*sub*martingale	$\mathbb{E}[M_\tau] \leq \mathbb{E}[M_\nu]$

Examples of application

In this section we note that, as an application of the stopping time theorem, a number of expectations can be computed in a simple and elegant way.

Brownian motion hitting a barrier

Given $a, b \in \mathbb{R}$, $a < b$, let the hitting[3] time $\tau_{a,b} : \Omega \longrightarrow \mathbb{R}_+$ be defined by

$$\tau_{a,b} = \inf\{t \geq 0 \ : \ B_t = a \text{ or } B_t = b\},$$

[3]A hitting time is a stopping time

which is the hitting time of the boundary $\{a,b\}$ of Brownian motion $(B_t)_{t\in\mathbb{R}_+}$, $a,b\in\mathbb{R}$, $a<b$.

Recall that Brownian motion $(B_t)_{t\in\mathbb{R}_+}$ is a martingale since it has independent increments, and those increments are centered:

$$\mathbb{E}[B_t - B_s] = 0, \qquad 0 \le s \le t.$$

Consequently, $(B_{\tau_{a,b}\wedge t})_{t\in\mathbb{R}_+}$ is still a martingale and by (9.6) we have

$$\mathbb{E}[B_{\tau_{a,b}} \mid B_0 = x] = \mathbb{E}[B_0 \mid B_0 = x] = x,$$

as the exchange between limit and expectation in (9.6) can be justified since

$$|B_{t\wedge\tau_{a,b}}| \le \max(|a|,|b|), \qquad t\in\mathbb{R}_+.$$

Hence we have

$$x = \mathbb{E}[B_{\tau_{a,b}} \mid B_0 = x] = a\times\mathbb{P}(B_{\tau_{a,b}} = a \mid B_0 = x) + b\times\mathbb{P}(B_{\tau_{a,b}} = b \mid B_0 = x),$$

under the additional condition

$$\mathbb{P}\left(X_{\tau_{a,b}} = a \mid X_0 = x\right) + \mathbb{P}(X_{\tau_{a,b}} = b \mid X_0 = x) = 1.$$

which yields

$$\mathbb{P}(B_{\tau_{a,b}} = b \mid B_0 = x) = \frac{x-a}{b-a}, \qquad a \le x \le b,$$

which also shows that

$$\mathbb{P}(B_{\tau_{a,b}} = a \mid B_0 = x) = \frac{b-x}{b-a}, \qquad a \le x \le b.$$

Note that the above result and its proof actually apply to any continuous martingale, and not only to Brownian motion.

Drifted Brownian motion hitting a barrier

Next, let us turn to the case of drifted Brownian motion

$$X_t = x + B_t + \mu t, \qquad t\in\mathbb{R}_+.$$

In this case the process $(X_t)_{t\in\mathbb{R}_+}$ is no longer a martingale and in order to use Theorem 9.1 we need to construct a martingale of a different type. Here we note that the process

$$M_t := e^{\sigma B_t - \sigma^2 t/2}, \qquad t\in\mathbb{R}_+,$$

is a martingale with respect to $(\mathcal{F}_t)_{t\in\mathbb{R}_+}$. Indeed, we have

$$\mathbb{E}[M_t \mid \mathcal{F}_s] = \mathbb{E}\left[e^{\sigma B_t - \sigma^2 t/2}\middle|\mathcal{F}_s\right] = e^{\sigma B_s - \sigma^2 s/2}, \qquad 0 \le s \le t.$$

By Theorem 9.1 we know that the stopped process $(M_{\tau_{a,b}\wedge t})_{t\in\mathbb{R}_+}$ is a martingale, hence its expectation is constant by Proposition 9.1, and (9.6) gives

$$1 = \mathbb{E}[M_0] = \mathbb{E}[M_{\tau_{a,b}}],$$

as the exchange between limit and expectation in (9.6) can be justified since

$$|M_{t\wedge\tau_{a,b}}| \leq \max(e^{\sigma|a|}, e^{\sigma|b|}), \qquad t \in \mathbb{R}_+.$$

Next we note that letting $\sigma = -2\mu$ we have

$$e^{\sigma X_t} = e^{\sigma x + \sigma B_t + \sigma\mu t} = e^{\sigma x + \sigma B_t - \sigma^2 t/2} = e^{\sigma x} M_t,$$

hence

$$\begin{aligned} 1 &= \mathbb{E}[M_{\tau_{a,b}}] \\ &= e^{-\sigma x}\,\mathbb{E}[e^{\sigma X_{\tau_{a,b}}}] \\ &= e^{\sigma(a-x)}\mathbb{P}\left(X_{\tau_{a,b}} = a \mid X_0 = x\right) + e^{\sigma(b-x)}\mathbb{P}(X_{\tau_{a,b}} = b \mid X_0 = x), \end{aligned}$$

under the additional condition

$$\mathbb{P}\left(X_{\tau_{a,b}} = a \mid X_0 = x\right) + \mathbb{P}(X_{\tau_{a,b}} = b \mid X_0 = x) = 1.$$

Finally this gives

$$\mathbb{P}(X_{\tau_{a,b}} = a \mid X_0 = x) = \frac{e^{\sigma x} - e^{\sigma b}}{e^{\sigma a} - e^{\sigma b}} = \frac{e^{-2\mu x} - e^{-2\mu b}}{e^{-2\mu a} - e^{-2\mu b}}, \tag{9.13}$$

$a \leq x \leq b$, and

$$\mathbb{P}(X_{\tau_{a,b}} = b \mid X_0 = x) = \frac{e^{-2\mu a} - e^{-2\mu x}}{e^{-2\mu a} - e^{-2\mu b}}, \qquad a \leq x \leq b.$$

Mean hitting time for Brownian motion

The martingale method also allows us to compute the expectation $\mathbb{E}[B_{\tau_{a,b}}]$, after checking that $(B_t^2 - t)_{t\in\mathbb{R}_+}$ is also a martingale. Indeed we have

$$\begin{aligned} &\mathbb{E}[B_t^2 - t \mid \mathcal{F}_s] = \mathbb{E}[(B_s + (B_t - B_s))^2 - t \mid \mathcal{F}_s] \\ &= \mathbb{E}[B_s^2 + (B_t - B_s)^2 + 2B_s(B_t - B_s) - t \mid \mathcal{F}_s] \\ &= \mathbb{E}[B_s^2 - s \mid \mathcal{F}_s] - (t-s) + \mathbb{E}[(B_t - B_s)^2 \mid \mathcal{F}_s] + 2\,\mathbb{E}[B_s(B_t - B_s) \mid \mathcal{F}_s] \\ &= B_s^2 - s - (t-s) + \mathbb{E}[(B_t - B_s)^2 \mid \mathcal{F}_s] + 2B_s\,\mathbb{E}[B_t - B_s \mid \mathcal{F}_s] \\ &= B_s^2 - s - (t-s) + \mathbb{E}[(B_t - B_s)^2] + 2B_s\,\mathbb{E}[B_t - B_s] \\ &= B_s^2 - s, \qquad 0 \leq s \leq t. \end{aligned}$$

Consequently the stopped process $(B^2_{\tau_{a,b}\wedge t} - \tau_{a,b}\wedge t)_{t\in\mathbb{R}_+}$ is still a martingale by Theorem 9.1 hence the expectation $\mathbb{E}[B^2_{\tau_{a,b}\wedge t} - \tau_{a,b}\wedge t]$ is constant in $t \in \mathbb{R}_+$, hence by (9.6) we get[4]

$$x^2 = \mathbb{E}[B_0^2 - 0 \mid B_0 = x]$$

[4]Here we note that it can be showed that $\mathbb{E}[\tau_{a,b}] < \infty$ in order to apply (9.6).

$$
\begin{aligned}
&= \mathbb{E}[B^2_{\tau_{a,b}} - \tau_{a,b} \mid B_0 = x] \\
&= \mathbb{E}[B^2_{\tau_{a,b}} \mid B_0 = x] - \mathbb{E}[\tau_{a,b} \mid B_0 = x] \\
&= b^2\mathbb{P}(B_{\tau_{a,b}} = b \mid B_0 = x) + a^2\mathbb{P}(B_{\tau_{a,b}} = a \mid B_0 = x) - \mathbb{E}[\tau_{a,b} \mid B_0 = x],
\end{aligned}
$$

i.e.,

$$
\begin{aligned}
\mathbb{E}[\tau_{a,b} \mid B_0 = x] &= b^2\mathbb{P}(B_{\tau_{a,b}} = b \mid B_0 = x) + a^2\mathbb{P}(B_{\tau_{a,b}} = a \mid B_0 = x) - x^2 \\
&= b^2\frac{x-a}{b-a} + a^2\frac{b-x}{b-a} - x^2 \\
&= (x-a)(b-x), \qquad a \leq x \leq b.
\end{aligned}
$$

Mean hitting time for drifted Brownian motion

Finally we show how to recover the value of the mean hitting time of drifted Brownian motion $X_t = B_t + \mu t$. As above, the process $X_t - \mu t$ is a martingale the stopped process $(X_{\tau_{a,b}\wedge t} - \mu(\tau_{a,b} \wedge t))_{t\in\mathbb{R}_+}$ is still a martingale by Theorem 9.1. Hence the expectation $\mathbb{E}[X_{\tau_{a,b}\wedge t} - \mu(\tau_{a,b} \wedge t)]$ is constant in $t \in \mathbb{R}_+$.

Since the stopped process $(X_{\tau_{a,b}\wedge t} - \mu t)_{t\in\mathbb{R}_+}$ is a martingale, we have

$$
x = \mathbb{E}[X_{\tau_{a,b}} - \mu\tau_{a,b} \mid X_0 = x],
$$

which gives

$$
\begin{aligned}
x &= \mathbb{E}[X_{\tau_{a,b}} - \mu\tau_{a,b} \mid X_0 = x] \\
&= \mathbb{E}[X_{\tau_{a,b}} \mid X_0 = x] - \mu\,\mathbb{E}[\tau_{a,b} \mid X_0 = x] \\
&= b\mathbb{P}(X_{\tau_{a,b}} = b \mid X_0 = x) + a\mathbb{P}(X_{\tau_{a,b}} = a \mid X_0 = x) - \mu\,\mathbb{E}[\tau_{a,b} \mid X_0 = x],
\end{aligned}
$$

i.e., by (9.13),

$$
\begin{aligned}
\mu\,\mathbb{E}[\tau_{a,b} \mid X_0 = x] &= b\mathbb{P}(X_{\tau_{a,b}} = b \mid X_0 = x) + a\mathbb{P}(X_{\tau_{a,b}} = a \mid X_0 = x) - x \\
&= b\frac{e^{-2\mu a} - e^{-2\mu x}}{e^{-2\mu a} - e^{-2\mu b}} + a\frac{e^{-2\mu x} - e^{-2\mu b}}{e^{-2\mu a} - e^{-2\mu b}} - x \\
&= \frac{b(e^{-2\mu a} - e^{-2\mu x}) + a(e^{-2\mu x} - e^{-2\mu b}) - x(e^{-2\mu a} - e^{-2\mu b})}{e^{-2\mu a} - e^{-2\mu b}},
\end{aligned}
$$

hence

$$
\mathbb{E}[\tau_{a,b} \mid X_0 = x] = \frac{b(e^{-2\mu a} - e^{-2\mu x}) + a(e^{-2\mu x} - e^{-2\mu b}) - x(e^{-2\mu a} - e^{-2\mu b})}{\mu(e^{-2\mu a} - e^{-2\mu b})},
$$

$a \leq x \leq b$.

9.4 Perpetual American Options

The price of an American put option expiring at time $T > 0$ with strike K can be expressed as the expected value of its discounted payoff:

$$f(t, S_t) = \sup_{\substack{t \leq \tau \leq T \\ \tau \text{ stopping time}}} \mathbb{E}^* \left[e^{-r(\tau - t)} (K - S_\tau)^+ \Big| S_t \right],$$

under the risk-neutral probability measure $\mathbb{P}^*$, where the supremum is taken over stopping times between t and a fixed maturity T. Similarly, the price of a finite expiration American call option with strike K is expressed as

$$f(t, S_t) = \sup_{\substack{t \leq \tau \leq T \\ \tau \text{ stopping time}}} \mathbb{E}^* \left[e^{-r(\tau - t)} (S_\tau - K)^+ \Big| S_t \right].$$

In this section we take $T = +\infty$, in which case we refer to these options as *perpetual* options, and the corresponding put and call are respectively priced as

$$f(t, S_t) = \sup_{\substack{\tau \geq t \\ \tau \text{ stopping time}}} \mathbb{E}^* \left[e^{-r(\tau - t)} (K - S_\tau)^+ \Big| S_t \right],$$

and

$$f(t, S_t) = \sup_{\substack{\tau \geq t \\ \tau \text{ stopping time}}} \mathbb{E}^* \left[e^{-r(\tau - t)} (S_\tau - K)^+ \Big| S_t \right].$$

Two-choice optimal stopping at a fixed price level for perpetual put options

In this section we consider the pricing of perpetual put options. Given $L \in (0, K)$ a fixed price, consider the following choices for the exercise of a *put* option with strike K:

1. If $S_t \leq L$, then exercise at time t.

2. Otherwise if $S_t > L$, wait until the first hitting time

$$\tau_L := \inf\{u \geq t \ : \ S_u \leq L\} \tag{9.14}$$

   and exercise the option at time τ_L.

Note that by definition of (9.14) we have $\tau_L = t$ if $S_t \leq L$.

In case $S_t \leq L$, the payoff will be

$$(K - S_t)^+ = K - S_t$$

since $K > L \geq S_t$, however in this case one would buy the option at price $K - S_t$ only to exercise it immediately for the same amount.

In case $S_t > L$, the price of the option will be

$$\begin{aligned} f_L(t, S_t) &= \mathbb{E}^* \left[e^{-r(\tau_L - t)} (K - S_{\tau_L})^+ \Big| S_t \right] \\ &= \mathbb{E}^* \left[e^{-r(\tau_L - t)} (K - L)^+ \Big| S_t \right] \\ &= (K - L)\, \mathbb{E}^* \left[e^{-r(\tau_L - t)} \Big| S_t \right]. \end{aligned} \tag{9.15}$$

We note that the starting date t does not matter when pricing perpetual options, hence $f_L(t, x)$ is actually independent of $t \in \mathbb{R}_+$, and the pricing of the perpetual put option can be performed by taking $t = 0$ and in the sequel we will work under

$$f_L(t, x) = f_L(x), \qquad x > 0.$$

Recall that the underlying asset price is written as

$$S_t = S_0 e^{\sigma \tilde{B}_t - \sigma^2 t/2 + rt}, \qquad t \in \mathbb{R}_+, \tag{9.16}$$

where $(\tilde{B}_t)_{t \in \mathbb{R}_+}$ is a standard Brownian motion under the risk-neutral probability measure $\mathbb{P}^*$, r is the risk-free interest rate, and $\sigma > 0$ is the volatility coefficient.

Proposition 9.3 *Assume that* $r \geq 0$. *We have*

$$\begin{aligned} f_L(x) &= \mathbb{E}^* \left[e^{-r(\tau_L - t)} (K - S_{\tau_L})^+ \Big| S_t = x \right] \\ &= \begin{cases} K - x, & 0 < x \leq L, \\ \\ (K - L) \left(\dfrac{x}{L} \right)^{-2r/\sigma^2}, & x \geq L. \end{cases} \end{aligned} \tag{9.17}$$

Proof. The result is obvious for $S_t = x \leq L$ since in this case we have $\tau_L = t$ and $S_{\tau_L} = S_t = x$, so that we only focus on the case $x \geq L$. In addition we take $t = 0$ without loss of generality. By the relation

$$\mathbb{E}^* \left[e^{-r(\tau_L - t)} (K - S_{\tau_L})^+ \Big| S_t = x \right] = \mathbb{E}^* \left[e^{-r(\tau_L - t)} (K - L) \Big| S_t = x \right], \tag{9.18}$$

we check that it suffices to compute $\mathbb{E}^* \left[e^{-r(\tau_L - t)} \Big| S_t = x \right]$. For this we note that from (9.16), for all $\lambda \in \mathbb{R}$ the process $(Z_t)_{t \in \mathbb{R}_+}$ defined as

$$Z_t := \left(\frac{S_t}{S_0} \right)^{\lambda} e^{-r\lambda t + \lambda \sigma^2 t/2 - \lambda^2 \sigma^2 t/2} = e^{\lambda \sigma \tilde{B}_t - \lambda^2 \sigma^2 t/2}, \qquad t \in \mathbb{R}_+,$$

is a martingale under the risk-neutral probability measure $\mathbb{P}^*$, hence we have[5]

$$\mathbb{E}^*[Z_{\tau_L}] = \mathbb{E}^*[Z_0] = 1,$$

with

$$\begin{aligned} Z_{\tau_L} &= \left(\frac{S_{\tau_L}}{S_0}\right)^{\lambda} e^{-(r\lambda - \lambda\sigma^2/2 + \lambda^2\sigma^2/2)\tau_L} \\ &= \left(\frac{L}{S_0}\right)^{\lambda} e^{-(r\lambda - \lambda\sigma^2/2 + \lambda^2\sigma^2/2)\tau_L}, \end{aligned}$$

which yields

$$\mathbb{E}^*\left[\left(\frac{L}{S_0}\right)^{\lambda} e^{-(r\lambda - \lambda\sigma^2/2 + \lambda^2\sigma^2/2)\tau_L}\right] = 1,$$

i.e.,

$$\mathbb{E}^*\left[e^{-(r\lambda - \lambda\sigma^2/2 + \lambda^2\sigma^2/2)\tau_L}\right] = \left(\frac{S_0}{L}\right)^{\lambda},$$

or

$$\mathbb{E}^*\left[e^{-r\tau_L}\right] = \left(\frac{S_0}{L}\right)^{\lambda}, \qquad (9.19)$$

provided we choose λ such that

$$-(r\lambda - \lambda\sigma^2/2 + \lambda^2\sigma^2/2) = -r,$$

i.e.,

$$0 = \lambda^2\sigma^2/2 + \lambda(r - \sigma^2/2) - r = \frac{\sigma^2}{2}(\lambda + 2r/\sigma^2)(\lambda - 1). \qquad (9.20)$$

This equation admits two solutions and we choose the negative solution $\lambda = -2r/\sigma^2$ since $S_0 = x > L$ and the expectation $\mathbb{E}^*[e^{-r\tau_L}] < 1$ in (9.19) is lower than 1 as $r \geq 0$. Consequently we have

$$\mathbb{E}^*\left[e^{-r(\tau_L - t)} \,\middle|\, S_t = x\right] = \left(\frac{x}{L}\right)^{-2r/\sigma^2} \qquad x \geq L, \qquad (9.21)$$

and we conclude by (9.18), which shows that

$$\begin{aligned} \mathbb{E}^*\left[e^{-r(\tau_L - t)}(K - S_{\tau_L})^+ \,\middle|\, S_t = x\right] &= (K - L)\,\mathbb{E}^*\left[e^{-r(\tau_L - t)} \,\middle|\, S_t = x\right] \\ &= (K - L)\left(\frac{x}{L}\right)^{-2r/\sigma^2}, \end{aligned}$$

when $S_t = x > L$. □

[5] Here the exchange of limit and expectation can be justified by monotone convergence, cf. p. 347 of [68].

We note that taking $L = K$ would yield a payoff always equal to 0 for the option holder, hence the value of L should be strictly lower than K. On the other hand, if $L = 0$ the value of τ_L will be infinite almost surely, hence the option price will be 0 when $r \geq 0$ from (9.15). Therefore there should be an optimal value L^*, which should be strictly comprised between 0 and K.

Figure 9.4 shows for $K = 100$ that there exists an optimal value $L^* = 85.71$ which maximizes the option price for all values of the underlying.

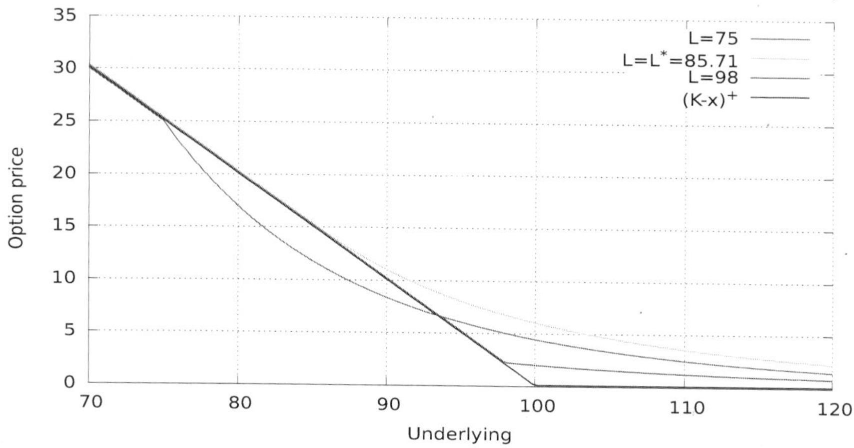

FIGURE 9.4: Graphs of the option price by exercise at τ_L for several values of L.

In order to compute L^* we observe that, geometrically, the slope of $f_L(x)$ at $x = L^*$ is equal to -1, i.e.,

$$f'_{L^*}(L^*) = -\frac{2r}{\sigma^2}(K - L^*)\frac{(L^*)^{-2r/\sigma^2-1}}{(L^*)^{-2r/\sigma^2}} = -1,$$

i.e.,

$$\frac{2r}{\sigma^2}(K - L^*) = L^*,$$

or

$$\boxed{L^* = \frac{2r}{2r + \sigma^2}K < K,}$$

cf. [68] page 351 for another derivation.

The next figure gives another view of the put option prices according to different values of L, with the optimal value $L^* = 85.71$.

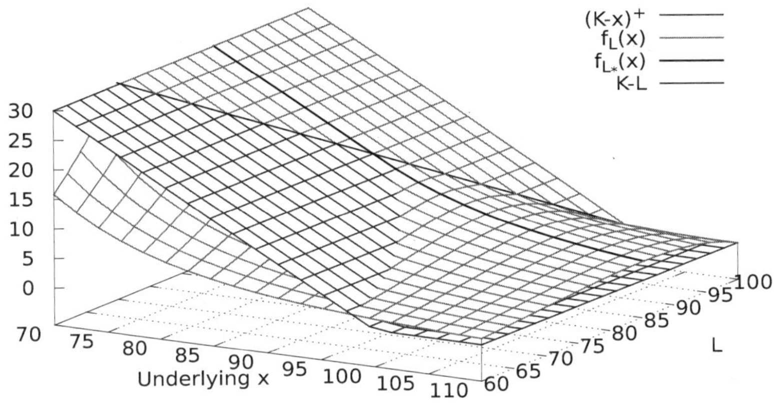

FIGURE 9.5: Graph of the option price as a function of L and of the underlying asset price.

In Figure 9.6 which is based on the stock price of HSBC Holdings (0005.HK) over year 2009, the optimal exercise strategy for an American put option with strike K=\$77.67 would have been to exercise whenever the underlying price goes above $L^* = \$62$, i.e., at approximately 54 days, for a payoff of \$38. Note that waiting a longer time, e.g., until 85 days, would have yielded a higher payoff of at least \$65. This is due to the fact that, here, optimization is done

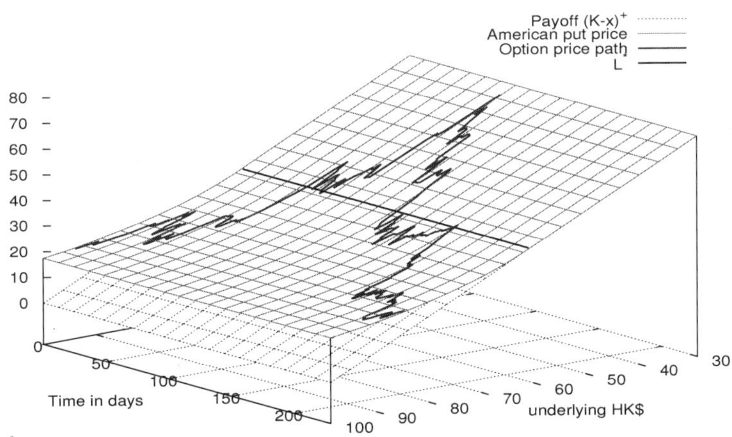

FIGURE 9.6: Path of the American put option price on the HSBC stock.

based on the past information only and makes sense in expectation (or average) over all possible future paths.

PDE approach

We can check by hand that

$$f_{L^*}(x) = \begin{cases} K - x, & 0 < x \leq L^* = \dfrac{2r}{2r+\sigma^2}K, \\ \\ \dfrac{K\sigma^2}{2r+\sigma^2}\left(\dfrac{2r+\sigma^2}{2r}\dfrac{x}{K}\right)^{-2r/\sigma^2}, & x \geq L^* = \dfrac{2r}{2r+\sigma^2}K, \end{cases}$$

satisfies the PDE

$$-rf_{L^*}(x) + rxf'_{L^*}(x) + \frac{1}{2}\sigma^2 x^2 f''_{L^*}(x) = \begin{cases} -rK < 0, & 0 < x \leq L^* < K, \\ \\ 0, & x > L^*. \end{cases} \tag{9.22}$$

in addition to the condition

$$\begin{cases} f_{L^*}(x) = K - x, & 0 < x < L^* < K, \\ \\ f_{L^*}(x) > (K-x)^+, & x \geq L^*. \end{cases}$$

This can be summarized in the following differential inequalities, or variational differential equation:

$$\begin{cases} f_{L^*}(x) \geq (K-x)^+, & (9.23a) \\ \\ rxf'_{L^*}(x) + \dfrac{1}{2}\sigma^2 x^2 f''_{L^*}(x) \leq rf_{L^*}(x), & (9.23b) \\ \\ \left(rf_{L^*}(x) - rxf'_{L^*}(x) - \dfrac{1}{2}\sigma^2 x^2 f''_{L^*}(x)\right)(f_{L^*}(x) - (K-x)^+) = 0, & (9.23c) \end{cases}$$

which admits an interpretation in terms of absence of arbitrage, as shown below.

By Itô's formula the discounted portfolio price $\tilde{f}_{L^*}(S_t) = e^{-rt}f_{L^*}(S_t)$ satisfies

$$d(\tilde{f}_{L^*}(S_t)) = \left(-rf_{L^*}(S_t) + rS_t f'_{L^*}(S_t) + \frac{1}{2}\sigma^2 S_t^2 f''_{L^*}(S_t)\right)e^{-rt}dt$$

$$+e^{-rt}\sigma S_t f'_{L^*}(S_t)d\tilde{B}_t, \tag{9.24}$$

hence from Equation (9.23c), $\tilde{f}_{L^*}(S_t)$ is a martingale when $f_{L^*}(S_t) > (K - S_t)^+$, i.e., $S_t > L^*$, and the expected rate of return of the option price $f_{L^*}(S_t)$ then equals the rate r of the risk-free asset as the investor prefers to wait.

On the other hand if $f_{L^*}(S_t) = (K - S_t)^+$, i.e., $0 < S_t < L^*$, it is not worth waiting as (9.23b) and (9.23c) show that the return of the option is lower than that of the risk-free asset, i.e.,:

$$-rf_{L^*}(S_t) + rS_t f'_{L^*}(S_t) + \frac{1}{2}\sigma^2 S_t^2 f''_{L^*}(S_t) < 0,$$

and exercise becomes immediate since the process $\tilde{f}_{L^*}(S_t)$ becomes a (strict) supermartingale. In this case, (9.23c) implies $f_{L^*}(x) = (K - x)^+$.

In view of the above derivation it should make sense to assert that $f_{L^*}(S_t)$ is the price at time t of the perpetual American put option. The next proposition shows that this is indeed the case, and that the optimal exercise time is $\tau^* = \tau_{L^*}$.

Proposition 9.4 *The price of the perpetual American put option is given for all $t \geq 0$ by*

$$\begin{aligned} f_{L^*}(S_t) &= \sup_{\substack{\tau \geq t \\ \tau \text{ stopping time}}} \mathbb{E}^*\left[e^{-r(\tau-t)}(K - S_\tau)^+ \Big| S_t\right] \\ &= \mathbb{E}^*\left[e^{-r(\tau_{L^*}-t)}(K - S_{\tau_{L^*}})^+ \Big| S_t\right] \\ &= \begin{cases} K - S_t, & 0 < S_t \leq L^*, \\ \\ \dfrac{K\sigma^2}{2r+\sigma^2}\left(\dfrac{2r+\sigma^2}{2r}\dfrac{S_t}{K}\right)^{-2r/\sigma^2}, & S_t \geq L^*. \end{cases} \end{aligned}$$

Proof. By Itô's formula (9.24) and the inequality (9.23b) one checks that the discounted portfolio price

$$u \mapsto e^{-ru} f_{L^*}(S_u), \qquad u \geq t,$$

is a supermartingale. As a consequence, for all stopping times τ we have, by (9.12),

$$e^{-rt} f_{L^*}(S_t) \;\geq\; \mathbb{E}^*\left[e^{-r\tau} f_{L^*}(S_\tau) \Big| S_t\right]$$

$$\geq \mathbb{E}^*\left[e^{-r\tau}(K - S_\tau)^+ \Big| S_t\right],$$

from (9.23a), which implies

$$e^{-rt} f_{L^*}(S_t) \geq \sup_{\substack{\tau \geq t \\ \tau \text{ stopping time}}} \mathbb{E}^*\left[e^{-r\tau}(K - S_\tau)^+ \Big| S_t\right]. \tag{9.25}$$

On the other hand, taking $\tau = \tau_{L^*}$ we note that the process

$$u \mapsto e^{-ru\wedge\tau_{L^*}} f_{L^*}(S_{u\wedge\tau_{L^*}}), \qquad u \geq t,$$

is not only a supermartingale, it is also a martingale until exercise at time τ_{L^*} by (9.22) since $S_{u\wedge\tau_{L^*}} \geq L^*$, hence we have

$$e^{-rt} f_{L^*}(S_t) = \mathbb{E}^*\left[e^{-r(u\wedge\tau_{L^*})} f_{L^*}(S_{u\wedge\tau_{L^*}}) \Big| S_t\right], \qquad u \geq t,$$

hence after letting u tend to infinity we obtain

$$\begin{aligned}
e^{-rt} f_{L^*}(S_t) &= \mathbb{E}^*\left[e^{-r\tau_{L^*}} f_{L^*}(S_{\tau_{L^*}}) \Big| S_t\right] \\
&= \mathbb{E}^*\left[e^{-r\tau_{L^*}} f_{L^*}(L^*) \Big| S_t\right] \\
&= \mathbb{E}^*\left[e^{-r\tau_{L^*}} (K - S_{\tau_{L^*}})^+ \Big| S_t\right] \\
&\leq \sup_{\substack{\tau \geq t \\ \tau \text{ stopping time}}} \mathbb{E}^*\left[e^{-r\tau_{L^*}} (K - S_{\tau_{L^*}})^+ \Big| S_t\right],
\end{aligned}$$

which shows that

$$e^{-rt} f_{L^*}(S_t) \leq \sup_{\substack{\tau \geq t \\ \tau \text{ stopping time}}} \mathbb{E}^*\left[e^{-r\tau}(K - S_\tau)^+ \Big| S_t\right], \qquad t \geq 0,$$

and the conclusion follows from (9.25). □

Two-choice optimal stopping at a fixed price level for perpetual call options

In this section we consider the pricing of perpetual call options. Given $L > K$ a fixed price, consider the following choices for the exercise of a *call* option with strike K:

1. If $S_t \geq L$, then exercise at time t.

2. Otherwise, wait until the first hitting time

$$\tau_L = \inf\{u \geq t \; : \; S_u = L\}$$

and exercise the option and time τ_L.

In case $S_t \geq L$, the payoff will be

$$(S_t - K)^+ = S_t - K$$

since $K < L \leq S_t$.

In case $S_t < L$, the price of the option will be

$$\begin{aligned} f_L(S_t) &= \mathbb{E}^* \left[e^{-r(\tau_L - t)} (S_{\tau_L} - K)^+ \Big| S_t \right] \\ &= \mathbb{E}^* \left[e^{-r(\tau_L - t)} (L - K)^+ \Big| S_t \right] \\ &= (L - K) \, \mathbb{E}^* \left[e^{-r(\tau_L - t)} \Big| S_t \right]. \end{aligned}$$

Proposition 9.5 *We have*

$$f_L(x) = \begin{cases} x - K, & x \geq L > K, \\ (L - K)\dfrac{x}{L}, & 0 < x \leq L. \end{cases} \tag{9.26}$$

Proof. We only need to consider the case $x < L$. Note that for all $\lambda \in \mathbb{R}$,

$$Z_t := \left(\frac{S_t}{S_0} \right)^{\lambda} e^{-r\lambda t + \lambda\sigma^2 t/2 - \lambda^2\sigma^2 t/2} = e^{\lambda\sigma\tilde{B}_t - \lambda^2\sigma^2 t/2}, \qquad t \in \mathbb{R}_+,$$

is a martingale under the risk-neutral probability measure $\tilde{\mathbb{P}}$. Hence the stopped process $(Z_{t\wedge\tau_L})_{t\in\mathbb{R}_+}$ is a martingale and it has constant expectation. Hence we have

$$\mathbb{E}^*[Z_{t\wedge\tau_L}] = \mathbb{E}^*[Z_0] = 1,$$

and by letting t go to infinity we get

$$\mathbb{E}^* \left[\left(\frac{S_{\tau_L}}{S_0} \right)^{\lambda} e^{-(r\lambda - \lambda\sigma^2/2 + \lambda^2\sigma^2/2)\tau_L} \right] = 1,$$

which yields

$$\mathbb{E}^* \left[e^{-(r\lambda - \lambda\sigma^2/2 + \lambda^2\sigma^2/2)\tau_L} \right] = \left(\frac{S_0}{L} \right)^{\lambda},$$

i.e.,

$$\mathbb{E}^* \left[e^{-r\tau_L} \right] = \left(\frac{S_0}{L} \right)^{\lambda}, \tag{9.27}$$

provided that λ is chosen such that

$$-(r\lambda - \lambda\sigma^2/2 + \lambda^2\sigma^2/2) = -r,$$

i.e.,

$$0 = \lambda^2\sigma^2/2 + \lambda(r - \sigma^2/2) - r = \frac{\sigma^2}{2}(\lambda + 2r/\sigma^2)(\lambda - 1).$$

Here we choose the positive solution $\lambda = 1$ since $S_0 = x < L$ and the expectation (9.27) is lower than 1. □

One sees from Figure 9.7 that the situation completely differs from the perpetual put option case, as there does not exist an optimal value L^* that would maximize the option price for all values of the underlying.

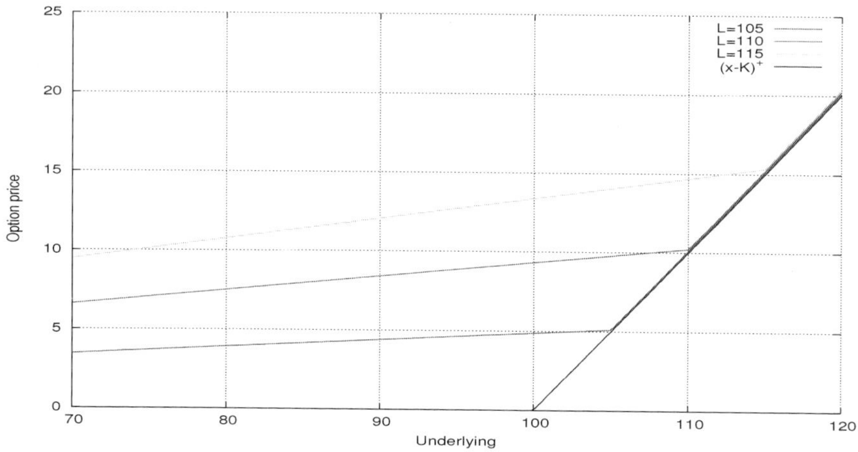

FIGURE 9.7: Graphs of the option price by exercising at τ_L for several values of L.

The intuition behind this picture is that there is no upper limit above which one should exercise the option, and in order to price the American perpetual call option we have to let L go to infinity, i.e., the "optimal" exercise strategy is to wait indefinitely.

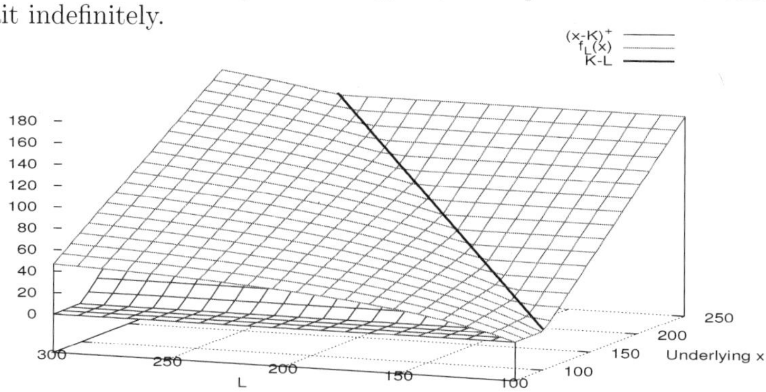

FIGURE 9.8: Graphs of the option prices parametrized by different values of L.

We check from (9.26) that

$$\lim_{L\to\infty} f_L(x) = x - \lim_{L\to\infty} K\frac{x}{L} = x, \qquad x > 0. \tag{9.28}$$

As a consequence we have the following proposition.

Proposition 9.6 *The price of the perpetual American call option is given by*

$$\sup_{\substack{\tau\geq t\\ \tau \text{ stopping time}}} \mathbb{E}^*\left[e^{-r(\tau-t)}(S_\tau - K)^+ \Big| S_t\right] = S_t, \qquad t \in \mathbb{R}_+. \tag{9.29}$$

Proof. For all $L > K$ we have

$$\begin{aligned} f_L(S_t) &= \mathbb{E}^*\left[e^{-r(\tau_L-t)}(S_{\tau_L} - K)^+ \Big| S_t\right] \\ &\leq \sup_{\substack{\tau\geq t\\ \tau \text{ stopping time}}} \mathbb{E}^*\left[e^{-r(\tau-t)}(S_\tau - K)^+ \Big| S_t\right], \qquad t \geq 0, \end{aligned}$$

hence taking the limit as $L \to \infty$ yields

$$S_t \leq \sup_{\substack{\tau\geq t\\ \tau \text{ stopping time}}} \mathbb{E}^*\left[e^{-r(\tau-t)}(S_\tau - K)^+ \Big| S_t\right] \tag{9.30}$$

from (9.28). On the other hand, for all stopping times $\tau \geq t$ we have, by (9.12),

$$\mathbb{E}^*\left[e^{-r(\tau-t)}(S_\tau - K)^+ \Big| S_t\right] \leq \mathbb{E}^*\left[e^{-r(\tau-t)}S_\tau \Big| S_t\right] \leq S_t, \qquad t \geq 0,$$

since $u \mapsto e^{-r(u-t)}S_u$ is a martingale, hence

$$\sup_{\substack{\tau\geq t\\ \tau \text{ stopping time}}} \mathbb{E}^*\left[e^{-r(\tau-t)}(S_\tau - K)^+ \Big| S_t\right] \leq S_t, \qquad t \geq 0,$$

which shows (9.29) by (9.30). □

We may also check that since $(e^{-rt}S_t)_{t\in\mathbb{R}_+}$ is a martingale, the process $t \mapsto (e^{-rt}S_t - K)^+$ is a *sub*martingale since the function $x \mapsto (x-K)^+$ is convex, hence for all bounded stopping times τ such that $t \leq \tau$ we have

$$(S_t - K)^+ \leq \mathbb{E}^*\left[(e^{-r(\tau-t)}S_\tau - K)^+ \Big| S_t\right] \leq \mathbb{E}^*\left[e^{-r(\tau-t)}(S_\tau - K)^+ \Big| S_t\right],$$

$t \geq 0$, showing that it is always better to wait than to exercise at time t, and the optimal exercise time is $\tau^* = +\infty$. This argument does not apply to American put options.

9.5 Finite Expiration American Options

In this section we consider finite expirations American put and call options with strike K, whose prices can be expressed as

$$f(t,S_t) = \sup_{\substack{t\leq\tau\leq T \\ \tau \text{ stopping time}}} \mathbb{E}^*\left[e^{-r(\tau-t)}(K-S_\tau)^+ \Big| S_t \right],$$

and

$$f(t,S_t) = \sup_{\substack{t\leq\tau\leq T \\ \tau \text{ stopping time}}} \mathbb{E}^*\left[e^{-r(\tau-t)}(S_\tau-K)^+ \Big| S_t \right].$$

Two-choice optimal stopping at fixed times with finite expiration

We start by considering the optimal stopping problem in a simplified setting where $\tau \in \{t, T\}$ is allowed at time t to take only *two* values t (which corresponds to immediate exercise) and T (wait until maturity).

Call options

Since $x \mapsto (x-K)^+$ is a non-decreasing convex function and the process $(e^{-rt}S_t)_{t\in\mathbb{R}_+}$ is a martingale under $\mathbb{P}^*$, we know that $t \mapsto (e^{-rt}S_t - e^{-rT}K)^+$ remains a submartingale by the Jensen inequality (9.2), hence

$$\begin{aligned}
(S_t-K)^+ &= e^{rt}(e^{-rt}S_t - e^{-rt}K)^+ \\
&\leq e^{rt}(e^{-rt}S_t - e^{-rT}K)^+ \\
&\leq e^{rt}\,\mathbb{E}^*[(e^{-rT}S_T - e^{-rT}K)^+ \mid \mathcal{F}_t] \\
&\leq e^{-r(T-t)}\,\mathbb{E}^*[(S_T-K)^+ \mid \mathcal{F}_t],
\end{aligned}$$

i.e.,

$$(x-K)^+ \leq e^{-r(T-t)}\,\mathbb{E}^*[(S_T-K)^+ | S_t = x], \qquad x,t>0,$$

as illustrated in Figure 9.9 using the Black–Scholes formula for European call options.

In other words, taking $x = S_t$, the payoff $(S_t-K)^+$ of immediate exercise at time t is always lower than the expected payoff $e^{-r(T-t)}\,\mathbb{E}^*[(S_T-K)^+|S_t=x]$ given by exercise at maturity T. As a consequence, the optimal strategy for the investor is to wait until time T to exercise an American call option, rather than exercising earlier at time t.

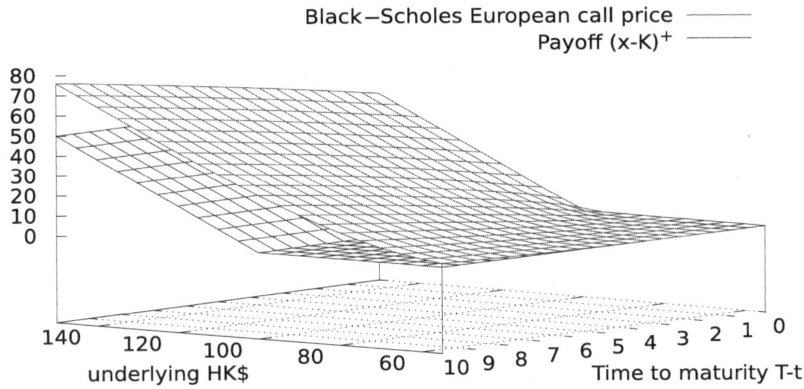

FIGURE 9.9: Expected Black–Scholes European call price vs $(x,t) \mapsto (x-K)^+$.

More generally it can be in fact shown that the price of an American call option equals the price of the corresponding European call option with maturity T, i.e.,

$$f(t,S_t) = e^{-r(T-t)}\,\mathbb{E}^*\left[(S_T-K)^+\Big|S_t\right],$$

i.e., T is the optimal exercise date, cf. e.g., §14.4 of [69] for a proof.

Put options

For put options the situation is entirely different. The Black–Scholes formula for European put options shows that the inequality

$$(K-x)^+ \leq e^{-r(T-t)}\,\mathbb{E}^*[(K-S_T)^+|S_t=x],$$

does not always hold, as illustrated in Figure 9.10.

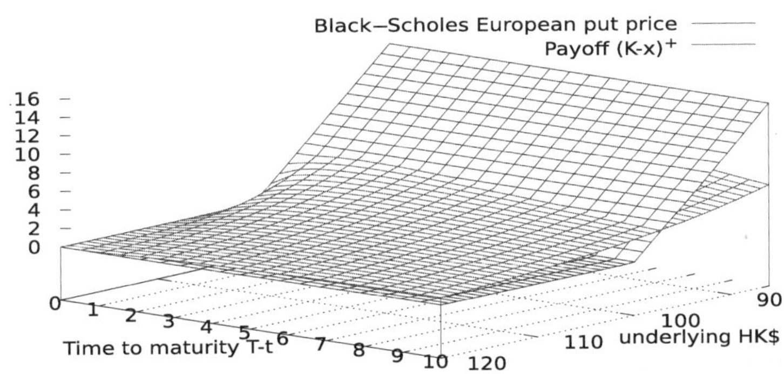

FIGURE 9.10: Black–Scholes put price function vs. $(x,t) \mapsto (K-x)^+$.

As a consequence, the optimal exercise decision for a put option depends on whether $(K - S_t)^+ \leq e^{-r(T-t)} \mathbb{E}^*[(K - S_T)^+|S_t]$ (in which case one chooses to exercise at time T) or $(K - S_t)^+ > e^{-r(T-t)} \mathbb{E}^*[(K - S_T)^+|S_t]$ (in which case one chooses to exercise at time t).

A view from above of the graph of Figure 9.10 shows the existence of an optimal frontier depending on time to maturity and on the value of the underlying asset instead of being given by a constant level L^* as in Section 9.4, cf. Figure 9.11:

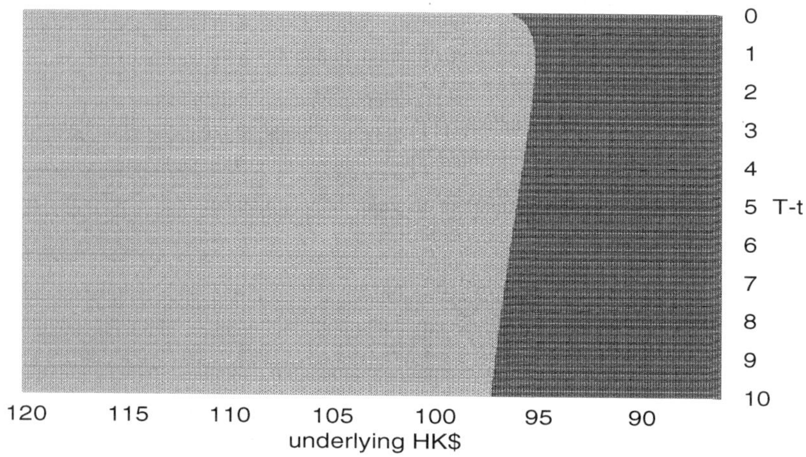

FIGURE 9.11: Optimal frontier for the exercise of a put option.

At a given time t, one will choose to exercise immediately if $(S_t, T-t)$ belongs to the area on the right, and to wait until maturity if $(S_t, T-t)$ belongs to the area on the left.

PDE characterization of the finite expiration American put price

Let us describe the PDE associated to American put options. After discretization $\{0 = t_0, t_1, \ldots, t_N = T\}$ of the time interval $[0, T]$, the optimal exercise strategy for an American put option can be described as follows at each time-step:

If $f(t, S_t) > (K - S_t)^+$, wait.

If $f(t, S_t) = (K - S_t)^+$, exercise the option at time t.

Note that we cannot have $f(t, S_t) < (K - S_t)^+$.

If $f(t, S_t) > (K - S_t)^+$ the expected return of the option equals that of the risk-free asset. This means that $f(t, S_t)$ follows the Black–Scholes PDE

$$rf(t, S_t) = \frac{\partial f}{\partial t}(t, S_t) + rS_t\frac{\partial f}{\partial x}(t, S_t) + \frac{1}{2}\sigma^2 S_t^2\frac{\partial^2 f}{\partial x^2}(t, S_t),$$

whereas if $f(t, S_t) = (K - S_t)^+$ it is not worth waiting as the return of the option is lower than that of the risk-free asset:

$$rf(t, S_t) \geq \frac{\partial f}{\partial t}(t, S_t) + rS_t\frac{\partial f}{\partial x}(t, S_t) + \frac{1}{2}\sigma^2 S_t^2\frac{\partial^2 f}{\partial x^2}(t, S_t).$$

As a consequence, $f(t, x)$ should solve the following variational PDE:

$$\begin{cases} f(t, x) \geq (K - x)^+, & \text{(9.31a)} \\ \dfrac{\partial f}{\partial t}(t, x) + rx\dfrac{\partial f}{\partial x}(t, x) + \dfrac{1}{2}\sigma^2 x^2\dfrac{\partial^2 f}{\partial x^2}(t, x) \leq rf(t, x), & \text{(9.31b)} \\ \left(\dfrac{\partial f}{\partial t}(t, x) + rx\dfrac{\partial f}{\partial x}(t, x) + \dfrac{1}{2}\sigma^2 x^2\dfrac{\partial^2 f}{\partial x^2}(t, x) - rf(t, x)\right) & \text{(9.31c)} \\ \qquad\qquad \times\, (f(t, x) - (K - x)^+) = 0, & \end{cases}$$

subject to the terminal condition $f(T, x) = (K - x)^+$. In other words, equality holds either in (9.31a) or in (9.31b) due to the presence of the term $(f(t, x) - (K - x)^+)$ in (9.31c).

The optimal exercise strategy consists in exercising the put option as soon as the equality $f(u, S_u) = (K - S_u)^+$ holds, i.e., at the time

$$\tau^* = \inf\{u \geq t \ : \ f(u, S_u) = (K - S_u)^+\},$$

after which the process $\tilde{f}_{L^*}(S_t)$ ceases to be a martingale and becomes a (strict) supermartingale.

A simple procedure to compute numerically the price of an American put option is to use a finite difference scheme while simply enforcing the condition $f(t, x) \geq (K - x)^+$ at iteration, adding the condition

$$f(t_i, x_j) := \max(f(t_i, x_j), (K - x_j)^+)$$

right after the computation of $f(t_i, x_j)$.

The next figure shows a numerical resolution of the variational PDE (9.31a)–(9.31c) using the above simplified (implicit) finite difference scheme. In comparison with Figure 9.6, one can check that the PDE solution becomes time-dependent in the finite expiration case.

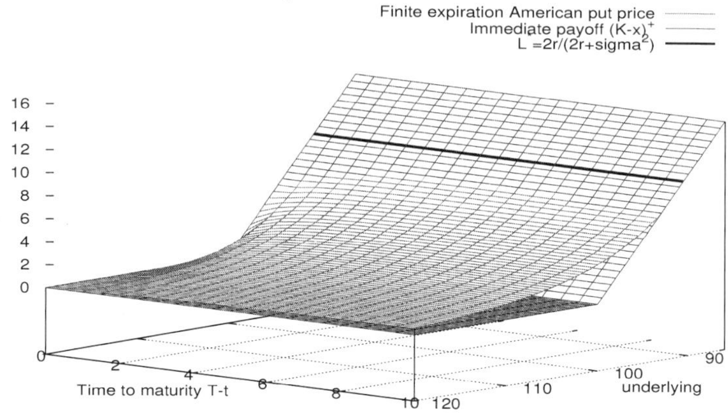

FIGURE 9.12: Numerical values of the finite expiration American put price.

In general, one will choose to exercise the put option when

$$f(t, S_t) = (K - S_t)^+,$$

i.e., within the upper area in Figure (9.12). We check that the optimal threshold $L^* = 90.64$ of the corresponding perpetual put option is within the exercise region, which is consistent since the perpetual optimal strategy should allow one to wait longer than in the finite expiration case.

The numerical computation of the put price

$$f(t, S_t) = \sup_{\substack{t \le \tau \le T \\ \tau \text{ stopping time}}} \mathbb{E}^* \left[e^{-r(\tau - t)} (K - S_\tau)^+ \Big| S_t \right]$$

can also be done by dynamic programming and backward optimization using the Longstaff–Schwartz (or Least Square Monte Carlo, LSM) algorithm [46], as in Figure 9.13.

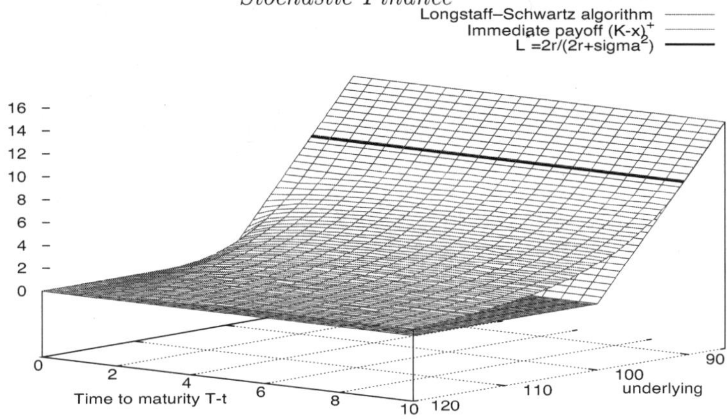

FIGURE 9.13: Longstaff–Schwartz algorithm for the finite expiration American put price.

In Figure 9.13 above and Figure 9.14 below the optimal threshold of the corresponding perpetual put option is again $L^* = 90.64$ and falls within the exercise region. Also, the optimal threshold is closer to L^* for large time to maturities, which shows that the perpetual option approximates the finite expiration option in that situation. In Figure 9.14 we compare the numerical computation of the American put price by the finite difference and Longstaff–Schwartz methods.

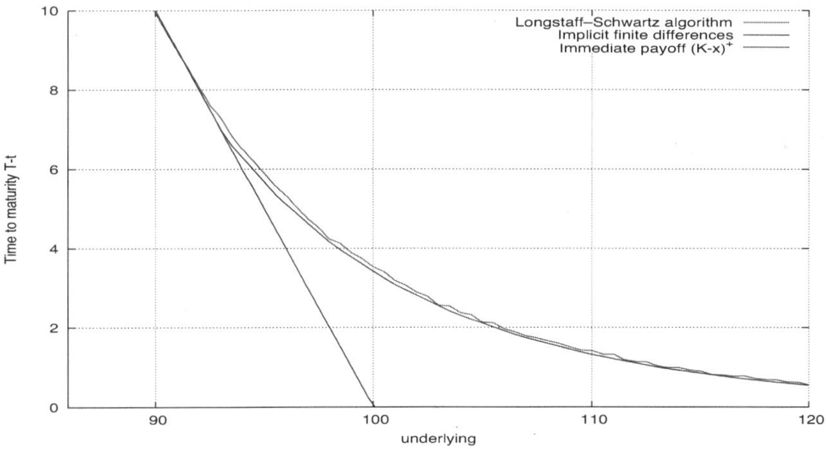

FIGURE 9.14: Comparison between Longstaff–Schwartz and finite differences.

It turns out that, although both results are very close, the Longstaff–Schwartz method performs better in the critical area close to exercise as it yields the expected continuously differentiable solution, and the simple numerical PDE solution tends to underestimate the optimal threshold. Also, a small error in the values of the solution translates into a large error on the value of the optimal exercise threshold.

The finite expiration American call option

In the next proposition we compute the price of a finite expiration American call option with an arbitrary convex payoff function ϕ.

Proposition 9.7 *Let $\phi : \mathbb{R} \longrightarrow \mathbb{R}$ be a nonnegative convex function such that $\phi(0) = 0$. The price of the finite expiration American call option with payoff function ϕ on the underlying asset $(S_t)_{t\in\mathbb{R}_+}$ is given by*

$$f(t, S_t) = \sup_{\substack{t\leq\tau\leq T \\ \tau \text{ stopping time}}} \mathbb{E}^* \left[e^{-r(\tau-t)}\phi(S_\tau) \Big| S_t \right] = e^{-r(T-t)}\, \mathbb{E}^* \left[\phi(S_T) \Big| S_t \right],$$

i.e., *the optimal strategy is to wait until the maturity time T to exercise the option, and $\tau^* = T$.*

Proof. Since the function ϕ is convex and $\phi(0) = 0$ we have

$$\phi(px) = \phi((1-p)\times 0 + px) \leq (1-p)\times\phi(0) + p\phi(x) \leq p\phi(x),$$

for all $p \in [0, 1]$ and $x \geq 0$. Hence the process $s \mapsto e^{-rs}\phi(S_{t+s})$ is a *sub*martingale since taking $p = e^{-r(\tau-t)}$ we have

$$\begin{aligned} \mathbb{E}^* \left[e^{-rs}\phi\left(S_{t+s}\right) \mid \mathcal{F}_t \right] &\geq e^{-rs}\phi\left(\mathbb{E}^*\left[S_{t+s} \mid \mathcal{F}_t\right]\right) \\ &\geq \phi\left(\mathbb{E}^*\left[e^{-rs}S_{t+s} \mid \mathcal{F}_t\right]\right) \\ &= \phi(S_t), \end{aligned}$$

where we used Jensen's inequality (9.2) applied to the convex function ϕ. Hence by the optional sampling theorem for *sub*martingales, cf (9.10), for all (bounded) stopping times τ comprised between t and T we have,

$$\mathbb{E}^*[e^{-r(\tau-t)}\phi(S_\tau) \mid \mathcal{F}_t] \leq e^{-r(T-t)}\, \mathbb{E}^*[\phi(S_T) \mid \mathcal{F}_t],$$

i.e., it is always better to wait until time T than to exercise at time $\tau \in [t, T]$, and this yields

$$\sup_{\substack{t\leq\tau\leq T \\ \tau \text{ stopping time}}} \mathbb{E}^* \left[e^{-r(\tau-t)}\phi(S_\tau) \Big| S_t \right] \leq e^{-r(T-t)}\, \mathbb{E}^* \left[\phi(S_T) \Big| S_t \right].$$

The converse inequality

$$e^{-r(T-t)}\, \mathbb{E}^* \left[\phi(S_T) \Big| S_t \right] \leq \sup_{\substack{t\leq\tau\leq T \\ \tau \text{ stopping time}}} \mathbb{E}^* \left[e^{-r(\tau-t)}\phi(S_\tau) \Big| S_t \right],$$

being obvious because T is a stopping time. □

As a consequence of Proposition 9.7 applied to the convex function $\phi(x) = (x - K)^+$, the price of the finite expiration American call option is given by

$$\begin{aligned} f(t, S_t) &= \sup_{\substack{t \le \tau \le T \\ \tau \text{ stopping time}}} \mathbb{E}^* \left[e^{-r(\tau - t)} (S_\tau - K)^+ \Big| S_t \right] \\ &= e^{-r(T-t)} \mathbb{E}^* \left[(S_T - K)^+ \Big| S_t \right], \end{aligned}$$

i.e., the optimal strategy is to wait until the maturity time T to exercise the option.

In the following table we summarize the optimal exercise strategies for the pricing of American options.

option type	perpetual	finite expiration
put option	$\begin{cases} K - S_t, & 0 < S_t \le L^*, \\ (K - L^*) \left(\frac{S_t}{L^*} \right)^{-2r/\sigma^2}, & S_t \ge L^*. \end{cases}$ $\tau^* = \tau_{L^*}$	Solve the PDE (9.31a)-(9.31c) for f(t,x) or use Longstaff–Schwartz [46] $\tau^* = T \wedge \inf\{u \ge t : f(u, S_u) = (K - S_u)^+\}$
call option	S_t $\tau^* = +\infty$	$e^{-r(T-t)} \mathbb{E}^*[(S_T - K)^+ \mid S_t]$, $\tau^* = T$

Exercises

Exercise 9.1 Stopping times. Let $(B_t)_{t \in \mathbb{R}_+}$ be a standard Brownian motion started at 0.

1. Consider the random time τ defined by

$$\tau := \inf\{t \in \mathbb{R}_+ \ : \ e^{B_t} = \alpha e^{-t/2}\},$$

which represents the first time the exponential of Brownian motion B_t crosses the path of $t \mapsto \alpha e^{-t/2}$, where $\alpha > 1$.

Is τ a stopping time?

2. If τ is a stopping time, compute $E[e^{-\tau}]$ by the stopping time theorem.

3. Consider the random time ν defined by

$$\nu := \inf\{t \in \mathbb{R}_+ \ : \ e^{B_t} = e^{B_1}\},$$

which represents the first time the exponential of Brownian motion B_t hits the level e^{B_1}.

Is ν a stopping time?

Exercise 9.2 (Doob–Meyer decomposition in discrete time). Let $(M_n)_{n\in\mathbb{N}}$ be a discrete-time *sub*martingale with respect to a filtration $(\mathcal{F}_n)_{n\in\mathbb{N}}$, with $\mathcal{F}_{-1} = \{\emptyset, \Omega\}$.

1. Show that there exist two processes $(N_n)_{n\in\mathbb{N}}$ and $(A_n)_{n\in\mathbb{N}}$ such that

   (i) $(N_n)_{n\in\mathbb{N}}$ is a martingale with respect to $(\mathcal{F}_n)_{n\in\mathbb{N}}$,

   (ii) $(A_n)_{n\in\mathbb{N}}$ is non-decreasing, i.e., $A_n \leq A_{n+1}$ a.s., $n \in \mathbb{N}$,

   (iii) $(A_n)_{n\in\mathbb{N}}$ is predictable in the sense that A_n is $\mathcal{F}_{n-1}$-measurable, $n \in \mathbb{N}$, and

   (iv) $M_n = N_n + A_n$, $n \in \mathbb{N}$.

   Hint: Let $A_0 = 0$,

$$A_{n+1} = A_n + \mathbb{E}[M_{n+1} - M_n \mid \mathcal{F}_n], \quad n \geq 0,$$

   and define $(N_n)_{n\in\mathbb{N}}$ in such a way that it satisfies the four required properties.

2. Show that for all bounded stopping times σ and τ such that $\sigma \leq \tau$ a.s., we have

$$\mathbb{E}[M_\sigma] \leq \mathbb{E}[M_\tau].$$

   Hint: Use the stopping time Theorem 9.1 for martingales and (9.11).

Exercise 9.3 American digital options. An American digital call (resp. put) option with maturity $T > 0$ can be exercised at any time $t \in [0, T]$, at the choice of the option holder.

The call (resp. put) option exercised at time t yields the payoff $\mathbf{1}_{[K,\infty)}(S_t)$ (resp. $\mathbf{1}_{[0,K]}(S_t)$), and the option holder wants to find an exercise strategy that will maximize his payoff.

1. Consider the following possible situations at time t:

(i) $S_t \geq K$,

(ii) $S_t < K$.

In each case (*i*) and (*ii*), tell whether you would choose to exercise the <u>call</u> option immediately or wait.

2. Consider the following possible situations at time t:

(i) $S_t > K$,

(ii) $S_t \leq K$.

In each case (*i*) and (*ii*), tell whether you would choose to exercise the <u>put</u> option immediately or wait.

3. The price $C_d^{\mathrm{Am}}(t, S_t)$ of an American digital call option is known to satisfy the Black–Scholes PDE

$$rC_d^{\mathrm{Am}}(t,x) = \frac{\partial}{\partial t} C_d^{\mathrm{Am}}(t,x) + rx\frac{\partial}{\partial x} C_d^{\mathrm{Am}}(t,x) + \frac{1}{2}\sigma^2 x^2 \frac{\partial^2}{\partial x^2} C_d^{\mathrm{Am}}(t,x).$$

Based on your answers to Question 1, how would you set the boundary conditions $C_d^{\mathrm{Am}}(t, K)$, $0 \leq t < T$, and $C_d^{\mathrm{Am}}(T, x)$, $0 \leq x < K$?

4. The price $P_d^{\mathrm{Am}}(t, S_t)$ of an American digital put option is known to satisfy the same Black–Scholes PDE

$$rP_d^{\mathrm{Am}}(t,x) = \frac{\partial}{\partial t} P_d^{\mathrm{Am}}(t,x) + rx\frac{\partial}{\partial x} P_d^{\mathrm{Am}}(t,x) + \frac{1}{2}\sigma^2 x^2 \frac{\partial^2}{\partial x^2} P_d^{\mathrm{Am}}(t,x). \tag{9.32}$$

Based on your answers to Question 2, how would you set the boundary conditions $P_d^{\mathrm{Am}}(t, K)$, $0 \leq t < T$, and $P_d^{\mathrm{Am}}(T, x)$, $x > K$?

5. Show that the optimal exercise strategy for the American digital call option with strike K is to exercise as soon as the underlying reaches the level K, at the time

$$\tau_K = \inf\{u \geq t \ : \ S_u = K\},$$

starting from any level $S_t \leq K$, and that the price $C_d^{\mathrm{Am}}(t, S_t)$ of the American digital call option is given by

$$C_d^{\mathrm{Am}}(t,x) = \mathbb{E}[e^{-r(\tau_K - t)} \mathbf{1}_{\{\tau_K < T\}} \mid S_t = x].$$

6. Show that the price $C_d^{\mathrm{Am}}(t, S_t)$ of the American digital call option is equal to

$$C_d^{\mathrm{Am}}(t,x) = \frac{x}{K}\Phi\left(\frac{(r + \sigma^2/2)\tau + \log(x/K)}{\sigma\sqrt{\tau}}\right)$$

$$+\left(\frac{x}{K}\right)^{-2r/\sigma^2}\Phi\left(\frac{-(r+\sigma^2/2)\tau+\log(x/K)}{\sigma\sqrt{\tau}}\right), \qquad 0\le x\le K,$$

where $\tau = T - t$. Show that this formula is consistent with your answer to Question 3.

7. Show that the optimal exercise strategy for the American digital put option with strike K is to exercise as soon as the underlying reaches the level K, at the time

$$\tau_K = \inf\{u \ge t \ : \ S_u = K\},$$

starting from any level $S_t \ge K$, and that the price $P_d^{\rm Am}(t, S_t)$ of the American digital put option is

$$P_d^{\rm Am}(t,x) = \mathbb{E}[e^{-r(\tau_K - t)}\mathbf{1}_{\{\tau_K<T\}} \mid S_t = x], \qquad x \ge K.$$

8. Show that the price $P_d^{\rm Am}(t, S_t)$ of the American digital put option is equal to

$$\begin{aligned} P_d^{\rm Am}(t,x) &= \frac{x}{K}\Phi\left(\frac{-(r+\sigma^2/2)\tau-\log(x/K)}{\sigma\sqrt{\tau}}\right) \\ &\quad +\left(\frac{x}{K}\right)^{-2r/\sigma^2}\Phi\left(\frac{(r+\sigma^2/2)\tau-\log(x/K)}{\sigma\sqrt{\tau}}\right), \qquad x\ge K, \end{aligned}$$

and that this formula is consistent with your answer to Question 4.

9. Does the call-put parity hold for American digital options?

Exercise 9.4 American forward contracts. Consider $(S_t)_{t\in\mathbb{R}_+}$ an asset price process given by

$$\frac{dS_t}{S_t} = rdt + \sigma dB_t,$$

where $(B_t)_{t\in\mathbb{R}_+}$ is a standard Brownian motion.

1. Compute the price

$$f(t,S_t) = \sup_{\substack{t\le\tau\le T \\ \tau \text{ stopping time}}} \mathbb{E}^*\left[e^{-r(\tau-t)}(K-S_\tau)\Big| S_t\right],$$

and optimal exercise strategy of a finite expiration American type put forward contract with strike K on the underlying asset $(S_t)_{t\in\mathbb{R}_+}$, with payoff $K - S_T$.

2. Compute the price

$$f(t,S_t) = \sup_{\substack{t\le\tau\le T \\ \tau \text{ stopping time}}} \mathbb{E}^*\left[e^{-r(\tau-t)}(S_\tau-K)\Big| S_t\right],$$

and optimal exercise strategy of a finite expiration American type call forward contract with strike K on the underlying asset $(S_t)_{t\in\mathbb{R}_+}$, with payoff $S_T - K$.

3. How are the answers to Questions 1 and 2 modified in the case of perpetual options?

Exercise 9.5 Let $p \geq 1$ and consider a power put option with payoff

$$((\kappa - S_\tau)^+)^p = \begin{cases} (\kappa - S_\tau)^p & \text{if } S_\tau \leq \kappa, \\ 0 & \text{if } S_\tau > \kappa, \end{cases}$$

when exercised at time τ, on an underlying asset whose price S_t is written as

$$S_t = S_0 e^{rt+\sigma B_t - \sigma^2 t/2}, \qquad t \in \mathbb{R}_+,$$

where $(B_t)_{t\in\mathbb{R}_+}$ is a standard Brownian motion under the risk-neutral probability measure $\mathbb{P}^*$, $r \geq 0$ is the risk-free interest rate, and $\sigma > 0$ is the volatility coefficient.

Given $L \in (0, \kappa)$ a fixed price, consider the following choices for the exercise of a *put* option with strike κ:

- If $S_t \leq L$, then exercise at time t.
- Otherwise, wait until the first hitting time $\tau_L := \inf\{u \geq t \ : \ S_u = L\}$, and exercise the option at time τ_L.

1. Under the above strategy, what is the option payoff equal to if $S_t \leq L$?
2. Show that in case $S_t > L$, the price of the option is equal to

$$f_L(S_t) = (\kappa - L)^p \, \mathbb{E}^* \left[e^{-r(\tau_L - t)} \middle| S_t \right].$$

3. Compute the price $f_L(S_t)$ of the option at time t.

   Hint. Recall that by (9.21) we have $\mathbb{E}^*[e^{-r(\tau_L-t)} \mid S_t = x] = (x/L)^{-2r/\sigma^2}$, $x \geq L$.

4. Compute the optimal value L^* that maximizes $L \mapsto f_L(x)$ for all fixed $x > 0$.

   Hint. Observe that, geometrically, the slope of $x \mapsto f_L(x)$ at $x = L^*$ is equal to $-p(\kappa - L^*)^{p-1}$.

5. How would you compute the American option price

$$f(t, S_t) = \sup_{\substack{\tau \geq t \\ \tau \text{ stopping time}}} \mathbb{E}^* \left[e^{-r(\tau - t)}((\kappa - S_\tau)^+)^p \Big| S_t \right] \ ?$$

Exercise 9.6 Same questions as in Exercise 9.5 for the option with payoff $\kappa - (S_\tau)^p$ when exercised at time τ, with $p > 0$.

Exercise 9.7 (cf. Exercise 8.5 page 372 of [68]). Consider an underlying asset price process written as

$$S_t = S_0 e^{(r-a)t + \sigma \tilde{B}_t - \sigma^2 t/2}, \qquad t \in \mathbb{R}_+,$$

where $(\tilde{B}_t)_{t\in\mathbb{R}_+}$ is a standard Brownian motion under the risk-neutral probability measure $\mathbb{P}^*$, $r > 0$ is the risk-free interest rate, $\sigma > 0$ is the volatility coefficient, and $a > 0$ is a constant dividend rate.

1. Show that for all $x \geq L$ and $\lambda \in \mathbb{R}$ the process $(Z_t)_{t\in\mathbb{R}_+}$ defined as

$$Z_t := \left(\frac{S_t}{S_0} \right)^\lambda e^{-(r-a)\lambda t + \lambda\sigma^2 t/2 - \lambda^2\sigma^2 t/2}, \qquad t \in \mathbb{R}_+,$$

   is a martingale under $\mathbb{P}^*$.

2. Let τ_L denote the hitting time

$$\tau_L = \inf\{u \in \mathbb{R}_+ \ : \ S_u \leq L\}.$$

   By application of the stopping time theorem to the martingale $(Z_t)_{t\in\mathbb{R}_+}$, show that

$$\mathbb{E}^* \left[e^{-r\tau_L} \right] = \left(\frac{S_0}{L} \right)^\lambda, \tag{9.33}$$

   with

$$\lambda = \frac{-(r - a - \sigma^2/2) - \sqrt{(r - a - \sigma^2/2)^2 + 4r\sigma^2/2}}{\sigma^2}. \tag{9.34}$$

3. Show that for all $L \in (0, K)$ we have

$$\mathbb{E}^* \left[e^{-r\tau_L}(K - S_{\tau_L})^+ \Big| S_0 = x \right] = \begin{cases} K - x, & 0 < x \leq L, \\ (K - L) \left(\dfrac{x}{L} \right)^{\frac{-(r-a-\sigma^2/2) - \sqrt{(r-a-\sigma^2/2)^2 + 4r\sigma^2/2}}{\sigma^2}}, & x \geq L. \end{cases}$$

4. Show that the value L^* of L that maximizes

$$f_L(x) := \mathbb{E}^*\left[e^{-r\tau_L}(K - S_{\tau_L})^+ \Big| S_0 = x\right]$$

for all x is given by

$$L^* = \frac{\lambda}{\lambda - 1}K.$$

5. Show that

$$f_{L^*}(x) = \begin{cases} K - x, & 0 < x \leq L^* = \dfrac{\lambda}{\lambda - 1}K, \\ \left(\dfrac{1-\lambda}{K}\right)^{\lambda - 1}\left(\dfrac{x}{-\lambda}\right)^{\lambda}, & x \geq L^* = \dfrac{\lambda}{\lambda - 1}K, \end{cases}$$

6. Show by hand computation that $f_{L^*}(x)$ satisfies the variational differential equation

$$\begin{cases} f_{L^*}(x) \geq (K - x)^+, & (9.35a) \\ (r - a)xf'_{L^*}(x) + \dfrac{1}{2}\sigma^2 x^2 f''_{L^*}(x) \leq rf_{L^*}(x), & (9.35b) \\ \left(rf_{L^*}(x) - (r - a)xf'_{L^*}(x) - \dfrac{1}{2}\sigma^2 x^2 f''_{L^*}(x)\right) & (9.35c) \\ \qquad \times (f_{L^*}(x) - (K - x)^+) = 0. \end{cases}$$

7. By Itô's formula, check that the discounted portfolio price

$$t \mapsto e^{-rt} f_{L^*}(S_t)$$

is a supermartingale.

8. Show that we have

$$f_{L^*}(S_0) \geq \sup_{\tau \text{ stopping time}} \mathbb{E}^*\left[e^{-r\tau}(K - S_\tau)^+ \Big| S_0\right].$$

9. Show that the stopped process

$$s \mapsto e^{-r(s \wedge \tau_{L^*})} f_{L^*}(S_{s \wedge \tau_{L^*}}), \qquad s \in \mathbb{R}_+,$$

is a martingale, and that

$$f_{L^*}(S_0) \leq \sup_{\tau \text{ stopping time}} \mathbb{E}^*\left[e^{-r\tau}(K - S_\tau)^+\right].$$

10. Fix $t \in \mathbb{R}_+$ and let τ_{L^*} denote the hitting time

$$\tau_{L^*} = \inf\{u \geq t \ : \ S_u = L^*\}.$$

Conclude that the price of the perpetual American put option with dividend is given for all $t \geq 0$ by

$$\begin{aligned} f_{L^*}(S_t) &= \mathbb{E}^*\left[e^{-r(\tau_{L^*}-t)}(K - S_{\tau_{L^*}})^+ \Big| S_t\right] \\ &= \begin{cases} K - S_t, & 0 < S_t \leq \dfrac{\lambda}{\lambda-1}K, \\ \left(\dfrac{1-\lambda}{K}\right)^{\lambda-1}\left(\dfrac{S_t}{-\lambda}\right)^{\lambda}, & S_t \geq \dfrac{\lambda}{\lambda-1}K, \end{cases} \end{aligned}$$

where λ is given by (9.34), and

$$\tau_{L^*} = \inf\{u \geq t \ : \ S_u \leq L\}.$$

Exercise 9.8 This exercise is a simplified adaptation of the paper [29].

We consider two risky assets S_1 and S_2 modeled by

$$S_1(t) = s_1(\theta)e^{\sigma_1 W_t + rt - \sigma_2^2 t/2} \quad \text{and} \quad S_2(t) = s_2(\theta)e^{\sigma_2 W_t + rt - \sigma_2^2 t/2}, \ t \in \mathbb{R}_+, \quad (9.36)$$

with $\sigma_2 > \sigma_1 \geq 0$, and the perpetual optimal stopping problem

$$\sup_{\tau \text{ stopping time}} \mathbb{E}[e^{-r\tau}(S_1(\tau) - S_2(\tau))^+],$$

where $(W_t)_{t\in\mathbb{R}_+}$ is a standard Brownian motion under $\mathbb{P}$.

1. Find $\alpha > 1$ such that the process

$$Z_t := e^{-rt}S_1(t)^\alpha S_2(t)^{1-\alpha}, \qquad t \in \mathbb{R}_+, \qquad (9.37)$$

is a martingale.

2. For some fixed $L \geq 1$, consider the hitting time

$$\tau_L = \inf\{t \in \mathbb{R}_+ \ : \ S_1(t) \geq L S_2(t)\},$$

and show that

$$\mathbb{E}[e^{-r\tau_L}(S_1(\tau_L) - S_2(\tau_L))^+] = (L-1)\,\mathbb{E}[e^{-r\tau_L}S_2(\tau_L)].$$

3. By an application of the stopping time theorem to the martingale (9.37), show that we have

$$\mathbb{E}[e^{-r\tau_L}(S_1(\tau_L) - S_2(\tau_L))^+] = \frac{L-1}{L^\alpha}S_1(0)^\alpha S_2(0)^{1-\alpha}.$$

4. Show that the price of the perpetual exchange option is given by

$$\sup_{\tau \text{ stopping time}} \mathbb{E}[e^{-r\tau}(S_1(\tau) - S_2(\tau))^+] = \frac{L^* - 1}{(L^*)^\alpha} S_1(0)^\alpha S_2(0)^{1-\alpha},$$

where

$$L^* = \frac{\alpha}{\alpha - 1}.$$

5. As an application of Question 4, compute the perpetual American put option price

$$\sup_{\tau \text{ stopping time}} \mathbb{E}[e^{-r\tau}(\kappa - S_2(\tau))^+]$$

when $r = \sigma_2^2/2$.

Chapter 10

Change of Numéraire and Forward Measures

In this chapter we introduce the notion of numéraire. This allows us to consider pricing under random discount rates using forward measures, with the pricing of exchange options (Margrabe formula) and foreign exchange options (Garman–Kohlagen formula) as main applications. A short introduction to the computation of self-financing hedging strategies under change of numéraire is also given in Section 10.5. The change of numéraire technique and associated forward measures will also be applied to the pricing of bonds and interest rate derivatives such as bond options in Chapter 12.

10.1 Notion of Numéraire

A *numéraire* is any strictly positive $\mathcal{F}_t$-adapted stochastic process $(N_t)_{t\in\mathbb{R}_+}$ that can be taken as a unit of reference when pricing an asset or a claim.

In general the price S_t of an asset in the numéraire N_t is given by

$$\hat{S}_t = \frac{S_t}{N_t}, \qquad t \in \mathbb{R}_+.$$

Deterministic numéraires transformations are easy to handle as a change of numéraire by a deterministic factor is a formal algebraic transformation that does not involve any risk. This can be the case for example when a currency is pegged to another currency, e.g., the exchange rate 6.55957 from Euro to French franc was fixed on January 1st, 1999.

On the other hand, a random numéraire may involve risk and allow for arbitrage opportunities.

Examples of numéraire include:

– the *money market account*

$$N_t = \exp\left(\int_0^t r_s ds\right),$$

where $(r_t)_{t\in\mathbb{R}_+}$ is a possibly random and time-dependent risk-free interest rate.

In this case,

$$\hat{S}_t = \frac{S_t}{N_t} = e^{-\int_0^t r_s ds} S_t, \qquad t \in \mathbb{R}_+,$$

represents the discounted price of the asset at time 0.

- an *exchange rate* $N_t = R_t$ with respect to a foreign currency.

  In this case,

$$\hat{S}_t = \frac{S_t}{R_t}, \qquad t \in \mathbb{R}_+,$$

  represents the price of the asset in units of the foreign currency. For example, if $R_t = 1.7$ is the exchange rate from Euro to Singapore dollar and $S_t = S\$1$, then $\hat{S}_t = S_t/R_t =$€ 0.59.

- *forward numéraire:* the price

$$P(t,T) = \mathbb{E}^*\left[e^{-\int_t^T r_s ds}\Big|\mathcal{F}_t\right], \qquad 0 \leq t \leq T,$$

  of a bond paying $P(T,T) = \$1$ at maturity T, in this case $R_t = P(t,T)$, $0 \leq t \leq T$. We check that

$$t \longmapsto e^{-\int_0^t r_s ds} P(t,T) = \mathbb{E}^*\left[e^{-\int_0^T r_s ds}\Big|\mathcal{F}_t\right], \qquad 0 \leq t \leq T,$$

  is a martingale.

- *annuity numéraire* of the form

$$R_t = \sum_{k=1}^{n} (T_k - T_{k-1}) P(t, T_k)$$

  where $P(t,T_1), \ldots, P(t,T_n)$ are bond prices with maturities $T_1 < \cdots < T_n$ arranged according to a *tenor structure.*

- combinations of the above, for example a foreign money market account $e^{\int_0^t r_s^{\mathrm{f}} ds} R_t$, expressed in local (or domestic) currency, where $(r_t^{\mathrm{f}})_{t\in\mathbb{R}_+}$ represents a short term interest rate on the foreign market.

When the numéraire is a random process, the pricing of a claim whose value has been transformed under change of numéraire, e.g., under a change of currency, has to take into account the risks existing on the foreign market.

In particular, in order to perform a fair pricing, one has to determine a probability measure (for example on the foreign market), under which the transformed process $\hat{S}_t = S_t/N_t$ will be a martingale.

For example in case $N_t = e^{\int_0^t r_s ds}$, the risk-neutral measure $\mathbb{P}^*$ is a measure under which the discounted price process

$$\hat{S}_t = \frac{S_t}{N_t} = e^{-\int_0^t r_s ds} S_t, \qquad t \in \mathbb{R}_+,$$

is a martingale.

In the next section we will see that this property can be extended to any kind of numéraire.

10.2 Change of Numéraire

In this section we review the pricing of options by a change of measure associated to a numéraire N_t, cf. e.g., [27] and references therein.

Most of the results of this chapter rely on the following assumption, which expresses absence of arbitrage. In the sequel, $(r_t)_{t\in\mathbb{R}_+}$ denotes an $\mathcal{F}_t$-adapted short term interest rate process.

Assumption (A) *Under the risk-neutral measure* $\mathbb{P}^*$, *the discounted numéraire*

$$t \longmapsto e^{-\int_0^t r_s ds} N_t$$

is an $\mathcal{F}_t$*-martingale.*

Taking the process $(N_t)_{t\in[0,T]}$ as a numéraire, we define the *forward measure* $\hat{\mathbb{P}}$ by

$$\frac{d\hat{\mathbb{P}}}{d\mathbb{P}^*} = e^{-\int_0^T r_s ds} \frac{N_T}{N_0}. \tag{10.1}$$

Recall that from Section 6.3 the above relation rewrites as

$$d\hat{\mathbb{P}} = e^{-\int_0^T r_s ds} \frac{N_T}{N_0} d\mathbb{P}^*,$$

which is equivalent to stating that

$$\int_\Omega \xi(\omega) d\hat{\mathbb{P}}(\omega) = \int_\Omega e^{-\int_0^T r_s ds} \frac{N_T}{N_0} \xi d\mathbb{P}^*,$$

or, under a different notation,

$$\hat{\mathbb{E}}[\xi] = \mathbb{E}^* \left[e^{-\int_0^T r_s ds} \frac{N_T}{N_0} \xi \right],$$

for all integrable $\mathcal{F}_T$-measurable random variable ξ.

More generally, (10.1) and the fact that

$$t \longmapsto e^{-\int_0^t r_s ds} N_t$$

is a martingale under $\mathbb{P}^*$ under Assumption (A), imply that

$$\mathbb{E}^* \left[\frac{d\hat{\mathbb{P}}}{d\mathbb{P}^*} \Big| \mathcal{F}_t \right] = \mathbb{E}^* \left[\frac{N_T}{N_0} e^{-\int_0^T r_s ds} \Big| \mathcal{F}_t \right] = \frac{N_t}{N_0} e^{-\int_0^t r_s ds}, \tag{10.2}$$

for all integrable $\mathcal{F}_t$-measurable random variables ξ, $0 \le t \le T$. We also have the following lemma. Note that (10.2) should not be confused with (10.3).

Lemma 10.1 *We have*

$$\frac{d\hat{\mathbb{P}}_{|\mathcal{F}_t}}{d\mathbb{P}^*_{|\mathcal{F}_t}} = e^{-\int_t^T r_s ds} \frac{N_T}{N_t}, \qquad 0 \le t \le T. \tag{10.3}$$

Proof. The proof of (10.3) relies on the abstract version of the Bayes formula. we start by noting that for all integrable $\mathcal{F}_t$-measurable random variable G we have

$$\begin{aligned}
\hat{\mathbb{E}}[G\hat{\mathbb{E}}[\hat{\xi}|\mathcal{F}_t]] &= \hat{\mathbb{E}}[\hat{\mathbb{E}}[G\hat{\xi}|\mathcal{F}_t]] \\
&= \hat{\mathbb{E}}[G\hat{\xi}] \\
&= \mathbb{E}^* \left[G\hat{\xi} e^{-\int_0^T r_s ds} \frac{N_T}{N_0} \right] \\
&= \mathbb{E}^* \left[G \frac{N_t}{N_0} e^{-\int_0^t r_s ds} \mathbb{E}^* \left[\hat{\xi} e^{-\int_t^T r_s ds} \frac{N_T}{N_t} \Big| \mathcal{F}_t \right] \right] \\
&= \hat{\mathbb{E}} \left[G \, \mathbb{E}^* \left[\hat{\xi} e^{-\int_t^T r_s ds} \frac{N_T}{N_t} \Big| \mathcal{F}_t \right] \right],
\end{aligned}$$

which shows that

$$\hat{\mathbb{E}}[\hat{\xi}|\mathcal{F}_t] = \mathbb{E}^* \left[\hat{\xi} e^{-\int_t^T r_s ds} \frac{N_T}{N_t} \Big| \mathcal{F}_t \right],$$

i.e., (10.3) holds. □

We note that in case the numéraire $N_t = e^{\int_0^t r_s ds}$ is equal to the money market account we simply have $\hat{\mathbb{P}} = \mathbb{P}^*$.

Pricing Using Change of Numéraire

The change of numéraire technique is specially useful for pricing under random interest rates.

The next proposition is the basic result of this section; it provides a way to price an option with arbitrary payoff ξ under a random discount factor $e^{-\int_t^T r_s ds}$ by use of the forward measure. It will be applied in Chapter 12 to the pricing of bond options and caplets, cf. Propositions 12.1, 12.2 and 12.3 below.

Proposition 10.1 *An option with payoff ξ is priced at time t as*

$$\mathbb{E}^*\left[e^{-\int_t^T r_s ds}\xi \,\Big|\mathcal{F}_t\right] = N_t \hat{\mathbb{E}}\left[\frac{\xi}{N_T}\Big|\mathcal{F}_t\right], \qquad 0 \leq t \leq T, \tag{10.4}$$

provided $\xi/N_T \in L^1(\hat{\mathbb{P}}, \mathcal{F}_T)$.

Each application of the formula (10.4) will require to

a) identify a suitable numéraire $(N_t)_{t\in\mathbb{R}_+}$, and to

b) make sure that the ratio ξ/N_T takes a sufficiently simple form,

in order to allow for the computation of the expectation in the right-hand side of (10.4).

Proof. Proposition 10.1 relies on Relation (10.3), which shows that

$$\begin{aligned} N_t \hat{\mathbb{E}}\left[\frac{\xi}{N_T}\Big|\mathcal{F}_t\right] &= N_t \mathbb{E}^*\left[\frac{\xi}{N_T}\frac{d\hat{\mathbb{P}}_{|\mathcal{F}_t}}{d\mathbb{P}^*_{|\mathcal{F}_t}}\Big|\mathcal{F}_t\right] \\ &= \mathbb{E}^*\left[e^{-\int_t^T r_s ds}\xi\Big|\mathcal{F}_t\right], \qquad 0 \leq t \leq T. \end{aligned}$$

□

Next we consider further examples of numéraires and associated examples of option prices.

Examples:

a) *Money market account:* $N_t = e^{\int_0^t r_s ds}$, where $(r_t)_{t\in\mathbb{R}_+}$ is a possibly random and time-dependent risk-free interest rate.

In this case we have $\hat{\mathbb{P}} = \mathbb{P}^*$ and (10.4) simply reads

$$\mathbb{E}^*\left[e^{-\int_t^T r_s ds}\xi \,\Big|\mathcal{F}_t\right] = e^{\int_0^t r_s ds}\,\mathbb{E}^*\left[e^{-\int_0^T r_s ds}\xi\Big|\mathcal{F}_t\right], \qquad 0 \leq t \leq T,$$

which yields no particular information.

b) *Forward numéraire:* $N_t = P(t,T)$ is the price $P(t,T)$ of a bond maturing at time T, $0 \le t \le T$. Here, the discounted bond price process $\left(e^{-\int_0^t r_s ds} P(t,T)\right)_{t\in[0,T]}$ is an $\mathcal{F}_t$-martingale under $\mathbb{P}^*$, i.e., Assumption (A) is satisfied and $N_t = P(t,T)$ can be taken as numéraire. In this case, (10.4) shows that a random claim ξ can be priced as

$$\mathbb{E}^*\left[e^{-\int_t^T r_s ds}\xi \,\Big|\mathcal{F}_t\right] = P(t,T)\hat{\mathbb{E}}\left[\xi\Big|\mathcal{F}_t\right], \qquad 0 \le t \le T, \tag{10.5}$$

since $P(T,T) = 1$, where the forward measure $\hat{\mathbb{P}}$ satisfies

$$\frac{d\hat{\mathbb{P}}}{d\mathbb{P}^*} = e^{-\int_0^T r_s ds}\frac{P(T,T)}{P(0,T)} = \frac{e^{-\int_0^T r_s ds}}{P(0,T)} \tag{10.6}$$

by (10.1).

c) *Annuity numéraire* of the form

$$N_t = \sum_{k=1}^n (T_k - T_{k-1})P(t,T_k)$$

where $P(t,T_1),\ldots,P(t,T_n)$ are bond prices with maturities $T_1 < \cdots < T_n$. Here, (10.4) shows that

$$\begin{aligned}&\mathbb{E}^*\left[e^{-\int_t^T r_s ds}(P(T,T_n) - P(T,T_1) - \kappa N_T)^+\Big|\mathcal{F}_t\right]\\ &= N_t\hat{\mathbb{E}}\left[\left(\frac{P(T,T_n) - P(T,T_1)}{N_T} - \kappa\right)^+\Big|\mathcal{F}_t\right],\end{aligned}$$

$0 \le t \le T$, where $(P(T,T_n) - P(T,T_1))/N_T$ is a *swap rate*.

In the sequel, given $(X_t)_{t\in\mathbb{R}_+}$ an asset price process, we define the process of forward prices

$$\hat{X}_t := \frac{X_t}{N_t}, \qquad 0 \le t \le T, \tag{10.7}$$

which represents the values at times t of X_t, expressed in units of the numéraire N_t. It will be useful to determine the dynamics of $(\hat{X}_t)_{t\in\mathbb{R}_+}$ under the forward measure $\hat{\mathbb{P}}$.

Proposition 10.2 *Let $(X_t)_{t\in\mathbb{R}_+}$ denote a continuous $\mathcal{F}_t$-adapted asset price process such that*

$$t \longmapsto e^{-\int_0^t r_s ds}X_t$$

is a martingale under $\mathbb{P}^*$. Then under change of numéraire, the process $(\hat{X}_t)_{t\in[0,T]}$ of forward prices is a martingale under $\hat{\mathbb{P}}$, provided it is integrable under $\hat{\mathbb{P}}$.

Proof. We need to show that

$$\hat{\mathbb{E}}\left[\frac{X_t}{N_t}\Big|\mathcal{F}_s\right] = \frac{X_s}{N_s}, \qquad 0 \leq s \leq t, \tag{10.8}$$

and we achieve this using a standard characterization of conditional expectation. Namely, for all bounded $\mathcal{F}_s$-measurable random variables G we note that under Assumption (A) we have

$$\begin{aligned}
\hat{\mathbb{E}}\left[G\frac{X_t}{N_t}\right] &= \mathbb{E}^*\left[G\frac{X_t}{N_t}\frac{d\hat{\mathbb{P}}}{d\mathbb{P}^*}\right] \\
&= \mathbb{E}^*\left[G\frac{X_t}{N_t}\,\mathbb{E}^*\left[\frac{d\hat{\mathbb{P}}}{d\mathbb{P}^*}\Big|\mathcal{F}_t\right]\right] \\
&= \mathbb{E}^*\left[Ge^{-\int_0^t r_u du}\frac{X_t}{N_0}\right] \\
&= \mathbb{E}^*\left[Ge^{-\int_0^s r_u du}\frac{X_s}{N_0}\right] \\
&= \hat{\mathbb{E}}\left[G\frac{X_s}{N_s}\right], \qquad 0 \leq s \leq t,
\end{aligned}$$

because

$$t \longmapsto e^{-\int_0^t r_s ds} X_t$$

is a martingale. Finally, the identity

$$\hat{\mathbb{E}}\left[G\frac{X_t}{N_t}\right] = \hat{\mathbb{E}}\left[G\frac{X_s}{N_s}\right], \qquad 0 \leq s \leq t,$$

for all bounded $\mathcal{F}_s$-measurable G, implies (10.8). □

Next we will rephrase Proposition 10.2 in Proposition 10.3 using the Girsanov theorem, which briefly recalled below.

Girsanov theorem

Recall that letting

$$\Phi_t = \mathbb{E}\left[\frac{d\hat{\mathbb{P}}}{d\mathbb{P}^*}\Big|\mathcal{F}_t\right], \qquad t \in [0, T],$$

the Girsanov theorem,[1] shows that the process $(\hat{W}_t)_{t\in\mathbb{R}_+}$ defined by

$$d\hat{W}_t = dW_t - \frac{1}{\Phi_t} d\Phi_t \cdot dW_t, \qquad t \in \mathbb{R}_+, \tag{10.9}$$

[1] See e.g., Theorem 40 in [59].

is a standard Brownian motion under $\hat{\mathbb{P}}$.

Next, Relation (10.2) shows that

$$\begin{aligned}\Phi_t &= \mathbb{E}\left[\frac{d\hat{\mathbb{P}}}{d\mathbb{P}^*}\Big|\mathcal{F}_t\right] \\ &= \mathbb{E}\left[\frac{N_T}{N_0}e^{-\int_0^T r_s ds}\Big|\mathcal{F}_t\right] \\ &= \frac{N_t}{N_0}e^{-\int_0^t r_s ds}, \qquad 0 \le t \le T,\end{aligned}$$

hence

$$d\Phi_t = -\Phi_t r_t dt + \frac{\Phi_t}{N_t}dN_t,$$

and Relation (10.9) becomes

$$d\hat{W}_t = dW_t - \frac{1}{N_t}dN_t \cdot dW_t, \qquad t \in \mathbb{R}_+. \tag{10.10}$$

The next proposition confirms the statement of Proposition 10.2, and in addition it determines the precise dynamics of $(\hat{X}_t)_{t\in\mathbb{R}_+}$ under $\hat{\mathbb{P}}$. See Exercise 10.1 for another calculation based on geometric Brownian motion, and Exercise 10.4 for an extension to correlated Brownian motions.

Proposition 10.3 *Assume that $(X_t)_{t\in\mathbb{R}_+}$ and $(N_t)_{t\in\mathbb{R}_+}$ satisfy the stochastic differential equations*

$$dX_t = r_t X_t dt + \sigma_t^X X_t dW_t, \quad \text{and} \quad dN_t = r_t N_t dt + \sigma_t^N N_t dW_t, \tag{10.11}$$

where $(\sigma_t^X)_{t\in\mathbb{R}_+}$ and $(\sigma_t^N)_{t\in\mathbb{R}_+}$ are $\mathcal{F}_t$-adapted volatility processes. Then we have

$$d\hat{X}_t = (\sigma_t^X - \sigma_t^N)\hat{X}_t d\hat{W}_t. \tag{10.12}$$

Proof. First we note that by (10.10) and (10.11),

$$d\hat{W}_t = dW_t - \frac{1}{N_t}dN_t \cdot dW_t = dW_t - \sigma_t^N dt, \qquad t \in \mathbb{R}_+,$$

is a standard Brownian motion under $\hat{\mathbb{P}}$. Next, by Itô's calculus we have

$$\begin{aligned}d\hat{X}_t &= d\left(\frac{X_t}{N_t}\right) \\ &= \frac{dX_t}{N_t} - \frac{X_t}{N_t^2}dN_t - \frac{1}{N_t^2}dN_t \cdot dX_t + X_t\frac{(dN_t)^2}{N_t^3} \\ &= \frac{1}{N_t}(r_t X_t dt + \sigma_t^X X_t dW_t) - \frac{X_t}{N_t^2}(r_t N_t dt + \sigma_t^N N_t dW_t)\end{aligned}$$

$$
\begin{aligned}
& -\frac{X_t N_t}{N_t^2}\sigma_t^X\sigma_t^N dt + X_t\frac{|\sigma_t^N|^2 N_t^2}{N_t^3}dt \\
= & \frac{X_t}{N_t}\sigma_t^X dW_t - \frac{X_t}{N_t}\sigma_t^N dW_t - \frac{X_t}{N_t}\sigma_t^X\sigma_t^N dt + X_t\frac{|\sigma_t^N|^2}{N_t}dt \\
= & \frac{X_t}{N_t}\left(\sigma_t^X dW_t - \sigma_t^N dW_t - \sigma_t^X\sigma_t^N dt + |\sigma_t^N|^2 dt\right) \\
= & \hat{X}_t(\sigma_t^X - \sigma_t^N)dW_t - \hat{X}_t(\sigma_t^X - \sigma_t^N)\sigma_t^N dt \\
= & \hat{X}_t(\sigma_t^X - \sigma_t^N)d\hat{W}_t,
\end{aligned}
$$

since $d\hat{W}_t = dW_t - \sigma_t^N dt$, $t \in \mathbb{R}_+$. □

We end this section with a comment on inverse changes of measure.

Inverse Change of Measure

In the next proposition we compute conditional inverse density $d\mathbb{P}/d\hat{\mathbb{P}}$.

Proposition 10.4 *We have*

$$
\hat{\mathbb{E}}\left[\frac{d\mathbb{P}}{d\hat{\mathbb{P}}}\Big|\mathcal{F}_t\right] = \frac{N_0}{N_t}\exp\left(\int_0^t r_s ds\right) \qquad 0 \le t \le T, \tag{10.13}
$$

and the process

$$
t \longmapsto \frac{N_0}{N_t}\exp\left(\int_0^t r_s ds\right), \qquad 0 \le t \le T,
$$

is an $\mathcal{F}_t$-martingale under $\hat{\mathbb{P}}$.

Proof. For all bounded and $\mathcal{F}_t$-measurable random variables F we have,

$$
\begin{aligned}
\hat{\mathbb{E}}\left[F\frac{d\mathbb{P}}{d\hat{\mathbb{P}}}\right] &= \mathbb{E}^*[F] \\
&= \mathbb{E}^*\left[F\frac{N_t}{N_t}\right] \\
&= \mathbb{E}^*\left[F\frac{N_T}{N_t}\exp\left(-\int_t^T r_s ds\right)\right] \\
&= \hat{\mathbb{E}}\left[F\frac{N_0}{N_t}\exp\left(\int_0^t r_s ds\right)\right].
\end{aligned}
$$

□

By Itô's calculus and (10.11) we also have

$$
\begin{aligned}
d\left(\frac{1}{N_t}\right) &= -\frac{1}{N_t^2}dN_t + \frac{1}{N_t^3}(dN_t)^2 \\
&= -\frac{1}{N_t^2}(r_t N_t dt + \sigma_t^N N_t dW_t) + \frac{|\sigma_t^N|^2}{N_t}dt
\end{aligned}
$$

$$= -\frac{1}{N_t^2}(r_t N_t dt + \sigma_t^N N_t(d\hat{W}_t + \sigma_t^N dt)) + \frac{|\sigma_t^N|^2}{N_t} dt$$
$$= -\frac{1}{N_t}(r_t dt + \sigma_t^N d\hat{W}_t),$$

and

$$d\left(\frac{1}{N_t}\exp\left(\int_0^t r_s ds\right)\right) = -\frac{1}{N_t}\exp\left(\int_0^t r_s ds\right)\sigma_t^N d\hat{W}_t,$$

which recovers the second part of Proposition 10.4, i.e., the martingale property of

$$t \longmapsto \frac{1}{N_t}\exp\left(\int_0^t r_s ds\right)$$

under $\hat{\mathbb{P}}$.

10.3 Foreign Exchange

Currency exchange is a typical application of change of numéraire that illustrate the principle of absence of arbitrage.

Let R_t denote the foreign exchange rate, i.e., R_t is the (possibly fractional) quantity of local currency that corresponds to one unit of foreign currency.

Consider an investor that intends to exploit an "overseas investment opportunity" by

a) at time 0, changing one unit of local currency into $1/R_0$ units of foreign currency,

b) investing $1/R_0$ on the foreign market at the rate r^{f} to make the amount $e^{tr^{\mathrm{f}}}/R_0$ until time t,

c) changing back $e^{tr^{\mathrm{f}}}/R_0$ into a quantity $e^{tr^{\mathrm{f}}}R_t/R_0$ of his local currency.

In other words, the foreign money market account $e^{tr^{\mathrm{f}}}$ is valued $e^{tr^{\mathrm{f}}}R_t$ on the local (or domestic) market, and its discounted value on the local market is

$$e^{-tr+tr^{\mathrm{f}}}R_t, \qquad t \in \mathbb{R}_+.$$

The outcome of this investment will be obtained by comparing $e^{tr^{\mathrm{f}}}R_t/R_0$ to the amount e^{rt} that could have been obtained by investing on the local market.

Taking

$$N_t := e^{tr^{\mathrm{f}}} R_t, \qquad t \in \mathbb{R}_+, \tag{10.14}$$

as *numéraire*, absence of arbitrage is expressed by stating that the discounted numéraire process

$$t \longmapsto e^{-rt} N_t = e^{-t(r-r^{\mathrm{f}})} R_t$$

is a *martingale* under $\mathbb{P}^*$, which is Assumption (A).

Next we find a characterization of this arbitrage condition under the parameters of the model.

Proposition 10.5 *Assume that the foreign exchange rate R_t satisfies a stochastic differential equation of the form*

$$dR_t = \mu R_t dt + \sigma R_t dW_t, \tag{10.15}$$

where $(W_t)_{t\in\mathbb{R}_+}$ is a standard Brownian motion under $\mathbb{P}^*$. Under the absence of arbitrage Assumption (A) for the numéraire (10.14), *we have*

$$\mu = r - r^{\mathrm{f}}, \tag{10.16}$$

hence the exchange rate process satisfies

$$dR_t = (r - r^{\mathrm{f}}) R_t dt + \sigma R_t dW_t. \tag{10.17}$$

under $\mathbb{P}^*$.

Proof. The equation (10.15) has solution

$$R_t = R_0 e^{\mu t + \sigma W_t - \sigma^2 t/2}, \qquad t \in \mathbb{R}_+,$$

hence the discounted value of the foreign money market account $e^{tr^{\mathrm{f}}}$ on the local market is

$$e^{-tr+tr^{\mathrm{f}}} R_t = R_0 e^{t(r^{\mathrm{f}}-r+\mu)+\sigma W_t - \sigma^2 t/2}, \qquad t \in \mathbb{R}_+.$$

Under absence of arbitrage, $e^{-t(r-r^{\mathrm{f}})} R_t = e^{-tr} N_t$ should be a martingale under $\mathbb{P}^*$ and this holds provided $r^{\mathrm{f}} - r + \mu = 0$, which yields (10.16) and (10.17). $\square$

As a consequence of Proposition 10.5, under absence of arbitrage a local investor who buys a unit of foreign currency in the hope of a higher return $r^{\mathrm{f}} >> r$ will have to face a lower (or even more negative) drift

$$\mu = r - r^{\mathrm{f}} << 0$$

in his exchange rate R_t,

The local money market account $X_t := e^{rt}$ is valued e^{rt}/R_t on the foreign market, and its discounted value on the foreign market is

$$
\begin{aligned}
t \longmapsto \frac{e^{t(r-r^{\mathrm{f}})}}{R_t} &= \frac{X_t}{N_t} \\
&= R_0 e^{t(r-r^{\mathrm{f}})-\mu t-\sigma W_t+\sigma^2 t/2} \\
&= R_0 e^{t(r-r^{\mathrm{f}})-\mu t-\sigma \hat{W}_t-\sigma^2 t/2},
\end{aligned} \tag{10.18}
$$

where

$$
\begin{aligned}
d\hat{W}_t &= dW_t - \frac{1}{N_t} dN_t \cdot dW_t \\
&= dW_t - \frac{1}{R_t} dR_t \cdot dW_t \\
&= dW_t - \sigma dt, \qquad t \in \mathbb{R}_+,
\end{aligned}
$$

is a standard Brownian motion under $\hat{\mathbb{P}}$ by (10.10). Under absence of arbitrage $e^{-t(r-r^{\mathrm{f}})}R_t$ is a martingale under $\mathbb{P}^*$ and (10.18) is a martingale under $\hat{\mathbb{P}}$ by Proposition 10.2, which recovers (10.16).

Proposition 10.6 *Under the absence of arbitrage condition* (10.16), *the exchange rate* $1/R_t$ *satisfies*

$$
d\left(\frac{1}{R_t}\right) = (r^{\mathrm{f}} - r)\frac{1}{R_t} dt - \frac{\sigma}{R_t} d\hat{W}_t, \tag{10.19}
$$

under $\hat{\mathbb{P}}$, *where* $(R_t)_{t\in\mathbb{R}_+}$ *is given by* (10.17).

Proof. By (10.16), the exchange rate $1/R_t$ is written by Itô's calculus as

$$
\begin{aligned}
d\left(\frac{1}{R_t}\right) &= -\frac{1}{R_t^2}(\mu R_t dt + \sigma R_t dW_t) + \frac{1}{R_t^3}\sigma^2 R_t^2 dt \\
&= -(\mu - \sigma^2)\frac{1}{R_t} dt - \frac{\sigma}{R_t} dW_t \\
&= -\frac{\mu}{R_t} dt - \frac{\sigma}{R_t} d\hat{W}_t \\
&= (r^{\mathrm{f}} - r)\frac{1}{R_t} dt - \frac{\sigma}{R_t} d\hat{W}_t,
\end{aligned}
$$

where $\hat{W}_t$ is a standard Brownian motion under $\hat{\mathbb{P}}$. □

Consequently, under absence of arbitrage, a foreign investor who buys a unit of the local currency in the hope of a higher return $r >> r^{\mathrm{f}}$ will have to face a lower (or even more negative) drift $-\mu = r^{\mathrm{f}} - r$ in his exchange rate $1/R_t$ as written in (10.19) under $\hat{\mathbb{P}}$.

Foreign exchange options

We now price a foreign exchange option with payoff $(R_T - \kappa)^+$ under $\mathbb{P}^*$ by the Black–Scholes formula as in the next proposition, also known as the Garman–Kohlagen [26] formula.

Proposition 10.7 *(Garman–Kohlagen formula). Assume that $(R_t)_{t\in\mathbb{R}_+}$ is given by* (10.17)*. The price of the foreign exchange call option on R_T with maturity T and strike κ is given by*

$$e^{-(T-t)r}\,\mathbb{E}^*[(R_T-\kappa)^+ \mid R_t] = e^{-(T-t)r^{\mathrm{f}}} R_t\Phi_+(t,R_t) - \kappa e^{-(T-t)r}\Phi_-(t,R_t),$$

$0 \le t \le T$, where

$$\Phi_+(t,x) = \Phi\left(\frac{\log(x/\kappa) + (T-t)(r - r^{\mathrm{f}} + \sigma^2/2)}{\sigma\sqrt{T-t}}\right),$$

and

$$\Phi_-(t,x) = \Phi\left(\frac{\log(x/\kappa) + (T-t)(r - r^{\mathrm{f}} - \sigma^2/2)}{\sigma\sqrt{T-t}}\right).$$

Proof. As a consequence of (10.17) we find the numéraire dynamics

$$\begin{aligned} dN_t &= d(e^{tr^{\mathrm{f}}}R_t) \\ &= r^{\mathrm{f}}e^{tr^{\mathrm{f}}}R_t dt + e^{tr^{\mathrm{f}}}dR_t \\ &= re^{tr^{\mathrm{f}}}R_t dt + \sigma e^{tr^{\mathrm{f}}}R_t dW_t \\ &= rN_t dt + \sigma N_t dW_t. \end{aligned}$$

Hence a standard application of the Black–Scholes formula yields

$$\begin{aligned} e^{-(T-t)r}\,\mathbb{E}^*[(R_T-\kappa)^+ \mid \mathcal{F}_t] &= e^{-(T-t)r}\,\mathbb{E}^*[(e^{-Tr^{\mathrm{f}}}N_T - \kappa)^+ \mid \mathcal{F}_t] \\ &= e^{-(T-t)r}e^{-Tr^{\mathrm{f}}}\,\mathbb{E}^*[(N_T - \kappa e^{Tr^{\mathrm{f}}})^+ \mid \mathcal{F}_t] \\ &= e^{-Tr^{\mathrm{f}}}\left(N_t\Phi\left(\frac{\log(N_t e^{-Tr^{\mathrm{f}}}/\kappa) + (r+\sigma^2/2)(T-t)}{\sigma\sqrt{T-t}}\right)\right. \\ &\quad \left. -\kappa e^{Tr^{\mathrm{f}}-(T-t)r}\Phi\left(\frac{\log(N_t e^{-Tr^{\mathrm{f}}}/\kappa) + (r-\sigma^2/2)(T-t)}{\sigma\sqrt{T-t}}\right)\right) \\ &= e^{-Tr^{\mathrm{f}}}\left(N_t\Phi\left(\frac{\log(R_t/\kappa) + (T-t)(r - r^{\mathrm{f}} + \sigma^2/2)}{\sigma\sqrt{T-t}}\right)\right. \\ &\quad \left. -\kappa e^{Tr^{\mathrm{f}}-(T-t)r}\Phi\left(\frac{\log(R_t/\kappa) + (T-t)(r - r^{\mathrm{f}} - \sigma^2/2)}{\sigma\sqrt{T-t}}\right)\right) \\ &= e^{-(T-t)r^{\mathrm{f}}}R_t\Phi_+(t,R_t) - \kappa e^{-(T-t)r}\Phi_-(t,R_t). \end{aligned}$$

□

Similarly, from (10.19) rewritten as

$$d\left(\frac{e^{rt}}{R_t}\right) = r^{\rm f}\frac{e^{rt}}{R_t}dt - \sigma\frac{e^{rt}}{R_t}d\hat{W}_t,$$

a foreign exchange option with payoff $(1/R_T - \kappa)^+$ can be priced under $\hat{\mathbb{P}}$ as a put option by the Black–Scholes formula, as

$$e^{-(T-t)r^{\rm f}}\hat{\mathbb{E}}\left[\left(\frac{1}{R_T} - \kappa\right)^+ \Big| R_t\right] \tag{10.20}$$

$$= e^{-(T-t)r^{\rm f}}e^{-rT}\hat{\mathbb{E}}\left[\left(\frac{e^{rT}}{R_T} - \kappa e^{rT}\right)^+ \Big| R_t\right] \tag{10.21}$$

$$= \frac{e^{-r(T-t)}}{R_t}\Phi_+\left(t, \frac{1}{R_t}\right) - \kappa e^{-(T-t)r^{\rm f}}\Phi_-\left(t, \frac{1}{R_t}\right),$$

where

$$\Phi_+(t,x) = \Phi\left(\frac{\log(x/\kappa) + (T-t)(r^{\rm f} - r + \sigma^2/2)}{\sigma\sqrt{T-t}}\right),$$

and

$$\Phi_-(t,x) = \Phi\left(\frac{\log(x/\kappa) + (T-t)(r^{\rm f} - r - \sigma^2/2)}{\sigma\sqrt{T-t}}\right).$$

Call/put duality for foreign exchange options

Let $N_t = e^{tr^{\rm f}}R_t$, where R_t is an exchange rate with respect to a foreign currency and r_f is the foreign market interest rate.

From Proposition 10.1 and (10.4) we have

$$e^{-(T-t)r}\,\mathbb{E}^*\left[\left(\frac{1}{\kappa} - R_T\right)^+ \Big| R_t\right] = N_t\hat{\mathbb{E}}\left[\frac{1}{e^{Tr^{\rm f}}R_T}\left(\frac{1}{\kappa} - R_T\right)^+ \Big| R_t\right],$$

and this yields the call/put duality

$$e^{-(T-t)r^{\rm f}}\hat{\mathbb{E}}\left[\left(\frac{1}{R_T} - \kappa\right)^+ \Big| R_t\right] = e^{-(T-t)r^{\rm f}}\hat{\mathbb{E}}\left[\frac{\kappa}{R_T}\left(\frac{1}{\kappa} - R_T\right)^+ \Big| R_t\right]$$

$$= e^{tr^{\rm f}}\hat{\mathbb{E}}\left[\frac{\kappa}{e^{Tr^{\rm f}}R_T}\left(\frac{1}{\kappa} - R_T\right)^+ \Big| R_t\right]$$

$$= \frac{\kappa}{N_t}e^{tr^{\rm f}-(T-t)r}\,\mathbb{E}^*\left[\left(\frac{1}{\kappa} - R_T\right)^+ \Big| R_t\right]$$

$$= \frac{\kappa}{R_t}e^{-(T-t)r}\,\mathbb{E}^*\left[\left(\frac{1}{\kappa} - R_T\right)^+ \Big| R_t\right], \tag{10.22}$$

between a call option with strike κ and a (possibly fractional) quantity κ/R_t of put option(s) with strike $1/\kappa$.

In the Black–Scholes case the duality (10.22) can be directly checked by verifying that (10.20) coincides with

$$\begin{aligned}
&\frac{\kappa}{R_t} e^{-(T-t)r} \mathbb{E}^* \left[\left(\frac{1}{\kappa} - R_T \right)^+ \Big| R_t \right] \\
&= \frac{\kappa}{R_t} e^{-(T-t)r} e^{-Tr^{\rm f}} \mathbb{E}^* \left[\left(\frac{e^{Tr^{\rm f}}}{\kappa} - e^{Tr^{\rm f}} R_T \right)^+ \Big| R_t \right] \\
&= \frac{\kappa}{R_t} e^{-(T-t)r} e^{-Tr^{\rm f}} \mathbb{E}^* \left[\left(\frac{e^{Tr^{\rm f}}}{\kappa} - N_T \right)^+ \Big| R_t \right] \\
&= \frac{\kappa}{R_t} \left(\frac{e^{-(T-t)r}}{\kappa} \Phi_-^p(t, R_t) - e^{-(T-t)r^{\rm f}} R_t \Phi_+^p(t, R_t) \right) \\
&= \frac{e^{-(T-t)r}}{R_t} \Phi_-^p(t, R_t) - \kappa e^{-(T-t)r^{\rm f}} \Phi_+^p(t, R_t) \\
&= \frac{e^{-r(T-t)}}{R_t} \Phi_+ \left(t, \frac{1}{R_t} \right) - \kappa e^{-(T-t)r^{\rm f}} \Phi_- \left(t, \frac{1}{R_t} \right),
\end{aligned}$$

where

$$\Phi_-^p(t, x) = \Phi \left(- \frac{\log(x\kappa) + (T-t)(r - r^{\rm f} - \sigma_t^2/2)}{\sigma\sqrt{T-t}} \right),$$

and

$$\Phi_+^p(t, x) = \Phi \left(- \frac{\log(x\kappa) + (T-t)(r - r^{\rm f} + \sigma^2/2)}{\sigma\sqrt{T-t}} \right).$$

10.4 Pricing of Exchange Options

Based on Proposition 10.2 we model the process $\hat{X}_t$ of forward prices as a continuous martingale under $\hat{\mathbb{P}}$, written as

$$d\hat{X}_t = \hat{\sigma}_t d\hat{W}_t, \qquad t \in \mathbb{R}_+, \tag{10.23}$$

where $(\hat{W}_t)_{t\in\mathbb{R}_+}$ is a standard Brownian motion under $\hat{\mathbb{P}}$ and $(\hat{\sigma}_t)_{t\in\mathbb{R}_+}$ is an $\mathcal{F}_t$-adapted process. The following lemma is a consequence of the Markov property of the process $(\hat{X}_t)_{t\in\mathbb{R}_+}$ and leads to the Margrabe formula of Proposition 10.8 below.

Lemma 10.2 *Assume that $(\hat{X}_t)_{t\in\mathbb{R}_+}$ has the dynamics*

$$d\hat{X}_t = \hat{\sigma}_t(\hat{X}_t) d\hat{W}_t, \tag{10.24}$$

where $x \longmapsto \hat{\sigma}_t(x)$ *is a Lipschitz function, uniformly in* $t \in \mathbb{R}_+$. *Then the option with payoff* $\xi = N_T \hat{g}(\hat{X}_T)$ *is priced at time* t *as*

$$N_t \hat{C}(t, \hat{X}_t) = N_t \hat{\mathbb{E}}\left[\hat{g}(\hat{X}_T)\Big|\mathcal{F}_t\right] = \mathbb{E}^*\left[e^{-\int_t^T r_s ds} N_T \hat{g}(\hat{X}_T)\Big|\mathcal{F}_t\right], \quad (10.25)$$

for some (measurable) function $\hat{C}(t, x)$ *on* $\mathbb{R}_+ \times \mathbb{R}_+$.

The next proposition states the Margrabe [47] formula for the pricing of exchange options by the zero interest rate Black–Scholes formula. It will be applied in particular in Proposition 12.2 below for the pricing of bond options.

Proposition 10.8 *(Margrabe formula). Assume that* $\hat{\sigma}_t(\hat{X}_t) = \hat{\sigma}(t)\hat{X}_t$, i.e., *the martingale* $(\hat{X}_t)_{t\in[0,T]}$ *is a geometric Brownian motion under* $\hat{\mathbb{P}}$ *with deterministic volatility* $(\hat{\sigma}(t))_{t\in[0,T]}$. *Then we have*

$$\mathbb{E}^*\left[e^{-\int_t^T r_s ds}(X_T - \kappa N_T)^+ \Big| \mathcal{F}_t\right] = X_t \Phi_+^0(t, \hat{X}_t) - \kappa N_t \Phi_-^0(t, \hat{X}_t), \quad (10.26)$$

$t \in [0, T]$, *where*

$$\Phi_+^0(t, x) = \Phi\left(\frac{\log(x/\kappa)}{v(t,T)} + \frac{v(t,T)}{2}\right), \qquad \Phi_-^0(t, x) = \Phi\left(\frac{\log(x/\kappa)}{v(t,T)} - \frac{v(t,T)}{2}\right),$$

and

$$v^2(t,T) = \int_t^T \hat{\sigma}^2(s) ds.$$

Proof. Taking $g(x) = (x - \kappa)^+$ in (10.25), the call option with payoff

$$(X_T - \kappa N_T)^+ = N_T(\hat{X}_T - \kappa)^+,$$

and floating strike κN_T is priced by (10.25) as

$$\begin{aligned}
\mathbb{E}^*\left[e^{-\int_t^T r_s ds}(X_T - \kappa N_T)^+\Big|\mathcal{F}_t\right] &= \mathbb{E}^*\left[e^{-\int_t^T r_s ds} N_T(\hat{X}_T - \kappa)^+\Big|\mathcal{F}_t\right] \\
&= N_t \hat{\mathbb{E}}\left[(\hat{X}_T - \kappa)^+\Big|\mathcal{F}_t\right] \\
&= N_t \hat{C}(t, \hat{X}_t),
\end{aligned}$$

where the function $\hat{C}(t, \hat{X}_t)$ is given by the Black–Scholes formula

$$\hat{C}(t, x) = x\Phi_+^0(t, x) - \kappa\Phi_-^0(t, x),$$

with zero interest rate, since $(\hat{X}_t)_{t\in[0,T]}$ is a geometric Brownian motion which is a martingale under $\hat{\mathbb{P}}$. Hence we have

$$\begin{aligned}
\mathbb{E}^*\left[e^{-\int_t^T r_s ds}(X_T - \kappa N_T)^+\Big|\mathcal{F}_t\right] &= N_t \hat{C}(t, \hat{X}_t) \\
&= N_t \hat{X}_t \Phi_+^0(t, \hat{X}_t) - \kappa N_t \Phi_-^0(t, \hat{X}_t),
\end{aligned}$$

$t \in \mathbb{R}_+$. □

In particular, from Proposition 10.3 and (10.12), we can take $\hat{\sigma}(t) = \sigma_t^X - \sigma_t^N$ when $(\sigma_t^X)_{t\in\mathbb{R}_+}$ and $(\sigma_t^N)_{t\in\mathbb{R}_+}$ are deterministic.

Examples:

a) When the short rate process $(r(t))_{t\in[0,T]}$ is a *deterministic* function and $N_t = e^{-\int_t^T r(s)ds}$, $0 \leq t \leq T$, Proposition 10.8 yields Merton's [49] "zero interest rate" version of the Black–Scholes formula

$$e^{-\int_t^T r(s)ds}\,\mathbb{E}^*\left[(X_T - \kappa)^+ \,\Big|\, \mathcal{F}_t\right]$$
$$= X_t\Phi_+^0\left(t, e^{\int_t^T r(s)ds}X_t\right) - \kappa e^{-\int_t^T r(s)ds}\Phi_-^0\left(t, e^{\int_t^T r(s)ds}X_t\right),$$

where $(X_t)_{t\in\mathbb{R}_+}$ satisfies the equation

$$\frac{dX_t}{X_t} = r(t)dt + \hat{\sigma}(t)dW_t, \qquad 0 \leq t \leq T.$$

b) In the case of pricing under a *forward numéraire*, i.e., $N_t = P(t,T)$, $0 \leq t \leq T$, we get

$$\mathbb{E}^*\left[e^{-\int_t^T r_s ds}(X_T - \kappa)^+ \,\Big|\, \mathcal{F}_t\right] = X_t\Phi_+(t, \hat{X}_t) - \kappa P(t,T)\Phi_-(t, \hat{X}_t),$$

$t \in \mathbb{R}_+$, since $P(T,T) = 1$. In particular, when $X_t = P(t,S)$ the above formula allows us to price a bond call option on $P(T,S)$ as

$$\mathbb{E}^*\left[e^{-\int_t^T r_s ds}(P(T,S) - \kappa)^+ \,\Big|\, \mathcal{F}_t\right] = P(t,S)\Phi_+(t,\hat{X}_t) - \kappa P(t,T)\Phi_-(t,\hat{X}_t),$$

$0 \leq t \leq T$, provided the martingale $\hat{X}_t = P(t,S)/P(t,T)$ under $\hat{\mathbb{P}}$ is given by a geometric Brownian motion, cf. Section 12.2.

10.5 Self-Financing Hedging by Change of Numéraire

In this section we reconsider and extend the Black–Scholes self-financing hedging strategies found in (6.20)–(6.21) and Proposition 6.7 of Chapter 6, and use the stochastic integral representation of the forward payoff ξ/N_T and change of numéraire to compute self-financing portfolio strategies. Our hedging portfolios will be built on the assets (X_t, N_t), not on X_t and the money market account $B_t = e^{\int_0^t r_s ds}$, extending the classical hedging portfolios that are available from the Black–Scholes formula, using a technique from [37], cf. also [57].

Assume that the forward claim $\xi/N_T \in L^2(\Omega)$ has the stochastic integral representation

$$\frac{\xi}{N_T} = \hat{\mathbb{E}}\left[\frac{\xi}{N_T}\right] + \int_0^T \hat{\phi}_t d\hat{X}_t, \tag{10.27}$$

where $(\hat{X}_t)_{t\in[0,T]}$ is given by (10.23) and $(\hat{\phi}_t)_{t\in[0,T]}$ is a square-integrable adapted process under $\hat{\mathbb{P}}$, from which it follows that the forward claim price

$$\hat{V}_t := \hat{\mathbb{E}}\left[\frac{\xi}{N_T}\Big|\mathcal{F}_t\right], \qquad 0 \le t \le T,$$

is a martingale that can be decomposed as

$$\hat{V}_t = \hat{\mathbb{E}}\left[\frac{\xi}{N_T}\right] + \int_0^t \hat{\phi}_s d\hat{X}_s, \qquad 0 \le t \le T. \tag{10.28}$$

The next proposition extends the argument of [37] to the general framework of pricing using change of numéraire. Note that this result differs from the standard formula that uses the money market account $B_t = e^{\int_0^t r_s ds}$ for hedging instead of N_t, cf. e.g., [27] pages 453–454. The notion of self-financing portfolio is similar to that of Definition 5.1.

Proposition 10.9 *Letting* $\hat{\eta}_t = \hat{V}_t - \hat{X}_t\hat{\phi}_t$, $0 \le t \le T$, *the portfolio* $(\hat{\phi}_t, \hat{\eta}_t)_{t\in[0,T]}$ *with value*

$$V_t = \hat{\phi}_t X_t + \hat{\eta}_t N_t, \qquad 0 \le t \le T,$$

is self-financing in the sense that

$$dV_t = \hat{\phi}_t dX_t + \hat{\eta}_t dN_t,$$

and it hedges the claim ξ, i.e.,

$$V_t = \hat{\phi}_t X_t + \hat{\eta}_t N_t = \mathbb{E}^*\left[e^{-\int_t^T r_s ds}\xi \,\Big|\mathcal{F}_t\right], \qquad 0 \le t \le T. \tag{10.29}$$

Proof. In order to check that the portfolio $(\hat{\phi}_t, \hat{\eta}_t)_{t\in[0,T]}$ hedges the claim ξ it suffices to check that (10.29) holds since by (10.4) the price V_t at time $t \in [0, T]$ of the hedging portfolio satisfies

$$V_t = N_t\hat{V}_t = \mathbb{E}^*\left[e^{-\int_t^T r_s ds}\xi \,\Big|\mathcal{F}_t\right], \qquad 0 \le t \le T.$$

Next, we show that the portfolio $(\hat{\phi}_t, \hat{\eta}_t)_{t\in[0,T]}$ is self-financing. By numéraire invariance, cf. e.g., page 184 of [60], we have

$$\begin{aligned}
dV_t &= d(N_t\hat{V}_t) \\
&= \hat{V}_t dN_t + N_t d\hat{V}_t + dN_t \cdot d\hat{V}_t \\
&= \hat{V}_t dN_t + N_t\hat{\phi}_t d\hat{X}_t + \hat{\phi}_t dN_t \cdot d\hat{X}_t \\
&= \hat{\phi}_t\hat{X}_t dN_t + N_t\hat{\phi}_t d\hat{X}_t + \hat{\phi}_t dN_t \cdot d\hat{X}_t + (\hat{V}_t - \hat{\phi}_t\hat{X}_t)dN_t \\
&= \hat{\phi}_t d(N_t\hat{X}_t) + (\hat{V}_t - \hat{\phi}_t\hat{X}_t)dN_t \\
&= \hat{\phi}_t dX_t + \hat{\eta}_t dN_t.
\end{aligned}$$

□

We now consider an application to the forward Delta hedging of European type options with payoff $\xi = \hat{g}(\hat{X}_T)$ where $\hat{g} : \mathbb{R} \longrightarrow \mathbb{R}$ and $(\hat{X}_t)_{t\in\mathbb{R}_+}$ has the Markov property as in (10.24), where $\hat{\sigma} : \mathbb{R}_+ \times \mathbb{R}$. Assuming that the function $\hat{C}(t,x)$ defined by

$$\hat{V}_t := \hat{\mathbb{E}}\left[g(\hat{X}_T)\middle|\mathcal{F}_t\right] = \hat{C}(t,\hat{X}_t)$$

is $\mathcal{C}^2$ on $\mathbb{R}_+$, we have the following corollary of Proposition 10.9.

Corollary 10.1 *Letting* $\hat{\eta}_t = \hat{C}(t,\hat{X}_t) - \hat{X}_t \dfrac{\partial \hat{C}}{\partial x}(t,\hat{X}_t)$, $0 \le t \le T$, *the portfolio* $\left(\dfrac{\partial \hat{C}}{\partial x}(t,\hat{X}_t), \hat{\eta}_t\right)_{t\in[0,T]}$ *with value*

$$V_t = \hat{\eta}_t N_t + X_t \frac{\partial \hat{C}}{\partial x}(t,\hat{X}_t), \qquad t \in \mathbb{R}_+,$$

is self-financing and hedges the claim $\xi = N_T \hat{g}(\hat{X}_T)$.

Proof. This result follows directly from Proposition 10.9 by noting that by Itô's formula, and the martingale property of $\hat{V}_t$ under $\hat{\mathbb{P}}$ the stochastic integral representation (10.28) is given by

$$\hat{\phi}_t = \frac{\partial \hat{C}}{\partial x}(t,\hat{X}_t), \qquad 0 \le t \le T.$$

□

In the case of a call option with payoff function $\xi = (X_T - \kappa N_T)^+ = N_T(\hat{X}_T - \kappa)^+$ on the geometric Brownian motion $(\hat{X}_t)_{t\in[0,T]}$ under $\hat{\mathbb{P}}$ with

$$\hat{\sigma}_t(\hat{X}_t) = \hat{\sigma}(t)\hat{X}_t,$$

where $(\hat{\sigma}(t))_{t\in[0,T]}$ is a deterministic function, we have the following corollary on the hedging of exchange options based on the Margrabe formula (10.26).

Corollary 10.2 *The decomposition*

$$\mathbb{E}^*\left[e^{-\int_t^T r_s ds}(X_T - \kappa N_T)^+ \middle| \mathcal{F}_t\right] = X_t \Phi_+^0(t,\hat{X}_t) - \kappa N_t \Phi_-^0(t,\hat{X}_t)$$

yields a self-financing portfolio $(\Phi_+^0(t,\hat{X}_t), -\kappa\Phi_-^0(t,\hat{X}_t))_{t\in[0,T]}$ *in* (X_t, N_t) *that hedges the claim* $\xi = (X_T - \kappa N_T)^+$.

Proof. We apply Corollary 10.1 and the classical relation

$$\frac{\partial \hat{C}}{\partial x}(t,x) = \Phi_+^0(t,x), \qquad x \in \mathbb{R},$$

for the function $\hat{C}(t,x) = x\Phi_+^0(t,x) - \kappa\Phi_-^0(t,x)$. □

Note that the Delta hedging method requires the computation of the function $\hat{C}(t, x)$ and that of the associated finite differences, and may not apply to path-dependent claims.

Examples:

a) When the short rate process $(r(t))_{t\in[0,T]}$ is a *deterministic* function and $N_t = e^{\int_t^T r(s)ds}$, Corollary 10.2 yields the usual Black–Scholes hedging strategy

$$\begin{aligned}&\left(\Phi_+(t, \hat{X}_t), -\kappa e^{\int_0^T r(s)ds}\Phi_-(t, X_t)\right)_{t\in[0,T]}\\ &= \left(\Phi_+^0(t, e^{\int_t^T r(s)ds}\hat{X}_t), -\kappa e^{\int_0^T r(s)ds}\Phi_-^0(t, e^{\int_t^T r(s)ds}X_t)\right)_{t\in[0,T]}\end{aligned}$$

in $(X_t, e^{\int_0^t r(s)ds})$, that hedges the claim $\xi = (X_T - \kappa)^+$.

b) In case $N_t = P(t, T)$ and $X_t = P(t, S)$, $0 \leq t \leq T < S$, Corollary 10.2 shows that a bond call option with payoff $(P(T, S) - \kappa)^+$ can be hedged as

$$\mathbb{E}^*\left[e^{-\int_t^T r_s ds}(P(T, S) - \kappa)^+ \middle| \mathcal{F}_t\right] = P(t, S)\Phi_+(t, \hat{X}_t) - \kappa P(t, T)\Phi_-(t, \hat{X}_t)$$

by the self-financing portfolio

$$(\Phi_+(t, \hat{X}_t), -\kappa\Phi_-(t, \hat{X}_t))_{t\in[0,T]}$$

in $(P(t, S), P(t, T))$, i.e., one needs to hold the quantity $\Phi_+(t, \hat{X}_t)$ of the bond maturing at time S, and to short a quantity $\kappa\Phi_-(t, \hat{X}_t)$ of the bond maturing at time T.

Exercises

Exercise 10.1 Let $(B_t)_{t\in\mathbb{R}_+}$ be a standard Brownian motion started at 0 under the risk-neutral measure $\mathbb{P}^*$. Consider a numéraire $(N_t)_{t\in\mathbb{R}_+}$ given by

$$N_t := N_0 e^{\eta B_t - \eta^2 t/2}, \qquad t \in \mathbb{R}_+,$$

and a risky asset $(X_t)_{t\in\mathbb{R}_+}$ given by

$$X_t := X_0 e^{\sigma B_t - \sigma^2 t/2}, \qquad t \in \mathbb{R}_+.$$

Let $\hat{\mathbb{P}}$ denote the forward measure relative to the numéraire $(N_t)_{t\in\mathbb{R}_+}$, under which the process $\hat{X}_t := X_t/N_t$ of forward prices is known to be a martingale.

1. Using the Itô formula, compute

$$d\hat{X}_t = d(X_t/N_t) = (X_0/N_0)d\left(e^{(\sigma-\eta)B_t-(\sigma^2-\eta^2)t/2}\right).$$

2. Explain why the exchange option price $\mathbb{E}[(X_T - \lambda N_T)^+]$ at time 0 has the Black–Scholes form

$$\mathbb{E}[(X_T - \lambda N_T)^+] \tag{10.30}$$
$$= X_0\Phi\left(\frac{\log(\hat{X}_0/\lambda)}{\hat{\sigma}\sqrt{T}} + \frac{\hat{\sigma}\sqrt{T}}{2}\right) - \lambda N_0\Phi\left(\frac{\log(\hat{X}_0/\lambda)}{\hat{\sigma}\sqrt{T}} - \frac{\hat{\sigma}\sqrt{T}}{2}\right).$$

Hint: Recall that the forward process $\hat{X}_t$ is a martingale under the forward measure $\hat{\mathbb{P}}$.

3. Give the value of $\hat{\sigma}$ in terms of σ and η.

Exercise 10.2 Bond options. Consider two bonds with maturities T and S, with prices $P(t,T)$ and $P(t,S)$ given by

$$\frac{dP(t,T)}{P(t,T)} = r_t dt + \zeta_t^T dW_t,$$

and

$$\frac{dP(t,S)}{P(t,S)} = r_t dt + \zeta_t^S dW_t,$$

where $(\zeta^T(s))_{s\in[0,T]}$ and $(\zeta^S(s))_{s\in[0,S]}$ are deterministic functions.

1. Show, using Itô's formula, that

$$d\left(\frac{P(t,S)}{P(t,T)}\right) = \frac{P(t,S)}{P(t,T)}(\zeta^S(t) - \zeta^T(t))d\hat{W}_t,$$

where $(\hat{W}_t)_{t\in\mathbb{R}_+}$ is a standard Brownian motion under $\hat{\mathbb{P}}$.

2. Show that

$$P(T,S) = \frac{P(t,S)}{P(t,T)}\exp\left(\int_t^T (\zeta^S(s) - \zeta^T(s))d\hat{W}_s - \frac{1}{2}\int_t^T |\zeta^S(s) - \zeta^T(s)|^2 ds\right).$$

Let $\hat{\mathbb{P}}$ denote the forward measure associated to the numéraire

$$N_t := P(t,T), \qquad 0 \le t \le T.$$

3. Show that for all $S,T > 0$ the price at time t

$$\mathbb{E}\left[e^{-\int_t^T r_s ds}(P(T,S) - \kappa)^+ \Big| \mathcal{F}_t\right]$$

of a bond call option on $P(T,S)$ with payoff $(P(T,S)-\kappa)^+$ is equal to

$$\mathbb{E}^*\left[e^{-\int_t^T r_s ds}(P(T,S)-\kappa)^+ \Big| \mathcal{F}_t\right] \tag{10.31}$$
$$= P(t,S)\Phi\left(\frac{v}{2}+\frac{1}{v}\log\frac{P(t,S)}{\kappa P(t,T)}\right) - \kappa P(t,T)\Phi\left(-\frac{v}{2}+\frac{1}{v}\log\frac{P(t,S)}{\kappa P(t,T)}\right),$$

where

$$v^2 = \int_t^T |\zeta^S(s)-\zeta^T(s)|^2 ds.$$

4. Compute the self-financing hedging strategy that hedges the bond option using a portfolio based on the assets $P(t,T)$ and $P(t,S)$.

Exercise 10.3 Consider two risky assets S_1 and S_2 modeled by the geometric Brownian motions

$$S_1(t) = e^{\sigma_1 W_t + \mu t} \qquad \text{and} \qquad S_2(t) = e^{\sigma_2 W_t + \mu t}, \qquad t \in \mathbb{R}_+, \tag{10.32}$$

where $(W_t)_{t\in\mathbb{R}_+}$ is a standard Brownian motion under $\mathbb{P}$.

1. Find a condition on r, μ and σ_2 so that the discounted price process $e^{-rt}S_2(t)$ is a martingale under $\mathbb{P}$.
2. Assume that $r-\mu=\sigma_2^2/2$, and let

$$X_t = e^{(\sigma_2^2-\sigma_1^2)t/2}S_1(t), \qquad t\in\mathbb{R}_+.$$

Show that the discounted process $e^{-rt}X_t$ is a martingale under $\mathbb{P}$.

3. Taking $N_t = S_2(t)$ as numéraire, show that the forward process $\hat{X}(t) = X_t/N_t$ is a martingale under the forward measure $\hat{\mathbb{P}}$ defined by

$$\frac{d\hat{\mathbb{P}}}{d\mathbb{P}} = e^{-rT}\frac{N_T}{N_0}.$$

Recall that

$$\hat{W}_t := W_t - \sigma_2 t$$

is a standard Brownian motion under $\hat{\mathbb{P}}$.

4. Using the relation

$$e^{-rT}\,\mathbb{E}[(S_1(T)-S_2(T))^+] = N_0\hat{\mathbb{E}}[(S_1(T)-S_2(T))^+/N_T],$$

compute the price

$$e^{-rT}\,\mathbb{E}[(S_1(T)-S_2(T))^+].$$

of the exchange option on the assets S_1 and S_2.

Exercise 10.4 Extension of Proposition 10.3 to correlated Brownian motions. Assume that $(S_t)_{t\in\mathbb{R}_+}$ and $(N_t)_{t\in\mathbb{R}_+}$ satisfy the stochastic differential equations

$$dS_t = r_t S_t dt + \sigma_t^S S_t dW_t^S, \quad \text{and} \quad dN_t = \eta_t N_t dt + \sigma_t^N N_t dW_t^N,$$

where $(W_t^S)_{t\in\mathbb{R}_+}$ and $(W_t^N)_{t\in\mathbb{R}_+}$ have the correlation

$$dW_t^S \cdot dW_t^N = \rho dt,$$

where $\rho \in [-1, 1]$.

1. Show that $(W_t^N)_{t\in\mathbb{R}_+}$ can be written as

$$W_t^N = \rho W_t^S + \sqrt{1-\rho^2} W_t, \qquad t \in \mathbb{R}_+,$$

   where $(W_t)_{t\in\mathbb{R}_+}$ is a standard Brownian motion under $\mathbb{P}^*$, independent of $(W_t^S)_{t\in\mathbb{R}_+}$.

2. Letting $X_t = S_t/N_t$, show that dX_t can be written as

$$dX_t = (r_t - \eta_t + (\sigma_t^N)^2 - \rho\sigma_t^N\sigma_t^S)X_t dt + \hat{\sigma}_t X_t dW_t^X,$$

   where W_t^X is a standard Brownian motion under $\mathbb{P}^*$ and $\hat{\sigma}_t$ is to be computed.

Exercise 10.5 Quanto options (Exercise 9.5 in [68]). Consider an asset priced S_t at time t, with

$$dS_t = r S_t dt + \sigma^S S_t dW_t^S,$$

and an exchange rate $(R_t)_{t\in\mathbb{R}_+}$ given by

$$dR_t = (r - r^{\mathrm{f}}) R_t dt + \sigma^R R_t dW_t^R,$$

from (10.16) in Proposition 10.5, where $(W_t^R)_{t\in\mathbb{R}_+}$ is written as

$$W_t^R = \rho W_t^S + \sqrt{1-\rho^2} W_t, \qquad t \in \mathbb{R}_+,$$

where $(W_t)_{t\in\mathbb{R}_+}$ is a standard Brownian motion under $\mathbb{P}^*$, 'independent of $(W_t^S)_{t\in\mathbb{R}_+}$, i.e., we have

$$dW_t^R \cdot dW_t^S = \rho dt,$$

where ρ is a correlation coefficient.

1. Let

$$a = r - r^{\mathrm{f}} + \rho\sigma^R\sigma^S - (\sigma^R)^2$$

   and $X_t = e^{at} S_t / R_t$, $t \in \mathbb{R}_+$, and show by Exercise 10.4 that dX_t can be written as

$$dX_t = r X_t dt + \hat{\sigma} X_t dW_t^X,$$

   where $(W_t^X)_{t\in\mathbb{R}_+}$ is a standard Brownian motion under $\mathbb{P}^*$ and $\hat{\sigma}$ is to be computed.

2. Compute the price

$$e^{-r(T-t)}\,\mathbb{E}^*\left[\left(\frac{S_T}{R_T}-\kappa\right)^+\Big|\mathcal{F}_t\right]$$

at time t of a quanto option.

Chapter 11

Forward Rate Modeling

This chapter is concerned with interest rate modeling, in which the mean reversion property plays an important role. We consider the main short rate models (Vasicek, CIR, CEV, affine models) and the computation of bond prices in such models. Next we consider the modeling of forward rates in the HJM and BGM models, as well as in two-factor models.

11.1 Short-Term Models

Mean Reversion

The first model to capture the mean reversion property of interest rates, a property not possessed by geometric Brownian motion, is the Vasicek [71] model, which is based on the Ornstein–Uhlenbeck process. Here, the short term interest rate process $(r_t)_{t\in\mathbb{R}_+}$ solves the equation

$$dr_t = (a - br_t)dt + \sigma dB_t, \tag{11.1}$$

where $(B_t)_{t\in\mathbb{R}_+}$ is a standard Brownian motion, with solution

$$r_t = r_0 e^{-bt} + \frac{a}{b}(1 - e^{-bt}) + \sigma \int_0^t e^{-b(t-s)} dB_s, \qquad t \in \mathbb{R}_+. \tag{11.2}$$

The law of r_t is Gaussian at all times t, with mean $r_0 e^{-bt} + \frac{a}{b}(1 - e^{-bt})$ and variance

$$\sigma^2 \int_0^t (e^{-b(t-s)})^2 ds = \sigma^2 \int_0^t e^{-2bs} ds = \frac{\sigma^2}{2b}.$$

This model has the interesting properties of being statistically stationary in time in the long run, and to admit a Gaussian $\mathcal{N}(a/b, \sigma^2/(2b))$ invariant distribution when $b > 0$; however, its drawback is to allow for negative values of r_t.

Figure 11.1 presents a random simulation of $t \longmapsto r_t$ in the Vasicek model with $r_0 = a/b = 5\%$, i.e., the reverting property of the process is with respect to its initial value $r_0 = 5\%$. Note that the interest rate in Figure 11.1 becomes negative for a short period of time, which is unusual for interest rates but may nevertheless happen.

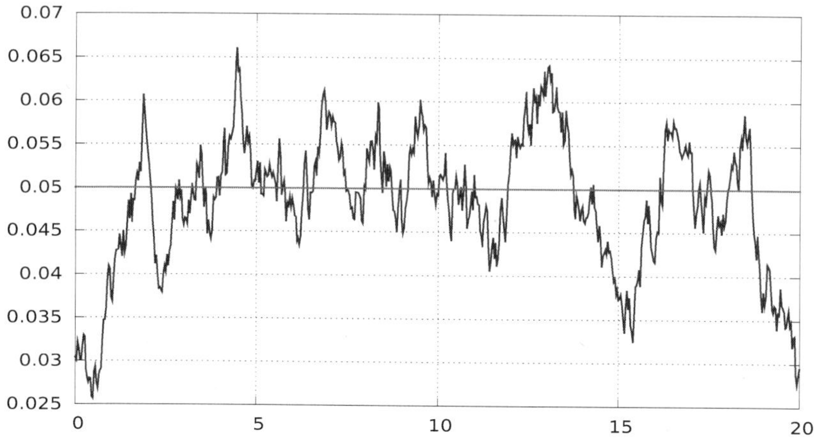

FIGURE 11.1: Graph of $t \longmapsto r_t$ in the Vasicek model.

The Cox–Ingersoll–Ross (CIR) [13] model brings a solution to the positivity problem encountered with the Vasicek model, by the use the nonlinear equation

$$dr_t = \beta(\alpha - r_t)dt + r_t^{1/2}\sigma dB_t.$$

Other classical mean reverting models include the Courtadon (1982) model

$$dr_t = \beta(\alpha - r_t)dt + \sigma r_t dB_t$$

where α, β, σ are nonnegative, and the exponential-Vasicek model

$$dr_t = r_t(\eta - a\log r_t)dt + \sigma r_t dB_t,$$

where a, η, σ are nonnegative.

Constant Elasticity of Variance (CEV)

Constant Elasticity of Variance models are designed to take into account non-constant volatilities that can vary as a power of the underlying asset. The Marsh–Rosenfeld (1983) model

$$dr_t = (\beta r_t^{-(1-\gamma)} + \alpha r_t)dt + \sigma r_t^{\gamma/2} dB_t$$

where α, β, σ, γ are nonnegative constants, covers most of the CEV models. For $\gamma = 1$ this is the CIR model, and for $\beta = 0$ we get the standard CEV model

$$dr_t = \alpha r_t dt + \sigma r_t^{\gamma/2} dB_t.$$

If $\gamma = 2$ this yields the Dothan [17] model

$$dr_t = \alpha r_t dt + \sigma r_t dB_t.$$

Affine Models

The class of short rate interest rate models admits a number of generalizations that can be found in the references quoted in the introduction of this chapter, among which is the class of affine models of the form

$$dr_t = (\eta(t) + \lambda(t)r_t)dt + \sqrt{\delta(t) + \gamma(t)r_t}dB_t. \tag{11.3}$$

Such models are called affine because the associated zero-coupon bonds can be priced using an affine PDE as will be seen in Proposition 11.2.

They also include the Ho-Lee model

$$dr_t = \theta(t)dt + \sigma dB_t,$$

where $\theta(t)$ is a deterministic function of time, and the Hull–White model

$$dr_t = (\theta(t) - \alpha(t)r_t)dt + \sigma(t)dB_t$$

which is a time-dependent extension of the Vasicek model.

11.2 Zero-Coupon Bonds

A zero-coupon bond is a contract priced $P_0(t,T)$ at time $t < T$ to deliver $P_0(T,T) = 1\$$ at time T. The computation of the arbitrage price $P_0(t,T)$ of a zero-coupon bond based on an underlying short term interest rate process $(r_t)_{t\in\mathbb{R}_+}$ is a basic and important issue in interest rate modeling.

In case the short term interest rate process $(r_t)_{t\in\mathbb{R}_+}$ is a deterministic function of time, a standard arbitrage argument shows that the price $P(t,T)$ of the bond is given by

$$P(t,T) = e^{-\int_t^T r_s ds}, \qquad 0 \le t \le T. \tag{11.4}$$

In case $(r_t)_{t\in\mathbb{R}_+}$ is an $\mathcal{F}_t$-adapted random process the formula (11.4) is no longer valid as it relies on future information, and we replace it with

$$P(t,T) = \mathbb{E}^*\left[e^{-\int_t^T r_s ds}\Big|\mathcal{F}_t\right], \qquad 0 \le t \le T, \tag{11.5}$$

under a risk-neutral measure $\mathbb{P}^*$. It is natural to write $P(t,T)$ as a conditional expectation under a martingale measure, as the use of conditional expectation helps to "filter out" the future information past time t contained in $\int_t^T r_s ds$. Expression (11.5) makes sense as the "best possible estimate" of the future quantity $e^{-\int_t^T r_s ds}$ in mean square sense, given information known up to time t.

Pricing bonds with a non-zero coupon is not difficult in the case of a deterministic continuous-time coupon yield at rate $c > 0$. In this case the price $P_c(t,T)$ of the coupon bond is given by

$$P_c(t,T) = e^{c(T-t)} P_0(t,T), \qquad 0 \le t \le T,$$

In the sequel we will only consider zero-coupon bonds, and let $P(t,T) = P_0(t,T)$, $0 \le t \le T$.

The following proposition shows that Assumption (A) of Chapter 10 is satisfied, i.e., the bond price process $t \longmapsto P(t,T)$ can be taken as a numéraire.

Proposition 11.1 *The discounted bond price process*

$$t \longmapsto e^{-\int_0^t r_s ds} P(t,T)$$

is a martingale under $\mathbb{P}^*$.

Proof. We have

$$\begin{aligned} e^{-\int_0^t r_s ds} P(t,T) &= e^{-\int_0^t r_s ds} \mathbb{E}^* \left[e^{-\int_t^T r_s ds} \Big| \mathcal{F}_t \right] \\ &= \mathbb{E}^* \left[e^{-\int_0^t r_s ds} e^{-\int_t^T r_s ds} \Big| \mathcal{F}_t \right] \\ &= \mathbb{E}^* \left[e^{-\int_0^T r_s ds} \Big| \mathcal{F}_t \right] \end{aligned}$$

and this suffices to conclude since by the "tower property" (A.20) of conditional expectations, any process of the form $t \longmapsto \mathbb{E}^*[F \mid \mathcal{F}_t]$, $F \in L^1(\Omega)$, is a martingale, cf. Relation (6.1). □

Bond pricing PDE

We assume from now on that the underlying short rate process is the solution to the stochastic differential equation

$$dr_t = \mu(t, r_t)dt + \sigma(t, r_t)dB_t \tag{11.6}$$

where $(B_t)_{t \in \mathbb{R}_+}$ is a standard Brownian motion under $\mathbb{P}^*$.

Since all solutions of stochastic differential equations such as (11.6) have the Markov property, cf e.g., Theorem V-32 of [61], the arbitrage price $P(t,T)$ can be rewritten as a function $F(t, r_t)$ of r_t, i.e.,

$$P(t,T) = \mathbb{E}^* \left[e^{-\int_t^T r_s ds} \Big| \mathcal{F}_t \right] = \mathbb{E}^* \left[e^{-\int_t^T r_s ds} \Big| r_t \right] = F(t, r_t),$$

and depends on r_t only instead of depending on all information available in $\mathcal{F}_t$ up to time t, meaning that the pricing problem can now be formulated as a search for the function $F(t,x)$.

Proposition 11.2 *(Bond pricing PDE) The bond pricing PDE for $P(t,T) = F(t,r_t)$ is written as*

$$xF(t,x) = \frac{\partial F}{\partial t}(t,x) + \mu(t,x)\frac{\partial F}{\partial x}(t,x) + \frac{1}{2}\sigma^2(t,x)\frac{\partial^2 F}{\partial x^2}(t,x), \quad (11.7)$$

$t \in \mathbb{R}_+$, $x \in \mathbb{R}$, subject to the terminal condition

$$F(T,x) = 1, \qquad x \in \mathbb{R}. \quad (11.8)$$

Proof. From Itô's formula we have

$$\begin{aligned}
d\left(e^{-\int_0^t r_s ds} P(t,T)\right) &= -r_t e^{-\int_0^t r_s ds} P(t,T)dt + e^{-\int_0^t r_s ds} dP(t,T) \\
&= -r_t e^{-\int_0^t r_s ds} F(t,r_t)dt + e^{-\int_0^t r_s ds} dF(t,r_t) \\
&= -r_t e^{-\int_0^t r_s ds} F(t,r_t)dt + e^{-\int_0^t r_s ds}\frac{\partial F}{\partial x}(t,r_t)(\mu(t,r_t)dt + \sigma(t,r_t)dB_t) \\
&\quad + e^{-\int_0^t r_s ds}\left(\frac{1}{2}\sigma^2(t,r_t)\frac{\partial^2 F}{\partial x^2}(t,r_t)dt + \frac{\partial F}{\partial t}(t,r_t)dt\right) \\
&= e^{-\int_0^t r_s ds}\sigma(t,r_t)\frac{\partial F}{\partial x}(t,r_t)dB_t \\
&\quad + e^{-\int_0^t r_s ds}\left(-r_t F(t,r_t) + \mu(t,r_t)\frac{\partial F}{\partial x}(t,r_t)\right. \\
&\quad \left. + \frac{1}{2}\sigma^2(t,r_t)\frac{\partial^2 F}{\partial x^2}(t,r_t) + \frac{\partial F}{\partial t}(t,r_t)\right) dt. \quad (11.9)
\end{aligned}$$

Given that $t \longmapsto e^{-\int_0^t r_s ds} P(t,T)$ is a martingale, the above expression (11.9) should only contain terms in dB_t (cf. Corollary 1, p. 72 of [61]), and all terms in dt should vanish inside (11.9). This leads to the identity

$$-r_t F(t,r_t) + \mu(t,r_t)\frac{\partial F}{\partial x}(t,r_t) + \frac{1}{2}\sigma^2(t,r_t)\frac{\partial^2 F}{\partial x^2}(t,r_t) + \frac{\partial F}{\partial t}(t,r_t) = 0.$$

Condition (11.8) is due to the fact that $P(T,T) = \$1$. □

The above proposition also shows that $t \longmapsto e^{-\int_0^t r_s ds} P(t,T)$ satisfies the stochastic differential equations

$$d\left(e^{-\int_0^t r_s ds} P(t,T)\right) = e^{-\int_0^t r_s ds}\sigma(t,r_t)\frac{\partial F}{\partial x}(t,r_t)dB_t.$$

Consequently we have

$$\frac{dP(t,T)}{P(t,T)} = r_t dt + \frac{\sigma(t,r_t)}{P(t,T)}\frac{\partial F}{\partial x}(t,r_t)dB_t = r_t dt + \sigma(t,r_t)\frac{\partial \log F}{\partial x}(t,r_t)dB_t.$$

In the Vasicek case

$$dr_t = (a - br_t)dt + \sigma dW_t,$$

we get

$$\frac{dP(t,T)}{P(t,T)} = r_t dt - \frac{\sigma}{b}(1 - e^{-b(T-t)})dW_t. \qquad (11.10)$$

Note that more generally, all affine short rate models as defined in Relation (11.3), including the Vasicek model, will yield a bond pricing formula of the form

$$P(t,T) = e^{A(T-t)+C(T-t)r_t},$$

cf. e.g., § 3.2.4. of [8].

Probabilistic PDE Solution

Next we solve the PDE (11.7) by direct computation of the conditional expectation

$$P(t,T) = \mathbb{E}^* \left[e^{-\int_t^T r_s ds} \Big| \mathcal{F}_t \right], \qquad (11.11)$$

in the Vasicek [71] model

$$dr_t = (a - br_t)dt + \sigma dB_t,$$

i.e., when the short rate $(r_t)_{t\in\mathbb{R}_+}$ has the expression

$$r_t = g(t) + \int_0^t h(t,s)dB_s = r_0 e^{-bt} + \frac{a}{b}(1 - e^{-bt}) + \sigma \int_0^t e^{-b(t-s)} dB_s,$$

where $g(t)$ and $h(t,s)$ are deterministic functions.

Letting $u \vee t = \max(u,t)$, using the fact that Wiener integrals are Gaussian random variables and the Gaussian characteristic function, we have

$$\begin{aligned}
P(t,T) &= \mathbb{E}^* \left[e^{-\int_t^T r_s ds} \Big| \mathcal{F}_t \right] \\
&= \mathbb{E}^* \left[e^{-\int_t^T (g(s)+\int_0^s h(s,u)dB_u)ds} \Big| \mathcal{F}_t \right] \\
&= e^{-\int_t^T g(s)ds} \, \mathbb{E}^* \left[e^{-\int_t^T \int_0^s h(s,u)dB_u ds} \Big| \mathcal{F}_t \right] \\
&= e^{-\int_t^T g(s)ds} \, \mathbb{E}^* \left[e^{-\int_0^T \int_{u\vee t}^T h(s,u)ds dB_u} \Big| \mathcal{F}_t \right] \\
&= e^{-\int_t^T g(s)ds} e^{-\int_0^t \int_{u\vee t}^T h(s,u)dsdB_u} \, \mathbb{E}^* \left[e^{-\int_t^T \int_{u\vee t}^T h(s,u)dsdB_u} \Big| \mathcal{F}_t \right] \\
&= e^{-\int_t^T g(s)ds} e^{-\int_0^t \int_t^T h(s,u)dsdB_u} \, \mathbb{E}^* \left[e^{-\int_t^T \int_u^T h(s,u)dsdB_u} \Big| \mathcal{F}_t \right] \\
&= e^{-\int_t^T g(s)ds} e^{-\int_0^t \int_t^T h(s,u)dsdB_u} \, \mathbb{E}^* \left[e^{-\int_t^T \int_u^T h(s,u)dsdB_u} \right] \\
&= e^{-\int_t^T g(s)ds} e^{-\int_0^t \int_t^T h(s,u)dsdB_u} e^{\frac{1}{2}\int_t^T \left(\int_u^T h(s,u)ds\right)^2 du} \\
&= e^{-\int_t^T (r_0 e^{-bs} + \frac{a}{b}(1-e^{-bs}))ds} e^{-\sigma \int_0^t \int_t^T e^{-b(s-u)} dsdB_u}
\end{aligned}$$

$$
\begin{aligned}
& \times e^{\frac{\sigma^2}{2}\int_t^T \left(\int_u^T e^{-b(s-u)}ds\right)^2 du} \\
= \;& e^{-\int_t^T (r_0 e^{-bs}+\frac{a}{b}(1-e^{-bs}))ds} e^{-\frac{\sigma}{b}(1-e^{-b(T-t)})\int_0^t e^{-b(t-u)}dB_u} \\
& \times e^{\frac{\sigma^2}{2}\int_t^T e^{2bu}\left(\frac{e^{-bu}-e^{-bT}}{b}\right)^2 du} \\
= \;& e^{-\frac{r_t}{b}(1-e^{-b(T-t)})+\frac{1}{b}(1-e^{-b(T-t)})(r_0 e^{-bt}+\frac{a}{b}(1-e^{-bt}))} \\
& \times e^{-\int_t^T (r_0 e^{-bs}+\frac{a}{b}(1-e^{-bs}))ds+\frac{\sigma^2}{2}\int_t^T e^{2bu}\left(\frac{e^{-bu}-e^{-bT}}{b}\right)^2 du} \\
= \;& e^{C(T-t)r_t+A(T-t)},
\end{aligned}
$$

where

$$C(T-t) = -\frac{1}{b}(1-e^{-b(T-t)}),$$

and

$$A(T-t) = \frac{4ab-3\sigma^2}{4b^3} + \frac{\sigma^2-2ab}{2b^2}(T-t) + \frac{\sigma^2-ab}{b^3}e^{-b(T-t)} - \frac{\sigma^2}{4b^3}e^{-2b(T-t)}.$$

Analytical PDE Solution

In order to solve the PDE (11.7) analytically we may look for a solution of the form

$$F(t,x) = e^{A(T-t)+xC(T-t)}, \tag{11.12}$$

where A and C are functions to be determined under the conditions $A(0)=0$ and $C(0)=0$. Plugging (11.12) into the PDE (11.7) yields the system of Riccati and linear differential equations

$$
\begin{cases}
-A'(s) = -aC(s) - \dfrac{\sigma^2}{2}C^2(s) \\[2ex]
-C'(s) = bC(s) + 1,
\end{cases}
$$

which can be solved to recover the above value of $P(t,T)$.

Some Bond Price Simulations

In this section we consider again the Vasicek model, in which the short rate $(r_t)_{t\in\mathbb{R}_+}$ is solution to (11.1). Figure 11.2 presents a random simulation of $t \longmapsto P(t,T)$ in the same Vasicek model. The graph of the corresponding deterministic bond price obtained for $a=b=\sigma=0$ is also shown on the Figure 11.2.

The above simulation can be compared to the actual market data of a coupon bond in Figure 11.4.

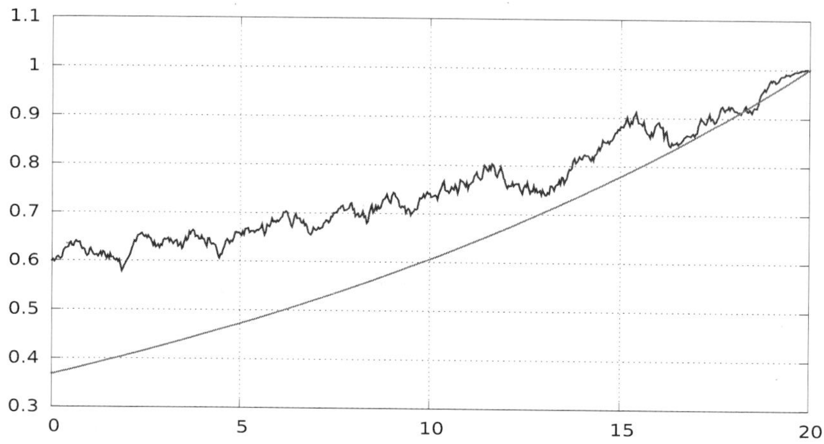

FIGURE 11.2: Graphs of $t \longmapsto P(t,T)$ and $t \longmapsto e^{-r_0(T-t)}$.

Figure 11.3 presents a random simulation of $t \longmapsto P(t,T)$ for a non-zero coupon bond with price $P_c(t,T) = e^{c(T-t)}P(t,T)$, and coupon rate $c > 0$, $0 \leq t \leq T$.

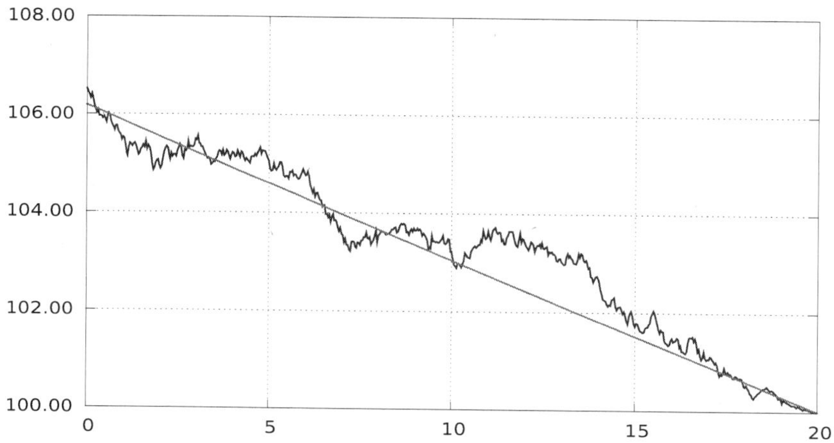

FIGURE 11.3: Graph of $t \longmapsto P(t,T)$ for a bond with a 2.3% coupon.

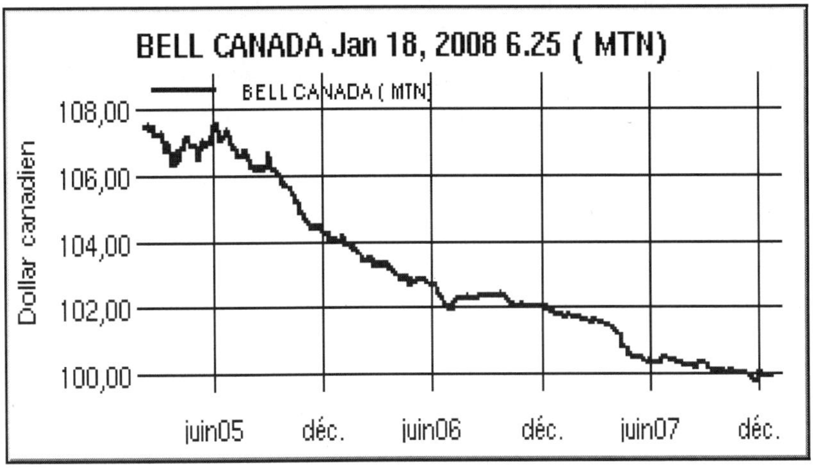

FIGURE 11.4: Bond price graph with maturity 01/18/08 and coupon rate 6.25%.

Bond pricing in the Dothan model

In the Dothan [17] model, the short term interest rate process $(r_t)_{t\in\mathbb{R}_+}$ is modeled according to a geometric Brownian motion

$$dr_t = \lambda r_t dt + \sigma r_t dB_t, \tag{11.13}$$

where the volatility $\sigma > 0$ and the drift $\lambda \in \mathbb{R}$ are constant parameters and $(B_t)_{t\in\mathbb{R}_+}$ is a standard Brownian motion. In this model the short term interest rate r_t remains always positive, while the proportional volatility term σr_t accounts for the sensitivity of the volatility of interest rate changes to the level of the rate r_t.

On the other hand, the Dothan model is the only lognormal short rate model that allows for an analytical formula for the zero coupon bond price

$$P(t,T) = \mathbb{E}\left[e^{-\int_t^T r_s ds} \Big| \mathcal{F}_t \right], \qquad 0 \le t \le T.$$

For convenience of notation we let $p = 1 - 2\lambda/\sigma^2$ and rewrite (11.13) as

$$dr_t = (1-p)\frac{1}{2}\sigma^2 r_t dt + \sigma r_t dB_t,$$

with solution

$$r_t = r_0 \exp\left(\sigma B_t - p\sigma^2 t/2\right), \qquad t \in \mathbb{R}_+,$$

where $p\sigma/2$ identifies to the market price of risk. By the Markov property of

$(r_t)_{t\in\mathbb{R}_+}$, the bond price $P(t,T)$ is a function $F(\tau, r_t)$ of r_t and of the time to maturity $\tau = T - t$:

$$P(t,T) = F(\tau, r_t) = \mathbb{E}\left[e^{-\int_t^T r_s ds} \Big| r_t \right], \qquad 0 \le t \le T. \tag{11.14}$$

By computation of the conditional expectation (11.14) using (8.38) we easily obtain the following result, cf. [52].

Proposition 11.3 *The zero-coupon bond price $P(t,T) = F(T-t, r_t)$ is given for all $p \in \mathbb{R}$ by*

$$F(\tau, r) = e^{-\sigma^2 p^2 \tau/8} \int_0^\infty \int_0^\infty e^{-ur} \exp\left(-2\frac{(1+z^2)}{\sigma^2 u}\right) \theta\left(\frac{4z}{\sigma^2 u}, \frac{\sigma^2 \tau}{4}\right) \frac{du}{u} \frac{dz}{z^{p+1}}. \tag{11.15}$$

Proof. We have

$$\begin{aligned} F(T-t, r_t) &= P(t,T) \\ &= \mathbb{E}\left[\exp\left(- \int_t^T r_s ds \right) \Big| \mathcal{F}_t \right] \\ &= \mathbb{E}\left[\exp\left(-r_t \int_t^T e^{\sigma(B_s - B_t) - \sigma^2 p(s-t)/2} ds \right) \Big| \mathcal{F}_t \right] \qquad (11.16) \\ &= \int_0^\infty \int_{-\infty}^\infty e^{-r_t u} \mathbb{P}\left(\int_0^\tau e^{\sigma B_s - p\sigma^2 s/2} ds \in du, B_\tau \in dy \right) \\ &= \int_0^\infty e^{-r_t u} \mathbb{P}\left(\int_0^\tau e^{\sigma B_s - p\sigma^2 s/2} ds \in du \right), \end{aligned}$$

and the conclusion follows from the change of variable $z = e^{\sigma y/2}$, using (8.38). □

See [52] and [51] for more results on bond pricing in the Dothan model, and [58] for numerical computations.

11.3 Forward Rates

A forward interest rate contract gives its holder a loan decided at present time t and to be delivered over a future period of time $[T, S]$ at a rate denoted by $f(t, T, S)$, $t \le T \le S$, and called a forward rate.

Let us determine the arbitrage or "fair" value of this rate using the instruments available in a bond market, which are bonds priced at $P(t,T)$ for various maturity dates $T > t$.

The loan can be realized using the bonds available on the market by proceeding in two steps:

1) at time t, borrow the amount $P(t,S)$ by shortselling one unit of bond with maturity S, which will mean refunding \$1 at time S.

2) since one only needs the money at time T, it makes sense to invest the amount $P(t,S)$ over the period $[t,T]$ by buying a (possibly fractional) quantity $P(t,S)/P(t,T)$ of a bond with maturity T priced $P(t,T)$ at time t. This will yield the amount

$$\$1 \times \frac{P(t,S)}{P(t,T)}$$

at time T.

As a consequence the investor will receive $P(t,S)/P(t,T)$ at time T, and will refund \$1 at time S.

The corresponding forward rate $f(t,T,S)$ is then given by the relation

$$\frac{P(t,S)}{P(t,T)} \exp\left((S-T)f(t,T,S)\right) = \$1, \qquad 0 \leq t \leq T \leq S, \tag{11.17}$$

where we used exponential compounding, which leads to the following definition (11.18).

Definition 11.1 *The forward rate $f(t,T,S)$ at time t for a loan on $[T,S]$ is given by*

$$f(t,T,S) = \frac{\log P(t,T) - \log P(t,S)}{S-T}. \tag{11.18}$$

The *spot* forward rate $f(t,t,T)$ is given by

$$f(t,t,T) = -\frac{\log P(t,T)}{T-t}, \quad \text{or} \quad P(t,T) = e^{-(T-t)f(t,t,T)}, \qquad 0 \leq t \leq T, \tag{11.19}$$

and is also called the *yield*.

Figure 11.5 presents a typical forward rate curve on the LIBOR (London Interbank Offered Rate) market with t =07 may 2003, δ = six months.

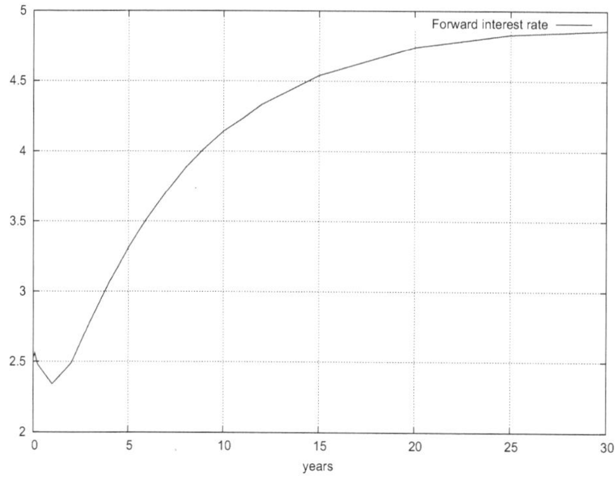

TimeSerieNb	505
AsOfDate	7-mai-03
2D	2,55
1W	2,53
1M	2,56
2M	2,52
3M	2,48
1Y	2,34
2Y	2,49
3Y	2,79
4Y	3,07
5Y	3,31
6Y	3,52
7Y	3,71
8Y	3,88
9Y	4,02
10Y	4,14
11Y	4,23
12Y	4,33
13Y	4,4
14Y	4,47
15Y	4,54
20Y	4,74
25Y	4,83
30Y	4,86

FIGURE 11.5: Graph of $T \longmapsto f(t,T,T+\delta)$.

The instantaneous forward rate $f(t,T)$ is defined by taking the limit of $f(t,T,S)$ as $S \searrow T$, i.e.,

$$\begin{aligned} f(t,T): &= \lim_{S\searrow T} f(t,T,S) \\ &= -\lim_{S\searrow T} \frac{\log P(t,S) - \log P(t,T)}{S-T} \\ &= -\lim_{\varepsilon\searrow 0} \frac{\log P(t,T+\varepsilon) - \log P(t,T)}{\varepsilon} \\ &= -\frac{\partial \log P(t,T)}{\partial T} \\ &= -\frac{1}{P(t,T)} \frac{\partial P(t,T)}{\partial T}. \end{aligned} \tag{11.20}$$

The above equation can be viewed as a differential equation to be solved for $\log P(t,T)$ under the initial condition $P(T,T)=1$, which yields the following proposition.

Proposition 11.4 *We have*

$$P(t,T) = \exp\left(-\int_t^T f(t,s)ds\right), \qquad 0 \le t \le T. \tag{11.21}$$

Proof. We check that

$$\log P(t,T) = \log P(t,T) - \log P(t,t) = \int_t^T \frac{\partial \log P(t,s)}{\partial s} ds = -\int_t^T f(t,s)ds.$$

□

As a consequence of (11.17) and (11.21) the forward rate $f(t,T,S)$ can be recovered from the instantaneous forward rate $f(t,s)$, as:

$$f(t,T,S) = \frac{1}{S-T}\int_T^S f(t,s)ds, \qquad 0 \le t \le T < S. \tag{11.22}$$

Forward Swap Rates

An interest rate swap makes it possible to exchange a variable forward rate $f(t,T,S)$ against a fixed rate κ over a time period $[T,S]$. Over a succession of time intervals $[T_1,T_2],\ldots,[T_{n-1},T_n]$, the sum of such exchanges will generate a cumulative discounted cash flow

$$\begin{aligned}
&\left(\sum_{k=1}^{n-1}(T_{k+1}-T_k)e^{-\int_t^{T_{k+1}} r_s ds} f(t,T_k,T_{k+1})\right) - \left(\sum_{k=1}^{n-1}(T_{k+1}-T_k)\kappa e^{-\int_t^{T_{k+1}} r_s ds}\right) \\
&\qquad = \sum_{k=1}^{n-1}(T_{k+1}-T_k)e^{-\int_t^{T_{k+1}} r_s ds}(f(t,T_k,T_{k+1})-\kappa),
\end{aligned}$$

at time t, in which we use linear interest rate compounding, and priced *at time* t as

$$\begin{aligned}
&\mathbb{E}\left[\sum_{k=1}^{n-1}(T_{k+1}-T_k)e^{-\int_t^{T_{k+1}} r_s ds}(f(t,T_k,T_{k+1})-\kappa)\,\Big|\,\mathcal{F}_t\right] \\
&\qquad = \sum_{k=1}^{n-1}(T_{k+1}-T_k)(f(t,T_k,T_{k+1})-\kappa)\,\mathbb{E}\left[e^{-\int_t^{T_{k+1}} r_s ds}\,\Big|\,\mathcal{F}_t\right] \\
&\qquad = \sum_{k=1}^{n-1}(T_{k+1}-T_k)P(t,T_{k+1})(f(t,T_k,T_{k+1})-\kappa).
\end{aligned}$$

The swap rate $S(t,T_1,T_n)$ is by definition the fair value of κ that cancels this cash flow and achieves equilibrium, i.e., $S(t,T_1,T_n)$ satisfies

$$\sum_{k=1}^{n-1}(T_{k+1}-T_k)P(t,T_{k+1})(f(t,T_k,T_{k+1})-S(t,T_1,T_n)) = 0, \tag{11.23}$$

and is given by

$$S(t,T_1,T_n) = \frac{1}{P(t,T_1,T_n)}\sum_{k=1}^{n-1}(T_{k+1}-T_k)P(t,T_{k+1})f(t,T_k,T_{k+1}), \tag{11.24}$$

where

$$P(t,T_1,T_n) = \sum_{k=1}^{n-1}(T_{k+1}-T_k)P(t,T_{k+1}), \qquad 0 \le t \le T_1,$$

is the annuity numéraire.

LIBOR Rates

Recall that the forward rate $f(t,T,S)$, $0 \leq t \leq T \leq S$, is defined using exponential compounding, from the relation

$$f(t,T,S) = -\frac{\log P(t,S) - \log P(t,T)}{S-T}. \tag{11.25}$$

In order to compute swaption prices one prefers to use forward rates as defined on the London InterBank Offered Rates (LIBOR) market instead of the standard forward rates given by (11.25).

The forward LIBOR $L(t,T,S)$ for a loan on $[T,S]$ is defined using linear compounding, i.e., by replacing (11.25) with the relation

$$1 + (S-T)L(t,T,S) = \frac{P(t,T)}{P(t,S)},$$

which yields the following definition.

Definition 11.2 *The forward LIBOR rate $L(t,T,S)$ at time t for a loan on $[T,S]$ is given by*

$$L(t,T,S) = \frac{1}{S-T}\left(\frac{P(t,T)}{P(t,S)} - 1\right), \qquad 0 \leq t \leq T < S. \tag{11.26}$$

Note that (11.26) above yields the same formula for the instantaneous forward rate

$$\begin{aligned} f(t,T) : &= \lim_{S \searrow T} L(t,T,S) \\ &= \lim_{S \searrow T} \frac{P(t,S) - P(t,T)}{(S-T)P(t,S)} \\ &= -\frac{1}{P(t,T)} \frac{\partial P(t,T)}{\partial T} \\ &= -\frac{\partial \log P(t,T)}{\partial T} \end{aligned}$$

as (11.20).

In addition, Relation (11.26) shows that the LIBOR rate can be viewed as a forward price $\hat{X}_t = X_t/N_t$ with numéraire $N_t = P(t,S)/(S-T)$ and $X_t = P(t,T) - P(t,S)$, according to Relation (10.7) of Chapter 10. As a consequence, from Proposition 10.2, the LIBOR rate $(L(t,T,S))_{t\in[T,S]}$ is a martingale under the forward measure $\hat{\mathbb{P}}$ defined by

$$\frac{d\hat{\mathbb{P}}}{d\mathbb{P}^*} = \frac{1}{P(0,S)} e^{-\int_0^S r_t dt}.$$

LIBOR Swap Rates

The LIBOR swap rate $S(t, T_1, T_n)$ satisfies the same relation as (11.23) with the forward rate $f(t, T_k, T_{k+1})$ replaced with the LIBOR rate $L(t, T_k, T_{k+1})$, i.e.,

$$\sum_{k=1}^{n-1}(T_{k+1} - T_k)P(t, T_{k+1})(L(t, T_k, T_{k+1}) - S(t, T_1, T_n)) = 0.$$

Proposition 11.5 *We have*

$$S(t, T_1, T_n) = \frac{P(t, T_1) - P(t, T_n)}{P(t, T_1, T_n)}, \qquad 0 \leq t \leq T_1, \quad 1 \leq i < j \leq n. \quad (11.27)$$

Proof. By (11.24) and (11.26) we have

$$\begin{aligned} S(t, T_1, T_n) &= \frac{1}{P(t, T_1, T_n)} \sum_{k=1}^{n-1}(T_{k+1} - T_k)P(t, T_{k+1})L(t, T_k, T_{k+1}) \\ &= \frac{1}{P(t, T_1, T_n)} \sum_{k=1}^{n-1} P(t, T_{k+1}) \left(\frac{P(t, T_k)}{P(t, T_{k+1})} - 1 \right) \\ &= \frac{1}{P(t, T_1, T_n)} \sum_{k=1}^{n-1} (P(t, T_k) - P(t, T_{k+1})) \\ &= \frac{P(t, T_1) - P(t, T_n)}{P(t, T_1, T_n)} \end{aligned} \quad (11.28)$$

by a telescoping summation. □

Clearly, a simple expression for the swap rate such as that of Proposition 11.5 cannot be obtained using the standard (i.e., non-LIBOR) rates defined in (11.25).

When $n = 2$, the swap rate $S(t, T_1, T_2)$ coincides with the forward rate $L(t, T_1, T_2)$:

$$S(t, T_1, T_2) = L(t, T_1, T_2), \qquad 1 \leq i \leq n - 1, \quad (11.29)$$

and the bond prices $P(t, T_1)$ can be recovered from the forward swap rates $S(t, T_1, T_n)$.

Similarly to the case of LIBOR rates, Relation (11.27) shows that the LIBOR swap rate can be viewed as a forward price with (annuity) numéraire $N_t = P(t, T_1, T_n)$ and $X_t = P(t, T_1) - P(t, T_n)$. Consequently the LIBOR swap rate $(S(t, T_1, T_n)_{t \in [T,S]}$ is a martingale under the forward measure $\hat{\mathbb{P}}$ defined from (10.1) by

$$\frac{d\hat{\mathbb{P}}}{d\mathbb{P}^*} = \frac{P(T_1, T_1, T_n)}{P(0, T_1, T_n)} e^{-\int_0^{T_1} r_t dt}.$$

11.4 HJM Model

In the previous chapter we have focused on the modeling of the short rate $(r_t)_{t\in\mathbb{R}_+}$ and on its consequences on the pricing of bonds $P(t,T)$, from which the forward rates $f(t,T,S)$ and $L(t,T,S)$ have been defined.

In this section we choose a different starting point and consider the problem of directly modeling the instantaneous forward rate $f(t,T)$. The graph given in Figure 11.4 presents a possible random evolution of a forward interest rate curve using the Musiela convention, i.e., we will write

$$g(x) = f(t,t+x) = f(t,T),$$

under the substitution $x = T-t$, $x \geq 0$, and represent a sample of the instantaneous forward curve $x \longmapsto f(t,t+x)$ for each $t \in \mathbb{R}_+$.

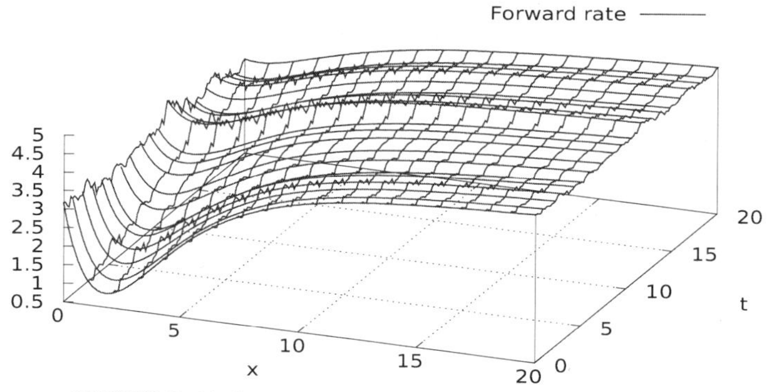

FIGURE 11.6: Stochastic process of forward curves.

In the HJM model, the instantaneous forward rate $f(t,T)$ is modeled under $\mathbb{P}$ by a stochastic differential equation of the form

$$d_t f(t,T) = \alpha(t,T)dt + \sigma(t,T)dB_t, \tag{11.30}$$

where $t \longmapsto \alpha(t,T)$ and $t \longmapsto \sigma(t,T)$, $0 \leq t \leq T$, are allowed to be random (adapted) processes. In the above equation, the date T is fixed and the differential d_t is with respect to t.

Under basic Markovianity assumptions, a HJM model with deterministic coefficients $\alpha(t,T)$ and $\sigma(t,T)$ will yield a short rate process $(r_t)_{t\in\mathbb{R}_+}$ of the form

$$dr_t = (a(t) - b(t)r_t)dt + \sigma(t)dB_t,$$

cf. § 6.6 of [56], which is the [33] Hull–White model, cf. Section 11.1, with explicit solution

$$r_t = r_s e^{-\int_s^t b(\tau)d\tau} + \int_s^t e^{-\int_u^t b(\tau)d\tau} a(u)du + \int_s^t \sigma(u)e^{-\int_u^t b(\tau)d\tau} dB_u,$$

$0 \leq s \leq t$.

The HJM Condition

How to "encode" absence of arbitrage in the defining equation (11.30) is an important question. Recall that under absence of arbitrage, the bond price $P(t,T)$ has been defined as

$$P(t,T) = \mathbb{E}^* \left[\exp\left(-\int_t^T r_s ds \right) \Big| \mathcal{F}_t \right], \qquad (11.31)$$

and the discounted bond price process

$$\begin{aligned} t \longmapsto \exp\left(-\int_0^t r_s ds\right) P(t,T) &= \exp\left(-\int_0^t r_s ds - \int_t^T f(t,s)ds\right) \\ &= \exp\left(-\int_0^t r_s ds - X_t\right) \qquad (11.32) \end{aligned}$$

is a martingale by Proposition 11.1 and Relation (11.21). This latter property will be used to characterize absence of arbitrage in the HJM model.

Proposition 11.6 *(HJM Condition [32]). Under the condition*

$$\alpha(t,T) = \sigma(t,T) \int_t^T \sigma(t,s)ds, \qquad t \in [0,T], \qquad (11.33)$$

which is known as the HJM absence of arbitrage condition, *the process* (11.32) *becomes a martingale.*

Proof. Consider the spot forward rate

$$f(t,t,T) = \frac{1}{T-t} \int_t^T f(t,s)ds,$$

and let

$$X_t = \int_t^T f(t,s)ds = -\log P(t,T), \qquad 0 \leq t \leq T,$$

with the relation

$$f(t,t,T) = \frac{1}{T-t} \int_t^T f(t,s)ds = \frac{X_t}{T-t}, \qquad 0 \leq t \leq T, \qquad (11.34)$$

where the dynamics of $t \longmapsto f(t,s)$ is given by (11.30). We have

$$\begin{aligned} d_t X_t &= -f(t,t)dt + \int_t^T d_t f(t,s)ds \\ &= -f(t,t)dt + \int_t^T \alpha(t,s)dsdt + \int_t^T \sigma(t,s)dsdB_t \\ &= -r_t dt + \left(\int_t^T \alpha(t,s)ds\right) dt + \left(\int_t^T \sigma(t,s)ds\right) dB_t, \end{aligned}$$

hence

$$|d_t X_t|^2 = \left(\int_t^T \sigma(t,s)ds\right)^2 dt.$$

Hence by Itô's calculus we have

$$\begin{aligned} d_t P(t,T) &= d_t e^{-X_t} \\ &= -e^{-X_t} d_t X_t + \frac{1}{2} e^{-X_t} (d_t X_t)^2 \\ &= -e^{-X_t} d_t X_t + \frac{1}{2} e^{-X_t} \left(\int_t^T \sigma(t,s)ds\right)^2 dt \\ &= -e^{-X_t} \left(-r_t dt + \int_t^T \alpha(t,s)dsdt + \int_t^T \sigma(t,s)dsdB_t\right) \\ &\quad + \frac{1}{2} e^{-X_t} \left(\int_t^T \sigma(t,s)ds\right)^2 dt, \end{aligned}$$

and the discounted bond price satisfies

$$\begin{aligned} & d_t \left(\exp\left(-\int_0^t r_s ds\right) P(t,T) \right) \\ &= -r_t \exp\left(-\int_0^t r_s ds - X_t\right) dt + \exp\left(-\int_0^t r_s ds\right) d_t P(t,T) \\ &= -r_t \exp\left(-\int_0^t r_s ds - X_t\right) dt - \exp\left(-\int_0^t r_s ds - X_t\right) d_t X_t \\ &\quad + \frac{1}{2} \exp\left(-\int_0^t r_s ds - X_t\right) \left(\int_t^T \sigma(t,s)ds\right)^2 dt \\ &= -r_t \exp\left(-\int_0^t r_s ds - X_t\right) dt \\ &\quad - \exp\left(-\int_0^t r_s ds - X_t\right) \left(-r_t dt + \int_t^T \alpha(t,s)dsdt + \int_t^T \sigma(t,s)dsdB_t\right) \\ &\quad + \frac{1}{2} \exp\left(-\int_0^t r_s ds - X_t\right) \left(\int_t^T \sigma(t,s)ds\right)^2 dt \end{aligned}$$

$$= -\exp\left(-\int_0^t r_s ds - X_t\right)\int_t^T \sigma(t,s)dsdB_t$$
$$-\exp\left(-\int_0^t r_s ds - X_t\right)\left(\int_t^T \alpha(t,s)dsdt - \frac{1}{2}\left(\int_t^T \sigma(t,s)ds\right)^2\right)dt.$$

Thus the process $P(t,T)$ will be a martingale provided that

$$\int_t^T \alpha(t,s)ds - \frac{1}{2}\left(\int_t^T \sigma(t,s)ds\right)^2 = 0, \qquad 0 \le t \le T. \tag{11.35}$$

Differentiating the above relation with respect to T, we get

$$\alpha(t,T) = \sigma(t,T)\int_t^T \sigma(t,s)ds,$$

which is in fact equivalent to (11.35). □

11.5 Forward Vasicek Rates

In this section we consider the Vasicek model, in which the short rate process is the solution (11.2) of (11.1) as illustrated in Figure 11.1.

In this model the forward rate is given by

$$f(t,T,S) = -\frac{\log P(t,S) - \log P(t,T)}{S-T}$$
$$= -\frac{r_t(C(S-t) - C(T-t)) + A(S-t) - A(T-t))}{S-T}$$
$$= -\frac{\sigma^2 - 2ab}{2b^2}$$
$$-\frac{1}{S-T}\left(\left(\frac{r_t}{b} + \frac{\sigma^2 - ab}{b^3}\right)(e^{-b(S-t)} - e^{-b(T-t)})\right.$$
$$\left.-\frac{\sigma^2}{4b^3}(e^{-2b(S-t)} - e^{-2b(T-t)})\right).$$

In this model the forward rate $t \longmapsto f(t,T,S)$ can be represented as in Figure 11.7, with here $b/a > r_0$.

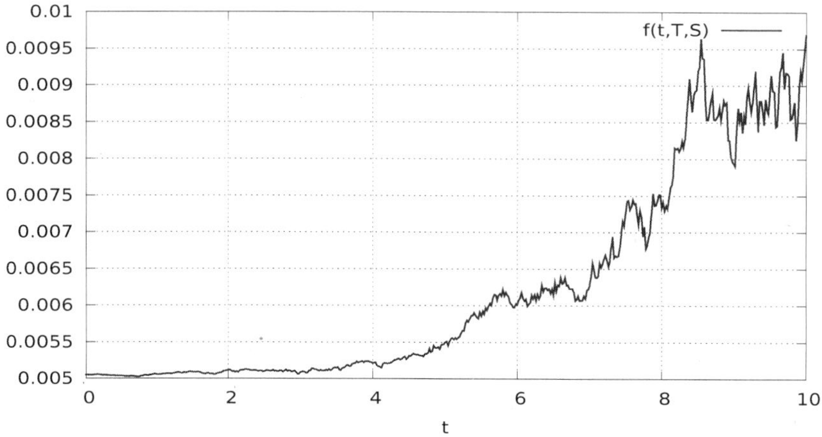

FIGURE 11.7: Forward rate process $t \longmapsto f(t,T,S)$.

Note that the forward rate cure $t \longmapsto f(t,T,S)$ is flat for small values of t.

The instantaneous short rate is given by

$$\begin{aligned} f(t,T) : &= -\frac{\partial \log P(t,T)}{\partial T} \\ &= r_t e^{-b(T-t)} + \frac{a}{b}(1 - e^{-b(T-t)}) - \frac{\sigma^2}{2b^2}(1 - e^{-b(T-t)})^2, \end{aligned} \tag{11.36}$$

and the relation $\lim_{T \searrow t} f(t,T) = r_t$ is easily recovered from this formula.

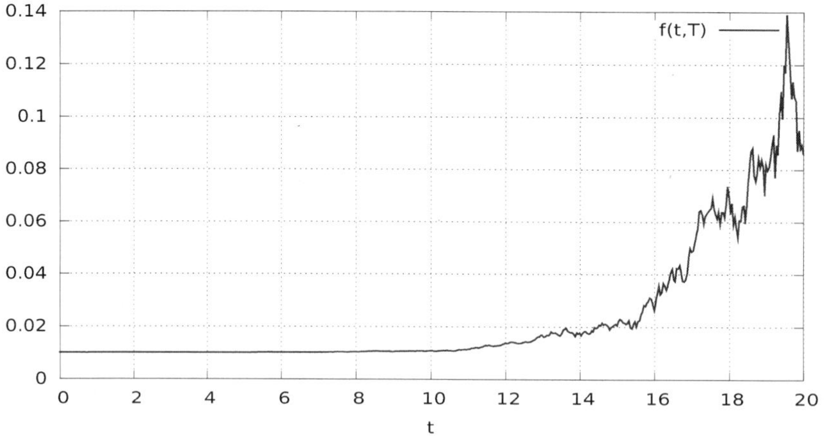

FIGURE 11.8: Instantaneous forward rate process $t \longmapsto f(t,T)$.

The instantaneous forward rate $t \longmapsto f(t,T)$ can be represented as in Figure 11.8, with here $t=0$ and $b/a > r_0$:

The HJM coefficients in the Vasicek model are in fact deterministic and taking $a=0$ we have

$$d_t f(t,T) = \sigma^2 e^{-b(T-t)} \int_t^T e^{b(t-s)} ds dt + \sigma e^{-b(T-t)} dB_t,$$

i.e.,

$$\alpha(t,T) = \sigma^2 e^{-b(T-t)} \int_t^T e^{-b(T-s)} ds, \qquad \text{and} \qquad \sigma(t,T) = \sigma e^{-b(T-t)},$$

and the HJM condition reads

$$\alpha(t,T) = \sigma^2 e^{-b(T-t)} \int_t^T e^{-b(s-t)} ds = \sigma(t,T) \int_t^T \sigma(t,s) ds. \tag{11.37}$$

Random simulations of the Vasicek instantaneous forward rates are provided in Figures 11.9 and 11.10.

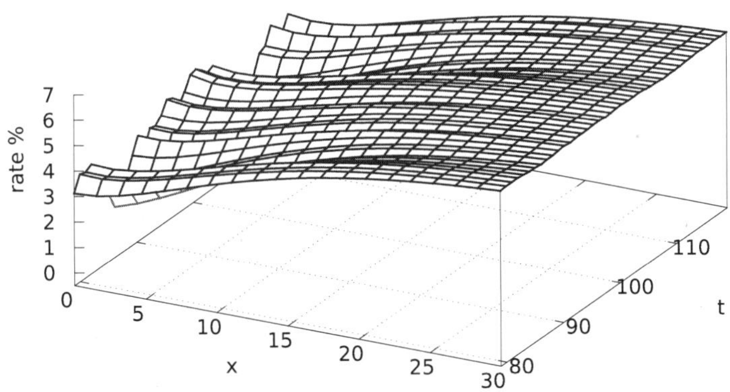

FIGURE 11.9: Forward instantaneous curve $(t,x) \longmapsto f(t,t+x)$ in the Vasicek model.

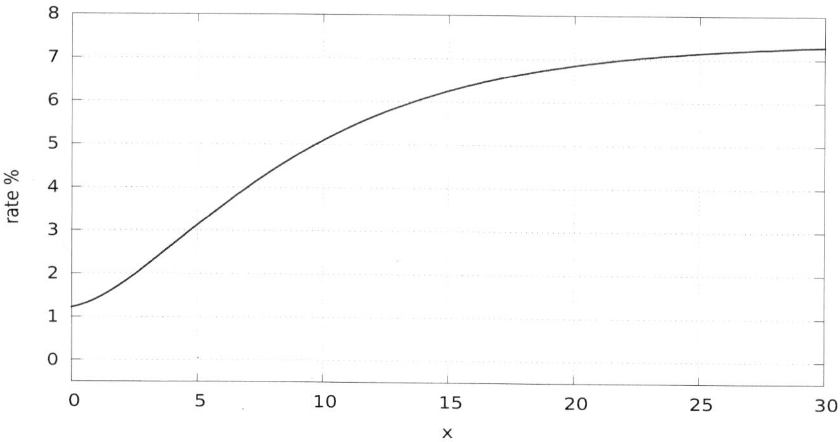

FIGURE 11.10: Forward instantaneous curve $x \longmapsto f(0,x)$ in the Vasicek model.

For $x = 0$ the first "slice" of this surface is actually the short rate Vasicek process $r_t = f(t,t) = f(t,t+0)$ which is represented in Figure 11.11 using another discretization.

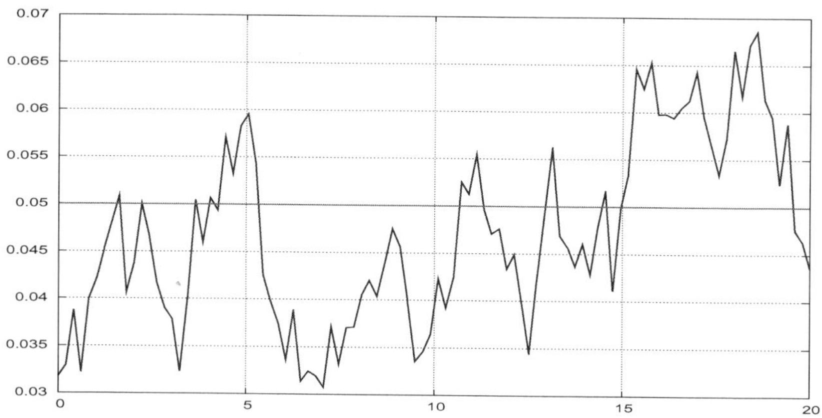

FIGURE 11.11: Short-term interest rate curve $t \longmapsto r_t$ in the Vasicek model.

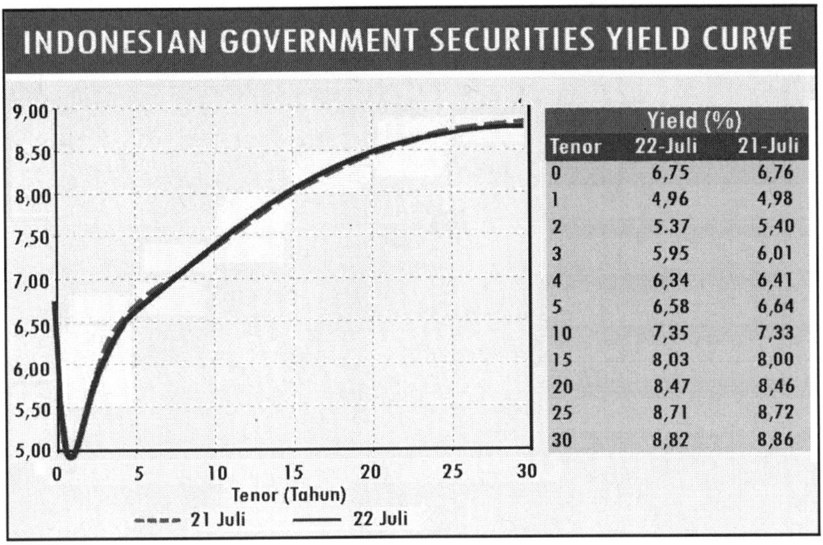

FIGURE 11.12: Market example of yield curves (11.19).

Another example of market data is given in Figure 11.12, in which the dashed and continuous curves refer respectively to July 21 and 22 of year 2011.

11.6 Modeling Issues

Parametrization of Forward Rates

In the Nelson–Siegel parametrization the forward interest rate curves are parametrized by 4 coefficients z_1, z_2, z_3, z_4, as

$$g(x) = z_1 + (z_2 + z_3 x)e^{-xz_4}, \qquad x \geq 0.$$

An example of a graph obtained by the Nelson–Siegel parametrization is given in Figure 11.13, for $z_1 = 1$, $z_2 = -10$, $z_3 = 100$, $z_4 = 10$.

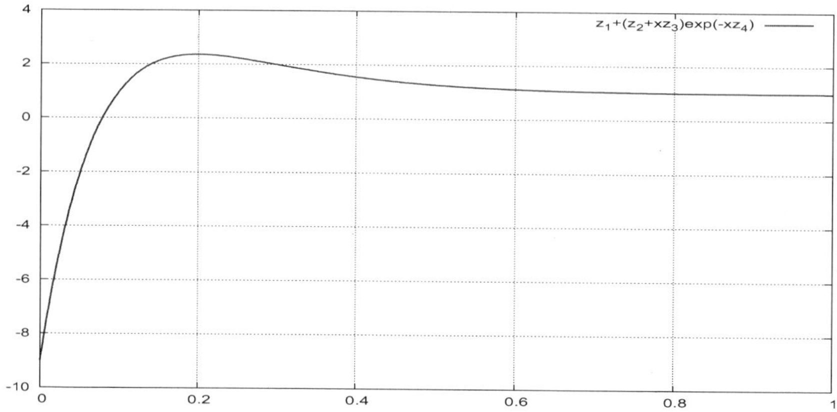

FIGURE 11.13: Graph of $x \longmapsto g(x)$ in the Nelson–Siegel model.

The Svensson parametrization has the advantage of reproducing two humps instead of one, the location and height of which can be chosen *via* 6 parameters z_1, z_2, z_3, z_4, z_5, z_6 as

$$g(x) = z_1 + (z_2 + z_3 x)e^{-xz_4} + z_5 x e^{-xz_6}, \qquad x \geq 0.$$

A typical graph of a Svensson parametrization is given in Figure 11.14, for $z_1 = 7$, $z_2 = -5$, $z_3 = -100$, $z_4 = 10$, $z_5 = -1/2$, $z_6 = -1$.

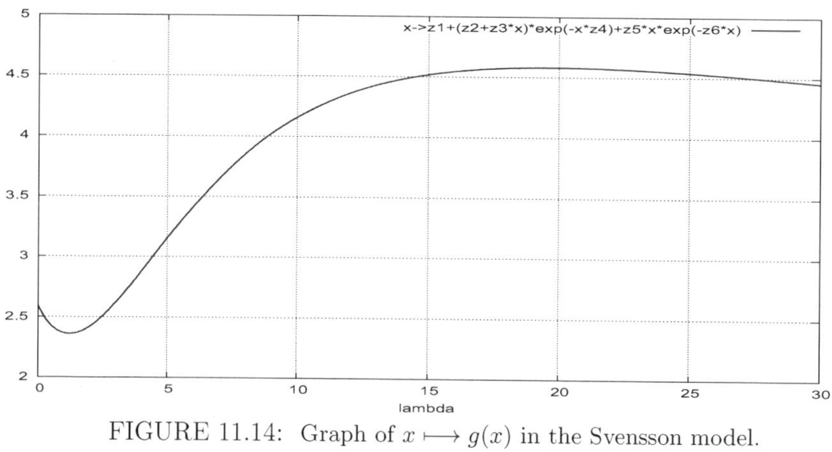

FIGURE 11.14: Graph of $x \longmapsto g(x)$ in the Svensson model.

Figure 11.15 presents a fit of the market data of Figure 11.5 using a Svensson curve.

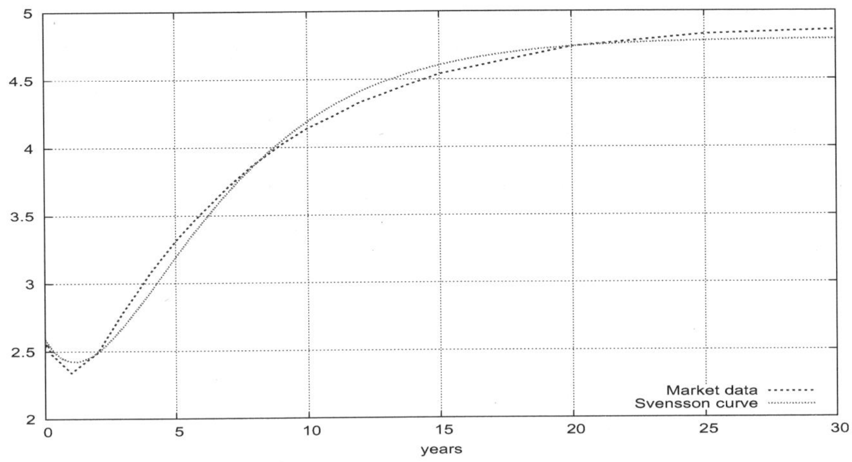

FIGURE 11.15: Comparison of market data *vs.* a Svensson curve.

One may think of constructing an instantaneous rate process taking values in the Svensson space, however this type of modelization is not consistent with absence of arbitrage, and it can be proved that the HJM curves cannot live in the Nelson–Siegel or Svensson spaces, cf. §3.5 of [5].

For example it can be easily shown that the forward curves of the Vasicek model are included neither in the Nelson–Siegel space, nor in the Svensson space, cf. [56] and §3.5 of [5]. In addition, such curves do not appear to correctly model the market forward curves considered above, cf. e.g., Figure 11.5.

In the Vasicek model we have

$$\frac{\partial f}{\partial T}(t,T) = \left(-br_t + a - \frac{\sigma^2}{b} + \frac{\sigma^2}{b}e^{-b(T-t)}\right) e^{-b(T-t)},$$

and one can check that the sign of the derivatives of f can only change once at most. As a consequence, the possible forward curves in the Vasicek model are limited to one change of "regime" per curve, as illustrated in Figure 11.16 for various values of r_t, and in Figure 11.17.

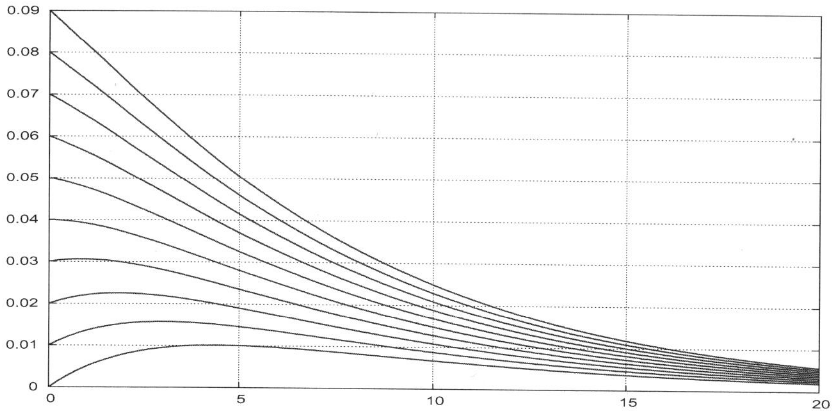

FIGURE 11.16: Graphs of forward rates.

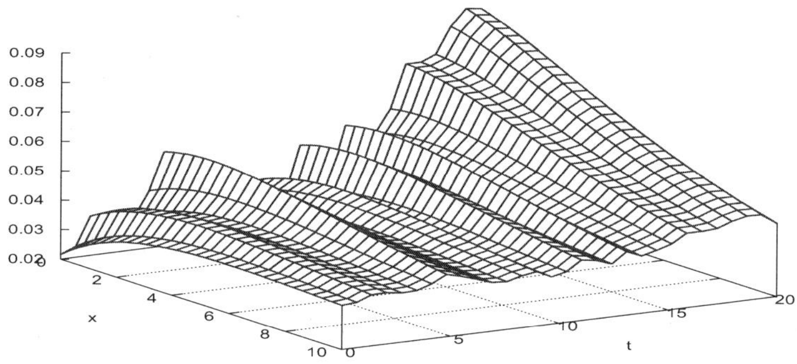

FIGURE 11.17: Forward instantaneous curve $(t, x) \longmapsto f(t, t + x)$ in the Vasicek model.

Another way to deal with the curve fitting problem is to use deterministic shifts for the fitting of one forward curve, such as the initial curve at $t = 0$, cf. e.g., § 8.2. of [56].

The Correlation Problem and a Two-Factor Model

The correlation problem is another issue of concern when using the affine models considered so far. Let us compare three bond price simulations with maturity $T_1 = 10$, $T_2 = 20$, and $T_3 = 30$ based on the same Brownian path, as given in Figure 11.18. Clearly, the bond prices $P(t, T_1)$ and $P(t, T_2)$ with maturities T_1 and T_2 are linked by the relation

$$P(t, T_2) = P(t, T_1) \exp(A(t, T_2) - A(t, T_1) + r_t(C(t, T_2) - C(t, T_1))), \quad (11.38)$$

meaning that bond prices with different maturities could be deduced from each other, which is unrealistic.

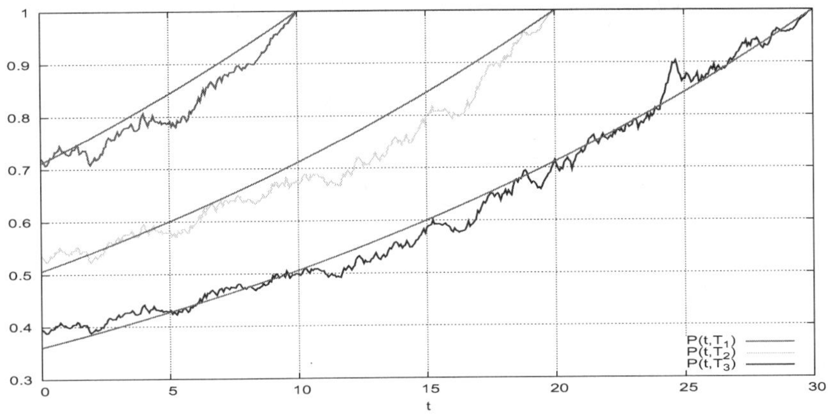

FIGURE 11.18: Graph of $t \longmapsto P(t, T_1)$.

For affine models of short rates we have the perfect correlation

$$\mathrm{Cor}(\log P(t, T_1), \log P(t, T_2)) = 1,$$

cf. § 8.3 of [56], since by (11.38), $\log P(t, T_1)$ and $\log P(t, T_2)$ are linked by the linear relation

$$\log P(t, T_2) = \log P(t, T_1) + A(t, T_2) - A(t, T_1) + r_t(C(t, T_2) - C(t, T_1)),$$

involving a same random variable $Z = r_t$. A solution to the correlation problem is to consider two control processes $(X_t)_{t\in\mathbb{R}_+}$, $(Y_t)_{t\in\mathbb{R}_+}$ which are the solution of

$$\begin{cases} dX_t = \mu_1(t, X_t)dt + \sigma_1(t, X_t)dB_t^{(1)}, \\ \\ dY_t = \mu_2(t, Y_t)dt + \sigma_2(t, Y_t)dB_t^{(2)}, \end{cases} \tag{11.39}$$

where $(B_t^{(1)})_{t\in\mathbb{R}_+}$, $(B_t^{(2)})_{t\in\mathbb{R}_+}$ have correlated Brownian motion with

$$\mathrm{Cov}(B_s^{(1)}, B_t^{(2)}) = \rho \min(s, t), \qquad s, t \in \mathbb{R}_+, \tag{11.40}$$

and

$$dB_t^{(1)} dB_t^{(2)} = \rho dt, \tag{11.41}$$

for some $\rho \in [-1, 1]$. In practice, $(B^{(1)})_{t\in\mathbb{R}_+}$ and $(B^{(2)})_{t\in\mathbb{R}_+}$ can be constructed from two independent Brownian motions $(W^{(1)})_{t\in\mathbb{R}_+}$ and $(W^{(2)})_{t\in\mathbb{R}_+}$, by letting

$$\begin{cases} B_t^{(1)} = W_t^{(1)}, \\ \\ B_t^{(2)} = \rho W_t^{(1)} + \sqrt{1-\rho^2} W_t^{(2)}, \qquad t \in \mathbb{R}_+, \end{cases}$$

and Relations (11.40) and (11.41) are easily satisfied from this construction.

In two-factor models one chooses to build the short term interest rate r_t *via*

$$r_t = X_t + Y_t, \qquad t \in \mathbb{R}_+.$$

By the previous standard arbitrage arguments we define the price of a bond with maturity T as

$$\begin{aligned} P(t,T) : &= \mathbb{E}\left[\exp\left(-\int_t^T r_s ds\right) \Big| \mathcal{F}_t\right] \\ &= \mathbb{E}\left[\exp\left(-\int_t^T r_s ds\right) \Big| X_t,\ Y_t\right] \\ &= F(t, X_t, Y_t), \end{aligned} \tag{11.42}$$

since the couple $(X_t, Y_t)_{t\in\mathbb{R}_+}$ is Markovian. Using the Itô formula with two variables and the fact that

$$t \longmapsto e^{-\int_0^t r_s ds} P(t,T) = e^{-\int_0^t r_s ds}\, \mathbb{E}\left[\exp\left(-\int_t^T r_s ds\right) \Big| \mathcal{F}_t\right]$$

is an $\mathcal{F}_t$-martingale under $\mathbb{P}^*$ we can derive a PDE

$$\begin{aligned} &-(x+y)F(t,x,y) + \mu_1(t,x)\frac{\partial F}{\partial x}(t,x,y) + \mu_2(t,y)\frac{\partial F}{\partial y}(t,x,y) \\ &+\frac{1}{2}\sigma_1^2(t,x)\frac{\partial^2 F}{\partial x^2}(t,x,y) + \frac{1}{2}\sigma_2^2(t,y)\frac{\partial^2 F}{\partial y^2}(t,x,y) \\ &+\rho\sigma_1(t,x)\sigma_2(t,y)\frac{\partial^2 F}{\partial x \partial y}(t,x,y) + \frac{\partial F}{\partial t}(t,X_t,Y_t) = 0, \end{aligned} \tag{11.43}$$

on $\mathbb{R}^2$ for the bond price $P(t,T)$. In the Vasicek model

$$\begin{cases} dX_t = -aX_t dt + \sigma dB_t^{(1)}, \\ dY_t = -bY_t dt + \eta dB_t^{(2)}, \end{cases}$$

this yields

$$P(t,T) = F_1(t,X_t)F_2(t,Y_t)\exp\left(U(t,T)\right), \tag{11.44}$$

where $F_1(t,X_t)$ and $F_2(t,Y_t)$ are the bond prices associated to X_t and Y_t in the Vasicek model, and

$$U(t,T) = \rho\frac{\sigma\eta}{ab}\left(T - t + \frac{e^{-a(T-t)}-1}{a} + \frac{e^{-b(T-t)}-1}{b} - \frac{e^{-(a+b)(T-t)}-1}{a+b}\right)$$

is a correlation term which vanishes when $(B_t^{(1)})_{t\in\mathbb{R}_+}$ and $(B_t^{(2)})_{t\in\mathbb{R}_+}$ are independent, i.e., when $\rho = 0$, cf [8], Chapter 4, Appendix A, and [56], [8].

Partial differentiation of $\log P(t,T)$ with respect to T leads to the instantaneous forward rate

$$f(t,T) = f_1(t,T) + f_2(t,T) - \rho\frac{\sigma\eta}{ab}(1 - e^{-a(T-t)})(1 - e^{-b(T-t)}), \quad (11.45)$$

where $f_1(t,T)$, $f_2(t,T)$ are the instantaneous forward rates corresponding to X_t and Y_t respectively, cf. § 8.4 of [56].

An example of a forward rate curve obtained in this way is given in Figure 11.19.

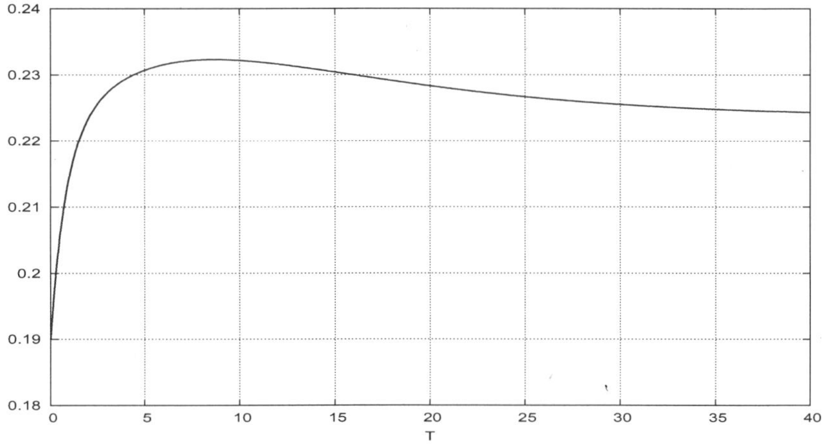

FIGURE 11.19: Graph of forward rates in a two-factor model.

Next in Figure 11.20 we present a graph of the evolution of forward curve in a two factor model.

11.7 BGM Model

The models (HJM, affine, etc.) considered in the previous chapter suffer from various drawbacks such as non-positivity of interest rates in Vasicek model, and lack of closed form solutions in more complex models. The BGM [7] model has the advantage of yielding positive interest rates, and to permit to derive explicit formulas for the computation of prices for interest rate derivatives such as caps and swaptions on the LIBOR market.

In the BGM model we work with a *tenor structure* $\{T_1, \ldots, T_n\}$ (see Section 12.1 for details) and consider the family $(\mathbb{P}_i)_{i=1,\ldots,n}$ of forward measures

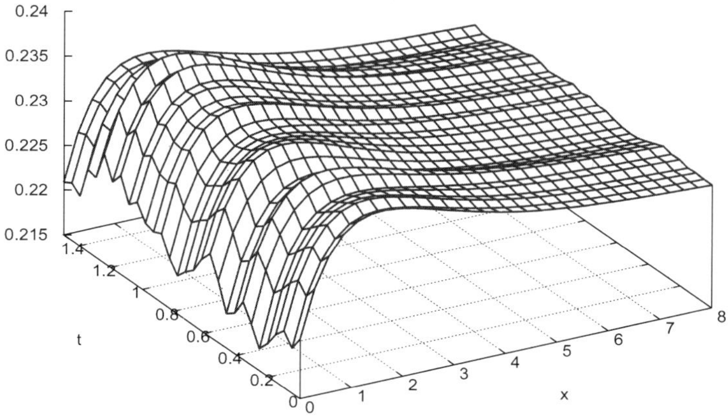

FIGURE 11.20: Random evolution of forward rates in a two-factor model.

defined by taking the bond prices $(P(t,T_1))_{t\in[0,T_1]}$, $i=1,\ldots,n$, as respective numéraires, i.e.,

$$\frac{d\mathbb{P}_i}{d\mathbb{P}_i^*} = \frac{e^{-\int_0^{T_1} r_s ds}}{P(0,T_1)},$$

cf. (10.6).

The forward LIBOR rate $L(t,T_1,T_2)$ is modeled as a geometric Brownian motion under $\mathbb{P}_2$, i.e.,

$$\frac{dL(t,T_1,T_2)}{L(t,T_1,T_2)} = \gamma_1(t)dB_t^{(2)}, \tag{11.46}$$

$0 \le t \le T_1$, $i=1,\ldots,n-1$, for some deterministic function $\gamma_1(t)$, with solution

$$L(u,T_1,T_2) = L(t,T_1,T_2)\exp\left(\int_t^u \gamma_1(s)dB_s^{(2)} - \frac{1}{2}\int_t^u |\gamma_1|^2(s)ds\right),$$

i.e., for $u = T_1$,

$$L(T_1,T_1,T_2) = L(t,T_1,T_2)\exp\left(\int_t^{T_1} \gamma_1(s)dB_s^{(2)} - \frac{1}{2}\int_t^{T_1} |\gamma_1|^2(s)ds\right).$$

Since $L(t,T_1,T_2)$ is a geometric Brownian motion under $\mathbb{P}_2$, standard caplets can be priced at time $t\in[0,T_1]$ from the Black–Scholes formula.

The following graph summarizes the notions introduced in this chapter.

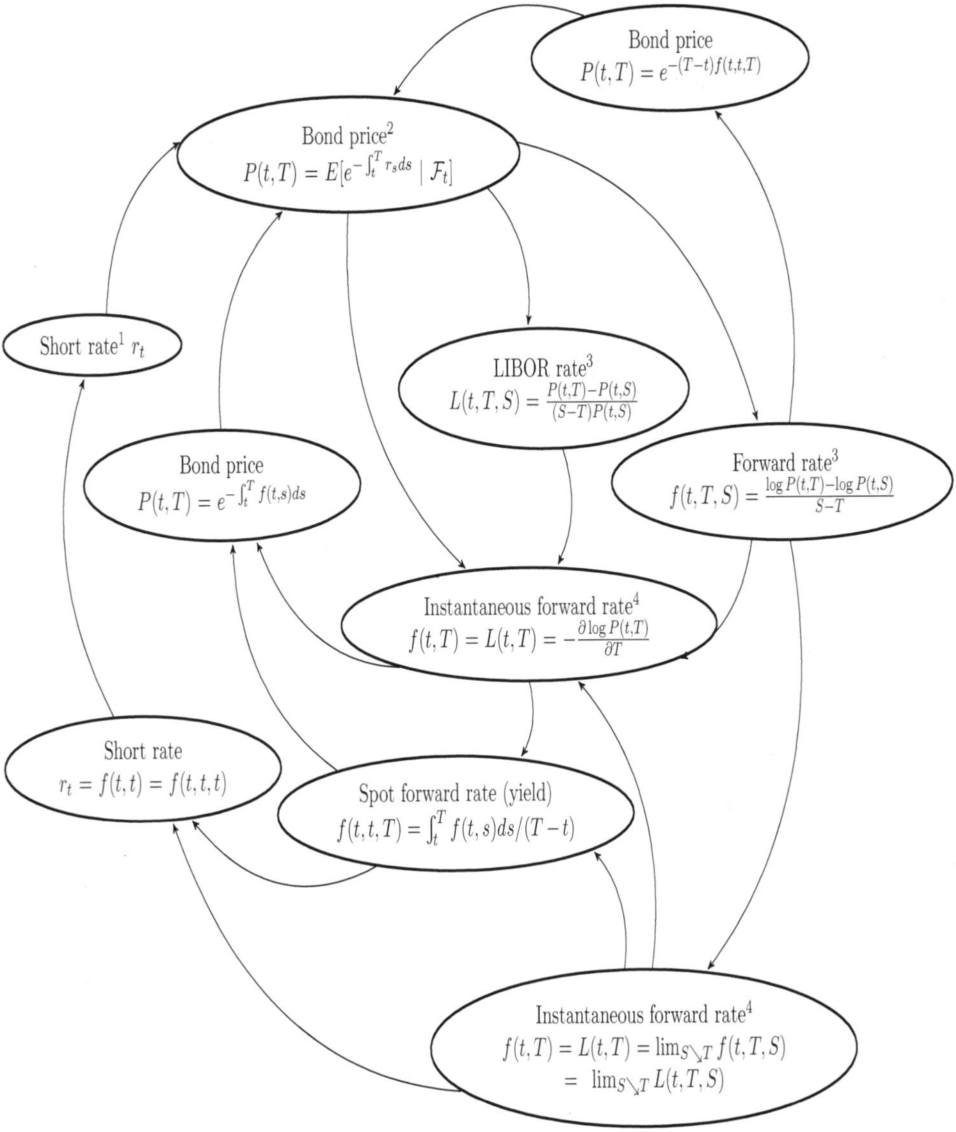

[1]Can be modeled by Vasiçek and other short rate models
[2]Can be modeled from $dP(t,T)/P(t,T)$.
[3]Can be modeled in the BGM model
[4]Can be modeled in the HJM model

FIGURE 11.21: Graph of stochastic interest rate modeling.

Exercises

Exercise 11.1 Let $(B_t)_{t\in\mathbb{R}_+}$ denote a standard Brownian motion started at 0 under the risk-neutral measure $\mathbb{P}^*$. We consider a short term interest rate process $(r_t)_{t\in\mathbb{R}_+}$ in a Ho–Lee model with constant deterministic volatility, defined by

$$dr_t = adt + \sigma dB_t,$$

where $a > 0$ and $\sigma > 0$. Let $P(t,T)$ will denote the arbitrage price of a zero-coupon bond in this model:

$$P(t,T) = \mathbb{E}^*\left[\exp\left(-\int_t^T r_s ds\right) \Big| \mathcal{F}_t\right], \qquad 0 \le t \le T. \tag{11.47}$$

1. State the bond pricing PDE satisfied by the function $F(t,x)$ defined *via*

$$F(t,x) := \mathbb{E}^*\left[\exp\left(-\int_t^T r_s ds\right) \Big| r_t = x\right], \qquad 0 \le t \le T.$$

2. Compute the arbitrage price $F(t,r_t) = P(t,T)$ from its expression (11.47) as a conditional expectation.

   Hint. One may use the *integration by parts* relation

$$\begin{aligned}\int_t^T B_s ds &= TB_T - tB_t - \int_t^T s dB_s \\ &= (T-t)B_t + T(B_T - B_t) - \int_t^T s dB_s \\ &= (T-t)B_t + \int_t^T (T-s) dB_s,\end{aligned}$$

   and the Laplace transform identity $\mathbb{E}[e^{\lambda X}] = e^{\lambda^2\eta^2/2}$ for $X \simeq \mathcal{N}(0,\eta^2)$.

3. Check that the function $F(t,x)$ computed in question 2 does satisfy the PDE derived in question 1.

4. Compute the forward rate $f(t,T,S)$ in this model.

   From now on we let $a = 0$.

5. Compute the instantaneous forward rate $f(t,T)$ in this model.

6. Derive the stochastic equation satisfied by the instantaneous forward rate $f(t,T)$.

7. Check that the HJM absence of arbitrage condition is satisfied in this equation.

Exercise 11.2 Let $(r_t)_{t\in\mathbb{R}_+}$ denote a short term interest rate process. For any $T > 0$, let $P(t,T)$ denote the price at time $t \in [0,T]$ of a zero coupon bond defined by the stochastic differential equation

$$\frac{dP(t,T)}{P(t,T)} = r_t dt + \sigma_t^T dB_t, \qquad 0 \le t \le T, \tag{11.48}$$

under the terminal condition $P(T,T) = 1$, where $(\sigma_t^T)_{t\in[0,T]}$ is an adapted process. Let the forward measure $\mathbb{P}_T$ be defined by

$$\mathbb{E}\left[\frac{d\mathbb{P}_T}{d\mathbb{P}}\Big|\mathcal{F}_t\right] = \frac{P(t,T)}{P(0,T)} e^{-\int_0^t r_s ds}, \qquad 0 \le t \le T.$$

Recall that

$$B_t^T := B_t - \int_0^t \sigma_s^T ds, \qquad 0 \le t \le T,$$

is a standard Brownian motion under $\mathbb{P}_T$.

1. Solve the stochastic differential equation (11.48).

2. Derive the stochastic differential equation satisfied by the discounted bond price process

$$t \longmapsto e^{-\int_0^t r_s ds} P(t,T), \qquad 0 \le t \le T,$$

and show that it is a martingale.

3. Show that

$$\mathbb{E}\left[e^{-\int_0^T r_s ds}\Big|\mathcal{F}_t\right] = e^{-\int_0^t r_s ds} P(t,T), \qquad 0 \le t \le T.$$

4. Show that

$$P(t,T) = \mathbb{E}\left[e^{-\int_t^T r_s ds}\Big|\mathcal{F}_t\right], \qquad 0 \le t \le T.$$

5. Compute $P(t,S)/P(t,T)$, $0 \le t \le T$, show that it is a martingale under $\mathbb{P}_T$ and that

$$P(T,S) = \frac{P(t,S)}{P(t,T)} \exp\left(\int_t^T (\sigma_s^S - \sigma_s^T) dB_s^T - \frac{1}{2}\int_t^T (\sigma_s^S - \sigma_s^T)^2 ds\right).$$

6. Assuming that $(\sigma_t^T)_{t\in[0,T]}$ and $(\sigma_t^S)_{t\in[0,S]}$ are deterministic functions, compute the price

$$\mathbb{E}\left[e^{-\int_t^T r_s ds}(P(T,S)-\kappa)^+\Big|\mathcal{F}_t\right] = P(t,T)\,\mathbb{E}_T\left[(P(T,S)-\kappa)^+\Big|\mathcal{F}_t\right]$$

of a bond option with strike κ.

Recall that if X is a centered Gaussian random variable with mean m_t and variance v_t^2 given $\mathcal{F}_t$, we have

$$\mathbb{E}[(e^X - K)^+ \mid \mathcal{F}_t] = e^{m_t+v_t^2/2}\Phi\left(\frac{v_t}{2} + \frac{1}{v_t}(m_t + v_t^2/2 - \log K)\right)$$
$$-K\Phi\left(-\frac{v_t}{2} + \frac{1}{v_t}(m_t + v_t^2/2 - \log K)\right)$$

where $\Phi(x)$, $x \in \mathbb{R}$, denotes the Gaussian distribution function.

Exercise 11.3 (Exercise 4.5 continued). Assume that the price $P(t,T)$ of a zero coupon bond is modeled as

$$P(t,T) = e^{-\mu(T-t)+X_t^T}, \qquad t \in [0,T],$$

where $\mu > 0$.

1. Show that the terminal condition $P(T,T) = 1$ is satisfied.
2. Compute the forward rate
$$f(t,T,S) = -\frac{1}{S-T}(\log P(t,S) - \log P(t,T)).$$
3. Compute the instantaneous forward rate
$$f(t,T) = -\lim_{S\searrow T}\frac{1}{S-T}(\log P(t,S) - \log P(t,T)).$$
4. Show that the limit $\lim_{T\searrow t} f(t,T)$ does not exist in $L^2(\Omega)$.
5. Show that $P(t,T)$ satisfies the stochastic differential equation
$$\frac{dP(t,T)}{P(t,T)} = \sigma dB_t + \frac{1}{2}\sigma^2 dt - \frac{\log P(t,T)}{T-t}dt, \qquad t \in [0,T].$$
6. Show, using the results of Exercise 11.2-(4), that
$$P(t,T) = \mathbb{E}\left[e^{-\int_t^T r_s^{\mathrm{T}} ds}\Big|\mathcal{F}_t\right],$$
where $(r_t^{\mathrm{T}})_{t\in[0,T]}$ is a process to be determined.
7. Compute the conditional density
$$\mathbb{E}\left[\frac{d\mathbb{P}_T}{d\mathbb{P}}\Big|\mathcal{F}_t\right] = \frac{P(t,T)}{P(0,T)}e^{-\int_0^t r_s^{\mathrm{T}} ds}$$
of the forward measure $\mathbb{P}_T$ with respect to $\mathbb{P}$.

8. Show that the process

$$\tilde{B}_t := B_t - \sigma t, \qquad 0 \le t \le T,$$

is a standard Brownian motion under $\mathbb{P}_T$.

9. Compute the dynamics of X_t^S and $P(t,S)$ under $\mathbb{P}_T$.
Hint: Show that

$$-\mu(S-T) + \sigma(S-T)\int_0^t \frac{1}{S-s} dB_s = \frac{S-T}{S-t}\log P(t,S).$$

10. Compute the bond option price

$$\mathbb{E}\left[e^{-\int_t^T r_s^T ds}(P(T,S)-K)^+ \Big| \mathcal{F}_t\right] = P(t,T)\,\mathbb{E}_T\left[(P(T,S)-K)^+ \Big| \mathcal{F}_t\right],$$

$0 \le t < T < S$.

Exercise 11.4 (Exercise 4.7 continued). Write down the bond pricing PDE for the function

$$F(t,x) = E\left[e^{-\int_t^T r_s ds} \Big| r_t = x\right]$$

and show that in case $\alpha = 0$ the corresponding bond price $P(t,T)$ equals

$$P(t,T) = e^{-B(T-t)r_t}, \qquad 0 \le t \le T,$$

where

$$B(x) = \frac{2(e^{\gamma x}-1)}{2\gamma + (\beta+\gamma)(e^{\gamma x}-1)},$$

with $\gamma = \sqrt{\beta^2 + 2\sigma^2}$.

Exercise 11.5 Let $(r_t)_{t\in\mathbb{R}_+}$ denote a short term interest rate process. For any $T>0$, let $P(t,T)$ denote the price at time $t \in [0,T]$ of a zero coupon bond defined by the stochastic differential equation

$$\frac{dP(t,T)}{P(t,T)} = r_t dt + \sigma_t^T dB_t, \qquad 0 \le t \le T,$$

under the terminal condition $P(T,T) = 1$, where $(\sigma_t^T)_{t\in[0,T]}$ is an adapted process. Let the forward measure $\mathbb{P}_T$ be defined by

$$\mathbb{E}\left[\frac{d\mathbb{P}_T}{d\mathbb{P}} \Big| \mathcal{F}_t\right] = \frac{P(t,T)}{P(0,T)} e^{-\int_0^t r_s ds}, \qquad 0 \le t \le T.$$

Recall that

$$B_t^T := B_t - \int_0^t \sigma_s^T ds, \qquad 0 \le t \le T,$$

is a standard Brownian motion under $\mathbb{P}_T$.

1. Solve the stochastic differential equation (11.48).

2. Derive the stochastic differential equation satisfied by the discounted bond price process

$$t \mapsto e^{-\int_0^t r_s ds} P(t,T), \qquad 0 \leq t \leq T,$$

and show that it is a martingale.

3. Show that

$$\mathbb{E}\left[e^{-\int_0^T r_s ds} \Big| \mathcal{F}_t \right] = e^{-\int_0^t r_s ds} P(t,T), \qquad 0 \leq t \leq T.$$

4. Show that

$$P(t,T) = \mathbb{E}\left[e^{-\int_t^T r_s ds} \Big| \mathcal{F}_t \right], \qquad 0 \leq t \leq T.$$

5. Compute $P(t,S)/P(t,T)$, $0 \leq t \leq T$, show that it is a martingale under $\mathbb{P}_T$ and that

$$P(T,S) = \frac{P(t,S)}{P(t,T)} \exp\left(\int_t^T (\sigma_s^S - \sigma_s^T) dB_s^T - \frac{1}{2} \int_t^T (\sigma_s^S - \sigma_s^T)^2 ds \right).$$

6. Assuming that $(\sigma_t^T)_{t\in[0,T]}$ and $(\sigma_t^S)_{t\in[0,S]}$ are deterministic functions, compute the price

$$\mathbb{E}\left[e^{-\int_t^T r_s ds} (P(T,S) - \kappa)^+ \Big| \mathcal{F}_t \right] = P(t,T)\, \mathbb{E}_T\left[(P(T,S) - \kappa)^+ \Big| \mathcal{F}_t \right]$$

of a bond option with strike κ.

Recall that if X is a centered Gaussian random variable with mean m_t and variance v_t^2 given $\mathcal{F}_t$, we have

$$\mathbb{E}[(e^X - K)^+ \mid \mathcal{F}_t] = e^{m_t + v_t^2/2} \Phi\left(\frac{v_t}{2} + \frac{1}{v_t}(m_t + v_t^2/2 - \log K) \right)$$
$$-K\Phi\left(-\frac{v_t}{2} + \frac{1}{v_t}(m_t + v_t^2/2 - \log K) \right)$$

where $\Phi(x)$, $x \in \mathbb{R}$, denotes the Gaussian distribution function.

Exercise 11.6 (Exercise 11.3 continued).

1. Compute the forward rate

$$f(t,T,S) = -\frac{1}{S-T}(\log P(t,S) - \log P(t,T)).$$

2. Compute the instantaneous forward rate

$$f(t,T) = -\lim_{S \searrow T} \frac{1}{S-T}(\log P(t,S) - \log P(t,T)).$$

3. Show that the limit $\lim_{T \searrow t} f(t,T)$ does not exist in $L^2(\Omega)$.

4. Show that $P(t,T)$ satisfies the stochastic differential equation

$$\frac{dP(t,T)}{P(t,T)} = \sigma dB_t + \frac{1}{2}\sigma^2 dt - \frac{\log P(t,T)}{T-t} dt, \qquad t \in [0,T].$$

5. Show, using the results of Exercise 11.5-(4), that

$$P(t,T) = \mathbb{E}\left[e^{-\int_t^T r_s^T ds} \Big| \mathcal{F}_t \right],$$

where $(r_t^T)_{t\in[0,T]}$ is a process to be determined.

6. Compute the conditional density

$$\mathbb{E}\left[\frac{d\mathbb{P}_T}{d\mathbb{P}} \Big| \mathcal{F}_t \right] = \frac{P(t,T)}{P(0,T)} e^{-\int_0^t r_s^T ds}$$

of the forward measure $\mathbb{P}_T$ with respect to $\mathbb{P}$.

7. Show that the process

$$\tilde{B}_t := B_t - \sigma t, \qquad 0 \le t \le T,$$

is a standard Brownian motion under $\mathbb{P}_T$.

8. Compute the dynamics of X_t^S and $P(t,S)$ under $\mathbb{P}_T$.
Hint: Show that

$$-\mu(S-T) + \sigma(S-T)\int_0^t \frac{1}{S-s} dB_s = \frac{S-T}{S-t}\log P(t,S).$$

9. Compute the bond option price

$$\mathbb{E}\left[e^{-\int_t^T r_s^T ds}(P(T,S)-K)^+ \Big| \mathcal{F}_t \right] = P(t,T)\,\mathbb{E}_T\left[(P(T,S)-K)^+ \Big| \mathcal{F}_t \right],$$

$0 \le t < T < S$.

Chapter 12

Pricing of Interest Rate Derivatives

In this chapter we consider the pricing of caplets, caps, and swaptions, using change of numéraire and forward swap measures.

12.1 Forward Measures and Tenor Structure

The maturity dates are arranged according to a discrete *tenor structure*

$$\{0 = T_0 < T_1 < T_2 < \cdots < T_n\}.$$

An example of forward interest rate curve data is given in the table of Figure 12.1, which contains the values of $(T_1, T_2, \ldots, T_{23})$ and of $\{f(t, t + T_i, t + T_i + \delta)\}_{i=1,\ldots,23}$, with $t = 07/05/2003$ and $\delta =$ six months.

2D	1W	1M	2M	3M	1Y	2Y	3Y	4Y	5Y	6Y	7Y
2.55	2.53	2.56	2.52	2.48	2.34	2.49	2.79	3.07	3.31	3.52	3.71
8Y	9Y	10Y	11Y	12Y	13Y	14Y	15Y	20Y	25Y	30Y	
3.88	4.02	4.14	4.23	4.33	4.40	4.47	4.54	4.74	4.83	4.86	

FIGURE 12.1: Forward rates arranged according to a tenor structure.

Recall that by definition of $P(t, T_i)$ and absence of arbitrage the process

$$t \longmapsto e^{-\int_0^t r_s ds} P(t, T_i), \qquad 0 \le t \le T_i, \quad i = 1, \ldots, n,$$

is an $\mathcal{F}_t$-martingale under $\mathbb{P}$, and as a consequence $(P(t, T_i))_{t \in [0, T_i]}$ can be taken as numéraire in the definition

$$\frac{d\hat{\mathbb{P}}_i}{d\mathbb{P}} = \frac{1}{P(0, T_i)} e^{-\int_0^{T_i} r_s ds} \tag{12.1}$$

of the *forward measure* $\hat{\mathbb{P}}_i$. The following proposition will allow us to price contingent claims using the forward measure $\hat{\mathbb{P}}_i$; it is a direct consequence of Proposition 10.1, noting that here we have $P(T_t, T_i) = 1$.

Proposition 12.1 *For all sufficiently integrable random variables F we have*

$$\mathbb{E}\left[Fe^{-\int_t^{T_i} r_s ds}\Big|\mathcal{F}_t\right] = P(t,T_i)\hat{\mathbb{E}}_i[F \mid \mathcal{F}_t], \qquad 0 \le t \le T, \quad i = 1,\ldots,n. \tag{12.2}$$

Recall that for all $T_i, T_j \ge 0$, the process

$$t \longmapsto \frac{P(t,T_j)}{P(t,T_i)}, \qquad 0 \le t \le \min(T_i,T_j),$$

is an $\mathcal{F}_t$-martingale under $\hat{\mathbb{P}}_i$, cf. Proposition 10.2.

Dynamics under the forward measure

In order to apply Proposition 12.1 and to compute the price

$$\mathbb{E}\left[e^{-\int_t^T r_s ds}F\Big|\mathcal{F}_t\right] = P(t,T_i)\hat{\mathbb{E}}_i[F \mid \mathcal{F}_t],$$

it can be useful to determine the dynamics of the underlying processes r_t, $f(t,T,S)$, and $P(t,T)$ under the forward measure $\hat{\mathbb{P}}_i$.

Let us assume that the dynamics of the bond price $P(t,T_i)$ is given by

$$\frac{dP(t,T_i)}{P(t,T_i)} = r_t dt + \zeta_i(t)dW_t, \tag{12.3}$$

for $i = 1,\ldots,n$, where $(W_t)_{t\in\mathbb{R}_+}$ is a standard Brownian motion under $\mathbb{P}$ and $(r_t)_{t\in\mathbb{R}_+}$ and $(\zeta_i(t))_{t\in\mathbb{R}_+}$ are adapted processes with respect to the filtration $(\mathcal{F}_t)_{t\in\mathbb{R}_+}$ generated by $(W_t)_{t\in\mathbb{R}_+}$.

By the Girsanov theorem,

$$\hat{W}_t^i := W_t - \int_0^t \zeta_i(s)ds, \qquad 0 \le t \le T_i, \tag{12.4}$$

is a standard Brownian motion under $\hat{\mathbb{P}}_i$ for all $i = 1,\ldots,n$, cf. e.g., (10.10), hence we have

$$d\hat{W}_t^j = dW_t - \zeta_j(t)dt, \qquad d\hat{W}_t^j = dW_t - \zeta_i(t)dt,$$

and

$$d\hat{W}_t^j = d\hat{W}_t^i - (\zeta_j(t) - \zeta_i(t))dt.$$

Hence the dynamics of $t \longmapsto P(t,T_i)$ under $\hat{\mathbb{P}}_i$ is given by

$$\frac{dP(t,T_i)}{P(t,T_i)} = r_t dt + |\zeta_i(t)|^2 dt + \zeta_t d\hat{W}_t^i, \tag{12.5}$$

where $(\hat{W}_t^i)_{t\in\mathbb{R}_+}$ is a standard Brownian motion under $\hat{\mathbb{P}}_i$.

In case the short rate process $(r_t)_{t\in\mathbb{R}_+}$ is Markovian and solution of

$$dr_t = \mu(t, r_t)dt + \sigma(t, r_t)dW_t,$$

its dynamics will be given under $\hat{\mathbb{P}}_i$ by

$$dr_t = \mu(t, r_t)dt + \sigma(t, r_t)\zeta_i(t)dt + \sigma(t, r_t)d\hat{W}_t^i. \tag{12.6}$$

In the Vasicek case we have

$$dr_t = (a - br_t)dt + \sigma dW_t,$$

and

$$\zeta_i(t) = -\frac{\sigma}{b}(1 - e^{-b(T_i - t)}), \qquad 0 \leq t \leq T_i,$$

by (11.10), hence from (12.6) we have

$$dr_t = (a - br_t)dt - \frac{\sigma^2}{b}(1 - e^{-b(T_i - t)})dt + \sigma d\hat{W}_t^i \tag{12.7}$$

and we obtain

$$\frac{dP(t, T_i)}{P(t, T_i)} = r_t dt + \frac{\sigma^2}{b^2}(1 - e^{-b(T_i - t)})^2 dt - \frac{\sigma}{b}(1 - e^{-b(T_i - t)})d\hat{W}_t^i,$$

from (11.10).

12.2 Bond Options

The next proposition can be obtained as an application of the Margrabe formula (10.26) of Proposition 10.8 by taking $X_t = P(t, T_j)$, $N_t = P(t, T_i)$, and $\hat{X}_t = X_t/N_t = P(t, T_j)/P(t, T_i)$. In the Vasicek model, this formula has been first obtained in [36].

Proposition 12.2 *The price of a bond call option on $P(T_i, T_j)$ with payoff $F = (P(T_i, T_j) - \kappa)^+$ can be computed as*

$$\begin{aligned}&\mathbb{E}\left[e^{-\int_t^{T_i} r_s ds}(P(T_i, T_j) - \kappa)^+ \middle| \mathcal{F}_t\right] \\ &= P(t, T_j)\Phi\left(\frac{v}{2} + \frac{1}{v}\log\frac{P(t, T_j)}{\kappa P(t, T_i)}\right) - \kappa P(t, T_i)\Phi\left(-\frac{v}{2} + \frac{1}{v}\log\frac{P(t, T_j)}{\kappa P(t, T_i)}\right),\end{aligned}$$

with

$$v^2 = \int_t^T |\zeta^j(s) - \zeta^i(s)|^2 ds.$$

Proof. First we note that using $N_t = P(t, T_i)$ as a numéraire the price of a bond call option on $P(T_i, T_j)$ with payoff $F = (P(T_i, T_j) - \kappa)^+$ can be written from Proposition 10.4 using the forward measure $\hat{\mathbb{P}}$, or directly by (10.5), as

$$\mathbb{E}\left[e^{-\int_t^{T_i} r_s ds}(P(T_i, T_j) - \kappa)^+ \Big| \mathcal{F}_t\right] = P(t, T_i)\hat{\mathbb{E}}_i\left[(P(T_i, T_j) - \kappa)^+ \Big| \mathcal{F}_t\right].$$

Next we use the Black–Scholes formula and the martingale property of the forward price $P(t, T_j)/P(t, T_i)$, which can be written as the geometric Brownian motion

$$P(T_i, T_j) = \frac{P(t, T_j)}{P(t, T_i)} \exp\left(\int_t^{T_i} (\zeta^i(s) - \zeta^j(s)) d\hat{W}_s^i - \frac{1}{2}\int_t^{T_i} |\zeta^i(s) - \zeta^j(s)|^2 ds\right),$$

under the forward measure $\hat{\mathbb{P}}$ when $(\zeta^i(s))_{s\in[0,T_i]}$ and $(\zeta^j(s))_{s\in[0,T_j]}$ in (12.3) are deterministic functions. The above relation can be obtained by solving (10.12) in Proposition 10.3. □

In the Vasicek case the above bond option price could also be computed from the joint law of $\left(r_T, \int_t^T r_s ds\right)$, which is Gaussian, or from the dynamics (12.5)-(12.7) of $P(t, T)$ and r_t under $\hat{\mathbb{P}}^i$, cf. § 7.3 of [56].

12.3 Caplet Pricing

The caplet on the spot forward rate $f(T, T, T_i)$ with strike κ is a contract with payoff

$$(f(T, T, T_i) - \kappa)^+,$$

priced at time $t \in [0, T]$ from Proposition 10.4 using the forward measure $\hat{\mathbb{P}}_i$ as

$$\mathbb{E}\left[e^{-\int_t^{T_i} r_s ds}(f(T, T, T_i) - \kappa)^+ \Big| \mathcal{F}_t\right] = P(t, T_i)\hat{\mathbb{E}}_i\left[(f(T, T, T_i) - \kappa)^+ \mid \mathcal{F}_t\right], \tag{12.8}$$

by taking $N_t = P(t, T_i)$ as a numéraire.

Next we consider the caplet with payoff

$$(L(T_i, T_i, T_{i+1}) - \kappa)^+$$

on the LIBOR rate

$$L(t, T_i, T_{i+1}) = \frac{1}{T_{i+1} - T_i}\left(\frac{P(t, T_i)}{P(t, T_{i+1})} - 1\right), \qquad 0 \le t \le T_i < T_{i+1},$$

which is a martingale under $\hat{\mathbb{P}}_{i+1}$ defined in (12.1), from Proposition 10.2.

We assume that $L(t, T_i, T_{i+1})$ is modeled in the BGM model of Section 11.7, i.e., we have

$$\frac{dL(t, T_i, T_{i+1})}{L(t, T_i, T_{i+1})} = \gamma_i(t) d\hat{B}_t^{i+1},$$

$0 \leq t \leq T_i$, $i = 1, \ldots, n-1$, where $t \longmapsto \gamma_i(t)$ is a deterministic function, $i = 1, \ldots, n-1$.

The next formula (12.9) is known as the Black caplet formula.

Proposition 12.3 *The caplet on $L(T_i, T_i, T_{i+1})$ is priced as time $t \in [0, T_i]$ as*

$$\begin{aligned} & \mathbb{E}\left[e^{-\int_t^{T_{i+1}} r_s ds} (L(T_i, T_i, T_{i+1}) - \kappa)^+ \Big| \mathcal{F}_t \right] \\ = \ & P(t, T_{i+1}) L(t, T_i, T_{i+1}) \Phi(d_+) - \kappa P(t, T_{i+1}) \Phi(d_-), \end{aligned} \tag{12.9}$$

where

$$d_+ = \frac{\log(L(t, T_i, T_{i+1})/\kappa) + \sigma_i^2(t)(T_i - t)/2}{\sigma_i(t)\sqrt{T_i - t}},$$

and

$$d_- = \frac{\log(L(t, T_i, T_{i+1})/\kappa) - \sigma_i^2(t)(T_i - t)/2}{\sigma_i(t)\sqrt{T_i - t}},$$

and

$$|\sigma_i(t)|^2 = \frac{1}{T_i - t} \int_t^{T_i} |\gamma_i|^2(s) ds.$$

Proof. By (12.8) we have

$$\begin{aligned} & \mathbb{E}\left[e^{-\int_t^{T_{i+1}} r_s ds} (L(T_i, T_i, T_{i+1}) - \kappa)^+ \Big| \mathcal{F}_t \right] \\ = \ & P(t, T_{i+1}) \mathbb{E}_{i+1} \left[(L(T_i, T_i, T_{i+1}) - \kappa)^+ \mid \mathcal{F}_t \right] \\ = \ & P(t, T_{i+1}) \mathrm{BS}(\kappa, L(t, T_i, T_{i+1}), \sigma_i(t), 0, T_i - t), \end{aligned}$$

$t \in [0, T_i]$, where

$$\mathrm{BS}(\kappa, x, \sigma, r, \tau) = x\Phi(d_+) - \kappa e^{-r\tau}\Phi(d_-)$$

is the Black–Scholes function with

$$|\sigma_i(t)|^2 = \frac{1}{T_i - t} \int_t^{T_i} |\gamma_i|^2(s) ds,$$

since $t \longmapsto L(t, T_i, T_{i+1})$ is a geometric Brownian motion with volatility $\gamma_i(t)$ under $\hat{\mathbb{P}}_{i+1}$. $\square$

We may also write

$$
\begin{aligned}
&(T_{i+1}-T_i)\,\mathbb{E}\left[e^{-\int_t^{T_{i+1}} r_s ds}(L(T_i,T_i,T_{i+1})-\kappa)^+\Big|\mathcal{F}_t\right] \\
&= P(t,T_{i+1})\left(\frac{P(t,T_i)}{P(t,T_{i+1})}-1\right)\Phi(d_+)-\kappa(T_{i+1}-T_i)P(t,T_{i+1})\Phi(d_-) \\
&= (P(t,T_i)-P(t,T_{i+1}))\Phi(d_+)-\kappa(T_{i+1}-T_i)P(t,T_{i+1})\Phi(d_-) \\
&= P(t,T_i)\Phi(d_+)-(1+\kappa(T_{i+1}-T_i))\,P(t,T_{i+1})\Phi(d_-),
\end{aligned}
$$

and this yields a self-financing hedging strategy

$$(\Phi(d_+),-(1+\kappa(T_{i+1}-T_i))\,\Phi(d_-))$$

in the bonds $(P(t,T_i),P(t,T_{i+1}))$ with maturities T_i and T_{i+1}, cf. Corollary 10.2 and [57]. Proposition 12.3 can also be proved by taking $P(t,T_{i+1})$ as numéraire and letting

$$\hat{X}_t = P(t,T_i)/P(t,T_{i+1}) = 1+(T_{i+1}-T_i)L(T_i,T_i,T_{i+1}).$$

Floorlets

Similarly, a floorlet on $f(T,T,T_i)$ with strike κ is a contract with payoff $(\kappa - f(T,T,T_i))^+$, priced at time $t\in[0,T]$ as

$$\mathbb{E}\left[e^{-\int_t^{T_i} r_s ds}(\kappa - f(T,T,T_i))^+\Big|\mathcal{F}_t\right] = P(t,T_i)\hat{\mathbb{E}}_i\left[(\kappa - f(T,T,T_i))^+ \mid \mathcal{F}_t\right].$$

Floorlets are analog to put options and can be similarly priced by the call/put parity in the Black–Scholes formula.

Cap Pricing

More generally one can consider caps that are relative to a given tenor structure $\{T_1,\ldots,T_n\}$, with discounted payoff

$$\sum_{k=1}^{n-1}(T_{k+1}-T_k)e^{-\int_t^{T_{k+1}} r_s ds}(f(T_k,T_k,T_{k+1})-\kappa)^+.$$

Pricing formulas for caps are easily deduced from analog formulas for caplets, since the payoff of a cap can be decomposed into a sum of caplet payoffs. Thus the price of a cap at time $t\in[0,T_1]$ is given by

$$
\begin{aligned}
&\mathbb{E}\left[\sum_{k=1}^{n-1}(T_{k+1}-T_k)e^{-\int_t^{T_{k+1}} r_s ds}(f(T_k,T_k,T_{k+1})-\kappa)^+\Big|\mathcal{F}_t\right] \\
&= \sum_{k=1}^{n-1}(T_{k+1}-T_k)\,\mathbb{E}\left[e^{-\int_t^{T_{k+1}} r_s ds}(f(T_k,T_k,T_{k+1})-\kappa)^+\Big|\mathcal{F}_t\right]
\end{aligned}
$$

$$= \sum_{k=1}^{n-1}(T_{k+1}-T_k)P(t,T_{k+1})\hat{\mathbb{E}}_{k+1}\left[(f(T_k,T_k,T_{k+1})-\kappa)^+\Big|\mathcal{F}_t\right].$$

In the above BGM model, the cap with payoff

$$\sum_{k=1}^{n-1}(T_{k+1}-T_k)(L(T_k,T_k,T_{k+1})-\kappa)^+$$

can be priced at time $t \in [0, T_1]$ as

$$\sum_{k=1}^{n-1}(T_{k+1}-T_k)P(t,T_{k+1})\mathrm{BS}(\kappa, L(t,T_k,T_{k+1}), \sigma_k(t), 0, T_k - t).$$

12.4 Forward Swap Measures

In this section we introduce the forward measures to be used for the pricing of swaptions, and we study their properties. We start with the definition of the *annuity numéraire*

$$P(t,T_i,T_j) = \sum_{k=i}^{j-1}(T_{k+1}-T_k)P(t,T_{k+1}), \qquad 0 \le t \le T_i, \tag{12.10}$$

with in particular

$$P(t,T_i,T_{i+1}) = (T_{i+1}-T_i)P(t,T_{i+1}), \qquad 0 \le t \le T_i.$$

$1 \le i < n$. The annuity numéraire satisfies the following martingale property, which can be proved by linearity and the fact that $t \longmapsto e^{-\int_0^t r_s ds}P(t,T_k)$ is a martingale for all $k = 1, \ldots, n$.

Proposition 12.4 *The discounted annuity numéraire*

$$t \longmapsto e^{-\int_0^t r_s ds}P(t,T_i,T_j) = e^{-\int_0^t r_s ds}\sum_{k=i}^{j-1}(T_{k+1}-T_k)P(t,T_{k+1}), \qquad 0 \le t \le T_i,$$

is a martingale under $\mathbb{P}$.

The forward swap measure $\hat{\mathbb{P}}_{i,j}$ is defined by

$$\frac{d\hat{\mathbb{P}}_{i,j}}{d\mathbb{P}} = e^{-\int_0^{T_i} r_s ds}\frac{P(T_i,T_i,T_j)}{P(0,T_i,T_j)}, \tag{12.11}$$

$1 \le i < j \le n$. We have

$$\mathbb{E}\left[\frac{d\hat{\mathbb{P}}_{i,j}}{d\mathbb{P}}\Big|\mathcal{F}_t\right] = \frac{1}{P(0,T_i,T_j)}\mathbb{E}\left[e^{-\int_0^{T_i} r_s ds}P(T_i,T_i,T_j)\Big|\mathcal{F}_t\right]$$

$$= \frac{P(t,T_i,T_j)}{P(0,T_i,T_j)} e^{-\int_0^t r_s ds},$$

$0 \leq t \leq T_i$, by Proposition 12.4, and

$$\frac{d\hat{\mathbb{P}}_{i,j|\mathcal{F}_t}}{d\mathbb{P}_{|\mathcal{F}_t}} = e^{-\int_t^{T_i} r_s ds} \frac{P(T_i,T_i,T_j)}{P(t,T_i,T_j)}, \qquad 0 \leq t \leq T_{i+1}, \tag{12.12}$$

by Proposition 10.3. We also know that the process

$$t \longmapsto v_k^{i,j}(t) := \frac{P(t,T_k)}{P(t,T_i,T_j)}$$

is an $\mathcal{F}_t$-martingale under $\hat{\mathbb{P}}_{i,j}$ by Proposition 10.2. It follows that the swap rate

$$S(t,T_i,T_j) := \frac{P(t,T_i) - P(t,T_j)}{P(t,T_i,T_j)} = v_i^{i,j}(t) - v_j^{i,j}(t), \qquad 0 \leq t \leq T_i,$$

defined in Proposition 11.5 is also a martingale under $\hat{\mathbb{P}}_{i,j}$.

Using the forward swap measure we obtain the following pricing formula for a given integrable claim with payoff of the form $P(T_i,T_i,T_j)F$:

$$\begin{aligned} \mathbb{E}\left[e^{-\int_t^{T_i} r_s ds} P(T_i,T_i,T_j) F \Big| \mathcal{F}_t \right] &= P(t,T_i,T_j)\, \mathbb{E}\left[F \frac{d\hat{\mathbb{P}}_{i,j|\mathcal{F}_t}}{d\mathbb{P}_{|\mathcal{F}_t}} \Big| \mathcal{F}_t \right] \\ &= P(t,T_i,T_j) \hat{\mathbb{E}}_{i,j}\left[F \Big| \mathcal{F}_t \right], \end{aligned} \tag{12.13}$$

after applying (12.11) and (12.12) on the last line, or Proposition 10.1.

12.5 Swaption Pricing on the LIBOR

A swaption on the forward rate $f(T_1,T_k,T_{k+1})$ is a contract meant to protect oneself against a risk based on an interest rate swap, and has payoff

$$\left(\sum_{k=i}^{j-1} (T_{k+1} - T_k) e^{-\int_{T_i}^{T_{k+1}} r_s ds} (f(T_i,T_k,T_{k+1}) - \kappa) \right)^+,$$

at time T_i.

This swaption can be priced at time $t \in [0,T_i]$ under a risk-neutral measure as

$$\mathbb{E}\left[e^{-\int_t^{T_i} r_s ds}\left(\sum_{k=i}^{j-1}(T_{k+1}-T_k)e^{-\int_{T_i}^{T_{k+1}} r_s ds}(f(T_i,T_k,T_{k+1})-\kappa)\right)^+ \Bigg| \mathcal{F}_t\right]. \tag{12.14}$$

In the sequel and in practice the price (12.14) of the swaption will be evaluated as

$$\mathbb{E}\left[e^{-\int_t^{T_i} r_s ds}\left(\sum_{k=i}^{j-1}(T_{k+1}-T_k)P(T_i,T_{k+1})(f(T_i,T_k,T_{k+1})-\kappa)\right)^+ \Bigg| \mathcal{F}_t\right], \tag{12.15}$$

where we approximate the discount factor $e^{-\int_{T_i}^{T_{k+1}} r_s ds}$ by its conditional expectation $P(T_i, T_{k+1})$ given $\mathcal{F}_{T_i}$.

Note that when $j = i+1$, the swaption price (12.15) coincides with the price at time t of a caplet on $[T_i, T_{i+1}]$ up to a factor $T_{i+1} - T_i$ since

$$\begin{aligned}
&\mathbb{E}\left[e^{-\int_t^{T_i} r_s ds}((T_{i+1}-T_i)P(T_i,T_{i+1})(f(T_i,T_i,T_{i+1})-\kappa))^+ \Big| \mathcal{F}_t\right]\\
&= (T_{i+1}-T_i)\,\mathbb{E}\left[e^{-\int_t^{T_i} r_s ds}P(T_i,T_{i+1})((f(T_i,T_i,T_{i+1})-\kappa))^+ \Big| \mathcal{F}_t\right]\\
&= (T_{i+1}-T_i)\,\mathbb{E}\left[e^{-\int_t^{T_i} r_s ds}\,\mathbb{E}\left[e^{-\int_{T_i}^{T_{i+1}} r_s ds}\Big|\mathcal{F}_{T_i}\right]((f(T_i,T_i,T_{i+1})-\kappa))^+ \Big| \mathcal{F}_t\right]\\
&= (T_{i+1}-T_i)\,\mathbb{E}\left[\mathbb{E}\left[e^{-\int_t^{T_i} r_s ds}e^{-\int_{T_i}^{T_{i+1}} r_s ds}((f(T_i,T_i,T_{i+1})-\kappa))^+ \Big|\mathcal{F}_{T_i}\right]\Big| \mathcal{F}_t\right]\\
&= (T_{i+1}-T_i)\,\mathbb{E}\left[e^{-\int_t^{T_{i+1}} r_s ds}(f(T_i,T_i,T_{i+1})-\kappa)^+ \Big| \mathcal{F}_t\right],
\end{aligned}$$

$0 \le t \le T_i$.

In case we replace the forward rate $f(t,T,S)$ with the LIBOR rate $L(t,T,S)$, the payoff of the swaption can be rewritten as in the following lemma which is a direct consequence of the definition of the swap rate $S(T_i,T_i,T_j)$.

Lemma 12.1 *The payoff of the swaption in* (12.15) *can be rewritten as*

$$\begin{aligned}
&\left(\sum_{k=i}^{j-1}(T_{k+1}-T_k)P(T_i,T_{k+1})(L(T_i,T_k,T_{k+1})-\kappa)\right)^+\\
&\qquad = (P(T_i,T_i)-P(T_i,T_j)-\kappa P(T_i,T_i,T_j))^+\\
&\qquad = P(T_i,T_i,T_j)\,(S(T_i,T_i,T_j)-\kappa)^+.
\end{aligned} \tag{12.16}$$

Proof. The relation

$$\sum_{k=i}^{j-1}(T_{k+1}-T_k)P(t,T_{k+1})(L(t,T_k,T_{k+1})-S(t,T_i,T_j)) = 0$$

that defines the forward swap rate $S(t, T_i, T_j)$ shows that

$$\begin{aligned}
\sum_{k=i}^{j-1} (T_{k+1} - T_k) P(t, T_{k+1}) L(t, T_k, T_{k+1}) \\
&= S(t, T_i, T_j) \sum_{k=i}^{j-1} (T_{k+1} - T_k) P(t, T_{k+1}) \\
&= P(t, T_i, T_j) S(t, T_i, T_j) \\
&= P(t, T_i) - P(t, T_j),
\end{aligned}$$

by the definition (12.10) of $P(t, T_i, T_j)$, hence

$$\begin{aligned}
\sum_{k=i}^{j-1} (T_{k+1} - T_k) P(t, T_{k+1}) (L(t, T_k, T_{k+1}) - \kappa) \\
&= P(t, T_i) - P(t, T_j) - \kappa P(t, T_i, T_j) \\
&= P(t, T_i, T_j) \left(S(t, T_i, T_j) - \kappa \right),
\end{aligned}$$

and for $t = T_i$ we get

$$\begin{aligned}
\left(\sum_{k=i}^{j-1} (T_{k+1} - T_k) P(T_i, T_{k+1}) (L(T_i, T_k, T_{k+1}) - \kappa) \right)^+ \\
&= P(T_i, T_i, T_j) \left(S(T_i, T_i, T_j) - \kappa \right)^+ .
\end{aligned}$$

□

The next proposition simply states that a swaption on the LIBOR rate can be priced as a European call option on the swap rate $S(T_i, T_i, T_j)$ under the forward swap measure $\hat{\mathbb{P}}_{i,j}$.

Proposition 12.5 *The price* (12.15) *of the swaption with payoff*

$$\left(\sum_{k=i}^{j-1} (T_{k+1} - T_k) P(T_i, T_{k+1}) (L(T_i, T_k, T_{k+1}) - \kappa) \right)^+ \tag{12.17}$$

on the LIBOR market can be written under the forward swap measure $\hat{\mathbb{P}}_{i,j}$ *as*

$$P(t, T_i, T_j) \hat{\mathbb{E}}_{i,j} \left[\left(S(T_i, T_i, T_j) - \kappa \right)^+ \Big| \mathcal{F}_t \right], \qquad 0 \le t \le T_i.$$

Proof. As a consequence of (12.13) and Lemma 12.1 we find

$$\mathbb{E} \left[e^{-\int_t^{T_i} r_s ds} \left(\sum_{k=i}^{j-1} (T_{k+1} - T_k) P(T_i, T_{k+1}) (L(T_i, T_k, T_{k+1}) - \kappa) \right)^+ \Bigg| \mathcal{F}_t \right]$$

$$
\begin{aligned}
&= \mathbb{E}\left[e^{-\int_t^{T_i} r_s ds}\left(P(T_i,T_i)-P(T_i,T_j)-\kappa P(T_i,T_i,T_j)\right)^+ \Big| \mathcal{F}_t\right] \\
&= \mathbb{E}\left[e^{-\int_t^{T_i} r_s ds} P(T_i,T_i,T_j)\left(S(T_i,T_i,T_j)-\kappa\right)^+ \Big| \mathcal{F}_t\right] \\
&= \frac{1}{P(t,T_i,T_j)}\mathbb{E}\left[\frac{d\hat{\mathbb{P}}_{i,j|\mathcal{F}_t}}{d\mathbb{P}_{|\mathcal{F}_t}}\left(S(T_i,T_i,T_j)-\kappa\right)^+ \Big| \mathcal{F}_t\right] \\
&= P(t,T_i,T_j)\hat{\mathbb{E}}_{i,j}\left[\left(S(T_1,T_1,T_n)-\kappa\right)^+ \Big| \mathcal{F}_t\right]. \qquad (12.18)
\end{aligned}
$$

□

In the next proposition we price a swaption with payoff (12.17) or equivalently (12.16).

Proposition 12.6 *Assume that the swap rate is modeled as a geometric Brownian motion under* $\mathbb{P}_{i,j}$, i.e.,

$$dS(t,T_i,T_j) = S(t,T_i,T_j)\hat{\sigma}(t)d\hat{W}_t^{i,j},$$

where $(\hat{\sigma}(t))_{t\in\mathbb{R}_+}$ *is a deterministic function. Then the swaption with payoff*

$$(P(T,T_i)-P(T,T_j)-\kappa P(T_i,T_i,T_j))^+ = P(T_i,T_i,T_j)\left(S(T_i,T_i,T_j)-\kappa\right)^+$$

can be priced using the Black–Scholes formula as

$$
\begin{aligned}
&\mathbb{E}\left[e^{-\int_t^{T_i} r_s ds} P(T_i,T_i,T_j)\left(S(T_i,T_i,T_j)-\kappa\right)^+ \Big| \mathcal{F}_t\right] \\
&\quad = (P(t,T_i)-P(t,T_j))\Phi_+(t,S(t,T_i,T_j)) \\
&\qquad -\kappa\Phi_-(t,S(t,T_i,T_j))\sum_{k=i}^{j-2}(T_{k+1}-T_k)P(t,T_{k+1}),
\end{aligned}
$$

where

$$d_+ = \frac{\log(S(t,T_i,T_j)/\kappa)+\sigma_{i,j}^2(t)(T_i-t)/2}{\sigma_{i,j}(t)\sqrt{T_i-t}},$$

and

$$d_- = \frac{\log(S(t,T_i,T_j)/\kappa)-\sigma_{i,j}^2(t)(T_i-t)/2}{\sigma_{i,j}(t)\sqrt{T_i-t}},$$

and

$$|\sigma_{i,j}(t)|^2 = \frac{1}{T_i-t}\int_t^{T_i}|\hat{\sigma}|^2(s)ds.$$

Proof. Since $S(t,T_i,T_j)$ is a geometric Brownian motion with variance $(\hat{\sigma}(t))_{t\in\mathbb{R}_+}$ under $\hat{\mathbb{P}}_{i,j}$, by (12.18) we have

$$
\begin{aligned}
&\mathbb{E}\left[e^{-\int_t^{T_i} r_s ds} P(T_i,T_i,T_j)\left(S(T_i,T_i,T_j)-\kappa\right)^+ \Big| \mathcal{F}_t\right] \\
&\quad = \mathbb{E}\left[e^{-\int_t^{T} r_s ds}(P(T,T_i)-P(T,T_j)-\kappa P(T_i,T_i,T_j))^+ \Big| \mathcal{F}_t\right]
\end{aligned}
$$

$$
\begin{aligned}
&= P(t,T_i,T_j)\hat{\mathbb{E}}_{i,j}\left[(S(T_i,T_i,T_j)-\kappa)^+ \Big| \mathcal{F}_t\right] \\
&= P(t,T_i,T_j)\mathrm{BS}(\kappa, S(T_i,T_i,T_j), \sigma_{i,j}(t), 0, T_i - t) \\
&= P(t,T_i,T_j)\left(S(t,T_i,T_j)\Phi_+(t,S(t,T_i,T_j)) - \kappa\Phi_-(t,S(t,T_i,T_j))\right) \\
&= (P(t,T_i)-P(t,T_j))\Phi_+(t,S(t,T_i,T_j)) - \kappa P(T_i,T_i,T_j)\Phi_-(t,S(t,T_i,T_j)) \\
&= (P(t,T_i)-P(t,T_j))\Phi_+(t,S(t,T_i,T_j)) \\
&\quad -\kappa\Phi_-(t,S(t,T_i,T_j))\sum_{k=i}^{j-2}(T_{k+1}-T_k)P(t,T_{k+1}).
\end{aligned}
$$

□

In addition the hedging strategy

$$
\begin{aligned}
&(\Phi_+(t,S(t,T_i,T_j)), -\kappa\Phi_-(t,S(t,T_i,T_j))(T_{i+1}-T_i), \ldots \\
&\qquad \ldots, -\kappa\Phi_-(t,S(t,T_i,T_j))(T_{j-1}-T_{j-2}), -\Phi_+(t,S(t,T_i,T_j)))
\end{aligned}
$$

based on the assets $(P(t,T_i),\ldots,P(t,T_j))$ is self-financing by Corollary 10.2, cf. also [57].

Swaption prices can also be computed by an approximation formula, from the exact dynamics of the swap rate $S(t,T_i,T_j)$ under $\hat{\mathbb{P}}_{i,j}$, based on the bond price dynamics of the form (12.3), cf. [66], page 17.

Exercises

Exercise 12.1 Given two bonds with maturities T, S and prices $P(t,T)$, $P(t,S)$, consider the LIBOR rate

$$
L(t,T,S) = \frac{P(t,T)-P(t,S)}{(S-T)P(t,S)}
$$

at time t, modeled as

$$
dL(t,T,S) = \mu_t L(t,T,S)dt + \sigma L(t,T,S)dW_t, \qquad 0 \le t \le T, \qquad (12.19)
$$

where $(W_t)_{t\in[0,T]}$ is a standard Brownian motion under the risk-neutral measure $\mathbb{P}^*$, $\sigma > 0$ is a constant, and $(\mu_t)_{t\in[0,T]}$ is an adapted process. Let

$$
F_t = \mathbb{E}^*\left[e^{-\int_t^S r_s ds}(\kappa - L(T,T,S))^+ \Big| \mathcal{F}_t\right]
$$

denote the price at time t of a floorlet option with strike κ, maturity T, and payment date S.

1. Rewrite the value of F_t using the forward measure $\hat{\mathbb{P}}_S$ with maturity S.
2. What is the dynamics of $L(t, T, S)$ under the forward measure $\hat{\mathbb{P}}_S$?
3. Write down the value of F_t using the Black–Scholes formula.

   Hint. Given X a centered Gaussian random variable with variance v^2 we have

   $$\mathbb{E}[(\kappa - e^{m+X})^+] = \kappa\Phi(-(m - \log\kappa)/v) - e^{m+\frac{v^2}{2}}\Phi(-v - (m - \log\kappa)/v),$$

   where Φ denotes the Gaussian cumulative distribution function.

Exercise 12.2 We work in the short rate model

$$dr_t = \sigma dB_t,$$

where $(B_t)_{t\in\mathbb{R}_+}$ is a standard Brownian motion under $\mathbb{P}$, and $\mathbb{P}_2$ is the forward measure defined by

$$\frac{d\mathbb{P}_2}{d\mathbb{P}} = \frac{1}{P(0, T_2)} e^{-\int_0^{T_2} r_s ds}.$$

1. State the expressions of ζ_t^1 and ζ_t^2 in

   $$\frac{dP(t, T_i)}{P(t, T_i)} = r_t dt + \zeta_t^i dB_t, \qquad i = 1, 2,$$

   and the dynamics of the $P(t, T_1)/P(t, T_2)$ under $\mathbb{P}_2$, where $P(t, T_1)$ and $P(t, T_2)$ are bond prices with maturities T_1 and T_2.
2. State the expression of the forward rate $f(t, T_1, T_2)$.
3. Compute the dynamics of $f(t, T_1, T_2)$ under the forward measure $\mathbb{P}_2$ with

   $$\frac{d\mathbb{P}_2}{d\mathbb{P}} = \frac{1}{P(0, T_2)} e^{-\int_0^{T_2} r_s ds}.$$

4. Compute the price

   $$(T_2 - T_1)\,\mathbb{E}\left[e^{-\int_t^{T_2} r_s ds}(f(T_1, T_1, T_2) - \kappa)^+ \Big| \mathcal{F}_t\right]$$

   of a cap at time $t \in [0, T_1]$, using the expectation under the forward measure $\mathbb{P}_2$.
5. Compute the dynamics of the swap rate process

   $$S(t, T_1, T_2) = \frac{P(t, T_1) - P(t, T_2)}{(T_2 - T_1)P(t, T_2)}, \qquad t \in [0, T_1],$$

   under $\mathbb{P}_2$.

6. Compute the swaption price

$$(T_2 - T_1)\,\mathbb{E}\left[e^{-\int_t^{T_1} r_s ds} P(T_1, T_2)(S(T_1, T_1, T_2) - \kappa)^+ \Big| \mathcal{F}_t\right]$$

on the swap rate $S(T_1, T_1, T_2)$ using the expectation under the forward swap measure $\mathbb{P}_{1,2}$.

Exercise 12.3 Consider three zero-coupon bonds $P(t, T_1)$, $P(t, T_2)$ and $P(t, T_3)$ with maturities $T_1 = \delta$, $T_2 = 2\delta$ and $T_3 = 3\delta$ respectively, and the forward LIBOR $L(t, T_1, T_2)$ and $L(t, T_2, T_3)$ defined by

$$L(t, T_i, T_{i+1}) = \frac{1}{\delta}\left(\frac{P(t, T_i)}{P(t, T_{i+1})} - 1\right), \qquad i = 1, 2.$$

Assume that $L(t, T_1, T_2)$ and $L(t, T_2, T_3)$ are modeled in the BGM model by

$$\frac{dL(t, T_1, T_2)}{L(t, T_1, T_2)} = e^{-at} d\hat{W}_t^2, \qquad 0 \leq t \leq T_1, \tag{12.20}$$

and $L(t, T_2, T_3) = b$, $0 \leq t \leq T_2$, for some constants $a, b > 0$, where $\hat{W}_t^2$ is a standard Brownian motion under the forward rate measure $\mathbb{P}_2$ defined by

$$\frac{d\mathbb{P}_2}{d\mathbb{P}} = \frac{e^{-\int_0^{T_2} r_s ds}}{P(0, T_2)}.$$

1. Compute $L(t, T_1, T_2)$, $0 \leq t \leq T_2$ by solving Equation (12.20).
2. Compute the price at time t:

$$\mathbb{E}\left[e^{-\int_t^{T_2} r_s ds}(L(T_1, T_1, T_2) - \kappa)^+ \Big| \mathcal{F}_t\right] = P(t, T_2)\hat{\mathbb{E}}_2\left[(L(T_1, T_1, T_2) - \kappa)^+ \mid \mathcal{F}_t\right],$$

of the caplet with strike κ, where $\hat{\mathbb{E}}_2$ denotes the expectation under the forward measure $\mathbb{P}_2$.

3. Compute

$$\frac{P(t, T_1)}{P(t, T_1, T_3)}, \quad 0 \leq t \leq T_1, \quad \text{and} \quad \frac{P(t, T_3)}{P(t, T_1, T_3)}, \qquad 0 \leq t \leq T_2,$$

in terms of b and $L(t, T_1, T_2)$, where $P(t, T_1, T_3)$ is the annuity numéraire

$$P(t, T_1, T_3) = \delta P(t, T_2) + \delta P(t, T_3), \qquad 0 \leq t \leq T_2.$$

4. Compute the dynamics of the swap rate

$$t \longmapsto S(t, T_1, T_3) = \frac{P(t, T_1) - P(t, T_3)}{P(t, T_1, T_3)}, \qquad 0 \leq t \leq T_1,$$

i.e., show that we have

$$dS(t, T_1, T_3) = \sigma_{1.3}(t) S(t, T_1, T_3) d\hat{W}_t^2,$$

where $\sigma_{1,3}(t)$ is a process to be determined.

5. Using the Black–Scholes formula, compute an approximation of the swaption price

$$\begin{aligned} & \mathbb{E}\left[e^{-\int_t^{T_1} r_s ds} P(T_1, T_1, T_3)(S(T_1, T_1, T_3) - \kappa)^+ \Big| \mathcal{F}_t\right] \\ = \; & P(t, T_1, T_3)\hat{\mathbb{E}}_2\left[(S(T_1, T_1, T_3) - \kappa)^+ \mid \mathcal{F}_t\right], \end{aligned}$$

at time $t \in [0, T_1]$. You will need to approximate $\sigma_{1,3}(s)$, $s \geq t$, by "freezing" all random terms at time t.

Hint. Given X a centered Gaussian random variable with variance v^2 we have

$$\mathbb{E}[(e^{m+X} - \kappa)^+] = e^{m+\frac{v^2}{2}}\Phi(v + (m - \log \kappa)/v) - \kappa\Phi((m - \log \kappa)/v),$$

where Φ denotes the Gaussian cumulative distribution function.

Exercise 12.4 Consider a portfolio $(\xi_t^T, \xi_t^S)_{t\in[0,T]}$ made of two bonds with maturities T, S, and value

$$V_t = \xi_t^T P(t, T) + \xi_t^S P(t, S), \qquad 0 \leq t \leq T,$$

at time t. We assume that the portfolio is self-financing, i.e.,

$$dV_t = \xi_t^T dP(t, T) + \xi_t^S dP(t, S), \qquad 0 \leq t \leq T, \tag{12.21}$$

and that it *hedges* the claim $(P(T, S) - \kappa)^+$, so that

$$\begin{aligned} V_t &= \mathbb{E}\left[e^{-\int_t^T r_s ds}(P(T, S) - \kappa)^+ \Big| \mathcal{F}_t\right] \\ &= P(t, T)\,\mathbb{E}_T\left[(P(T, S) - \kappa)^+ \Big| \mathcal{F}_t\right], \qquad 0 \leq t \leq T. \end{aligned}$$

1. Show that

$$\begin{aligned} & \mathbb{E}\left[e^{-\int_t^T r_s ds}(P(T, S) - K)^+ \Big| \mathcal{F}_t\right] \\ = \; & P(0, T)\,\mathbb{E}_T\left[(P(T, S) - K)^+\right] + \int_0^t \xi_s^T dP(s, T) + \int_0^t \xi_s^S dP(s, S). \end{aligned}$$

2. Show that under the self-financing condition (12.21), the discounted portfolio price $\tilde{V}_t = e^{-\int_0^t r_s ds}V_t$ satisfies

$$d\tilde{V}_t = \xi_t^T d\tilde{P}(t, T) + \xi_t^S d\tilde{P}(t, S),$$

where $\tilde{P}(t, T) = e^{-\int_0^t r_s ds}P(t, T)$ and $\tilde{P}(t, S) = e^{-\int_0^t r_s ds}P(t, S)$ denote the discounted bond prices.

3. Show that

$$\mathbb{E}_T\left[(P(T,S)-K)^+ \,|\mathcal{F}_t\right] = \mathbb{E}_T\left[(P(T,S)-K)^+\right] + \int_0^t \frac{\partial C}{\partial x}(X_u, T-u, v(u,T))dX_u.$$

Hint: use the martingale property and the Itô formula.

4. Show that the discounted portfolio price $\hat{V}_t = V_t/P(t,T)$ satisfies

$$\begin{aligned} d\hat{V}_t &= \frac{\partial C}{\partial x}(X_t, T-t, v(t,T))dX_t \\ &= \frac{P(t,S)}{P(t,T)}\frac{\partial C}{\partial x}(X_t, T-t, v(t,T))(\sigma_t^S - \sigma_t^T)d\hat{B}_t^T. \end{aligned}$$

5. Show that

$$dV_t = P(t,S)\frac{\partial C}{\partial x}(X_t, T-t, v(t,T))(\sigma_t^S - \sigma_t^T)dB_t + \hat{V}_t dP(t,T).$$

6. Show that

$$d\tilde{V}_t = \tilde{P}(t,S)\frac{\partial C}{\partial x}(X_t, T-t, v(t,T))(\sigma_t^S - \sigma_t^T)dB_t + \hat{V}_t d\tilde{P}(t,T).$$

7. Compute the hedging strategy $(\xi_t^T, \xi_t^S)_{t\in[0,T]}$ of the bond option.

8. Show that

$$\frac{\partial C}{\partial x}(x,\tau,v) = \Phi\left(\frac{\log(x/K) + \tau v^2/2}{\sqrt{\tau}v}\right).$$

Exercise 12.5 Consider a bond market with tenor structure $\{T_i, \ldots, T_j\}$ and bonds with maturities $T_i, \ldots, T_j$, whose prices $P(t,T_i), \ldots P(t,T_j)$ at time t are given by

$$\frac{dP(t,T_k)}{P(t,T_k)} = r_t dt + \zeta_k(t)dB_t, \qquad k = i, \ldots, j,$$

where $(r_t)_{t\in\mathbb{R}_+}$ is a short term interest rate process and $(B_t)_{t\in\mathbb{R}_+}$ denotes a standard Brownian motion generating a filtration $(\mathcal{F}_t)_{t\in\mathbb{R}_+}$, and $\zeta_i(t), \ldots, \zeta_j(t)$ are volatility processes.

The swap rate $S(t,T_i,T_j)$ is defined by

$$S(t,T_i,T_j) = \frac{P(t,T_i) - P(t,T_j)}{P(t,T_i,T_j)},$$

where

$$P(t, T_i, T_j) = \sum_{k=i}^{j-1} (T_{k+1} - T_k) P(t, T_{k+1})$$

is the annuity numéraire. Recall that a swaption on the LIBOR market can be priced at time $t \in [0, T_i]$ as

$$\mathbb{E}\left[e^{-\int_t^{T_i} r_s ds} \left(\sum_{k=i}^{j-1} (T_{k+1} - T_k) P(T_i, T_{k+1}) (L(T_i, T_k, T_{k+1}) - \kappa) \right)^+ \Big| \mathcal{F}_t \right]$$

$$= \quad P(t, T_i, T_j) \, \mathbb{E}_{i,j} \left[(S(T_i, T_i, T_j) - \kappa)^+ \Big| \mathcal{F}_t \right], \qquad (12.22)$$

under the forward swap measure $\mathbb{P}_{i,j}$ defined by

$$\frac{d\mathbb{P}_{i,j}}{d\mathbb{P}} = e^{-\int_0^{T_i} r_s ds} \frac{P(T_i, T_i, T_j)}{P(0, T_i, T_j)}, \qquad 1 \leq i < j \leq n,$$

under which

$$\hat{B}_t^{i,j} := B_t - \sum_{k=i}^{j-1} (T_{k+1} - T_k) \frac{P(t, T_{k+1})}{P(t, T_i, T_j)} \zeta_{k+1}(t) dt \qquad (12.23)$$

is a standard Brownian motion. We assume that the swap rate is modeled as a geometric Brownian motion

$$dS(t, T_i, T_j) = S(t, T_i, T_j) \sigma_{i,j}(t) d\hat{B}_t^{i,j}, \qquad 0 \leq t \leq T_i, \qquad (12.24)$$

where the swap rate volatility is a deterministic function $\sigma_{i,j}(t)$. In the sequel we denote $S_t = S(t, T_i, T_j)$ for simplicity of notation.

1. Solve the Equation (12.24) on the interval $[t, T_i]$, and compute $S(T_i, T_i, T_j)$ from the initial condition $S(t, T_i, T_j)$.

2. Show that the price (12.18) of the swaption can be written as

$$P(t, T_i, T_j) C(S_t, v(t, T_i)),$$

where

$$v^2(t, T_i) = \int_t^{T_i} |\sigma_{i,j}(s)|^2 ds,$$

and $C(x, v)$ is a function to be specified using the Black–Scholes formula $\text{BS}(K, x, \sigma, r, \tau)$, with

$$\mathbb{E}[(xe^{m+X} - K)^+] = \Phi(v + (m + \log(x/K))/v) - K\Phi((m + \log(x/K))/v),$$

where X is a centered Gaussian random variable with mean $m = r\tau - v^2/2$ and variance v^2.

3. Consider a portfolio $(\xi_t^i, \ldots, \xi_t^j)_{t\in[0,T_i]}$ made of bonds with maturities $T_i, \ldots, T_j$ and value

$$V_t = \sum_{k=i}^{j} \xi_t^k P(t, T_k),$$

at time $t \in [0, T_i]$. We assume that the portfolio is self-financing, i.e.,

$$dV_t = \sum_{k=i}^{j} \xi_t^k dP(t, T_k), \qquad 0 \le t \le T_i, \tag{12.25}$$

and that it *hedges* the claim $(S(T_i, T_i, T_j) - \kappa)^+$, so that

$$\begin{aligned} V_t &= \mathbb{E}\left[e^{-\int_t^{T_i} r_s ds} \left(\sum_{k=i}^{j-1} (T_{k+1} - T_k) P(T_i, T_{k+1}) (L(T_i, T_k, T_{k+1}) - \kappa) \right)^+ \Bigg| \mathcal{F}_t \right] \\ &= P(t, T_i, T_j)\, \mathbb{E}_{i,j}\left[(S(T_i, T_i, T_j) - \kappa)^+ \Big| \mathcal{F}_t \right], \end{aligned}$$

$0 \le t \le T_i$. Show that

$$\begin{aligned} &\mathbb{E}\left[e^{-\int_t^{T_i} r_s ds} \left(\sum_{k=i}^{j-1} (T_{k+1} - T_k) P(T_i, T_{k+1}) (L(T_i, T_k, T_{k+1}) - \kappa) \right)^+ \Bigg| \mathcal{F}_t \right] \\ &= P(0, T_i, T_j)\, \mathbb{E}_{i,j}\left[(S(T_i, T_i, T_j) - \kappa)^+ \right] + \sum_{k=i}^{j} \int_0^t \xi_s^k dP(s, T_i), \end{aligned}$$

$0 \le t \le T_i$.

4. Show that under the self-financing condition (12.25), the discounted portfolio price $\tilde{V}_t = e^{-\int_0^t r_s ds} V_t$ satisfies

$$d\tilde{V}_t = \sum_{k=i}^{j} \xi_t^k d\tilde{P}(t, T_k),$$

where $\tilde{P}(t, T_k) = e^{-\int_0^t r_s ds} P(t, T_k)$, $k = i, \ldots, j$, denote the discounted bond prices.

5. Show that

$$\begin{aligned} &\mathbb{E}_{i,j}\left[(S(T_i, T_i, T_j) - \kappa)^+ \Big| \mathcal{F}_t \right] \\ &= \mathbb{E}_{i,j}\left[(S(T_i, T_i, T_j) - \kappa)^+ \right] + \int_0^t \frac{\partial C}{\partial x}(S_u, v(u, T_i)) dS_u. \end{aligned}$$

Hint: use the martingale property and the Itô formula.

6. Show that the discounted portfolio price $\hat{V}_t = V_t/P(t,T_i,T_j)$ satisfies

$$d\hat{V}_t = \frac{\partial C}{\partial x}(S_t, v(t,T_i))dS_t = S_t\frac{\partial C}{\partial x}(S_t, v(t,T_i))\sigma_t^{i,j}d\hat{B}_t^{i,j}.$$

7. Show that

$$dV_t = (P(t,T_i) - P(t,T_j))\frac{\partial C}{\partial x}(S_t, v(t,T_i))\sigma_t^{i,j}dB_t + \hat{V}_t dP(t,T_i,T_j).$$

8. Show that

$$\begin{aligned} dV_t &= S_t\zeta_i(t)\frac{\partial C}{\partial x}(S_t, v(t,T_i))\sum_{k=i}^{j-1}(T_{k+1} - T_k)P(t,T_{k+1})dB_t \\ &+(\hat{V}_t - S_t\frac{\partial C}{\partial x}(S_t, v(t,T_i)))\sum_{k=i}^{j-1}(T_{k+1} - T_k)P(t,T_{k+1})\zeta_{k+1}(t)dB_t \\ &+\frac{\partial C}{\partial x}(S_t, v(t,T_i))P(t,T_j)(\zeta_i(t) - \zeta_j(t))dB_t. \end{aligned}$$

9. Show that

$$\begin{aligned} d\tilde{V}_t &= \frac{\partial C}{\partial x}(S_t, v(t,T_i))d(\tilde{P}(t,T_i) - \tilde{P}(t,T_j)) \\ &+(\hat{V}_t - S_t\frac{\partial C}{\partial x}(S_t, v(t,T_i)))d\tilde{P}(t,T_i,T_j). \end{aligned}$$

10. Show that

$$\frac{\partial C}{\partial x}(x, v(t,T_i)) = \Phi\left(\frac{\log(x/K)}{v(t,T_i)} + \frac{v(t,T_i)}{2}\right).$$

11. Show that we have

$$\begin{aligned} d\tilde{V}_t &= \Phi\left(\frac{\log(S_t/K)}{v(t,T_i)} + \frac{v(t,T_i)}{2}\right)d(\tilde{P}(t,T_i) - \tilde{P}(t,T_j)) \\ &-\kappa\Phi\left(\frac{\log(S_t/K)}{v(t,T_i)} - \frac{v(t,T_i)}{2}\right)d\tilde{P}(t,T_i,T_j). \end{aligned}$$

12. Show that the hedging strategy is given by

$$\xi_t^i = \Phi\left(\frac{\log(S_t/K)}{v(t,T_i)} + \frac{v(t,T_i)}{2}\right),$$

$$\xi_t^j = -\Phi\left(\frac{\log(S_t/K)}{v(t,T_i)} + \frac{v(t,T_i)}{2}\right) - \kappa(T_{j+1} - T_j)\Phi\left(\frac{\log(S_t/K)}{v(t,T_i)} - \frac{v(t,T_i)}{2}\right),$$

and

$$\xi_t^k = -\kappa(T_{k+1} - T_k)\Phi\left(\frac{\log(S_t/K)}{v(t,T_i)} - \frac{v(t,T_i)}{2}\right), \qquad i+1 \le k \le j-1.$$

Chapter 13

Default Risk in Bond Markets

The bond pricing model of Chapter 11 is based on the terminal condition $P(T,T) = \$1$, i.e., the bond payoff at maturity is always equal to \$1, and default never occurs. In this chapter we allow for the possibility of default at a random time τ, in which case the terminal payoff of a bond vanishes at maturity. We also consider the credit default options (swaps) that can act as a protection against default.

13.1 Survival Probabilities and Failure Rate

Given $t > 0$, let $\mathbb{P}(\tau \geq t)$ denote the probability that a random system with lifetime τ survives at least t years. Assuming that survival probabilities $\mathbb{P}(\tau \geq t)$ are strictly positive for all $t > 0$, we can compute the conditional probability for that system to survive up to time T, given that it was still functioning at time $t \in [0, T]$, as

$$\mathbb{P}(\tau > T \mid \tau > t) = \frac{\mathbb{P}(\tau > T \text{ and } \tau > t)}{\mathbb{P}(\tau > t)} = \frac{\mathbb{P}(\tau > T)}{\mathbb{P}(\tau > t)}, \qquad 0 \leq t \leq T,$$

with

$$\begin{aligned}
\mathbb{P}(\tau < T \mid \tau > t) &= 1 - \mathbb{P}(\tau > T \mid \tau > t) \\
&= \frac{\mathbb{P}(\tau > t) - \mathbb{P}(\tau > T)}{\mathbb{P}(\tau > t)} \\
&= \frac{\mathbb{P}(\tau < T) - \mathbb{P}(\tau < t)}{\mathbb{P}(\tau > t)} \\
&= \frac{\mathbb{P}(t < \tau < T)}{\mathbb{P}(\tau > t)}, \qquad 0 \leq t \leq T,
\end{aligned}$$

and the conditional survival probability law

$$\begin{aligned}
\mathbb{P}(\tau \in dx \mid \tau > t) &= \mathbb{P}(x < \tau < x + dx \mid \tau > t) \\
&= \mathbb{P}(\tau < x + dx \mid \tau > t) - \mathbb{P}(\tau < x \mid \tau > t) \\
&= \frac{\mathbb{P}(\tau < x + dx) - \mathbb{P}(\tau < x)}{\mathbb{P}(\tau > t)}
\end{aligned}$$

$$= \frac{1}{\mathbb{P}(\tau > t)} d\mathbb{P}(\tau < x)$$
$$= -\frac{1}{\mathbb{P}(\tau > t)} d\mathbb{P}(\tau > x), \qquad x > t.$$

From this we can deduce the *failure rate* function

$$\begin{aligned}
\lambda(t) &:= \frac{\mathbb{P}(\tau < t + dt \mid \tau > t)}{dt} \\
&= \frac{1}{\mathbb{P}(\tau > t)} \frac{\mathbb{P}(t < \tau < t + dt)}{dt} \\
&= \frac{1}{\mathbb{P}(\tau > t)} \frac{\mathbb{P}(\tau > t) - \mathbb{P}(\tau > t + dt)}{dt} \\
&= -\frac{d}{dt} \log \mathbb{P}(\tau > t) \\
&= -\frac{1}{\mathbb{P}(\tau > t)} \frac{d}{dt} \mathbb{P}(\tau > t), \qquad t > 0,
\end{aligned}$$

which satisfies the differential equation

$$\frac{d}{dt}\mathbb{P}(\tau > t) = -\lambda(t)\mathbb{P}(\tau > t),$$

which can be solved as

$$\mathbb{P}(\tau > t) = \exp\left(- \int_0^t \lambda(u) du \right), \qquad t \in \mathbb{R}_+, \tag{13.1}$$

under the initial condition $\mathbb{P}(\tau > 0) = 1$. This allows us to rewrite the survival probability as

$$\mathbb{P}(\tau > T \mid \tau > t) = \frac{\mathbb{P}(\tau > T)}{\mathbb{P}(\tau > t)} = \exp\left(- \int_t^T \lambda(u) du \right), \qquad 0 \leq t \leq T,$$

with

$$\mathbb{P}(\tau > t + h \mid \tau > t) = e^{-\lambda(t)h} \simeq 1 - \lambda(t)h, \tag{13.2}$$

and

$$\mathbb{P}(\tau < t + h \mid \tau > t) = 1 - e^{-\lambda(t)h} \simeq \lambda(t)h, \tag{13.3}$$

as h tends to 0. When the failure rate $\lambda(t) = \lambda > 0$ is a constant function of time, Relation (13.1) shows that

$$\mathbb{P}(\tau > T) = e^{-\lambda T}, \qquad T \geq 0,$$

i.e., τ has the exponential distribution with parameter λ. Note that given $(\tau_n)_{n\geq 1}$ a sequence of *i.i.d.* exponentially distributed random variables, letting

$$T_n = \tau_1 + \cdots + \tau_n, \qquad n \geq 1,$$

defines the sequence of jump times of a standard Poisson process with intensity $\lambda > 0$, cf. Section 14.1 below for details.

13.2 Stochastic Default

We now model the failure rate function $(\lambda_t)_{t\in\mathbb{R}_+}$ as a random process adapted to a filtration $(\mathcal{F}_t)_{t\in\mathbb{R}_+}$.

In case the random time τ is a *stopping time* with respect to $(\mathcal{F}_t)_{t\in\mathbb{R}_+}$, i.e., the knowledge of whether default already occurred at time t is contained in $\mathcal{F}_t$, $t \in \mathbb{R}_+$, and we have

$$\{\tau > t\} \in \mathcal{F}_t, \qquad t \in \mathbb{R}_+,$$

cf. Section 9.3, we have

$$\mathbb{P}(\tau > t \mid \mathcal{F}_t) = \mathbb{E}\left[\mathbf{1}_{\{\tau>t\}} \mid \mathcal{F}_t\right] = \mathbf{1}_{\{\tau>t\}}, \qquad t \in \mathbb{R}_+.$$

In the sequel we will not assume that τ is an $\mathcal{F}_t$-stopping time, and by analogy with (13.1) we will write $\mathbb{P}(\tau > t \mid \mathcal{F}_t)$ as

$$\mathbb{P}(\tau > t \mid \mathcal{F}_t) = \exp\left(-\int_0^t \lambda_u du\right), \qquad t > 0. \tag{13.4}$$

This is the case in particular in [45] when λ_u has the form $\lambda_u = h(X_u)$, and τ is given by

$$\tau = \inf\left\{t \in \mathbb{R}_+ \ : \ \int_0^t h(X_u)du \geq L\right\},$$

where h is a non-negative function, $(X_t)_{t\in\mathbb{R}_+}$ is a process generating a filtration $(\mathcal{F}_t)_{t\in\mathbb{R}_+}$, and L is an independent exponentially distributed random variable.

We let $(\mathcal{G}_t)_{t\in\mathbb{R}_+}$ be the filtration defined by

$$\mathcal{G}_t = \mathcal{F}_t \vee \sigma(\{\tau \leq u\} \ : \ 0 \leq u \leq t), \qquad t \in \mathbb{R}_+,$$

i.e., $\mathcal{G}_t$ contains the additional information on whether default at time τ has occurred or not before time t. The process λ_t can also be chosen among the classical mean-reverting diffusion processes, including jump-diffusion processes.

Taking $F = 1$ in the next Lemma 13.1 shows that the survival probability up to time T, given information known up to t, is given by

$$\begin{aligned}\mathbb{P}(\tau > T \mid \mathcal{G}_t) &= \mathbb{E}\left[\mathbf{1}_{\{\tau>T\}} \mid \mathcal{G}_t\right] \\ &= \mathbf{1}_{\{\tau>t\}}\,\mathbb{E}\left[\exp\left(-\int_t^T \lambda_u du\right) \Big| \mathcal{F}_t\right], \qquad 0 \leq t \leq T.\end{aligned} \tag{13.5}$$

Lemma 13.1 *([31]) For any $\mathcal{F}_T$-measurable integrable random variable F we have*

$$\mathbb{E}\left[F\mathbf{1}_{\{\tau>T\}} \mid \mathcal{G}_t\right] = \mathbf{1}_{\{\tau>t\}}\,\mathbb{E}\left[F\exp\left(-\int_t^T \lambda_u du\right)\Big|\mathcal{F}_t\right].$$

Proof. By (13.4) we have

$$\frac{\mathbb{P}(\tau > T \mid \mathcal{F}_T)}{\mathbb{P}(\tau > t \mid \mathcal{F}_t)} = \frac{\exp\left(-\int_0^T \lambda_u du\right)}{\exp\left(-\int_0^t \lambda_u du\right)} = \exp\left(-\int_t^T \lambda_u du\right),$$

hence, since F is $\mathcal{F}_T$-measurable,

$$\begin{aligned}
\mathbf{1}_{\{\tau>t\}}\,\mathbb{E}\left[F\exp\left(-\int_t^T \lambda_u du\right)\Big|\mathcal{F}_t\right] &= \mathbf{1}_{\{\tau>t\}}\,\mathbb{E}\left[F\frac{\mathbb{P}(\tau > T \mid \mathcal{F}_T)}{\mathbb{P}(\tau > t \mid \mathcal{F}_t)}\Big|\mathcal{F}_t\right] \\
&= \frac{\mathbf{1}_{\{\tau>t\}}}{\mathbb{P}(\tau > t \mid \mathcal{F}_t)}\,\mathbb{E}\left[F\,\mathbb{E}[\mathbf{1}_{\{\tau>T\}} \mid \mathcal{F}_T]\Big|\mathcal{F}_t\right] \\
&= \frac{\mathbf{1}_{\{\tau>t\}}}{\mathbb{P}(\tau > t \mid \mathcal{F}_t)}\,\mathbb{E}\left[\mathbb{E}[F\mathbf{1}_{\{\tau>T\}} \mid \mathcal{F}_T]\Big|\mathcal{F}_t\right] \\
&= \mathbf{1}_{\{\tau>t\}}\frac{\mathbb{E}\left[F\mathbf{1}_{\{\tau>T\}}\Big|\mathcal{F}_t\right]}{\mathbb{P}(\tau > t \mid \mathcal{F}_t)} \\
&= \mathbf{1}_{\{\tau>t\}}\,\mathbb{E}\left[F\mathbf{1}_{\{\tau>T\}}\Big|\mathcal{F}_t \vee \{\tau > t\}\right] \\
&= \mathbf{1}_{\{\tau>t\}}\,\mathbb{E}\left[F\mathbf{1}_{\{\tau>T\}}\Big|\mathcal{G}_t\right], \qquad 0 \le t \le T.
\end{aligned}$$

□

The computation of $\mathbb{P}(\tau > T \mid \mathcal{G}_t)$ according to (13.5) is then similar to that of a bond price, by considering the failure rate $\lambda(t)$ as a virtual short term interest rate. In particular the failure rate $\lambda(t, T)$ can be modeled in the HJM framework of Section 11.4, and

$$\mathbb{P}(\tau > T \mid \mathcal{G}_t) = \mathbb{E}\left[\exp\left(-\int_t^T \lambda(t,u)du\right)\Big|\mathcal{F}_t\right]$$

can then be computed by applying HJM bond pricing techniques.

13.3 Defaultable Bonds

The price of a default bond with maturity T, (random) default time τ and (possibly random) recovery rate $\xi \in [0, 1]$ is given by

$$P(t,T) \;=\; \mathbb{E}\left[\mathbf{1}_{\{\tau>T\}}\exp\left(-\int_t^T r_u du\right)\Big|\mathcal{G}_t\right]$$

$$+\mathbb{E}\left[\xi\mathbf{1}_{\{\tau\le T\}}\exp\left(-\int_t^T r_u du\right)\Big|\mathcal{G}_t\right], \qquad 0\le t\le T,$$

Taking $F=\exp\left(-\int_t^T r_u du\right)$ in Lemma 13.1, we get

$$\mathbb{E}\left[\mathbf{1}_{\{\tau>T\}}\exp\left(-\int_t^T r_u du\right)\Big|\mathcal{G}_t\right] = \mathbf{1}_{\{\tau>t\}}\mathbb{E}\left[\exp\left(-\int_t^T (r_u+\lambda_u)du\right)\Big|\mathcal{F}_t\right],$$

cf. e.g., [45], [31], [19], hence

$$\begin{aligned} P(t,T) &= \mathbf{1}_{\{\tau>t\}}\mathbb{E}\left[\exp\left(-\int_t^T (r_u+\lambda_u)du\right)\Big|\mathcal{F}_t\right] \\ &\quad +\mathbb{E}\left[\xi\mathbf{1}_{\{\tau\le T\}}\exp\left(-\int_t^T r_u du\right)\Big|\mathcal{G}_t\right], \qquad 0\le t\le T. \end{aligned}$$

In the case of complete default (zero-recovery) we have $\xi=0$ and

$$P(t,T)=\mathbf{1}_{\{\tau>t\}}\mathbb{E}\left[\exp\left(-\int_t^T (r_s+\lambda_s)ds\right)\Big|\mathcal{F}_t\right], \qquad 0\le t\le T. \quad (13.6)$$

From the above expression (13.6) we note that the effect of the presence of a default time τ is to decrease the bond price, which can be viewed as an increase of the short rate by the amount λ_u.

This treatment of default risk has some similarity with that of coupon bonds which can be priced as

$$P(t,T)=e^{c(T-t)}\mathbb{E}\left[\exp\left(-\int_t^T r_s ds\right)\Big|\mathcal{G}_t\right],$$

where $c>0$ is a continuous-time deterministic coupon rate.

Finally, from Proposition 12.1 the bond price (13.6) can also be expressed under the forward measure $\tilde{\mathbb{P}}$ with maturity T, as

$$\begin{aligned} P(t,T) &= \mathbf{1}_{\{\tau>t\}}\mathbb{E}\left[\exp\left(-\int_t^T (r_s+\lambda_s)ds\right)\Big|\mathcal{F}_t\right] \\ &= \mathbf{1}_{\{\tau>t\}}\mathbb{E}\left[\exp\left(-\int_t^T r_s ds\right)\Big|\mathcal{F}_t\right]\mathbb{E}_{\tilde{\mathbb{P}}}\left[\exp\left(-\int_t^T \lambda_s ds\right)\Big|\mathcal{F}_t\right] \\ &= \mathbf{1}_{\{\tau>t\}}\mathbb{E}\left[\exp\left(-\int_t^T r_s ds\right)\Big|\mathcal{F}_t\right]Q(t,T), \end{aligned}$$

where

$$Q(t,T)=\mathbb{E}_{\tilde{\mathbb{P}}}\left[\exp\left(-\int_t^T \lambda_s ds\right)\Big|\mathcal{F}_t\right]$$

denotes the survival probability under the forward measure $\tilde{\mathbb{P}}$, cf. [11], [10].

13.4 Credit Default Swaps

We work with a tenor structure $\{t = T_i < \cdots < T_j = T\}$.

A Credit Default Swap (CDS) is a contract consisting of

- **a premium leg:** the buyer is purchasing protection at time t against default at time T_k, $k = i+1, \ldots, j$, and has to make a fixed payment S_t at times $T_{i+1}, \ldots, T_j$ between t and T in compensation.

  The discounted value at time t of the premium leg is

$$\begin{aligned} V(t,T) &= \mathbb{E}\left[\sum_{k=i}^{j-1} S_t \delta_k \mathbf{1}_{\{\tau > T_{k+1}\}} \exp\left(-\int_t^{T_{k+1}} r_s ds\right) \Big| \mathcal{G}_t\right] \\ &= \sum_{k=i}^{j-1} S_t \delta_k \, \mathbb{E}\left[\mathbf{1}_{\{\tau > T_{k+1}\}} \exp\left(-\int_t^{T_{k+1}} r_s ds\right) \Big| \mathcal{G}_t\right] \\ &= S_t \sum_{k=i}^{j-1} \delta_k P(t, T_{k+1}) \\ &= S_t P(t, T_i, T_j), \end{aligned}$$

  where $\delta_k = T_{k+1} - T_k$, $P(t, T_i, T_j)$ is the annuity numéraire (12.10), and

$$P(t, T_k) = \mathbf{1}_{\{\tau > t\}} \, \mathbb{E}\left[\exp\left(-\int_t^{T_k} (r_s + \lambda_s) ds\right) \Big| \mathcal{F}_t\right], \qquad 0 \le t \le T_k,$$

  is the defaultable bond price with maturity T_k, $k = i, \ldots, j-1$. For simplicity we have ignored a possible accrual interest term over the time period $[T_k, \tau]$ when $\tau \in [T_k, T_{k+1}]$ in the above value of the premium leg.

- **a protection leg:** the seller or issuer of the contract makes a payment $1 - \xi_{k+1}$ to the buyer in case default occurs at time T_{k+1}, $k = i, \ldots, j-1$.

  The value at time t of the protection leg is

$$\mathbb{E}\left[\sum_{k=i}^{j-1} \mathbf{1}_{(T_k, T_{k+1}]}(\tau)(1 - \xi_{k+1}) \exp\left(-\int_t^{T_{k+1}} r_s ds\right) \Big| \mathcal{G}_t\right],$$

  where ξ_{k+1} is the recovery rate associated with the maturity T_{k+1}, $k = i, \ldots, j-1$.

In the case of a non-random recovery rate ξ_k the value of the protection leg becomes

$$\sum_{k=i}^{j-1}(1-\xi_{k+1})\,\mathbb{E}\left[\mathbf{1}_{(T_k,T_{k+1}]}(\tau)\exp\left(-\int_t^{T_{k+1}} r_s ds\right)\Big|\mathcal{G}_t\right].$$

The spread S_t is computed by equating the values of the protection and premium legs, i.e., from the relation

$$\begin{aligned} V(t,T) &= S_t P(t,T_i,T_j) \\ &= \mathbb{E}\left[\sum_{k=i}^{j-1}\mathbf{1}_{(T_k,T_{k+1}]}(\tau)(1-\xi_{k+1})\exp\left(-\int_t^{T_{k+1}} r_s ds\right)\Big|\mathcal{G}_t\right], \end{aligned}$$

which yields

$$S_t = \frac{1}{P(t,T_i,T_j)}\sum_{k=i}^{j-1}\mathbb{E}\left[\mathbf{1}_{(T_k,T_{k+1}]}(\tau)(1-\xi_{k+1})\exp\left(-\int_t^{T_{k+1}} r_s ds\right)\Big|\mathcal{G}_t\right].$$

In the case of a constant recovery rate ξ we find

$$S_t = \frac{1-\xi}{P(t,T_i,T_j)}\sum_{k=i}^{j-1}\mathbb{E}\left[\mathbf{1}_{(T_k,T_{k+1}]}(\tau)\exp\left(-\int_t^{T_{k+1}} r_s ds\right)\Big|\mathcal{G}_t\right],$$

and if τ is constrained to take values in the tenor structure $\{T_i,\ldots,T_j\}$ we get

$$S_t = \frac{1-\xi}{P(t,T_i,T_j)}\,\mathbb{E}\left[\mathbf{1}_{[t,T]}(\tau)\exp\left(-\int_t^{\tau} r_s ds\right)\Big|\mathcal{G}_t\right].$$

13.5 Exercises

Exercise 13.1 Defaultable bonds. Consider a (random) default time τ with law

$$\mathbb{P}(\tau > t \mid \mathcal{F}_t) = \exp\left(-\int_0^t \lambda_u du\right),$$

where λ_t is a (random) default rate process which is adapted to the filtration $(\mathcal{F}_t)_{t\in\mathbb{R}_+}$. Recall that the probability of survival up to time T, given information known up to time t, is given by

$$\mathbb{P}(\tau > T \mid \mathcal{G}_t) = \mathbf{1}_{\{\tau>t\}}E\left[\exp\left(-\int_t^T \lambda_u du\right)\Big|\mathcal{F}_t\right],$$

where $\mathcal{G}_t = \mathcal{F}_t \vee \sigma(\{\tau < u\} : 0 \leq u \leq t)$, $t \in \mathbb{R}_+$, is the filtration defined by adding the default time information to the history $(\mathcal{F}_t)_{t\in\mathbb{R}_+}$. In this framework, the price $P(t,T)$ of defaultable bond with maturity T, short term interest rate r_t and (random) default time τ is given by

$$\begin{aligned} P(t,T) &= E\left[\mathbf{1}_{\{\tau>T\}} \exp\left(-\int_t^T r_u du\right) \Big| \mathcal{G}_t\right] \qquad (13.7) \\ &= \mathbf{1}_{\{\tau>t\}} E\left[\exp\left(-\int_t^T (r_u+\lambda_u)du\right) \Big| \mathcal{F}_t\right]. \end{aligned}$$

In the sequel we assume that the processes $(r_t)_{t\in\mathbb{R}_+}$ and $(\lambda_t)_{t\in\mathbb{R}_+}$ are modeled according to the Vasicek processes

$$\begin{cases} dr_t = -ar_t dt + \sigma dB_t^{(1)}, \\ \\ d\lambda_t = -b\lambda_t dt + \eta dB_t^{(2)}, \end{cases}$$

where $(B_t^{(1)})_{t\in\mathbb{R}_+}$ and $(B_t^{(2)})_{t\in\mathbb{R}_+}$ are two standard $\mathcal{F}_t$-Brownian motions with correlation $\rho \in [-1,1]$, and $dB_t^{(1)} dB_t^{(2)} = \rho dt$.

1. Give a justification for the fact that

$$E\left[\exp\left(-\int_t^T (r_u+\lambda_u)du\right) \Big| \mathcal{F}_t\right]$$

can be written as a function $F(t, r_t, \lambda_t)$ of t, r_t and λ_t, $t \in [0,T]$.

2. Show that

$$t \longmapsto \exp\left(-\int_0^t (r_s+\lambda_s)ds\right) E\left[\exp\left(-\int_t^T (r_u+\lambda_u)du\right) \Big| \mathcal{F}_t\right]$$

is an $\mathcal{F}_t$-martingale under $\mathbb{P}$.

3. Use the Itô formula with two variables to derive a PDE on $\mathbb{R}^2$ for the function $F(t,x,y)$.

4. Show that we have

$$\int_t^T r_s ds = C(a,t,T) r_t + \sigma \int_t^T C(a,s,T) dB_s^{(1)},$$

and

$$\int_t^T \lambda_s ds = C(b,t,T)\lambda_t + \eta \int_t^T C(b,s,T) dB_s^{(2)},$$

where

$$C(a,t,T) = -\frac{1}{a}(e^{-a(T-t)} - 1).$$

5. Show that the random variable

$$\int_t^T r_s ds + \int_t^T \lambda_s ds$$

is Gaussian and compute its mean

$$\mathbb{E}\left[\int_t^T r_s ds + \int_t^T \lambda_s ds \Big| \mathcal{F}_t\right]$$

and variance

$$\mathrm{Var}\left[\int_t^T r_s ds + \int_t^T \lambda_s ds \Big| \mathcal{F}_t\right],$$

conditionally to $\mathcal{F}_t$.

6. Compute $P(t,T)$ from its expression (13.7) as a conditional expectation.

7. Show that the solution $F(t,x,y)$ to the 2-dimensional PDE of Question 3 is

$$\begin{aligned} F(t,x,y) &= \exp\left(-C(a,t,T)x - C(b,t,T)y\right) \\ &\quad \times \exp\left(\frac{\sigma^2}{2}\int_t^T C^2(a,s,T)ds + \frac{\eta^2}{2}\int_t^T C^2(b,s,T)ds\right) \\ &\quad \times \exp\left(\rho\sigma\eta\int_t^T C(a,s,T)C(b,s,T)ds\right). \end{aligned}$$

8. Show that the defaultable bond price $P(t,T)$ can also be written as

$$P(t,T) = e^{U(t,T)}\mathbb{P}(\tau > T \mid \mathcal{G}_t)\,\mathbb{E}\left[\exp\left(-\int_t^T r_s ds\right)\Big|\mathcal{F}_t\right],$$

where

$$U(t,T) = \rho\frac{\sigma\eta}{ab}\left(T - t - C(a,t,T) - C(b,t,T) + C(a+b,t,T)\right).$$

9. By partial differentiation of $\log P(t,T)$ with respect to T, compute the corresponding instantaneous short rate $f(t,T) = -\dfrac{\partial \log P(t,T)}{\partial T}$.

10. Show that $\mathbb{P}(\tau > T \mid \mathcal{G}_t)$ can be written using an HJM type default rate as

$$\mathbb{P}(\tau > T \mid \mathcal{G}_t) = \mathbf{1}_{\{\tau > t\}}\exp\left(-\int_t^T f_2(t,u)du\right),$$

where

$$f_2(t,u) = \lambda_t e^{-b(u-t)} - \frac{\eta^2}{2}C^2(b,t,u).$$

11. Show how the result of Question 8 can be simplified when $(B_t^{(1)})_{t\in\mathbb{R}_+}$ and $(B_t^{(2)})_{t\in\mathbb{R}_+}$ are independent.

Chapter 14

Stochastic Calculus for Jump Processes

The modeling of a risky asset by stochastic processes with continuous paths, based on Brownian motions, suffers from several defects. First, the path continuity assumption does not seem reasonable in view of the possibility of sudden price variations (jumps) resulting from market crashes. Secondly, the modeling of risky asset prices by Brownian motion relies on the use of the Gaussian distribution which tends to underestimate the probabilities of extreme events.

A solution is to use stochastic processes with jumps, that will account for sudden variations of the asset prices. On the other hand, such jump models are generally based on the Poisson distribution which has a slower tail decay than the Gaussian distribution. This allows one to assign higher probabilities to extreme events, resulting in a more realistic modeling of asset prices.

14.1 Poisson Process

The most elementary and useful jump process is the *standard Poisson process* which is a stochastic process $(N_t)_{t\in\mathbb{R}_+}$ with jumps of size $+1$ only, and whose paths are constant in between two jumps, i.e., at time t, the value N_t of the process is given by

$$N_t = \sum_{k=1}^{\infty} \mathbf{1}_{[T_k,\infty)}(t), \qquad t \in \mathbb{R}_+, \tag{14.1}$$

where

$$\mathbf{1}_{[T_k,\infty)}(t) = \begin{cases} 1 & \text{if } t \geq T_k, \\ 0 & \text{if } 0 \leq t < T_k, \end{cases}$$

$k \geq 1$, and $(T_k)_{k\geq 1}$ is the increasing family of jump times of $(N_t)_{t\in\mathbb{R}_+}$ such that

$$\lim_{k\to\infty} T_k = +\infty.$$

In addition, $(N_t)_{t\in\mathbb{R}_+}$ satisfies the following conditions:

1. Independence of increments: for all $0 \leq t_0 < t_1 < \cdots < t_n$ and $n \geq 1$ the random variables

$$N_{t_1} - N_{t_0}, \ldots, N_{t_n} - N_{t_{n-1}},$$

are independent.

2. Stationarity of increments: $N_{t+h} - N_{s+h}$ has the same distribution as $N_t - N_s$ for all $h > 0$ and $0 \leq s \leq t$.

The meaning of the above stationarity condition is that for all fixed $k \in \mathbb{N}$ we have

$$\mathbb{P}(N_{t+h} - N_{s+h} = k) = \mathbb{P}(N_t - N_s = k),$$

for all $h > 0$, i.e., the value of the probability

$$\mathbb{P}(N_{t+h} - N_{s+h} = k)$$

does not depend on $h > 0$, for all fixed $0 \leq s \leq t$ and $k \in \mathbb{N}$.

The next figure represents a sample path of a Poisson process.

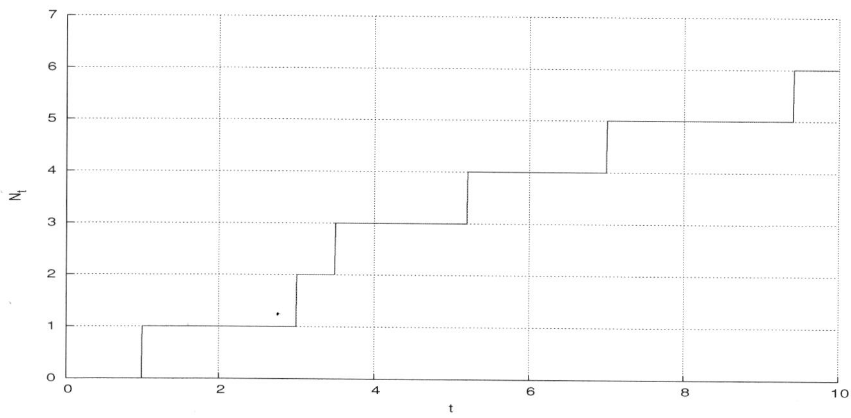

FIGURE 14.1: Sample path of a Poisson process $(N_t)_{t \in \mathbb{R}_+}$.

Based on the above assumption, given a time value $T > 0$ a natural question arises:

what is the probability distribution of the random variable N_T?

We already know that N_t takes values in $\mathbb{N}$ and therefore it has a discrete distribution for all $t \in \mathbb{R}_+$.

It is a remarkable fact that the distribution of the increments of $(N_t)_{t\in\mathbb{R}_+}$, can be completely determined from the above conditions, as shown in the following theorem.

As seen in the next result, $N_t - N_s$ has the Poisson distribution with parameter $\lambda(t-s)$.

Theorem 14.1 *Assume that the counting process $(N_t)_{t\in\mathbb{R}_+}$ satisfies the above Conditions 1 and 2. Then for all fixed $0 \le s \le t$ we have*

$$P(N_t - N_s = k) = e^{-\lambda(t-s)}\frac{(\lambda(t-s))^k}{k!}, \qquad k \in \mathbb{N}, \tag{14.2}$$

for some constant $\lambda > 0$.

The parameter $\lambda > 0$ is called the intensity of the Poisson process $(N_t)_{t\in\mathbb{R}_+}$ and it is given by

$$\lambda := \lim_{h\to 0}\frac{1}{h}\mathbb{P}(N_h = 1). \tag{14.3}$$

The proof of the above Theorem 14.1 is technical and not included here, cf. e.g., [6] for details, and we could in fact take this distribution property (14.2) as one of the hypotheses that define the Poisson process.

Precisely, we could restate the definition of the standard Poisson process $(N_t)_{t\in\mathbb{R}_+}$ with intensity $\lambda > 0$ as being a process defined by (14.1), which is assumed to have independent increments distributed according to the Poisson distribution, in the sense that for all $0 \le t_0 \le t_1 < \cdots < t_n$,

$$(N_{t_1} - N_{t_0}, \ldots, N_{t_n} - N_{t_{n-1}})$$

is a vector of independent Poisson random variables with respective parameters

$$(\lambda(t_1 - t_0), \ldots, \lambda(t_n - t_{n-1})).$$

In particular, N_t has the Poisson distribution with parameter λt, i.e.,

$$\mathbb{P}(N_t = k) = \frac{(\lambda t)^k}{k!}e^{-\lambda t}, \qquad t > 0.$$

The *expected value* $E[N_t]$ of N_t can be computed as

$$\mathbb{E}[N_t] = \lambda t, \tag{14.4}$$

cf. Exercise A.1.

Short Time Behavior

From (14.3) above we deduce the short time asymptotics [1]

$$\mathbb{P}(N_h = 1) = h\lambda e^{-h\lambda} \simeq h\lambda, \qquad h \to 0,$$

and

$$\mathbb{P}(N_h = 0) = e^{-h\lambda} \simeq 1 - h\lambda, \qquad h \to 0.$$

By stationarity of the Poisson process we find more generally that

$$\mathbb{P}(N_{t+h} - N_t = 1) = h\lambda e^{-h\lambda} \simeq h\lambda, \qquad h \to 0,$$

and

$$\mathbb{P}(N_{t+h} - N_t = 0) = e^{-h\lambda} \simeq 1 - h\lambda, \qquad h \to 0,$$

for all $t > 0$.

This means that within a "short" interval $[t, t+h]$ of length h, the increment $N_{t+h} - N_t$ behaves like a Bernoulli random variable with parameter λh. This fact can be used for the random simulation of Poisson process paths.

We also find that

$$\mathbb{P}(N_{t+h} - N_t = 2) \simeq h^2 \frac{\lambda^2}{2}, \qquad h \to 0, \qquad t > 0,$$

and more generally

$$\mathbb{P}(N_{t+h} - N_t = k) \simeq h^k \frac{\lambda^k}{k!}, \qquad h \to 0, \qquad t > 0.$$

The intensity of the Poisson process can in fact be made time-dependent (e.g., by a time change), in which case we have

$$\mathbb{P}(N_t - N_s = k) = \exp\left(-\int_s^t \lambda(u)du \right) \frac{\left(\int_s^t \lambda(u)du \right)^k}{k!}, \qquad k \geq 0.$$

In particular,

$$\mathbb{P}(N_{t+dt} - N_t = k) = \begin{cases} e^{-\lambda(t)dt} \simeq 1 - \lambda(t)dt, & k = 0, \\ \lambda(t)e^{-\lambda(t)dt}dt \simeq \lambda(t)dt, & k = 1, \\ o(dt), & k \geq 2, \end{cases}$$

and $\mathbb{P}(N_{t+dt} - N_t = 0)$, $\mathbb{P}(N_{t+dt} - N_t = 1)$ coincide respectively with (13.2) and (13.3) above. The intensity process $(\lambda(t))_{t\in\mathbb{R}_+}$ can also be made random in the case of Cox processes.

[1] We use the notation $f(h) \simeq h^k$ to mean that $\lim_{h\to 0} f(h)/h^k = 1$.

Poisson Process Jump Times

In order to prove the next proposition we note that we have the equivalence

$$\{T_1 > t\} \Longleftrightarrow \{N_t = 0\},$$

and more generally

$$\{T_n > t\} \Longleftrightarrow \{N_t \leq n - 1\},$$

for all $n \geq 1$.

In the next proposition we compute the distribution of T_n with its density. It coincides with the *gamma* distribution with integer parameter $n \geq 1$, also known as the Erlang distribution in queueing theory.

Proposition 14.1 *For all $n \geq 1$ the probability distribution of T_n has the density function*

$$t \longmapsto \lambda^n e^{-\lambda t} \frac{t^{n-1}}{(n-1)!}$$

on $\mathbb{R}_+$, i.e., *for all $t > 0$ the probability $\mathbb{P}(T_n \geq t)$ is given by*

$$\mathbb{P}(T_n \geq t) = \lambda^n \int_t^\infty e^{-\lambda s} \frac{s^{n-1}}{(n-1)!} ds.$$

Proof. We have

$$\mathbb{P}(T_1 > t) = \mathbb{P}(N_t = 0) = e^{-\lambda t}, \qquad t \in \mathbb{R}_+,$$

and by induction, assuming that

$$\mathbb{P}(T_{n-1} > t) = \lambda \int_t^\infty e^{-\lambda s} \frac{(\lambda s)^{n-2}}{(n-2)!} ds, \qquad n \geq 2,$$

we obtain

$$\begin{aligned}
\mathbb{P}(T_n > t) &= \mathbb{P}(T_n > t \geq T_{n-1}) + \mathbb{P}(T_{n-1} > t) \\
&= \mathbb{P}(N_t = n-1) + \mathbb{P}(T_{n-1} > t) \\
&= e^{-\lambda t} \frac{(\lambda t)^{n-1}}{(n-1)!} + \lambda \int_t^\infty e^{-\lambda s} \frac{(\lambda s)^{n-2}}{(n-2)!} ds \\
&= \lambda \int_t^\infty e^{-\lambda s} \frac{(\lambda s)^{n-1}}{(n-1)!} ds, \qquad t \in \mathbb{R}_+,
\end{aligned}$$

where we applied an integration by parts to derive the last line. □

In particular, for all $n \in \mathbb{Z}$ and $t \in \mathbb{R}_+$, we have

$$\mathbb{P}(N_t = n) = p_n(t) = e^{-\lambda t} \frac{(\lambda t)^n}{n!},$$

i.e., $p_{n-1} : \mathbb{R}_+ \to \mathbb{R}_+$, $n \geq 1$, is the density function of T_n.

Similarly we could show that the time

$$\tau_k := T_{k+1} - T_k$$

spent in state $k \in \mathbb{N}$, with $T_0 = 0$, forms a sequence of independent identically distributed random variables having the exponential distribution with parameter $\lambda > 0$, i.e.,

$$\mathbb{P}(\tau_0 > t_0, \ldots, \tau_n > t_n) = e^{-\lambda(t_0+t_1+\cdots+t_n)}, \qquad t_0, \ldots, t_n \in \mathbb{R}_+.$$

Since the expectation of the exponentially distributed random variable τ_k with parameter $\lambda > 0$ is given by

$$\mathbb{E}[\tau_k] = \frac{1}{\lambda},$$

we can check that the higher the intensity λ (i.e., the higher the probability of having a jump within a small interval), the smaller is the time spent in each state $k \in \mathbb{N}$ on average.

In addition, given that $\{N_T = n\}$, the n jump times on $[0, T]$ of the Poisson process $(N_t)_{t\in\mathbb{R}_+}$ are independent uniformly distributed random variables on $[0, T]^n$.

Compensated Poisson Martingale

From (14.4) above we deduce that

$$\mathbb{E}[N_t - \lambda t] = 0, \tag{14.5}$$

i.e., the compensated Poisson process $(N_t - \lambda t)_{t\in\mathbb{R}_+}$ has *centered increments.*

Since in addition $(N_t - \lambda t)_{t\in\mathbb{R}_+}$ also has independent increments we get the following proposition.

Proposition 14.2 *The compensated Poisson process*

$$(N_t - \lambda t)_{t\in\mathbb{R}_+}$$

is a martingale *with respect to its own filtration* $(\mathcal{F}_t)_{t\in\mathbb{R}_+}$.

Extensions of the Poisson process include Poisson processes with time-dependent intensity, and with random time-dependent intensity (Cox processes). Renewal processes are counting processes

$$N_t = \sum_{n\geq 1} \mathbf{1}_{[T_n,\infty)}(t), \qquad t \in \mathbb{R}_+,$$

in which $\tau_k = T_{k+1} - T_k$, $k \in \mathbb{N}$, is a sequence of independent identically distributed random variables. In particular, Poisson processes are renewal processes.

14.2 Compound Poisson Processes

The Poisson process itself appears to be too limited to develop realistic price models as its jumps are of constant size. Therefore there is some interest in considering jump processes that can have random jump sizes.

Let $(Z_k)_{k\geq 1}$ denote an *i.i.d.* sequence of square-integrable random variables with probability distribution $\nu(dy)$ on $\mathbb{R}$, independent of the Poisson process $(N_t)_{t\in\mathbb{R}_+}$. We have

$$\mathbb{P}(Z_k \in [a,b]) = \nu([a,b]) = \int_a^b \nu(dy), \qquad -\infty < a \leq b < \infty.$$

Definition 14.1 *The process*

$$Y_t = \sum_{k=1}^{N_t} Z_k, \qquad t \in \mathbb{R}_+, \tag{14.6}$$

is called a compound Poisson process.

The next figure represents a sample path of a compound Poisson process, with here $Z_1 = 0.9$, $Z_2 = -0.7$, $Z_3 = 1.4$, $Z_4 = 0.6$, $Z_5 = -2.5$, $Z_6 = 1.5$, $Z_7 = -1.2$.

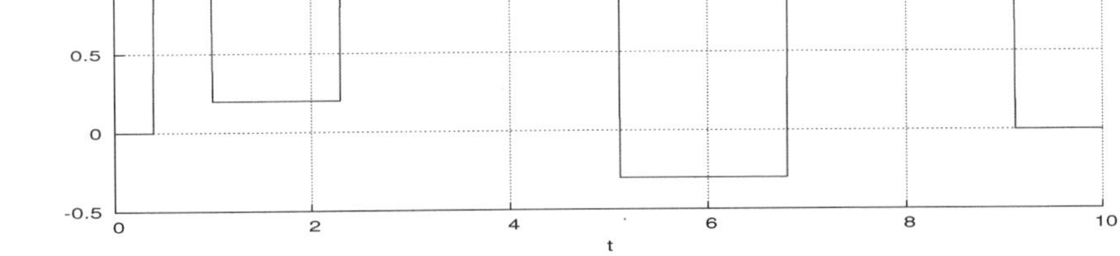

FIGURE 14.2: Sample path of a compound Poisson process $(Y_t)_{t\in\mathbb{R}_+}$.

Given that $\{N_T = n\}$, the n jump sizes of $(Y_t)_{t\in\mathbb{R}_+}$ on $[0,T]$ are independent random variables which are distributed on $\mathbb{R}$ according to $\nu(dx)$. Based on this fact, the next proposition allows us to compute the characteristic function of the increment $Y_T - Y_t$.

Proposition 14.3 *For any $t \in [0,T]$ we have*

$$\mathbb{E}\left[\exp\left(i\alpha(Y_T - Y_t)\right)\right] = \exp\left(\lambda(T-t)\int_{-\infty}^{\infty}(\mathrm{e}^{iy\alpha}-1)\nu(dy)\right),$$

$\alpha \in \mathbb{R}$.

Proof. Since N_t has a Poisson distribution with parameter $t > 0$ and is independent of $(Z_k)_{k\geq 1}$, for all $\alpha \in \mathbb{R}$ we have by conditioning:

$$\begin{aligned}
\mathbb{E}\left[\exp\left(i\alpha(Y_T - Y_t)\right)\right] &= \sum_{n=0}^{\infty}\mathbb{E}\left[\exp\left(i\alpha\sum_{k=1}^{n}Z_k\right)\right]\mathbb{P}(N_T - N_t = n) \\
&= \mathrm{e}^{-\lambda(T-t)}\sum_{n=0}^{\infty}\frac{\lambda^n}{n!}(T-t)^n\,\mathbb{E}\left[\exp\left(i\alpha\sum_{k=1}^{n}Z_k\right)\right] \\
&= \mathrm{e}^{-\lambda(T-t)}\sum_{n=0}^{\infty}\frac{\lambda^n}{n!}(T-t)^n\left(\mathbb{E}\left[\exp\left(i\alpha Z_1\right)\right]\right)^n \\
&= \exp\left(\lambda(T-t)\,\mathbb{E}\left[\exp\left(i\alpha Z_1\right)\right]\right) \\
&= \exp\left(\lambda(T-t)\int_{-\infty}^{\infty}(\mathrm{e}^{i\alpha y}-1)\nu(dy)\right),
\end{aligned}$$

since $\nu(dy)$ is the probability distribution of Z_1 and $\int_{-\infty}^{\infty}\nu(dy) = 1$. □

From the characteristic function we can compute the expectation and variance of Y_t for fixed t, as

$$\mathbb{E}[Y_t] = \lambda t\,\mathbb{E}[Z_1] \quad \text{and} \quad \mathrm{Var}\,[Y_t] = \lambda t\,\mathbb{E}[|Z_1|^2].$$

For the expectation we have

$$\mathbb{E}[Y_t] = -i\frac{d}{d\alpha}\,\mathbb{E}[\mathrm{e}^{i\alpha Y_t}]_{|\alpha=0} = \lambda t\int_{-\infty}^{\infty}y\nu(dy) = \lambda t\,\mathbb{E}[Z_1].$$

This relation can also be directly recovered as

$$\begin{aligned}
\mathbb{E}[Y_t] &= \mathbb{E}\left[\mathbb{E}\left[\sum_{k=1}^{N_t}Z_k\Big|N_t\right]\right] \\
&= \mathrm{e}^{-\lambda t}\sum_{n=0}^{\infty}\frac{\lambda^n t^n}{n!}\,\mathbb{E}\left[\sum_{k=1}^{n}Z_k\Big|N_t = n\right] \\
&= \mathrm{e}^{-\lambda t}\sum_{n=0}^{\infty}\frac{\lambda^n t^n}{n!}\,\mathbb{E}\left[\sum_{k=1}^{n}Z_k\right] \\
&= \lambda t\mathrm{e}^{-\lambda t}\,\mathbb{E}[Z_1]\sum_{n=1}^{\infty}\frac{(\lambda t)^{n-1}}{(n-1)!}
\end{aligned}$$

$$= \lambda t \, \mathbb{E}[Z_1].$$

More generally one can show that for all $0 \leq t_0 \leq t_1 \cdots \leq t_n$ and $\alpha_1, \ldots, \alpha_n \in \mathbb{R}$ we have

$$\begin{aligned}
\mathbb{E}\left[\prod_{k=1}^{n} e^{i\alpha_k (Y_{t_k} - Y_{t_{k-1}})}\right] &= \exp\left(\lambda \sum_{k=1}^{n} (t_k - t_{k-1}) \int_{-\infty}^{\infty} (e^{i\alpha_k y} - 1)\nu(dy)\right) \\
&= \prod_{k=1}^{n} \exp\left(\lambda (t_k - t_{k-1}) \int_{-\infty}^{\infty} (e^{i\alpha_k y} - 1)\nu(dy)\right) \\
&= \prod_{k=1}^{n} \mathbb{E}\left[e^{i\alpha (Y_{t_k} - Y_{t_{k-1}})}\right].
\end{aligned}$$

This shows in particular that the compound Poisson process $(Y_t)_{t \in \mathbb{R}_+}$ has independent increments, as the standard Poisson process $(N_t)_{t \in \mathbb{R}_+}$.

Since the compensated Poisson process also has centered increments by (14.5), we have the following proposition.

Proposition 14.4 *The compensated compound Poisson process*

$$M_t := Y_t - \lambda t \, \mathbb{E}[Z_1], \qquad t \in \mathbb{R}_+,$$

is a martingale.

By construction, compound Poisson processes only have a *finite* number of jumps on any interval. They belong to the family of *Lévy processes* which may have an infinite number of jumps on any finite time interval, cf. [12].

14.3 Stochastic Integrals with Jumps

Given $(\phi_t)_{t \in \mathbb{R}_+}$ a stochastic process we let the stochastic integral of $(\phi_t)_{t \in \mathbb{R}_+}$ with respect to $(Y_t)_{t \in \mathbb{R}_+}$ be defined by

$$\int_0^T \phi_t dY_t := \sum_{k=1}^{N_T} \phi_{T_k} Z_k.$$

Note that this expression $\int_0^T \phi_t dY_t$ has a natural financial interpretation as the value at time T of a portfolio containing a (possibly fractional) quantity

ϕ_t of a risky asset at time t, whose price evolves according to random returns Z_k at random times T_k.

In particular the compound Poisson process $(Y_t)_{t\in\mathbb{R}_+}$ in (14.1) admits the stochastic integral representation

$$Y_t = Y_0 + \int_0^t Z_{N_s} dN_s.$$

Next, given $(W_t)_{t\in\mathbb{R}_+}$ a standard Brownian motion independent of $(Y_t)_{t\in\mathbb{R}_+}$ and $(X_t)_{t\in\mathbb{R}_+}$ a jump-diffusion process of the form

$$X_t = \int_0^t u_s dW_s + \int_0^t v_s ds + Y_t, \qquad t \in \mathbb{R}_+,$$

where $(\phi_t)_{t\in\mathbb{R}_+}$ is a process which is adapted to the filtration $(\mathcal{F}_t)_{t\in\mathbb{R}_+}$ generated by $(W_t)_{t\in\mathbb{R}_+}$ and $(Y_t)_{t\in\mathbb{R}_+}$, and such that

$$\mathbb{E}\left[\int_0^\infty \phi_s^2 |u_s|^2 ds\right] < \infty \quad \text{and} \quad \mathbb{E}\left[\int_0^\infty |\phi_s v_s| ds\right] < \infty,$$

we let the stochastic integral of $(\phi_s)_{s\in\mathbb{R}_+}$ with respect to $(X_s)_{s\in\mathbb{R}_+}$ be defined by

$$\int_0^T \phi_s dX_s := \int_0^T \phi_s u_s dW_s + \int_0^T \phi_s v_s ds + \sum_{k=1}^{N_T} \phi_{T_k} Z_k, \qquad T > 0.$$

The compound Poisson compensated stochastic integral can be shown to satisfy the Itô isometry

$$\mathbb{E}\left[\left(\int_0^T \phi_{s^-}(dY_s - \lambda\,\mathbb{E}[Z_1]dt)\right)^2\right] = \lambda\,\mathbb{E}[|Z_1|^2]\,\mathbb{E}\left[\int_0^T |\phi|_s^2 ds\right], \tag{14.7}$$

provided the process $(\phi_s)_{s\in\mathbb{R}_+}$ is adapted to the filtration generated by $(Y_t)_{t\in\mathbb{R}_+}$, which makes the left limit process $(\phi_{s^-})_{s\in\mathbb{R}_+}$ *predictable.* The proof of (14.7) is similar to that of Proposition 4.2 in the case of simple predictable processes.

For the mixed continuous-jump martingale

$$X_t = \int_0^t u_s dW_s + Y_t - \lambda t\,\mathbb{E}[Z_1], \qquad t \in \mathbb{R}_+,$$

we have the isometry

$$\mathbb{E}\left[\left(\int_0^T \phi_{s^-} dX_s\right)^2\right] = \mathbb{E}\left[\int_0^T |\phi_s|^2 |u_s|^2 ds\right] + \lambda\,\mathbb{E}[|Z_1|^2]\,\mathbb{E}\left[\int_0^T |\phi_s|^2 ds\right]. \tag{14.8}$$

provided $(\phi_s)_{s\in\mathbb{R}_+}$ is adapted to the filtration $(\mathcal{F}_t)_{t\in\mathbb{R}_+}$ generated by $(W_t)_{t\in\mathbb{R}_+}$ and $(Y_t)_{t\in\mathbb{R}_+}$.

This isometry formula will be used in Section 15.5 for the computation of hedging strategies in jump models.

When $(X_t)_{t\in\mathbb{R}_+}$ takes the form

$$X_t = X_0 + \int_0^t u_s dW_s + \int_0^t v_s ds + \int_0^t \eta_s dY_s, \qquad t \in \mathbb{R}_+,$$

the stochastic integral of $(\phi_t)_{t\in\mathbb{R}_+}$ with respect to $(X_t)_{t\in\mathbb{R}_+}$ satisfies

$$\begin{aligned}\int_0^T \phi_s dX_s &:= \int_0^T \phi_s u_s dW_s + \int_0^T \phi_s v_s ds + \int_0^T \eta_s \phi_s dY_s \\ &= \int_0^T \phi_s u_s dW_s + \int_0^T \phi_s v_s ds + \sum_{k=1}^{N_T} \phi_{T_k} \eta_{T_k} Z_k, \qquad T > 0.\end{aligned}$$

14.4 Itô Formula with Jumps

Let us first consider the case of a standard Poisson process $(N_t)_{t\in\mathbb{R}_+}$ with intensity λ. We have the *telescoping sum*

$$\begin{aligned} f(N_t) &= f(0) + \sum_{k=1}^{N_t} (f(k) - f(k-1)) \\ &= f(0) + \int_0^t (f(1 + N_{s^-}) - f(N_{s^-}))dN_s \\ &= f(0) + \int_0^t (f(N_s) - f(N_{s^-}))dN_s. \end{aligned}$$

Here, N_{s^-} denotes the left limit of the Poisson process at time s, i.e.,

$$N_{s^-} = \lim_{h \searrow 0} N_{s-h}.$$

In particular we have

$$k = N_{T_k} = 1 + N_{T_k^-}, \qquad k \geq 1.$$

By the same argument we find, in the case of the compound Poisson process $(Y_t)_{t\in\mathbb{R}_+}$,

$$\begin{aligned} f(Y_t) &= f(0) + \sum_{k=1}^{N_t}(f(Y_{T_k^-} + Z_k) - f(Y_{T_k^-})) \\ &= f(0) + \int_0^t (f(Z_{N_s} + Y_{s^-}) - f(Y_{s^-}))dN_s \\ &= f(0) + \int_0^t (f(Y_s) - f(Y_{s^-}))dN_s, \end{aligned}$$

which can be decomposed using a compensated Poisson stochastic integral as

$$f(Y_t) = f(0) + \int_0^t (f(Y_s) - f(Y_{s^-}))(dN_s - \lambda ds) + \lambda\int_0^t (f(Y_s) - f(Y_{s^-}))ds.$$

More generally, for a process of the form

$$X_t = X_0 + \int_0^t u_s dW_s + \int_0^t v_s ds + \int_0^t \eta_s dY_s, \qquad t\in\mathbb{R}_+,$$

we find, by combining the Itô formula for Brownian motion with the above argument we get

$$\begin{aligned} f(X_t) &= f(X_0) + \int_0^t u_s f'(X_s)dW_s + \frac{1}{2}\int_0^t f''(X_s)|u_s|^2 ds \\ &\quad + \int_0^t v_s f'(X_s)ds + \sum_{k=1}^{N_T}(f(X_{T_k^-} + \eta_{T_k}Z_k) - f(X_{T_k^-})) \\ &= f(X_0) + \int_0^t u_s f'(X_s)dW_s + \frac{1}{2}\int_0^t f''(X_s)|u_s|^2 ds + \int_0^t v_s f'(X_s)ds \\ &\quad + \int_0^t (f(X_{s^-} + \eta_s Z_{N_s}) - f(X_{s^-}))dN_s \qquad t\in\mathbb{R}_+. \end{aligned}$$

i.e.,

$$\begin{aligned} f(X_t) &= f(X_0) + \int_0^t u_s f'(X_s)dW_s + \frac{1}{2}\int_0^t f''(X_s)|u_s|^2 ds + \int_0^t v_s f'(X_s)ds \\ &\quad + \int_0^t (f(X_s) - f(X_{s^-}))dN_s, \qquad t\in\mathbb{R}_+. \end{aligned} \tag{14.9}$$

For example, in case

$$X_t = \int_0^t u_s dW_s + \int_0^t v_s ds + \int_0^t \eta_s dN_s, \qquad t\in\mathbb{R}_+,$$

we get

$$\begin{aligned} f(X_t) &= f(0) + \int_0^t u_s f'(X_s) dW_s + \frac{1}{2}\int_0^t |u_s|^2 f''(X_s) dW_s \\ &+ \int_0^t v_s f'(X_s) ds + \int_0^t (f(X_{s^-} + \eta_s) - f(X_{s^-})) dN_s \\ &= f(0) + \int_0^t u_s f'(X_s) dW_s + \frac{1}{2}\int_0^t |u_s|^2 f''(X_s) dW_s \\ &+ \int_0^t v_s f'(X_s) ds + \int_0^t (f(X_s) - f(X_{s^-})) dN_s. \end{aligned}$$

For a process of the form

$$X_t = X_0 + \int_0^t u_s dW_s + \int_0^t \eta_s dY_t, \qquad t \in \mathbb{R}_+,$$

the Itô formula with jumps can be rewritten as

$$\begin{aligned} f(X_t) = f(X_0) &+ \int_0^t u_s f'(X_{s^-}) dW_s + \frac{1}{2}\int_0^t f''(X_s)|u_s|^2 ds + \int_0^t \eta_s f'(X_{s^-}) dY_s \\ &+ \int_0^t (f(X_s) - f(X_{s^-}) - \Delta X_s f'(X_{s^-}))\, d(N_s - s) \\ &+ \int_0^t (f(X_s) - f(X_{s^-}) - \Delta X_s f'(X_{s^-}))\, ds, \qquad t \in \mathbb{R}_+, \end{aligned}$$

and this formulation is at the basis of the extension of Itô's formula to Lévy processes with an infinite number of jumps using the bound

$$|f(x+y) - f(x) - y f'(x)| \leq C y^2,$$

for f a $\mathcal{C}_b^2(\mathbb{R})$ function. Such processes, also called "infinite activity Lévy processes" [12] are also useful in financial modeling and include the gamma process, stable processes, variance gamma processes, etc.

14.5 Stochastic Differential Equations with Jumps

Let us start with the simplest example

$$dS_t = \eta S_{t^-} dN_t, \tag{14.10}$$

of a stochastic differential equation with respect to the standard Poisson process, with constant coefficient $\eta \in \mathbb{R}$.

When

$$\Delta N_t = N_t - N_{t^-} = 1,$$

i.e., when the Poisson process has a jump at time t, the equation (14.10) reads

$$dS_t = S_t - S_{t^-} = \eta S_{t^-}, \qquad t > 0.$$

which can be solved to yield

$$S_t = (1+\eta)S_{t^-}, \qquad t > 0.$$

By induction, applying this procedure for each jump time gives us the solution

$$S_t = S_0(1+\eta)^{N_t}, \qquad t \in \mathbb{R}_+.$$

Next, consider the case where η is time-dependent, i.e.,

$$dS_t = \eta_t S_{t^-} dN_t. \tag{14.11}$$

At each jump time T_k, Relation (14.11) reads

$$dS_{T_k} = S_{T_k} - S_{T_k^-} = \eta_{T_k} S_{T_k^-},$$

i.e.,

$$S_{T_k} = (1+\eta_{T_k})S_{T_k^-},$$

and repeating this argument for all $k = 1, \ldots, N_t$ yields the product solution

$$S_t = S_0 \prod_{k=1}^{N_t}(1+\eta_{T_k}) = S_0 \prod_{\substack{\Delta N_s = 1\\ 0\le s\le t}} (1+\eta_s), \qquad t \in \mathbb{R}_+.$$

The equation

$$dS_t = \mu_t S_t dt + \eta_t S_{t^-}(dN_t - \lambda dt), \tag{14.12}$$

is then solved as

$$S_t = S_0 \exp\left(\int_0^t \mu_s ds - \lambda \int_0^t \eta_s ds\right) \prod_{k=1}^{N_t}(1+\eta_{T_k}), \qquad t \in \mathbb{R}_+.$$

A random simulation of the numerical solution of the above equation (14.12) is given in Figure 14.3.

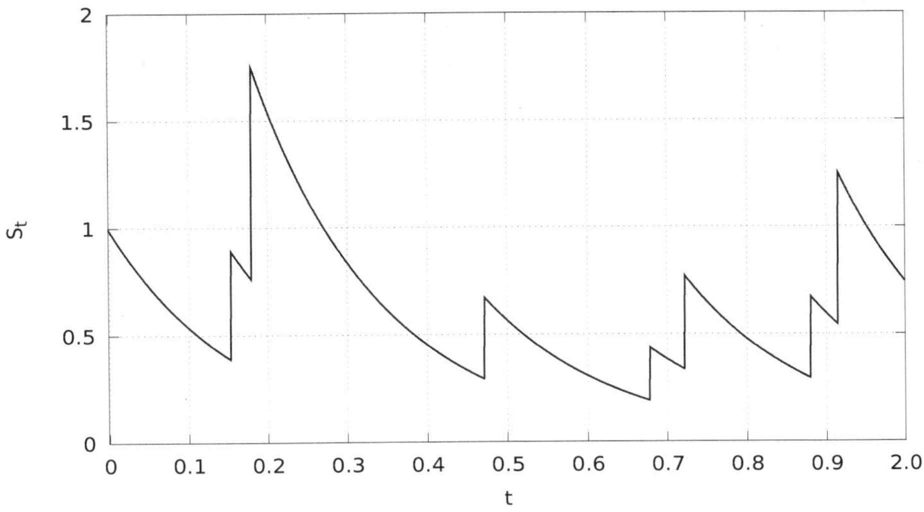

FIGURE 14.3: Geometric Poisson process.

The above simulation can be compared to the real sales ranking data of Figure 14.4. A random simulation of the geometric compound Poisson process

$$S_t = S_0 \exp\left(\int_0^t \mu_s ds - \lambda \mathbb{E}[Z_1] \int_0^t \eta_s ds\right) \prod_{k=1}^{N_t} (1 + \eta_{T_k} Z_k) \qquad t \in \mathbb{R}_+,$$

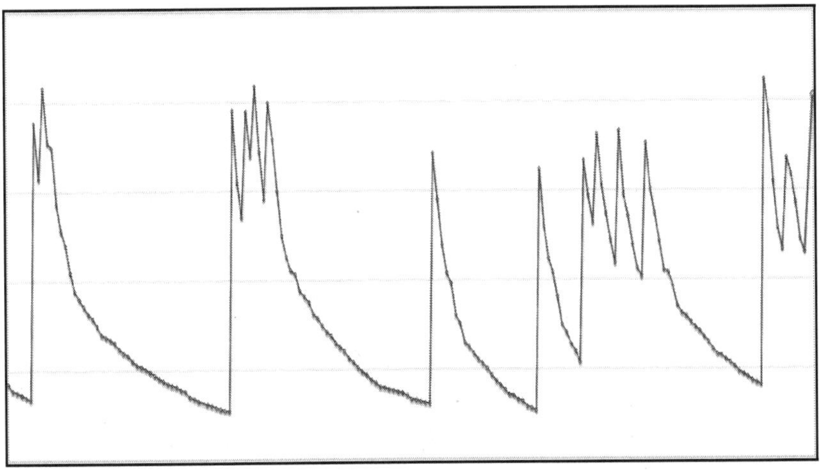

FIGURE 14.4: Ranking data.

solution of

$$dS_t = \mu_t S_t dt + \eta_t S_{t^-}(dY_t - \lambda\,\mathbb{E}[Z_1]dt),$$

is given in Figure 14.5.

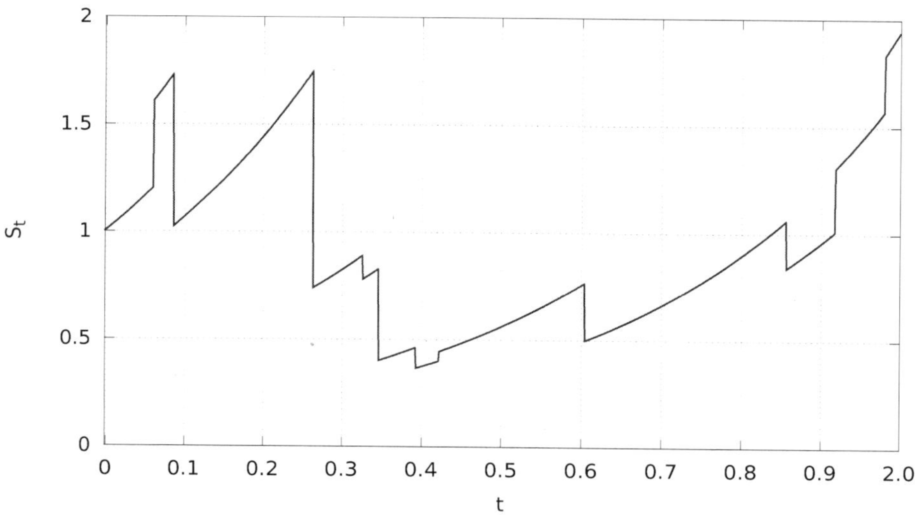

FIGURE 14.5: Geometric compound Poisson process.

In the case of a jump-diffusion stochastic differential equation of the form

$$dS_t = \mu_t S_t dt + \eta_t S_{t^-}(dY_t - \lambda\,\mathbb{E}[Z_1]dt) + \sigma_t S_t dW_t,$$

we get

$$\begin{aligned} S_t &= S_0 \exp\left(\int_0^t \mu_s ds - \lambda\,\mathbb{E}[Z_1]\int_0^t \eta_s ds + \int_0^t \sigma_s dW_s - \frac{1}{2}\int_0^t |\sigma_s|^2 ds\right) \\ &\quad \times \prod_{k=1}^{N_t}(1 + \eta_{T_k} Z_k), \end{aligned}$$

$t \in \mathbb{R}_+$. A random simulation of the geometric Brownian motion with compound Poisson jumps is given in Figure 14.6.

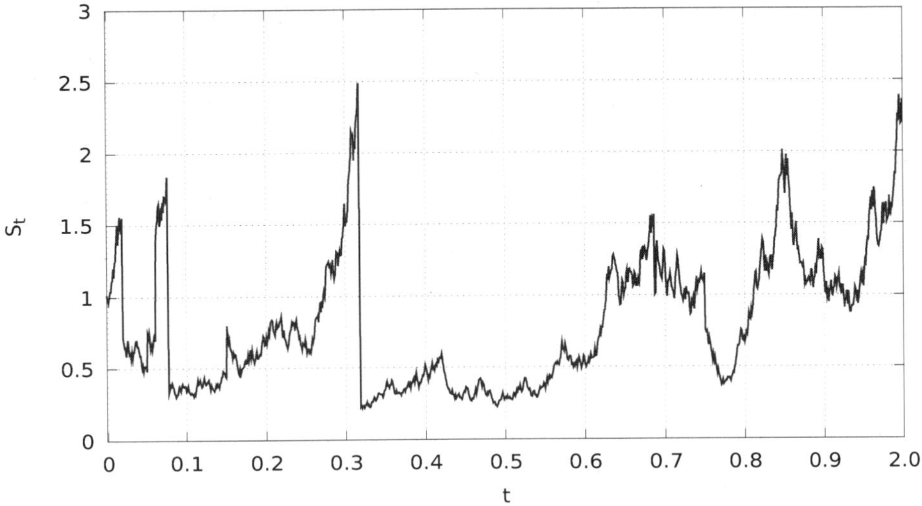

FIGURE 14.6: Geometric Brownian motion with compound Poisson jumps.

By rewriting S_t as

$$\begin{aligned} S_t &= S_0 \exp\left(\int_0^t \mu_s ds + \int_0^t \eta_s (dY_s - \lambda \mathbb{E}[Z_1] ds) + \int_0^t \sigma_s dW_s - \frac{1}{2} \int_0^t |\sigma_s|^2 ds \right) \\ &\quad \times \prod_{k=1}^{N_t} (e^{-\eta_{T_k}} (1 + \eta_{T_k} Z_k)), \end{aligned}$$

$t \in \mathbb{R}_+$, one can extend this jump model to processes with an infinite number of jumps on any finite time interval, cf. [12].

14.6 Girsanov Theorem for Jump Processes

Recall that in its simplest form, the Girsanov theorem for Brownian motion follows from the calculation

$$\begin{aligned} \mathbb{E}[f(W_T - \mu T)] &= \frac{1}{\sqrt{2\pi T}} \int_{-\infty}^{\infty} f(x - \mu T) e^{-x^2/(2T)} dx \\ &= \frac{1}{\sqrt{2\pi T}} \int_{-\infty}^{\infty} f(x) e^{-(x+\mu T)^2/(2T)} dx \\ &= \frac{1}{\sqrt{2\pi T}} \int_{-\infty}^{\infty} f(x) e^{-\mu x - \mu^2 T/2} e^{-x^2/(2T)} dx \\ &= \mathbb{E}[f(W_T) e^{-\mu W_T - \mu^2 T/2}] \\ &= \tilde{\mathbb{E}}[f(W_T)], \end{aligned} \tag{14.13}$$

for any bounded measurable function f on $\mathbb{R}$, which shows that W_T is a Gaussian random variable with mean $-\mu T$ under the probability measure $\tilde{\mathbb{P}}$ defined by

$$d\tilde{\mathbb{P}} = e^{-\mu W_T - \mu^2 T/2} d\mathbb{P},$$

cf. Section 6.2. Equivalently we have

$$\mathbb{E}[f(W_T)] = \tilde{\mathbb{E}}[f(W_T + \mu T)], \tag{14.14}$$

hence

under the probability measure

$$d\tilde{\mathbb{P}} = e^{-\mu W_T - \mu^2 T/2} d\mathbb{P},$$

the random variable $W_T + \mu T$ has a centered Gaussian distribution.

More generally, the Girsanov theorem states that $(W_t + \mu t)_{t\in[0,T]}$ is a standard Brownian motion under $\tilde{\mathbb{P}}$.

When Brownian motion is replaced with a standard Poisson process $(N_t)_{t\in\mathbb{R}_+}$, the above space shift

$$W_t \longmapsto W_t + \mu t$$

may not be used because $N_t + \mu t$ cannot be a Poisson process, whatever the change of probability applied, since by construction, the paths of the standard Poisson process has jumps of unit size and remain constant between jump times.

The correct way to proceed in order to extend (14.14) to the Poisson case is to replace the space shift with a *time contraction* (or dilation) by a certain factor $1 + c$ with $c > -1$, i.e.,

$$N_t \longmapsto N_{t/(1+c)}.$$

By analogy with (14.13) we write

$$\begin{aligned} \mathbb{E}[f(N_{T(1+c)})] &= \sum_{k=0}^{\infty} f(k)\mathbb{P}(N_{T(1+c)} = k) \qquad (14.15)\\ &= e^{-\lambda(1+c)T} \sum_{k=0}^{\infty} f(k) \frac{(\lambda T(1+c))^k}{k!} \\ &= e^{-\lambda T} e^{-\lambda c T} \sum_{k=0}^{\infty} f(k)(1+c)^k \frac{(\lambda T)^k}{k!} \end{aligned}$$

$$\begin{aligned}
&= e^{-\lambda cT} \sum_{k=0}^{\infty} f(k)(1+c)^k \mathbb{P}(N_T = k) \\
&= e^{-\lambda cT}\, \mathbb{E}[f(N_T)(1+c)^{N_T}] \\
&= e^{-\lambda cT} \int_\Omega (1+c)^{N_T} f(N_T) d\mathbb{P} \\
&= \int_\Omega f(N_T) d\tilde{\mathbb{P}} \\
&= \tilde{\mathbb{E}}[f(N_T)],
\end{aligned}$$

for f any bounded function on $\mathbb{N}$, where the probability measure $\tilde{\mathbb{P}}$ is defined by

$$d\tilde{\mathbb{P}} = e^{-\lambda cT}(1+c)^{N_T} d\mathbb{P}.$$

Consequently,

> under the probability measure
>
> $$d\tilde{\mathbb{P}} = e^{-\lambda cT}(1+c)^{N_T} d\mathbb{P},$$
>
> the law of the random variable N_T is that of $N_{T(1+c)}$ under $\mathbb{P}$, i.e., it is a Poisson random variable with intensity $\lambda(1+c)T$.

Equivalently we have

$$\mathbb{E}[f(N_T)] = \tilde{\mathbb{E}}[f(N_{T/(1+c)})],$$

i.e., under $\tilde{\mathbb{P}}$ the law of $N_{T/(1+c)}$ is that of a standard Poisson random variable with parameter λT.

In addition we have

$$\begin{aligned}
N_{t/(1+c)} &= \sum_{n\geq 1} \mathbf{1}_{[T_n,\infty)}(t/(1+c)) \\
&= \sum_{n\geq 1} \mathbf{1}_{[(1+c)T_n,\infty)}(t), \qquad t \in \mathbb{R}_+,
\end{aligned}$$

which shows that under $\tilde{\mathbb{P}}$, the jump times of $(N_{t/(1+c)})_{t\in[0,T]}$ are given by

$$((1+c)T_n)_{n\geq 1},$$

and we know that they are distributed as the jump times of a Poisson process with intensity λ.

Next taking $\tilde{\lambda} > 0$ and

$$c = -1 + \frac{\tilde{\lambda}}{\lambda},$$

we can rewrite the above by saying that

under the probability measure

$$d\tilde{\mathbb{P}} = e^{-\lambda cT}(1+c)^{N_T} d\mathbb{P} = e^{-(\tilde{\lambda}-\lambda)T} \left(\frac{\tilde{\lambda}}{\lambda}\right)^{N_T} d\mathbb{P},$$

the law of N_T is that of a Poisson random variable with intensity

$$\tilde{\lambda}T = \lambda(1+c)T.$$

Consequently, since both $(N_t)\lambda t)_{t\in\mathbb{R}_+}$ and $(N_t-(1+c)\lambda t)_{t\in\mathbb{R}_+}$ are processes with independent increments, the compensated Poisson process

$$N_t - (1+c)\lambda t = N_t - \tilde{\lambda}t$$

is a martingale under $\tilde{\mathbb{P}}$ by (6.2), although when $c \neq 0$ it is not a martingale under $\mathbb{P}$.

In the case of compound Poisson processes the Girsanov theorem can be extended to variations in jump sizes in addition to time variations, and we have the following more general result.

Theorem 14.2 *Let $(Y_t)_{t\geq 0}$ be a compound Poisson process with intensity $\lambda > 0$ and jump distribution $\nu(dx)$. Consider another jump distribution $\tilde{\nu}(dx)$, and let*

$$\phi(x) = \frac{\tilde{\lambda}}{\lambda}\frac{d\tilde{\nu}}{d\nu}(x) - 1, \qquad x \in \mathbb{R}.$$

Then,

under the probability measure

$$d\tilde{\mathbb{P}}_{\tilde{\lambda},\tilde{\nu}} := e^{-(\tilde{\lambda}-\lambda)T} \prod_{k=1}^{N_T} (1+\phi(Z_k)) d\tilde{\mathbb{P}}_{\lambda,\nu},$$

the process

$$Y_t = \sum_{k=1}^{N_t} Z_k, \qquad t \in \mathbb{R}_+,$$

is a compound Poisson process with

- *modified intensity $\tilde{\lambda} > 0$, and*
- *modified jump distribution $\tilde{\nu}(dx)$.*

Proof. For any bounded measurable function f on $\mathbb{R}$, we extend (14.15) to the following change of variable

$$\begin{aligned}
\mathbb{E}_{\tilde{\lambda},\tilde{\nu}}\left[f(Y_T)\right] &= e^{-(\tilde{\lambda}-\lambda)T}\,\mathbb{E}_{\lambda,\nu}\left[f(Y_T)\prod_{i=1}^{N_T}(1+\phi(Z_i))\right] \\
&= e^{-(\tilde{\lambda}-\lambda)T}\sum_{k=0}^{\infty}\mathbb{E}_{\lambda,\nu}\left[f\left(\sum_{i=1}^{k}Z_i\right)\prod_{i=1}^{k}(1+\phi(Z_i))\Big| N_T=k\right]\mathbb{P}(N_T=k) \\
&= e^{-\tilde{\lambda}T}\sum_{k=0}^{\infty}\frac{(\lambda T)^k}{k!}\,\mathbb{E}_{\lambda,\nu}\left[f\left(\sum_{i=1}^{k}Z_i\right)\prod_{i=1}^{k}(1+\phi(Z_i))\right] \\
&= e^{-\tilde{\lambda}T}\sum_{k=0}^{\infty}\frac{(\lambda T)^k}{k!}\int_{-\infty}^{\infty}\cdots\int_{-\infty}^{\infty}f(z_1+\cdots+z_k)\prod_{i=1}^{k}(1+\phi(z_i))\nu(dz_1)\cdots\nu(dz_k) \\
&= e^{-\tilde{\lambda}T}\sum_{k=0}^{\infty}\frac{(\tilde{\lambda}T)^k}{k!}\int_{-\infty}^{\infty}\cdots\int_{-\infty}^{\infty}f(z_1+\cdots+z_k)\left(\prod_{i=1}^{k}\frac{d\tilde{\nu}}{d\nu}(z_i)\right)\nu(dz_1)\cdots\nu(dz_k) \\
&= e^{-\tilde{\lambda}T}\sum_{k=0}^{\infty}\frac{(\tilde{\lambda}T)^k}{k!}\int_{-\infty}^{\infty}\cdots\int_{-\infty}^{\infty}f(z_1+\cdots+z_k)\tilde{\nu}(dz_1)\cdots\tilde{\nu}(dz_k).
\end{aligned}$$

This shows that under $\mathbb{P}_{\tilde{\lambda},\tilde{\nu}}$, Y_T has the distribution of a compound Poisson process with intensity $\tilde{\lambda}$ and jump distribution $\tilde{\nu}$. We refer to Proposition 9.6 of [12] for the independence of increments of $(Y_t)_{t\in\mathbb{R}_+}$ under $\tilde{\mathbb{P}}_{\tilde{\lambda},\tilde{\nu}}$. □

Note that the compound Poisson process with intensity $\tilde{\lambda} > 0$ and jump distribution $\tilde{\nu}$ can be built as

$$X_t := \sum_{k=1}^{N_{\tilde{\lambda}t/\lambda}} h(Z_k),$$

provided $\tilde{\nu}$ is the image measure of ν by the function $h : \mathbb{R} \to \mathbb{R}$, i.e.,

$$\mathbb{P}(h(Z_k) \in A) = \mathbb{P}(Z_k \in h^{-1}(A)) = \nu(h^{-1}(A)) = \tilde{\nu}(A),$$

for all measurable subset A of $\mathbb{R}$.

Compensated Compound Poisson Martingale

As a consequence of Theorem 14.2, the compensated process

$$Y_t - \tilde{\lambda}t\,\mathbb{E}_{\tilde{\nu}}[Z_1]$$

becomes a martingale under the probability measure $\tilde{\mathbb{P}}_{\tilde{\lambda},\tilde{\nu}}$ defined by

$$d\tilde{\mathbb{P}}_{\tilde{\lambda},\tilde{\nu}} = e^{-(\tilde{\lambda}-\lambda)T}\prod_{k=1}^{N_T}(1+\phi(Z_k))d\tilde{\mathbb{P}}_{\lambda,\nu}.$$

Finally, the Girsanov theorem can be extended to the linear combination of a standard Brownian motion $(W_t)_{t\in\mathbb{R}_+}$ and an independent compound Poisson process $(Y_t)_{t\in\mathbb{R}_+}$, as in the following result which is a particular case of Theorem 33.2 of [65].

Theorem 14.3 *Let $(Y_t)_{t\geq 0}$ be a compound Poisson process with intensity $\lambda > 0$ and jump distribution $\nu(dx)$. Consider another jump distribution $\tilde{\nu}(dx)$ and intensity parameter $\tilde{\lambda} > 0$, and let*

$$\phi(x) = \frac{\tilde{\lambda}}{\lambda}\frac{d\tilde{\nu}}{d\nu}(x) - 1, \qquad x \in \mathbb{R},$$

and let $(u_t)_{t\in\mathbb{R}_+}$ be a bounded adapted process. Then the process

$$\left(W_t + \int_0^t u_s ds + Y_t - \tilde{\lambda}\,\mathbb{E}_{\tilde{\nu}}[Z_1]t\right)_{t\in\mathbb{R}_+}$$

is a martingale under the probability measure

$$d\tilde{\mathbb{P}}_{u,\tilde{\lambda},\tilde{\nu}} = \exp\left(-(\tilde{\lambda}-\lambda)T - \int_0^T u_s dW_s - \frac{1}{2}\int_0^T |u_s|^2 ds\right)\prod_{k=1}^{N_T}(1+\phi(Z_k))d\tilde{\mathbb{P}}_{\lambda,\nu}. \tag{14.16}$$

As a consequence of Theorem 14.3, if

$$W_t + \int_0^t v_s ds + Y_t$$

is not a martingale under $\tilde{\mathbb{P}}_{\lambda,\nu}$, it will become a martingale under $\tilde{\mathbb{P}}_{u,\tilde{\lambda},\tilde{\nu}}$ provided $u, \tilde{\lambda}$ and $\tilde{\nu}$ are chosen in such a way that

$$v_s = u_s - \tilde{\lambda}\,\mathbb{E}_{\tilde{\nu}}[Z_1], \qquad s \in \mathbb{R}. \tag{14.17}$$

Moreover, under $\tilde{\mathbb{P}}_{u,\tilde{\lambda},\tilde{\nu}}$, the process $W_t + \int_0^t u_s ds$ is a standard Brownian motion, and $Y_t - \tilde{\lambda}\,\mathbb{E}_{\tilde{\nu}}[Z_1]t$ is a martingale.

When $\tilde{\lambda} = \lambda = 0$, Theorem 14.3 coincides with the usual Girsanov theorem for Brownian motion, in which case (14.17) admits only one solution given by $u = v$ and there is uniqueness of $\tilde{\mathbb{P}}_{u,0,0}$. Note that uniqueness occurs also when $u = 0$ in the absence of Brownian motion with Poisson jumps of fixed size a (i.e., $\tilde{\nu}(dx) = \nu(dx) = \delta_a(dx)$) since in this case (14.17) also admits only one solution $\tilde{\lambda} = v$ and there is uniqueness of $\tilde{\mathbb{P}}_{0,\tilde{\lambda},\delta_a}$. These remarks will be of importance for arbitrage pricing in jump models in Chapter 15.

Exercises

Exercise 14.1 Let $(N_t)_{t\in\mathbb{R}_+}$ be a standard Poisson process with intensity $\lambda > 0$, started at $N_0 = 0$.

1. Solve the stochastic differential equation

$$dS_t = \eta S_{t^-} dN_t - \eta\lambda S_t dt.$$

2. Consider now the compound Poisson process $Y_t := \sum_{k=1}^{N_t} Z_k$, where $(Z_k)_{k\geq 1}$ is an *i.i.d.* sequence of $\mathcal{N}(0,1)$ Gaussian random variables. Let $r > 0$. Solve the stochastic differential equation

$$dS_t = rS_t dt + \eta S_{t^-} dY_t.$$

Exercise 14.2 Show, by direct computation or using the characteristic function, that the variance of the compound Poisson process Y_t with intensity $\lambda > 0$ satisfies

$$\operatorname{Var}[Y_t] = \lambda t\, \mathbb{E}[|Z_1|^2] = \lambda t \int_{-\infty}^{\infty} x^2 \nu(dx).$$

Exercise 14.3 Consider an exponential compound Poisson process of the form

$$S_t = S_0 e^{\mu t + \sigma W_t + Y_t}, \qquad t \in \mathbb{R}_+,$$

where $(Y_t)_{t\in\mathbb{R}_+}$ is a compound Poisson process of the form (14.6).

1. Derive the stochastic differential equation with jumps satisfied by $(S_t)_{t\in\mathbb{R}_+}$.

2. Let $r > 0$. Find a family $(\tilde{\mathbb{P}}_{u,\tilde{\lambda},\tilde{\nu}})$ of probability measures under which the discounted asset price $e^{-rt}S_t$ is a martingale.

Exercise 14.4 Consider $(N_t)_{t\in\mathbb{R}_+}$ a standard Poisson process with intensity $\lambda > 0$, independent of $(W_t)_{t\in\mathbb{R}_+}$, under a probability measure $\mathbb{P}$. Let $(S_t)_{t\in\mathbb{R}_+}$ be defined by the stochastic differential equation

$$dS_t = \mu S_t dt + Y_{N_t} S_{t^-} dN_t, \tag{14.18}$$

where $(Y_k)_{k\geq 1}$ is an i.i.d. sequence of random variables of the form $Y_k = e^{X_k} - 1$ where $X_k \simeq \mathcal{N}(0, \sigma^2)$, $k \geq 1$.

1. Solve the equation (14.18).

2. We assume that μ and the risk-free rate $r > 0$ are chosen such that the discounted process $(e^{-rt}S_t)_{t\in\mathbb{R}_+}$ is a martingale under $\mathbb{P}$. What relation does this impose on μ and r?

3. Under the relation of Question (2), compute the price at time t of a European call option on S_T with strike κ and maturity T, using a series expansion of Black–Scholes functions.

Chapter 15

Pricing and Hedging in Jump Models

In this chapter we consider the problem of option pricing and hedging in jump-diffusion models. In comparison with the continuous case the situation is further complicated by the existence of multiple risk-neutral measures. As a consequence, perfect replicating hedging strategies cannot be computed in general.

15.1 Risk-Neutral Measures

Consider an asset price modeled by the equation,

$$dS_t = \mu S_t dt + \sigma S_t dW_t + S_{t^-} dY_t, \tag{15.1}$$

where $(Y_t)_{t\in\mathbb{R}_+}$ is the compound Poisson process defined in Section 14.2, with jump size distribution $\nu(dx)$ under $\mathbb{P}_\nu$. The equation (15.1) has for a solution

$$S_t = S_0 \exp\left(\mu t + \sigma W_t - \frac{1}{2}\sigma^2 t\right) \prod_{k=1}^{N_t} (1 + Z_k), \tag{15.2}$$

$t \in \mathbb{R}_+$. An important issue for non-abitrage pricing is to determine a risk-neutral probability measure $\mathbb{P}^*$ under which the discounted process $(e^{-rt}S_t)_{t\in\mathbb{R}_+}$ is a martingale, and this goal can be achieved using the Girsanov theorem for jump processes, cf. Section 14.6.

We have

$$\begin{aligned} d(e^{-rt}S_t) &= -re^{-rt}S_t dt + e^{-rt}dS_t \\ &= (\mu - r)e^{-rt}S_t dt + \sigma e^{-rt}S_t dW_t + e^{-rt}S_{t^-} dY_t \\ &= (\mu - r + \lambda\,\mathbb{E}_\nu[Z_1])e^{-rt}S_t dt + \sigma e^{-rt}S_t dW_t + e^{-rt}S_{t^-}(dY_t - \lambda\,\mathbb{E}_\nu[Z_1]dt), \end{aligned}$$

which yields a martingale under $\mathbb{P}$ provided

$$\mu - r + \lambda\,\mathbb{E}_\nu[Z_1] = 0;$$

however, that condition may not be satisfied under $\mathbb{P}_\nu$ by the market parameters.

In that case a change of measure might be needed. In order for the discounted process $(e^{-rt}S_t)_{t\in\mathbb{R}_+}$ to be a martingale, we may choose a drift parameter $u \in \mathbb{R}$, and intensity $\tilde{\lambda} > 0$, and a jump distribution $\tilde{\nu}$ satisfying

$$\mu - r = \sigma u - \tilde{\lambda}\,\mathbb{E}_{\tilde{\nu}}[Z_1]. \tag{15.3}$$

The Girsanov theorem for jump processes then shows that

$$dW_t + udt + dY_t - \tilde{\lambda}\,\mathbb{E}_{\tilde{\nu}}[Z_1]dt$$

is a martingale under the probability measure $\mathbb{P}_{u,\tilde{\lambda},\tilde{\nu}}$ defined in (14.16). Consequently the discounted asset price

$$\begin{aligned} d(e^{-rt}S_t) &= (\mu - r)e^{-rt}S_t dt + \sigma e^{-rt}S_t dW_t + e^{-rt}S_{t^-}dY_t \\ &= \sigma e^{-rt}S_t(dW_t + udt) + e^{-rt}S_{t^-}(dY_t - \tilde{\lambda}\,\mathbb{E}_{\tilde{\nu}}[Z_1]dt), \end{aligned}$$

is a martingale under $\mathbb{P}_{u,\tilde{\lambda},\tilde{\nu}}$.

In this setting the non-uniqueness of the risk neutral measure is apparent since additional degrees of freedom are involved in the choices of u, λ and the measure $\tilde{\nu}$, whereas in the continuous case the choice of $u = (\mu - r)/\sigma$ in (6.4) was unique.

15.2 Pricing in Jump Models

Recall that a market is without arbitrage if and only if it admits at least one risk-neutral measure.

Consider the probability measure $\mathbb{P}_{u,\tilde{\lambda},\tilde{\nu}}$ built in the previous section, under which the discounted asset price

$$d(e^{-rt}S_t) = e^{-rt}S_{t^-}(dY_t - \tilde{\lambda}\,\mathbb{E}_\nu[Z_1]dt) + \sigma e^{-rt}S_t d\tilde{W}_t,$$

is a martingale, and $\tilde{W}_t = W_t + udt$ is a standard Brownian motion under $\mathbb{P}_{u,\tilde{\lambda},\tilde{\nu}}$.

Then the arbitrage price of a claim with payoff C is given by

$$e^{-r(T-t)}\,\mathbb{E}_{u,\tilde{\lambda},\tilde{\nu}}[C \mid \mathcal{F}_t] \tag{15.4}$$

under $\mathbb{P}_{u,\tilde{\lambda},\tilde{\nu}}$.

Clearly the price (15.4) of C is no longer unique in the presence of jumps due to the infinity of choices satisfying the martingale condition (15.3), and such a market is not complete, except if either $\tilde{\lambda} = \lambda = 0$, or ($\sigma = 0$ and $\tilde{\nu} = \nu = \delta_1$).

Pricing of Vanilla Options

The price of a vanilla option with payoff of the form $\phi(S_T)$ on the underlying asset S_T can be written from (15.4) as

$$e^{-r(T-t)} \mathbb{E}_{u,\tilde{\lambda},\tilde{\nu}}[\phi(S_T) \mid \mathcal{F}_t], \tag{15.5}$$

where the expectation can be computed as

$$\begin{aligned}
&\mathbb{E}_{u,\tilde{\lambda},\tilde{\nu}}[\phi(S_T) \mid \mathcal{F}_t] \\
&= \mathbb{E}_{u,\tilde{\lambda},\tilde{\nu}}\left[\phi\left(S_0 \exp\left(\mu T + \sigma W_T - \frac{1}{2}\sigma^2 T\right)\prod_{k=1}^{N_T}(1+Z_k)\right) \Big| \mathcal{F}_t\right] \\
&= \mathbb{E}_{u,\tilde{\lambda},\tilde{\nu}}\left[\phi\left(S_t \exp\left(\mu(T-t) + \sigma(W_T - W_t) - \frac{1}{2}\sigma^2(T-t)\right)\prod_{k=N_t}^{N_T}(1+Z_k)\right) \Big| \mathcal{F}_t\right] \\
&= \mathbb{E}_{u,\tilde{\lambda},\tilde{\nu}}\left[\phi\left(x \exp\left(\mu(T-t) + \sigma(W_T - W_t) - \frac{1}{2}\sigma^2(T-t)\right)\prod_{k=N_t}^{N_T}(1+Z_k)\right)\right]_{x=S_t} \\
&= \sum_{n=0}^{\infty} \mathbb{P}_{u,\tilde{\lambda},\tilde{\nu}}(N_T - N_t = n) \\
&\quad \mathbb{E}_{u,\tilde{\lambda},\tilde{\nu}}\left[\phi\left(x e^{\mu(T-t)+\sigma(W_T-W_t)-\frac{1}{2}\sigma^2(T-t)}\prod_{k=N_t}^{N_T}(1+Z_k)\right) \Big| N_T - N_t = n\right]_{x=S_t} \\
&= e^{-\tilde{\lambda}(T-t)} \sum_{n=0}^{\infty} \frac{(\tilde{\lambda}(T-t))^n}{n!} \\
&\quad \times \mathbb{E}_{u,\tilde{\lambda},\tilde{\nu}}\left[\phi\left(x e^{\mu(T-t)+\sigma(W_T-W_t)-\frac{1}{2}\sigma^2(T-t)}\prod_{k=1}^{n}(1+Z_k)\right)\right]_{x=S_t} \\
&= e^{-\tilde{\lambda}(T-t)} \sum_{n=0}^{\infty} \frac{(\tilde{\lambda}(T-t))^n}{n!} \int_{-\infty}^{\infty} \cdots \int_{-\infty}^{\infty} \\
&\quad \mathbb{E}_{u,\tilde{\lambda},\tilde{\nu}}\left[\phi\left(x e^{\mu(T-t)+\sigma(W_T-W_t)-\frac{1}{2}\sigma^2(T-t)}\prod_{k=1}^{n}(1+z_k)\right)\right]_{x=S_t} \tilde{\nu}(dz_1)\cdots\tilde{\nu}(dz_n),
\end{aligned}$$

hence

$$\begin{aligned}
&e^{-r(T-t)} \mathbb{E}_{0,\tilde{\lambda},\tilde{\nu}}[\phi(S_T) \mid \mathcal{F}_t] \\
&= \frac{1}{\sqrt{2\pi(T-t)}} e^{-(r+\tilde{\lambda})(T-t)} \sum_{n=0}^{\infty} \frac{(\tilde{\lambda}(T-t))^n}{n!} \int_{-\infty}^{\infty} \cdots \int_{-\infty}^{\infty} \\
&\quad \phi\left(S_t e^{\mu(T-t)+\sigma(x-u)-\frac{1}{2}\sigma^2(T-t)}\prod_{k=1}^{n}(1+z_k)\right) e^{-\frac{x^2}{2(T-t)}} \tilde{\nu}(dz_1)\cdots\tilde{\nu}(dz_n)dx.
\end{aligned}$$

15.3 Black–Scholes PDE with Jumps

Recall that by the Markov property of $(S_t)_{t\in\mathbb{R}_+}$ the price (15.5) at time t of the option with payoff $\phi(S_T)$ can be written as a function $f(t,S_t)$ of t and S_t, i.e.,

$$f(t,S_t) = e^{-r(T-t)}\,\mathbb{E}_{u,\tilde\lambda,\tilde\nu}[\phi(S_T)\mid\mathcal{F}_t], \tag{15.6}$$

with the terminal condition $f(T,x)=\phi(x)$. In addition,

$$t \longmapsto e^{r(T-t)} f(t,S_t)$$

is a martingale under $\mathbb{P}_{u,\tilde\lambda,\tilde\nu}$ by the same argument as in (6.1).

In this section we derive a partial integro-differential equation (PIDE) for the function $(t,x)\longmapsto f(t,x)$. We have

$$dS_t = rS_t dt + \sigma S_t d\tilde W_t + S_{t^-}(dY_t - \tilde\lambda\,\mathbb{E}_{\tilde\nu}[Z_1]dt), \tag{15.7}$$

where $\tilde W_t = W_t + ut$ is a standard Brownian motion under $\mathbb{P}_{u,\tilde\lambda,\tilde\nu}$.

Hence the Itô formula with jumps (14.9) shows that

$$\begin{aligned}
df(t,S_t) &= \frac{\partial f}{\partial t}(t,S_t)dt + rS_t\frac{\partial f}{\partial x}(t,S_t)dt + \sigma S_t\frac{\partial f}{\partial x}(t,S_t)d\tilde W_t\sigma^2 S_t^2\frac{\partial^2 f}{\partial x^2}(t,S_t)dt\\
&\quad +\frac{1}{2} - \tilde\lambda\,\mathbb{E}_{\tilde\nu}[Z_1]S_t\frac{\partial f}{\partial x}(t,S_t)dt + (f(t,S_{t^-}(1+Z_{N_t})) - f(t,S_{t^-}))dN_t\\
&= \sigma S_t\frac{\partial f}{\partial x}(t,S_t)d\tilde W_t\\
&\quad +(f(t,S_{t^-}(1+Z_{N_t})) - f(t,S_{t^-}))dN_t - \tilde\lambda\,\mathbb{E}_{\tilde\nu}[(f(t,x(1+Z_1))\\
&\qquad -f(t,x))]_{x=S_t}dt + \frac{\partial f}{\partial t}(t,S_t)dt + rS_t\frac{\partial f}{\partial x}(t,S_t)dt + \frac{1}{2}\sigma^2 S_t^2\frac{\partial^2 f}{\partial x^2}(t,S_t)dt\\
&\quad +\tilde\lambda\,\mathbb{E}_{\tilde\nu}[(f(t,x(1+Z_1)) - f(t,x))]_{x=S_t}dt - \tilde\lambda\,\mathbb{E}_{\tilde\nu}[Z_1]S_t\frac{\partial f}{\partial x}(t,S_t)dt.
\end{aligned}$$

Based on the relation

$$d(e^{r(T-t)}f(t,S_t)) = -re^{r(T-t)}f(t,S_t) + e^{r(T-t)}df(t,S_t)$$

and the facts that the Brownian motion $(\tilde W_t)_{t\in\mathbb{R}_+}$, the differential

$$(f(t,S_{t^-}(1+Z_{N_t})) - f(t,S_{t^-}))dN_t - \tilde\lambda\,\mathbb{E}_{\tilde\nu}[(f(t,x(1+Z_1)) - f(t,x))]_{x=S_t}dt$$

and the process $t\mapsto e^{r(T-t)}f(t,S_t)$ all represent martingales under $\mathbb{P}_{u,\tilde\lambda,\tilde\nu}$, we conclude to the vanishing

$$-rf(t,S_t) + \frac{\partial f}{\partial t}(t,S_t) + rS_t\frac{\partial f}{\partial x}(t,S_t) + \frac{1}{2}\sigma^2 S_t^2\frac{\partial^2 f}{\partial x^2}(t,S_t)$$

$$+\tilde{\lambda}\,\mathbb{E}_{\tilde{\nu}}[(f(t,x(1+Z_1))-f(t,x))]_{x=S_t}-\tilde{\lambda}\,\mathbb{E}_{\tilde{\nu}}[Z_1]S_t\frac{\partial f}{\partial x}(t,S_t)=0,$$

or

$$-rf(t,x)+\frac{\partial f}{\partial t}(t,x)+rx\frac{\partial f}{\partial x}(t,x)+\frac{1}{2}\sigma^2x^2\frac{\partial^2 f}{\partial x^2}(t,x)$$
$$+\tilde{\lambda}\,\mathbb{E}_{\tilde{\nu}}[(f(t,x(1+Z_1))-f(t,x))]-\tilde{\lambda}\,\mathbb{E}_{\tilde{\nu}}[Z_1]x\frac{\partial f}{\partial x}(t,x)=0,$$

which leads to the Partial *Integro-Differential* Equation (PIDE)

$$\begin{aligned} rf(t,x) &= \frac{\partial f}{\partial t}(t,x)+rx\frac{\partial f}{\partial x}(t,x)+\frac{1}{2}\sigma^2x^2\frac{\partial^2 f}{\partial x^2}(t,x)\\ &\quad +\tilde{\lambda}\int_{-\infty}^{\infty}\left(f(t,x(1+y))-f(t,x)-yx\frac{\partial f}{\partial x}(t,x)\right)\tilde{\nu}(dy), \end{aligned} \tag{15.8}$$

under the terminal condition $f(T,x)=\phi(x)$.

In addition we found that the change $df(t,S_t)$ in the portfolio price (15.6) is given by

$$\begin{aligned} df(t,S_t) &= \sigma S_t\frac{\partial f}{\partial x}(t,S_t)d\tilde{W}_t+rf(t,S_t)dt+(f(t,S_{t^-}(1+Z_{N_t}))\\ &\quad -f(t,S_{t^-}))dN_t-\tilde{\lambda}\,\mathbb{E}_{\tilde{\nu}}[(f(t,x(1+Z_1))-f(t,x))]_{x=S_t}dt. \end{aligned} \tag{15.9}$$

In the case of Poisson jumps with fixed size a, i.e., $Y_t=aN_t$, $\nu(dx)=\delta_a(dx)$, the PIDE (15.8) reads

$$\begin{aligned} rf(t,x) &= \frac{\partial f}{\partial t}(t,x)+rx\frac{\partial f}{\partial x}(t,x)+\frac{1}{2}\sigma^2x^2\frac{\partial^2 f}{\partial x^2}(t,x)\\ &\quad +\tilde{\lambda}\left(f(t,x(1+a))-f(t,x)-ax\frac{\partial f}{\partial x}(t,x)\right), \end{aligned}$$

and we have

$$\begin{aligned} df(t,S_t) &= \sigma S_t\frac{\partial f}{\partial x}(t,S_t)d\tilde{W}_t+rf(t,S_t)dt\\ &\quad +(f(t,S_{t^-}(1+a))-f(t,S_{t^-}))dN_t-\tilde{\lambda}(f(t,S_t(1+a))-f(t,S_t))dt. \end{aligned}$$

15.4 Exponential Models

Instead of modeling the asset price $(S_t)_{t\in\mathbb{R}_+}$ through a stochastic exponential (15.2) solution of the stochastic differential equation with jumps of the form

(15.1), we may consider an exponential price process of the form

$$S_t = S_0 e^{\mu t + \sigma W_t + Y_t} = S_0 \exp\left(\mu t + \sigma W_t + \sum_{k=1}^{N_t} Z_k\right), \qquad t \in \mathbb{R}_+,$$

and choose a risk-neutral measure $\mathbb{P}_{u,\tilde{\lambda},\tilde{\nu}}$ under which $(e^{-rt}S_t)_{t\in\mathbb{R}_+}$ is a martingale. Then the expectation

$$e^{-r(T-t)}\,\mathbb{E}_{u,\tilde{\lambda},\tilde{\nu}}[\phi(S_T) \mid \mathcal{F}_t]$$

also becomes a (non-unique) arbitrage price at time $t \in [0,T]$ for the contingent claim with payoff $\phi(S_T)$.

Such an arbitrage price can be expressed as

$$\begin{aligned}
e^{-r(T-t)}\,\mathbb{E}_{u,\tilde{\lambda},\tilde{\nu}}[\phi(S_T) \mid \mathcal{F}_t] &= e^{-r(T-t)}\,\mathbb{E}_{u,\tilde{\lambda},\tilde{\nu}}[\phi(S_0 e^{\mu T + \sigma W_T + Y_T}) \mid \mathcal{F}_t] \\
&= e^{-r(T-t)}\,\mathbb{E}_{u,\tilde{\lambda},\tilde{\nu}}[\phi(S_t e^{\mu(T-t)+\sigma(W_T-W_t)+Y_T-Y_t}) \mid \mathcal{F}_t] \\
&= e^{-r(T-t)}\,\mathbb{E}_{u,\tilde{\lambda},\tilde{\nu}}[\phi(x e^{\mu(T-t)+\sigma(W_T-W_t)+Y_T-Y_t})]_{x=S_t} \\
&= e^{-r(T-t)-\tilde{\lambda}(T-t)} \\
&\quad \times \sum_{n=0}^{\infty} \frac{(\tilde{\lambda}T)^n}{n!}\,\mathbb{E}_{u,\tilde{\lambda},\tilde{\nu}}\left[\phi\left(x e^{\mu(T-t)+\sigma(W_T-W_t)} \exp\left(\sum_{k=1}^{n} Z_k\right)\right)\right]_{x=S_t}.
\end{aligned}$$

In the exponential model

$$S_t = S_0 e^{\tilde{\mu} t + \sigma W_t + Y_t} = S_0 e^{(\tilde{\mu}+\sigma^2/2)t + \sigma W_t - \sigma^2 t/2 + Y_t}$$

the process S_t satisfies

$$dS_t = \left(\tilde{\mu} + \frac{1}{2}\sigma^2\right) S_t dt + \sigma S_t dW_t + S_{t^-}(e^{\Delta Y_t} - 1) dN_t,$$

hence S_t has jumps of size $S_{T_k^-}(e^{Z_k} - 1)$, $k \geq 1$, and (15.3) reads

$$\tilde{\mu} + \frac{1}{2}\sigma^2 - r = \sigma u - \tilde{\lambda}\,\mathbb{E}_{\tilde{\nu}}[e^{Z_1} - 1].$$

The Merton Model

We assume that $(Z_k)_{k\geq 1}$ is a family of independent identically distributed Gaussian $\mathcal{N}(\delta, \eta^2)$ random variables under $\mathbb{P}_{u,\tilde{\lambda},\tilde{\nu}}$ with

$$\tilde{\mu} + \frac{1}{2}\sigma^2 - r = \sigma u - \tilde{\lambda}\,\mathbb{E}_{\tilde{\nu}}[e^{Z_1} - 1] = \sigma u - \tilde{\lambda}(e^{\delta + \eta^2/2} - 1),$$

from (15.3), hence is a standard Brownian motion under $\mathbb{P}_{u,\tilde{\lambda},\tilde{\nu}}$. For simplicity we choose $u = 0$, i.e.,

$$\tilde{\mu} = r - \frac{1}{2}\sigma^2 - \tilde{\lambda}(e^{\delta + \eta^2/2} - 1),$$

Hence we have

$$\begin{aligned}
& e^{-r(T-t)} \mathbb{E}_{\tilde{\lambda},\tilde{\nu}}[\phi(S_T) \mid \mathcal{F}_t] \\
&= e^{-r(T-t)-\tilde{\lambda}(T-t)} \sum_{n=0}^{\infty} \frac{(\tilde{\lambda}(T-t))^n}{n!} \\
&\quad \times \mathbb{E}_{\tilde{\lambda},\tilde{\nu}} \left[\phi \left(x e^{\tilde{\mu}(T-t)+\sigma(W_T-W_t)} \exp \left(\sum_{k=1}^{n} Z_k \right) \right) \right]_{x=S_t} \\
&= e^{-r(T-t)-\tilde{\lambda}(T-t)} \sum_{n=0}^{\infty} \frac{(\tilde{\lambda}(T-t))^n}{n!} \mathbb{E}\left[\phi(x e^{\tilde{\mu}(T-t)+n\delta+X}) \right]_{x=S_t},
\end{aligned}$$

where

$$X = \sigma(W_T - W_t) + \sum_{k=1}^{n} (Z_k - \delta) \simeq \mathcal{N}(0, \sigma^2(T-t) + n\eta^2)$$

is a centered Gaussian random variable with variance

$$v^2 = \sigma^2(T-t) + \sum_{k=1}^{n} \operatorname{Var} Z_k = \sigma^2(T-t) + n\eta^2.$$

Hence when $\phi(x) = (x-\kappa)^+$, using the relation

$$\mathrm{BS}(x, \kappa, v^2/\tau, r, \tau) = e^{-r\tau} \mathbb{E}[(x e^{X - v^2/2 + r\tau} - K)^+]$$

we get

$$\begin{aligned}
& e^{-r(T-t)-\tilde{\lambda}(T-t)} \mathbb{E}_{\tilde{\lambda},\tilde{\nu}}[(S_T - \kappa)^+ \mid \mathcal{F}_t] \\
&= e^{-r(T-t)-\tilde{\lambda}(T-t)} \sum_{n=0}^{\infty} \frac{(\tilde{\lambda}(T-t))^n}{n!} \mathbb{E}\left[(x e^{\tilde{\mu}(T-t)+n\delta+X} - \kappa)^+ \right]_{x=S_t} \\
&= e^{-r(T-t)-\tilde{\lambda}(T-t)} \sum_{n=0}^{\infty} \frac{(\tilde{\lambda}(T-t))^n}{n!} \\
&\quad \times \mathbb{E}\left[(x e^{(r-\frac{1}{2}\sigma^2-\tilde{\lambda}(e^{\delta+\eta^2/2}-1))(T-t)+n\delta+X} - \kappa)^+ \right]_{x=S_t} \\
&= e^{-r(T-t)-\tilde{\lambda}(T-t)} \sum_{n=0}^{\infty} \frac{(\tilde{\lambda}(T-t))^n}{n!} \\
&\quad \times \mathbb{E}\left[(x e^{n\delta+n\eta^2/2-\tilde{\lambda}(e^{\delta+\eta^2/2}-1)(T-t)+X-v^2/2+r(T-t)} - \kappa)^+ \right]_{x=S_t} \\
&= e^{-\tilde{\lambda}(T-t)} \sum_{n=0}^{\infty} \frac{(\tilde{\lambda}(T-t))^n}{n!} \\
&\quad \times \mathrm{BS}(S_t e^{n\delta+\frac{1}{2}n\eta^2-\tilde{\lambda}(e^{\delta+\eta^2/2}-1)(T-t)}, \kappa, \sigma^2 + n\eta^2/(T-t), r, T-t).
\end{aligned}$$

We may also write

$$
\begin{aligned}
& e^{-r(T-t)-\tilde{\lambda}(T-t)} \mathbb{E}_{\tilde{\lambda},\tilde{\nu}}[(S_T-\kappa)^+ \mid \mathcal{F}_t] \\
= & \ e^{-\tilde{\lambda}(T-t)} \sum_{n=0}^{\infty} \frac{(\tilde{\lambda}(T-t))^n}{n!} e^{n\delta+\frac{1}{2}n\eta^2-\tilde{\lambda}(e^{\delta+\eta^2/2}-1)(T-t)} \\
& \times \mathrm{BS}\left(S_t, \kappa e^{-n\delta-\frac{1}{2}n\eta^2+\tilde{\lambda}(e^{\delta+\eta^2/2}-1)(T-t)}, \sigma^2+n\eta^2/(T-t), r, T-t\right) \\
= & \ e^{-\tilde{\lambda}e^{\delta+\eta^2/2}(T-t)} \sum_{n=0}^{\infty} \frac{(\tilde{\lambda}e^{\delta+\frac{1}{2}n\eta^2}(T-t))^n}{n!} \\
& \times \mathrm{BS}\left(S_t, \kappa, \sigma^2+n\eta^2/(T-t), r+n\frac{\delta+\eta^2/2}{T-t}-\tilde{\lambda}(e^{\delta+\eta^2/2}-1), T-t\right).
\end{aligned}
$$

15.5 Self-Financing Hedging with Jumps

Consider a portfolio with value

$$V_t = \eta_t e^{rt} + \xi_t S_t$$

at time t, and satisfying the self-financing condition

$$dV_t = r\eta_t e^{rt} dt + \xi_t dS_t,$$

cf. Relation (5.1). When the portfolio hedges the claim $\phi(S_T)$ we must have $V_t = f(t, S_t)$ for all times $t \in [0,T]$ hence, by (15.7),

$$
\begin{aligned}
dV_t &= df(t, S_t) \\
&= r\eta_t e^{rt} dt + \xi_t dS_t \\
&= r\eta_t e^{rt} dt + \xi_t (rS_t dt + \sigma S_t d\tilde{W}_t + S_{t^-}(dY_t - \tilde{\lambda}\,\mathbb{E}_{\tilde{\nu}}[Z_1]dt)) \\
&= rV_t dt + \sigma \xi_t S_t d\tilde{W}_t + \xi_t S_{t^-}(dY_t - \tilde{\lambda}\,\mathbb{E}_{\tilde{\nu}}[Z_1]dt) \\
&= rf(t, S_t)dt + \sigma \xi_t S_t d\tilde{W}_t + \xi_t S_{t^-}(dY_t - \tilde{\lambda}\,\mathbb{E}_{\tilde{\nu}}[Z_1]dt), \qquad (15.10)
\end{aligned}
$$

has to match

$$
\begin{aligned}
df(t, S_t) &= rf(t, S_t)dt + \sigma S_t \frac{\partial f}{\partial x}(t, S_t) d\tilde{W}_t \qquad (15.11) \\
& + (f(t, S_{t^-}(1+Z_{N_t})) - f(t, S_{t^-}))dN_t - \tilde{\lambda}\,\mathbb{E}_{\tilde{\nu}}[(f(t, x(1+Z_1)) - f(t,x))]_{x=S_t} dt,
\end{aligned}
$$

which is obtained from (15.9).

In such a situation we say that the claim C can be exactly replicated.

Exact replication is possible in essentially only two situations:

(i) Continuous market, $\lambda = \tilde{\lambda} = 0$. In this case we find the usual Black–Scholes Delta:

$$\xi_t = \frac{\partial f}{\partial x}(t, S_t). \tag{15.12}$$

(ii) Poisson jump market, $\sigma = 0$ and $Y_t = aN_t$, $\nu(dx) = \delta_a(dx)$. In this case we find

$$\xi_t = \frac{1}{aS_{t^-}}(f(t, S_{t^-}(1+a)) - f(t, S_{t^-})). \tag{15.13}$$

Note that in the limit $a \to 0$ this expression recovers the Black–Scholes Delta formula (15.12).

When Conditions (i) or (ii) above are not satisfied, exact replication is not possible and this results into an hedging error given from (15.10) and (15.11) by

$$\begin{aligned}
V_T - \phi(S_T) &= V_T - f(T, S_T) \\
&= V_0 - f(0, S_0) + \int_0^T dV_t - \int_0^T df(t, S_t) \\
&= V_0 - f(0, S_0) + \sigma \int_0^T S_t \left(\xi_t - \frac{\partial f}{\partial x}(t, S_t) \right) d\tilde{W}_t \\
&\quad + \int_0^T \xi_t S_{t^-}(Z_{N_t} dN_t - \tilde{\lambda} \mathbb{E}_{\tilde{\nu}}[Z_1] dt) \\
&\quad - \int_0^T (f(t, S_{t^-}(1 + Z_{N_t})) - f(t, S_{t^-})) dN_t \\
&\quad + \tilde{\lambda} \int_0^T \mathbb{E}_{\tilde{\nu}}[(f(t, x(1+Z_1)) - f(t, x))]_{x=S_t} dt.
\end{aligned}$$

Assuming for simplicity that $Y_t = aN_t$, i.e., $\nu(dx) = \delta_a(dx)$, we get

$$\begin{aligned}
V_T - f(T, S_T) &= V_0 - f(0, S_0) + \sigma \int_0^T S_t \left(\xi_t - \frac{\partial f}{\partial x}(t, S_t) \right) d\tilde{W}_t \\
&\quad - \int_0^T (f(t, S_{t^-}(1+a)) - f(t, S_{t^-}) - a\xi_t S_{t^-})(dN_t - \tilde{\lambda} dt),
\end{aligned}$$

hence the mean square hedging error is given from the Itô isometry (14.8) by

$$\begin{aligned}
&\mathbb{E}_{u,\tilde{\lambda}}[(V_T - f(T, S_T))^2] \\
&= (V_0 - f(0, S_0))^2 + \sigma^2 \mathbb{E}_{u,\tilde{\lambda}} \left[\left(\int_0^T S_t \left(\xi_t - \frac{\partial f}{\partial x}(t, S_t) \right) d\tilde{W}_t \right)^2 \right] \\
&\quad + \mathbb{E}_{u,\tilde{\lambda}} \left[\left(\int_0^T (f(t, S_{t^-}(1+a)) - f(t, S_{t^-}) - a\xi_t S_{t^-})(dN_t - \tilde{\lambda} dt) \right)^2 \right] \\
&= (V_0 - f(0, S_0))^2 + \sigma^2 \mathbb{E}_{u,\tilde{\lambda}} \left[\int_0^T S_t^2 \left(\xi_t - \frac{\partial f}{\partial x}(t, S_t) \right)^2 dt \right]
\end{aligned}$$

$$+\tilde{\lambda}\,\mathbb{E}_{u,\tilde{\lambda}}\left[\int_0^T ((f(t,S_t(1+a)) - f(t,S_t) - a\xi_t S_t))^2\,dt\right].$$

Clearly, the initial portfolio value V_0 that minimizes the above quantity is

$$V_0 = f(0,S_0) = e^{-rT}\,\mathbb{E}_{u,\tilde{\lambda},\tilde{\nu}}[\phi(S_T)].$$

When hedging only the risk generated by the Brownian part we let

$$\xi_t = \frac{\partial f}{\partial x}(t,S_t)$$

as in the Black–Scholes model, and in this case the hedging error due to the presence of jumps becomes

$$\mathbb{E}_{u,\tilde{\lambda}}[(V_T - f(T,S_T))^2] = \tilde{\lambda}\,\mathbb{E}_{u,\tilde{\lambda}}\left[\int_0^T ((f(t,S_t(1+a)) - f(t,S_t) - a\xi_t S_t))^2\,dt\right].$$

Next, let us find the optimal strategy $(\xi_t)_{t\in\mathbb{R}_+}$ that minimizes the remaining hedging error

$$\mathbb{E}_{u,\tilde{\lambda}}\left[\int_0^T \left(\sigma^2 S_t^2\left(\xi_t - \frac{\partial f}{\partial x}(t,S_t)\right)^2 + \tilde{\lambda}\,((f(t,S_t(1+a)) - f(t,S_t) - a\xi_t S_t))^2\right)dt\right].$$

For all $t\in[0,T]$, the almost-sure minimum of

$$\xi_t \longmapsto \sigma^2 S_t^2\left(\xi_t - \frac{\partial f}{\partial x}(t,S_t)\right)^2 + \tilde{\lambda}\,((f(t,S_t(1+a)) - f(t,S_t) - a\xi_t S_t))^2$$

is given by differentiation with respect to ξ_t, as the solution of

$$\sigma^2 S_t^2\left(\xi_t - \frac{\partial f}{\partial x}(t,S_t)\right) - a\tilde{\lambda} S_t\,((f(t,S_t(1+a)) - f(t,S_t) - a\xi_t S_t)) = 0,$$

i.e.,

$$\xi_t = \frac{\sigma^2 \dfrac{\partial f}{\partial x}(t,S_t) + \dfrac{a\tilde{\lambda}}{S_t}(f(t,S_{t^-}(1+a)) - f(t,S_{t^-}))}{\sigma^2 + a^2\tilde{\lambda}}, \qquad t\in[0,T]. \tag{15.14}$$

We note that the optimal strategy (15.14) is a weighted average of the Brownian and jump hedging strategies (15.12) and (15.13) according to the respective variance parameters σ^2 and $a^2\tilde{\lambda}$ of the continuous and jump components.

Clearly, if $a\tilde{\lambda} = 0$ we get

$$\xi_t = \frac{\partial f}{\partial x}(t, S_t), \qquad t \in [0, T],$$

which is the Black–Scholes perfect replication strategy, and when $\sigma = 0$ we recover

$$\xi_t = \frac{f(t, S_{t^-}(1+a)) - f(t, S_{t^-})}{aS_{t^-}}, \qquad t \in [0, T].$$

which is (15.13).

Note that the fact that perfect replication is not possible in a jump-diffusion model can be interpreted as a more realistic feature of the model, as perfect replication is not possible in the real world.

See [38] for an example of a complete market model with jumps, in which continuous and jump noise are mutually excluding each other over time.

Exercises

Exercise 15.1 Consider a forward call contract with payoff $S_T - K$ on a jump diffusion risky asset $(S_t)_{t\in\mathbb{R}_+}$ given by

$$dS_t = \mu S_t dt + \sigma S_t dW_t + S_{t^-} dY_t.$$

1. Show that the forward claim admits a unique arbitrage price to be computed in a market with risk-free rate $r > 0$.

2. Show that the forward claim admits an exact replicating portfolio strategy based on the two assets S_t and e^{rt}.

3. Show that the portfolio strategy of Question 2 coincides with the optimal portfolio strategy (15.14).

Exercise 15.2 Consider $(W_t)_{t\in\mathbb{R}_+}$ a standard Brownian motion and $(N_t)_{t\in\mathbb{R}_+}$ a standard Poisson process with intensity $\lambda > 0$, independent of $(W_t)_{t\in\mathbb{R}_+}$, under a probability measure $\mathbb{P}^*$. Let $(S_t)_{t\in\mathbb{R}_+}$ be defined by the stochastic differential equation

$$dS_t = \mu S_t dt + \eta S_{t^-} dN_t + \sigma S_t dW_t. \qquad (15.15)$$

(i) Solve the equation (15.15).

(ii) We assume that μ, η and the risk-free rate $r > 0$ are chosen such that the discounted process $(e^{-rt}S_t)_{t\in\mathbb{R}_+}$ is a martingale under $\mathbb{P}^*$. What relation does this impose on μ, η, λ and r ?

(iii) Under the relation of Question (ii), compute the price at time t of a European call option on S_T with strike κ and maturity T, using a series expansion of Black–Scholes functions.

Exercise 15.3 Consider $(N_t)_{t\in\mathbb{R}_+}$ a standard Poisson process with intensity $\lambda > 0$ under a probability measure $\mathbb{P}$. Let $(S_t)_{t\in\mathbb{R}_+}$ be defined by the stochastic differential equation

$$dS_t = rS_t dt + Y_{N_t} S_{t^-} dN_t,$$

where $(Y_k)_{k\geq 1}$ is an *i.i.d.* sequence of uniformly distributed random variables on $[-1, 1]$.

1. Show that the discounted process $(e^{-rt}S_t)_{t\in\mathbb{R}_+}$ is a martingale under $\mathbb{P}$.

2. Compute the price at time 0 of a European call option on S_T with strike κ and maturity T, using a series of multiple integrals.

Chapter 16

Basic Numerical Methods

This chapter is an elementary introduction to finite difference methods for the resolution of PDEs and stochastic differential equations. We cover the explicit and implicit finite difference schemes for the heat equations and the Black–Scholes PDE, as well as the Euler and Milshtein schemes for stochastic differential equations.

16.1 The Heat Equation

Consider the heat equation

$$\frac{\partial \phi}{\partial t}(t,x) = \frac{\partial^2 \phi}{\partial x^2}(t,x) \tag{16.1}$$

with initial condition

$$\phi(0,x) = f(x)$$

on a compact interval $[0,T] \times [0,X]$ divided into the grid points

$$(t_i, x_j) = (i\Delta t, j\Delta x), \qquad i = 0,\ldots,N, \quad j = 0,\ldots,M,$$

with $\Delta t = T/N$ and $\Delta x = X/M$. Our goal is to obtain a discrete approximation

$$(\phi(t_i,x_j))_{0\leq i\leq N,\ 0\leq j\leq M}$$

of the solution to (16.1), by evaluating derivatives using finite differences.

Explicit method

Using the *forward* time difference approximation of (16.1) we get

$$\frac{\phi(t_{i+1},x_j) - \phi(t_i,x_j)}{\Delta t} = \frac{\phi(t_i,x_{j+1}) + \phi(t_i,x_{j-1}) - 2\phi(t_i,x_j)}{(\Delta x)^2}$$

and letting $\rho = (\Delta t)/(\Delta x)^2$ this yields

$$\phi(t_{i+1},x_j) = \rho\phi(t_i,x_{j+1}) + (1-2\rho)\phi(t_i,x_j) + \rho\phi(t_i,x_{j-1}),$$

$1 \leq j \leq M-1$, i.e

$$\Phi_{i+1} = A\Phi_i + \rho \begin{bmatrix} \phi(t_i, x_0) \\ 0 \\ \vdots \\ 0 \\ \phi(t_i, x_M) \end{bmatrix}, \qquad i = 0, 1, \ldots, N-1,$$

with

$$\Phi_i = \begin{bmatrix} \phi(t_i, x_1) \\ \vdots \\ \phi(t_i, x_{M-1}) \end{bmatrix}, \qquad i = 0, 1, \ldots, N,$$

and

$$A = \begin{bmatrix} 1-2\rho & \rho & 0 & \cdots & 0 & 0 & 0 \\ \rho & 1-2\rho & \rho & \cdots & 0 & 0 & 0 \\ 0 & \rho & 1-2\rho & \cdots & 0 & 0 & 0 \\ \vdots & \vdots & \vdots & \ddots & \vdots & \vdots & \vdots \\ 0 & 0 & 0 & \cdots & 1-2\rho & \rho & 0 \\ 0 & 0 & 0 & \cdots & \rho & 1-2\rho & \rho \\ 0 & 0 & 0 & \cdots & 0 & \rho & 1-2\rho \end{bmatrix}.$$

The vector

$$\begin{bmatrix} \phi(t_i, x_0) \\ 0 \\ \vdots \\ 0 \\ \phi(t_i, x_M) \end{bmatrix}, \qquad i = 0, \ldots, N,$$

can be given by the lateral boundary conditions $\phi(t, 0)$ and $\phi(t, X)$.

Implicit method

Using the *backward* time difference approximation

$$\frac{\phi(t_i, x_j) - \phi(t_{i-1}, x_j)}{\Delta t} = \frac{\phi(t_i, x_{j+1}) + \phi(t_i, x_{j-1}) - 2\phi(t_i, x_j)}{(\Delta x)^2}$$

and letting $\rho = (\Delta t)/(\Delta x)^2$ we get

$$\phi(t_{i-1}, x_j) = -\rho\phi(t_i, x_{j+1}) + (1+2\rho)\phi(t_i, x_j) - \rho\phi(t_i, x_{j-1}),$$

$1 \leq j \leq M-1$, i.e.,

$$\Phi_{i-1} = B\Phi_i + \rho \begin{bmatrix} \phi(t_i, x_0) \\ 0 \\ \vdots \\ 0 \\ \phi(t_i, x_M) \end{bmatrix}, \qquad i = 1, 2, \ldots, N,$$

with

$$B = \begin{bmatrix} 1+2\rho & -\rho & 0 & \cdots & 0 & 0 & 0 \\ -\rho & 1+2\rho & -\rho & \cdots & 0 & 0 & 0 \\ 0 & -\rho & 1+2\rho & \cdots & 0 & 0 & 0 \\ \vdots & \vdots & \vdots & \ddots & \vdots & \vdots & \vdots \\ 0 & 0 & 0 & \cdots & 1+2\rho & -\rho & 0 \\ 0 & 0 & 0 & \cdots & -\rho & 1+2\rho & -\rho \\ 0 & 0 & 0 & \cdots & 0 & -\rho & 1+2\rho \end{bmatrix}.$$

By inversion of the matrix B, Φ_i is given in terms of Φ_{i-1} as

$$\Phi_i = B^{-1}\Phi_{i-1} - \rho B^{-1} \begin{bmatrix} \phi(t_i, x_0) \\ 0 \\ \vdots \\ 0 \\ \phi(t_i, x_M) \end{bmatrix}, \qquad i = 1, \ldots, N.$$

16.2 Black–Scholes PDE

Consider the Black–Scholes PDE

$$r\phi(t,x) = \frac{\partial \phi}{\partial t}(t,x) + rx\frac{\partial \phi}{\partial x}(t,x) + \frac{1}{2}x^2\sigma^2\frac{\partial^2 \phi}{\partial x^2}(t,x), \tag{16.2}$$

under the terminal condition $\phi(T,x) = (x-K)^+$, resp. $\phi(T,x) = (K-x)^+$, for a European call, resp. put, option.

Note that here time runs *backwards* as we start from a terminal condition at time T. Thus here the explicit method uses *backward* differences while the implicit method uses *forward* differences.

Explicit method

Using the *backward* time difference approximation of (16.2) we get

$$\begin{aligned} r\phi(t_i, x_j) &= \frac{\phi(t_i, x_j) - \phi(t_{i-1}, x_j)}{\Delta t} + rx_j\frac{\phi(t_i, x_{j+1}) - \phi(t_i, x_{j-1})}{2\Delta x} \\ &\quad + \frac{1}{2}x_j^2\sigma^2\frac{\phi(t_i, x_{j+1}) + \phi(t_i, x_{j-1}) - 2\phi(t_i, x_j)}{(\Delta x)^2}, \end{aligned}$$

$1 \le j \le M-1$, i.e.,

$$\phi(t_{i-1}, x_j) = \frac{1}{2}\Delta t(\sigma^2 j^2 - rj)\phi(t_i, x_{j-1}) + (1 - \Delta t(\sigma^2 j^2 + r))\phi(t_i, x_j)$$

$$+\frac{1}{2}\Delta t(\sigma^2 j^2 + rj)\phi(t_i, x_{j+1}),$$

$1 \leq j \leq M-1$, where the boundary conditions $\phi(t_i, x_0)$ and $\phi(t_i, x_M)$ are given by

$$\phi(t_i, x_0) = 0, \qquad \phi(t_i, x_M) = x_M - Ke^{-r(T-t_i)}, \qquad 0 \leq i \leq N,$$

for a European call option, and

$$\phi(t_i, x_0) = Ke^{-r(T-t_i)}, \qquad \phi(t_i, x_M) = 0 \qquad 0 \leq i \leq N,$$

for a European put option.

The explicit finite difference method is known to have a divergent behavior when time runs backwards, as illustrated in Figure 16.1.

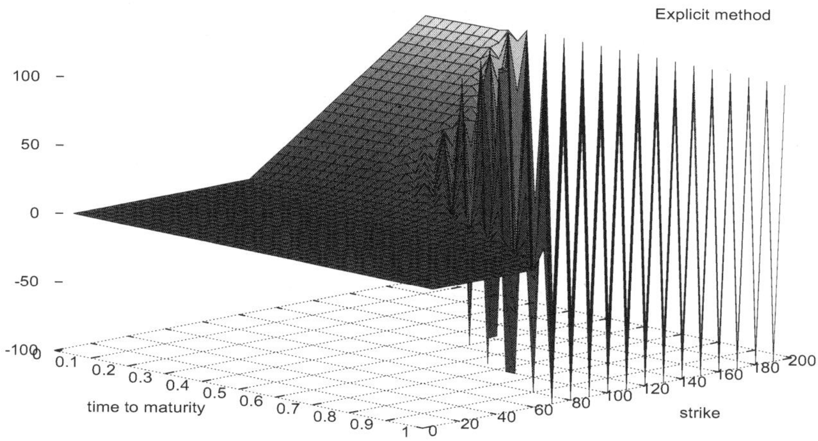

FIGURE 16.1: Divergence of the explicit finite difference method.

Implicit method

Using the *forward* time difference approximation of (16.2) we get

$$\begin{aligned} r\phi(t_i, x_j) &= \frac{\phi(t_{i+1}, x_j) - \phi(t_i, x_j)}{\Delta t} + rx_j \frac{\phi(t_i, x_{j+1}) - \phi(t_i, x_{j-1})}{2\Delta x} \\ &\quad + \frac{1}{2} x_j^2 \sigma^2 \frac{\phi(t_i, x_{j+1}) + \phi(t_i, x_{j-1}) - 2\phi(t_i, x_j)}{(\Delta x)^2}, \end{aligned}$$

$1 \leq j \leq M-1$, i.e.,

$$\begin{aligned} \phi(t_{i+1}, x_j) &= -\frac{1}{2}\Delta t(\sigma^2 j^2 - rj)\phi(t_i, x_{j-1}) + (1 + \Delta t(\sigma^2 j^2 + r))\phi(t_i, x_j) \\ &\quad -\frac{1}{2}\Delta t(\sigma^2 j^2 + rj)\phi(t_i, x_{j+1}), \end{aligned}$$

$1 \leq j \leq M-1$, i.e.,

$$\Phi_{i+1} = B\Phi_i + \begin{bmatrix} \frac{1}{2}\Delta t \left(r - \sigma^2\right) \phi(t_i, x_0) \\ 0 \\ \vdots \\ 0 \\ -\frac{1}{2}\Delta t \left(r(M-1) + \sigma^2 (M-1)^2\right) \phi(t_i, x_M) \end{bmatrix},$$

$i = 0, 1, \ldots, N-1$, with

$$B_{j,j-1} = \frac{1}{2}\Delta t \left(rj - \sigma^2 j^2\right), \quad B_{j,j} = 1 + \sigma^2 j^2 \Delta t + r\Delta t,$$

and

$$B_{j,j+1} = -\frac{1}{2}\Delta t \left(rj + \sigma^2 j^2\right),$$

for $j = 1, \ldots, M-1$, and $B(i,j) = 0$ otherwise.

By inversion of the matrix B, Φ_i is given in terms of Φ_{i+1} as

$$\Phi_i = B^{-1}\Phi_{i+1} - B^{-1} \begin{bmatrix} \frac{1}{2}\Delta t \left(r - \sigma^2\right) \phi(t_i, x_0) \\ 0 \\ \vdots \\ 0 \\ -\frac{1}{2}\Delta t \left(r(M-1) + \sigma^2 (M-1)^2\right) \phi(t_i, x_M) \end{bmatrix},$$

$i = 0, 1, \ldots, N-1$, where the boundary conditions $\phi(t_i, x_0)$ and $\phi(t_i, x_M)$ can be provided as in the case of the explicit method.

Note that for all $j = 1, \ldots, M-1$ we have

$$B_{j,j-1} + B_{j,j} + B_{j,j+1} = 1 + r\Delta t,$$

hence when the terminal condition is a constant $\phi(T, x) = c > 0$ we get

$$\phi(t_i, x) = c(1 + r\Delta t)^{-(N-i)} = c(1 + rT/N)^{-(N-i)}, \qquad i = 0, \ldots, N,$$

hence for all $s \in [0, T]$,

$$\begin{aligned} \phi(s, x) &= \lim_{N\to\infty} \phi(t_{[Ns/T]}, x) \\ &= c \lim_{N\to\infty} (1 + rT/N)^{-(N-[Ns/T])} \\ &= c \lim_{N\to\infty} (1 + rT/N)^{-[N(T-s)/T]} \\ &= c \lim_{N\to\infty} (1 + rT/N)^{-(T-s)/T} \\ &= ce^{-r(T-s)}, \end{aligned}$$

as expected, where $[x]$ denotes the integer part of $x \in \mathbb{R}$. The implicit finite difference method is known to be more stable than the explicit method, as illustrated in Figure 16.2, in which the discretization parameters have been taken to be the same as in Figure 16.1.

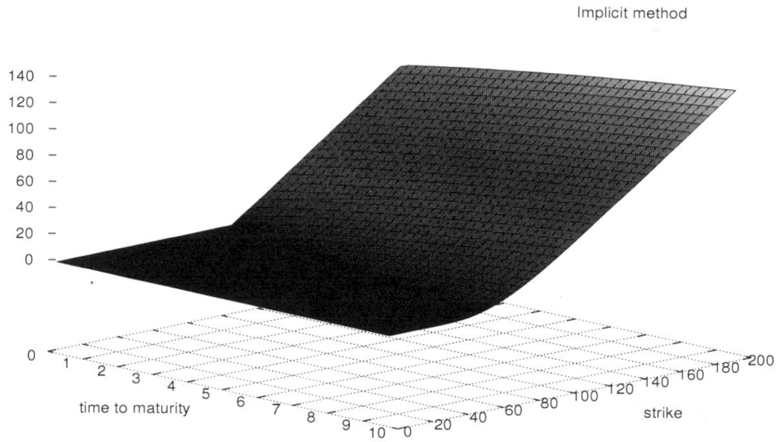

FIGURE 16.2: Stability of the implicit finite difference method.

16.3 Euler Discretization

In order to apply the Monte Carlo method in option pricing, we need to generate random samples whose empirical means are used for the evaluation of expectations. This can be done by discretizing the solutions of stochastic differential equations. Despite its apparent simplicity, the Monte Carlo method can be delicate to implement and the optimization of Monte Carlo algorithms and random number generation have been the object of numerous works which are outside the scope of this text, cf. e.g., [30], [43].

The Euler discretization scheme for the stochastic differential equation

$$dX_t = b(X_t)dt + a(X_t)dW_t$$

is given by

$$\begin{aligned}\hat{X}^N_{t_{k+1}} &= \hat{X}^N_{t_k} + \int_{t_k}^{t_{k+1}} b(X_s)ds + \int_{t_k}^{t_{k+1}} a(X_s)dW_s \\ &\simeq \hat{X}^N_{t_k} + b(\hat{X}^N_{t_k})(t_{k+1} - t_k) + a(\hat{X}^N_{t_k})(W_{t_{k+1}} - W_{t_k}).\end{aligned}$$

In particular, when X_t is the geometric Brownian motion given by

$$dX_t = rX_t dt + \sigma X_t dW_t$$

we get

$$\hat{X}^N_{t_{k+1}} = \hat{X}^N_{t_k} + r\hat{X}^N_{t_k}(t_{k+1} - t_k) + \sigma \hat{X}^N_{t_k}(W_{t_{k+1}} - W_{t_k}),$$

which can be computed as

$$\hat{X}^N_{t_k} = \hat{X}^N_{t_0} \prod_{i=1}^{k} \left(1 + r(t_i - t_{i-1}) + \sigma(W_{t_i} - W_{t_{i-1}})\right).$$

16.4 Milshtein Discretization

In the Milshtein scheme we expand $a(X_s)$ as

$$a(X_s) \simeq a(X_{t_k}) + a'(X_{t_k})b(X_{t_k})(s - t_k) + a'(X_{t_k})a(X_{t_k})(W_s - W_{t_k}).$$

As a consequence we get

$$\begin{aligned}
\hat{X}^N_{t_{k+1}} &= \hat{X}^N_{t_k} + \int_{t_k}^{t_{k+1}} b(X_s)ds + \int_{t_k}^{t_{k+1}} a(X_s)dW_s \\
&\simeq \hat{X}^N_{t_k} + \int_{t_k}^{t_{k+1}} b(X_s)ds + a(X_{t_k})(W_{t_{k+1}} - W_{t_k}) \\
&\quad + a'(X_{t_k})b(X_{t_k}) \int_{t_k}^{t_{k+1}} (s - t_k)dW_s + a'(X_{t_k})a(X_{t_k}) \int_{t_k}^{t_{k+1}} (W_s - W_{t_k})dW_s \\
&= \hat{X}^N_{t_k} + \int_{t_k}^{t_{k+1}} b(X_s)ds + a(X_{t_k})(W_{t_{k+1}} - W_{t_k}) \\
&\quad + \frac{1}{2}a'(X_{t_k})b(X_{t_k})(t_{k+1} - t_k)^2 + a'(X_{t_k})a(X_{t_k}) \int_{t_k}^{t_{k+1}} (W_s - W_{t_k})dW_s \\
&\simeq \hat{X}^N_{t_k} + \int_{t_k}^{t_{k+1}} b(X_s)ds + a(X_{t_k})(W_{t_{k+1}} - W_{t_k}) \\
&\quad + a'(X_{t_k})a(X_{t_k}) \int_{t_k}^{t_{k+1}} (W_s - W_{t_k})dW_s.
\end{aligned}$$

Next using Itô's formula we note that

$$(W_{t_{k+1}} - W_{t_k})^2 = 2\int_{t_k}^{t_{k+1}} (W_s - W_{t_k})dW_s + \int_{t_k}^{t_{k+1}} ds,$$

hence

$$\int_{t_k}^{t_{k+1}} (W_s - W_{t_k})dW_s = \frac{1}{2}((W_{t_{k+1}} - W_{t_k})^2 - (t_{k+1} - t_k)),$$

and

$$\hat{X}^N_{t_{k+1}} \simeq \hat{X}^N_{t_k} + \int_{t_k}^{t_{k+1}} b(X_s)ds + a(X_{t_k})(W_{t_{k+1}} - W_{t_k})$$

$$
\begin{aligned}
& +\frac{1}{2}a'(X_{t_k})a(X_{t_k})((W_{t_{k+1}} - W_{t_k})^2 - (t_{k+1} - t_k)) \\
= \ & \hat{X}^N_{t_k} + \int_{t_k}^{t_{k+1}} b(X_s)ds + a(X_{t_k})(W_{t_{k+1}} - W_{t_k}) \\
& +\frac{1}{2}a'(X_{t_k})a(X_{t_k})((W_{t_{k+1}} - W_{t_k})^2 - (t_{k+1} - t_k)) \\
= \ & \hat{X}^N_{t_k} + b(X_{t_k})(t_{k+1} - t_k) + a(X_{t_k})(W_{t_{k+1}} - W_{t_k}) \\
& +\frac{1}{2}a'(X_{t_k})a(X_{t_k})((W_{t_{k+1}} - W_{t_k})^2 - (t_{k+1} - t_k)).
\end{aligned}
$$

As a consequence the Milshtein scheme is written as

$$
\begin{aligned}
\hat{X}^N_{t_{k+1}} \ \simeq \ & \hat{X}^N_{t_k} + b(\hat{X}^N_{t_k})(t_{k+1} - t_k) + a(\hat{X}^N_{t_k})(W_{t_{k+1}} - W_{t_k}) \\
& +\frac{1}{2}a'(\hat{X}^N_{t_k})a(\hat{X}^N_{t_k})((W_{t_{k+1}} - W_{t_k})^2 - (t_{k+1} - t_k)),
\end{aligned}
$$

i.e., in the Milshtein scheme we take into account the "small" difference

$$
(W_{t_{k+1}} - W_{t_k})^2 - (t_{k+1} - t_k)
$$

existing between $(\Delta W_t)^2$ and Δt. Taking $(\Delta W_t)^2$ equal to Δt brings us back to the Euler scheme.

When X_t is the geometric Brownian motion given by

$$
dX_t = rX_t dt + \sigma X_t dW_t
$$

we get

$$
\begin{aligned}
\hat{X}^N_{t_{k+1}} \ = \ & \hat{X}^N_{t_k} + (r - \sigma^2/2)\hat{X}^N_{t_k}(t_{k+1} - t_k) + \sigma \hat{X}^N_{t_k}(W_{t_{k+1}} - W_{t_k}) \\
& +\frac{1}{2}\sigma^2 \hat{X}^N_{t_k}(W_{t_{k+1}} - W_{t_k})^2,
\end{aligned}
$$

which can be computed as

$$
\begin{aligned}
\hat{X}^N_{t_k} \ = \ & \hat{X}^N_{t_0} \prod_{i=1}^{k} \Bigg(1 + (r - \sigma^2/2)(t_i - t_{i-1}) + \sigma(W_{t_i} - W_{t_{i-1}}) \\
& +\frac{1}{2}\sigma^2 (W_{t_i} - W_{t_{i-1}})^2 \Bigg).
\end{aligned}
$$

Appendix

Background on Probability Theory

In this appendix we review a number of basic probabilistic tools that are needed in option pricing and hedging. We refer to [35], [16], [53] for more on the needed probability background.

Probability Spaces and Events

We will need the following notation coming from set theory. Given A and B to abstract sets, "$A \subset B$" means that A is contained in B, and the property that ω belongs to the set A is denoted by "$\omega \in A$." The finite set made of n elements $\omega_1, \ldots, \omega_n$ is denoted by $\{\omega_1, \ldots, \omega_n\}$, and we will usually make a distinction between the element ω and its associated singleton set $\{\omega\}$.

A probability space is an abstract set Ω that contains the possible outcomes of a random experiment.

Examples:

i) Coin tossing: $\Omega = \{H, T\}$.

ii) Rolling one die: $\Omega = \{1, 2, 3, 4, 5, 6\}$.

iii) Picking on card at random in a pack of 52: $\Omega = \{1, 2, 3, , \ldots, 52\}$.

iv) An integer-valued random outcome: $\Omega = \mathbb{N}$.

In this case the outcome $\omega \in \mathbb{N}$ can be the random number of trials needed until some event occurs.

v) A non-negative, real-valued outcome: $\Omega = \mathbb{R}_+$.

In this case the outcome $\omega \in \mathbb{R}_+$ may represent the (non-negative) value of a stock price, or a continuous random time.

vi) A random continuous parameter (such as time, weather, price or wealth, temperature, ...): $\Omega = \mathbb{R}$.

vii) Random choice of a continuous path in the space $\Omega = \mathcal{C}(\mathbb{R}_+)$ of all continuous functions on $\mathbb{R}_+$.

In this case, $\omega \in \Omega$ is a function $\omega : \mathbb{R}_+ \longrightarrow \mathbb{R}$ and a typical example is the graph $t \longmapsto \omega(t)$ of a stock price over time.

Product spaces:

Probability spaces can be built as product spaces and used for the modeling of repeated random experiments.

i) Rolling two dice: $\Omega = \{1, 2, 3, 4, 5, 6\} \times \{1, 2, 3, 4, 5, 6\}$.

In this case a typical element of Ω is written as $\omega = (k, l)$ with $k, l \in \{1, 2, 3, 4, 5, 6\}$.

ii) A finite number n of real-valued samples: $\Omega = \mathbb{R}^n$.

In this case the outcome ω is a vector $\omega = (x_1, \ldots, x_n) \in \mathbb{R}^n$ with n components.

Note that to some extent, the more complex Ω is, the better it fits a practical and useful situation, e.g., $\Omega = \{H, T\}$ corresponds to a simple coin tossing experiment while $\Omega = \mathcal{C}(\mathbb{R}_+)$ the space of continuous functions on $\mathbb{R}_+$ can be applied to the modeling of stock markets. On the other hand, in many cases and especially in the most complex situations, we will *not* attempt to specify Ω explicitly.

Events

An event is a collection of outcomes, which is represented by a subset of Ω.

The collections $\mathcal{G}$ of events that we will consider are called σ-algebras, and assumed to satisfy the following conditions.

(i) $\emptyset \in \mathcal{G}$,

(ii) For all countable sequences $A_n \in \mathcal{G}$, $n \geq 1$, we have $\bigcup_{n\geq 1} A_n \in \mathcal{G}$,

(iii) $A \in \mathcal{G} \Longrightarrow (\Omega \setminus A) \in \mathcal{G}$,

where $\Omega \setminus A := \{\omega \in \Omega \ : \ \omega \notin A\}$.

The collection of all events in Ω will be generally denoted by $\mathcal{F}$. The empty set $\emptyset$ and the full space Ω are considered as events but they are of less importance because Ω corresponds to "any outcome may occur" while $\emptyset$ corresponds to an absence of outcome, or no experiment.

In the context of stochastic processes, two σ-algebras $\mathcal{G}$ and $\mathcal{F}$ such that $\mathcal{G} \subset \mathcal{F}$ will refer to two different amounts of information, the amount of information associated to $\mathcal{G}$ being here lower than the one associated to $\mathcal{F}$.

The formalism of σ-algebras helps in describing events in a short and precise way.

Examples:

i) $\Omega = \{1, 2, 3, 4, 5, 6\}$.

The event $A = \{2, 4, 6\}$ corresponds to

"the result of the experiment is an even number."

ii) Taking again $\Omega = \{1, 2, 3, 4, 5, 6\}$,

$$\mathcal{F} := \{\Omega, \emptyset, \{2, 4, 6\}, \{1, 3, 5\}\}$$

defines a σ-algebra on Ω which corresponds to the knowledge of parity of an integer picked at random from 1 to 6.

Note that in the set-theoretic notation, an event A is a subset of Ω, i.e., $A \subset \Omega$, while it is an element of $\mathcal{F}$, i.e., $A \in \mathcal{F}$. For example, we have $\Omega \supset \{2, 4, 6\} \in \mathcal{F}$, while $\{\{2, 4, 6\}, \{1, 3, 5\}\} \subset \mathcal{F}$.

Taking

$$\mathcal{G} := \{\Omega, \emptyset, \{2, 4, 6\}, \{2, 4\}, \{6\}, \{1, 2, 3, 4, 5\}, \{1, 3, 5, 6\}, \{1, 3, 5\}\} \supset \mathcal{F},$$

defines a σ-algebra on Ω which is bigger than $\mathcal{F}$ and corresponds to the knowledge whether the outcome is equal to 6 or not, in addition to the parity information contained in $\mathcal{F}$.

iii) Take

$$\Omega = \{H,T\} \times \{H,T\} = \{(H,H),(H.T),(T,H),(T,T)\}.$$

In this case, the collection $\mathcal{F}$ of all possible events is given by

$$\begin{aligned}\mathcal{F} = \{\emptyset, &\{(H,H)\},\{(T,T)\},\{(H,T)\},\{(T,H)\},\\ &\{(T,T),(H,H)\},\{(H,T),(T,H)\},\{(H,T),(T,T)\},\\ &\{(T,H),(T,T)\},\{(H,T),(H,H)\},\{(T,H),(H,H)\},\\ &\{(H,H),(T,T),(T,H)\},\{(H,H),(T,T),(H,T)\},\\ &\{(H,T),(T,H),(H,H)\},\{(H,T),(T,H),(T,T)\},\Omega\}.\end{aligned} \tag{A.1}$$

Note that the set $\mathcal{F}$ of all events considered in (A.1) above has altogether

$1 = \binom{n}{0}$ event of cardinal 0,

$4 = \binom{n}{1}$ events of cardinal 1,

$6 = \binom{n}{2}$ events of cardinal 2,

$4 = \binom{n}{3}$ events of cardinal 3,

$1 = \binom{n}{4}$ event of cardinal 4,

with $n = 4$, for a total of

$$16 = 2^n = \sum_{k=0}^{4} \binom{4}{k} = 1 + 4 + 6 + 4 + 1$$

events.

The collection of events

$$\mathcal{G} := \{\emptyset, \{(T,T),(H,H)\},\{(H,T),(T,H)\},\Omega\}$$

defines a sub σ-algebra of $\mathcal{F}$, associated to the information "the results of two coin tossings are different."

Exercise: Write down the set of all events on $\Omega = \{H,T\}$.

Note also that (H,T) is different from (T,H), whereas $\{(H,T),(T,H)\}$ is equal to $\{(T,H),(H,T)\}$.

In addition we will usually make a distinction between the outcome $\omega \in \Omega$ and its associated event $\{\omega\} \in \mathcal{F}$, which satisfies $\{\omega\} \subset \Omega$.

Probability Measures

A probability measure is a mapping $\mathbb{P} : \mathcal{F} \longrightarrow [0,1]$ that assigns a probability $\mathbb{P}(A) \in [0,1]$ to any event A, with the properties

a) $\mathbb{P}(\Omega) = 1$, and

b) $\mathbb{P}\left(\bigcup_{n=1}^{\infty} A_n\right) = \sum_{n=1}^{\infty} \mathbb{P}(A_n)$, whenever $A_k \cap A_l = \emptyset$, $k \neq l$.

A property or event is said to hold $\mathbb{P}$-almost surely (also written $\mathbb{P}$-*a.s.*) if it holds with probability equal to one.

In particular we have

$$\mathbb{P}(A_1 \cup \cdots \cup A_n) = \mathbb{P}(A_1) + \cdots + \mathbb{P}(A_n)$$

when the subsets $A_1, \ldots, A_n$ of Ω are disjoints, and

$$\mathbb{P}(A \cup B) = \mathbb{P}(A) + \mathbb{P}(B)$$

if $A \cap B = \emptyset$. In the general case we can write

$$\mathbb{P}(A \cup B) = \mathbb{P}(A) + \mathbb{P}(B) - \mathbb{P}(A \cap B).$$

The triple

$$(\Omega, \mathcal{F}, \mathbb{P}) \tag{A.2}$$

was introduced by A.N. Kolmogorov (1903–1987), and is generally referred to as the Kolmogorov framework.

In addition we have the following convergence properties.

1. Let $(A_n)_{n\in\mathbb{N}}$ be a *non-decreasing* sequence of events, i.e., $A_n \subset A_{n+1}$, $n \in \mathbb{N}$. Then we have

$$\mathbb{P}\left(\bigcup_{n\in\mathbb{N}} A_n\right) = \lim_{n\to\infty} \mathbb{P}(A_n).$$

2. Let $(A_n)_{n\in\mathbb{N}}$ be a *non-increasing* sequence of events, i.e., $A_{n+1} \subset A_n$, $n \in \mathbb{N}$. Then we have

$$\mathbb{P}\left(\bigcap_{n\in\mathbb{N}} A_n\right) = \lim_{n\to\infty} \mathbb{P}(A_n). \tag{A.3}$$

Conditional Probabilities and Independence

Given any two events $A, B \subset \Omega$ with $\mathbb{P}(B) \neq 0$, we call

$$\mathbb{P}(A \mid B) := \frac{\mathbb{P}(A \cap B)}{\mathbb{P}(B)}$$

the probability of A *given* B, or *conditionally to* B. Note that if $\mathbb{P}(B) = 1$ we have $\mathbb{P}(A \cap B^c) \leq \mathbb{P}(B^c) = 0$ hence $\mathbb{P}(A \cap B) = \mathbb{P}(A)$ and $\mathbb{P}(A \mid B) = \mathbb{P}(A)$.

We also recall the following property:

$$\mathbb{P}\left(B \cap \bigcup_{n=1}^{\infty} A_n\right) = \sum_{n=1}^{\infty} \mathbb{P}(B \cap A_n) = \sum_{n=1}^{\infty} \mathbb{P}(B \mid A_n)\mathbb{P}(A_n) = \sum_{n=1}^{\infty} \mathbb{P}(A_n \mid B)\mathbb{P}(B),$$

for any family of events $(A_n)_{n\geq 1}$, B, provided $A_i \cap A_j = \emptyset$, $i \neq j$, and $\mathbb{P}(A_n) > 0$, $n \geq 1$. This also shows that conditional probability measures are probability measures, in the sense that whenever $\mathbb{P}(B) > 0$ we have

a) $\mathbb{P}(\Omega \mid B) = 1$, and

b) $\mathbb{P}\left(\bigcup_{n=1}^{\infty} A_n \middle| B\right) = \sum_{n=1}^{\infty} \mathbb{P}(A_n \mid B)$, whenever $A_k \cap A_l = \emptyset$, $k \neq l$.

In particular if $\bigcup_{n=1}^{\infty} A_n = \Omega$ we get the *law of total probability*

$$\mathbb{P}(B) = \sum_{n=1}^{\infty} \mathbb{P}(B \cap A_n) = \sum_{n=1}^{\infty} \mathbb{P}(A_n \mid B)\mathbb{P}(B) = \sum_{n=1}^{\infty} \mathbb{P}(B \mid A_n)\mathbb{P}(A_n), \quad \text{(A.4)}$$

provided $A_i \cap A_j = \emptyset$, $i \neq j$, and $\mathbb{P}(A_n) > 0$, $n \geq 1$. However we have in general

$$\mathbb{P}\left(A \middle| \bigcup_{n=1}^{\infty} B_n\right) \neq \sum_{n=1}^{\infty} \mathbb{P}(A \mid B_n),$$

even when $B_k \cap B_l = \emptyset$, $k \neq l$. Indeed, taking for example $A = \Omega = B_1 \cup B_2$ with $B_1 \cap B_2 = \emptyset$ and $\mathbb{P}(B_1) = \mathbb{P}(B_2) = 1/2$, we have

$$1 = \mathbb{P}(\Omega \mid B_1 \cup B_2) \neq \mathbb{P}(\Omega \mid B_1) + \mathbb{P}(\Omega \mid B_2) = 2.$$

Finally, two events A and B are said to be independent if

$$\mathbb{P}(A \mid B) = \mathbb{P}(A),$$

i.e., if

$$\mathbb{P}(A \cap B) = \mathbb{P}(A)\mathbb{P}(B).$$

In this case we find

$$\mathbb{P}(A \mid B) = \mathbb{P}(A).$$

Random Variables

A real-valued random variable is a mapping

$$\begin{aligned} X \ : \ \Omega &\longrightarrow \mathbb{R} \\ \omega &\longmapsto X(\omega) \end{aligned}$$

from a probability space Ω into the state space $\mathbb{R}$.

Given $X : \Omega \longrightarrow \mathbb{R}$ a random variable and A a (measurable)[1] subset of $\mathbb{R}$, we denote by $\{X \in A\}$ the event

$$\{X \in A\} = \{\omega \in \Omega \ : \ X(\omega) \in A\}.$$

Given $\mathcal{G}$ a σ-algebra on $\mathcal{G}$, the mapping $X : \Omega \longrightarrow \mathbb{R}$ is said to be $\mathcal{G}$-measurable if

$$\{X \leq x\} = \{\omega \in \Omega \ : \ X(\omega) \leq x\} \in \mathcal{G},$$

for all $x \in \mathbb{R}$. In this case we will also say that X depends only on the information contained in $\mathcal{G}$.

Examples:

i) Let $\Omega = \{1, 2, 3, 4, 5, 6\} \times \{1, 2, 3, 4, 5, 6\}$, and consider the mapping

$$\begin{aligned} X \ : \ \Omega &\longrightarrow \mathbb{R} \\ (k, l) &\longmapsto k + l. \end{aligned}$$

Then X is a random variable giving the sum of the two numbers appearing on each die.

ii) The time needed every day to travel from home to work or school is a random variable, as the precise value of this time may change from day to day under unexpected circumstances.

---

[1] Measurability of subsets of $\mathbb{R}$ refers to *Borel measurability*, a concept which will not be defined in this text.

iii) The price of a risky asset is a random variable.

In the sequel we will often use the notion of indicator function $\mathbf{1}_A$ of an event A. The indicator function $\mathbf{1}_A$ is the random variable

$$\begin{aligned} \mathbf{1}_A \quad : \quad \Omega &\longrightarrow \{0,1\} \\ \omega &\longmapsto \mathbf{1}_A(\omega) \end{aligned}$$

defined by

$$\mathbf{1}_A(\omega) = \begin{cases} 1 & \text{if } \omega \in A, \\ 0 & \text{if } \omega \notin A, \end{cases}$$

with the property

$$\mathbf{1}_{A\cap B}(\omega) = \mathbf{1}_A(\omega)\mathbf{1}_B(\omega), \tag{A.5}$$

since

$$\begin{aligned} \omega \in A \cap B &\iff \{\omega \in A \text{ and } \omega \in B\} \\ &\iff \{\mathbf{1}_A(\omega) = 1 \text{ and } \mathbf{1}_B(\omega) = 1\} \\ &\iff \mathbf{1}_A(\omega)\mathbf{1}_B(\omega) = 1. \end{aligned}$$

We also have

$$\mathbf{1}_{A\cup B} = \mathbf{1}_A + \mathbf{1}_B - \mathbf{1}_{A\cap B} = \mathbf{1}_A + \mathbf{1}_B - \mathbf{1}_A\mathbf{1}_B,$$

and

$$\mathbf{1}_{A\cup B} = \mathbf{1}_A + \mathbf{1}_B, \tag{A.6}$$

if $A \cap B = \emptyset$. In addition, any Bernoulli random variable $X : \Omega \longrightarrow \{0,1\}$ can be written as an indicator function

$$X = \mathbf{1}_A$$

on Ω with $A = \{X = 1\} = \{\omega \in \Omega \ : \ X(\omega) = 1\}$. For example if $\Omega = \mathbb{N}$ and $A = \{k\}$, for all $l \in \mathbb{N}$ we have

$$\mathbf{1}_{\{k\}}(l) = \begin{cases} 1 & \text{if } k = l, \\ 0 & \text{if } k \neq l. \end{cases}$$

If X is a random variable we also let

$$\mathbf{1}_{\{X=n\}} = \begin{cases} 1 & \text{if } X = n, \\ 0 & \text{if } X \neq n, \end{cases}$$

and

$$\mathbf{1}_{\{X<n\}} = \begin{cases} 1 & \text{if } X < n, \\ 0 & \text{if } X \geq n. \end{cases}$$

Probability Distributions

The probability distribution of a random variable $X : \Omega \longrightarrow \mathbb{R}$ is the collection

$$\{\mathbb{P}(X \in A) \ : \ A \text{ measurable subset of } \mathbb{R}\}.$$

In fact the distributions of X can be reduced to the knowledge of either

$$\{\mathbb{P}(a < X \leq b) \ : \ a < b \in \mathbb{R}\},$$

or

$$\{\mathbb{P}(X \leq a) \ : \ a \in \mathbb{R}\},$$

or

$$\{\mathbb{P}(X \geq a) \ : \ a \in \mathbb{R}\}.$$

Two random variables X and Y are said to be independent under the probability $\mathbb{P}$ if their probability distributions satisfy

$$\mathbb{P}(X \in A \ , \ Y \in B) = \mathbb{P}(X \in A)\mathbb{P}(Y \in B)$$

for all (measurable) subsets A and B of $\mathbb{R}$.

Distributions Admitting a Density

In this case the distribution of X is given by

$$\mathbb{P}(a \leq X \leq b) = \int_a^b f(x)dx$$

where the function $f : \mathbb{R} \longrightarrow \mathbb{R}_+$ is called the density of the distribution of X. We also say that the distribution of X is absolutely continuous, or that X is an absolutely continuous random variable. This, however, does *not* imply that the density function $f : \mathbb{R} \longrightarrow \mathbb{R}_+$ is continuous.

In particular we always have

$$\int_{-\infty}^{\infty} f(x)dx = \mathbb{P}(-\infty \leq X \leq \infty) = 1$$

for all probability density functions $f : \mathbb{R} \longrightarrow \mathbb{R}_+$.

The density f_X can be recovered from the distribution functions

$$x \longmapsto \mathbb{P}(X \leq x) = \int_{-\infty}^{x} f_X(s)ds, \qquad x \in \mathbb{R},$$

and

$$x \longmapsto \mathbb{P}(X \geq x) = \int_{x}^{\infty} f_X(s)ds, \qquad x \in \mathbb{R},$$

as

$$f_X(x) = \frac{\partial}{\partial x}\int_{-\infty}^{x} f_X(s)ds = -\frac{\partial}{\partial x}\int_{x}^{\infty} f_X(s)ds, \qquad x \in \mathbb{R}.$$

Examples:

i) The uniform distribution on an interval.

The density of the uniform distribution on the interval $[a, b]$ is given by

$$f(x) = \frac{1}{b-a}\mathbf{1}_{[a,b]}(x), \qquad x \in \mathbb{R}.$$

ii) The Gaussian distribution.

The density of the standard normal distribution is given by

$$f(x) = \frac{1}{\sqrt{2\pi}}e^{-x^2/2}, \qquad x \in \mathbb{R}.$$

More generally, X has a Gaussian distribution with mean $\mu \in \mathbb{R}$ and variance $\sigma^2 > 0$ (in this case we write $X \simeq \mathcal{N}(\mu, \sigma^2)$) if

$$f(x) = \frac{1}{\sqrt{2\pi\sigma^2}}e^{-(x-\mu)^2/(2\sigma^2)}, \qquad x \in \mathbb{R}.$$

iii) The exponential distribution with parameter $\lambda > 0$.

In this case we have

$$f(x) = \lambda \mathbf{1}_{[0,\infty)}(x)e^{-\lambda x} = \begin{cases} \lambda e^{-\lambda x}, & x \geq 0 \\ 0, & x < 0. \end{cases}$$

We also have

$$\mathbb{P}(X > t) = e^{-\lambda t}, \qquad t \in \mathbb{R}_+. \tag{A.7}$$

In addition, if $X_1, \ldots, X_n$ are independent exponentially distributed random variables with parameters $\lambda_1, \ldots, \lambda_n$ we have

$$\begin{aligned} \mathbb{P}(\min(X_1, \ldots, X_n) > t) &= \mathbb{P}(X_1 > t, \ldots, X_n > t) \\ &= \mathbb{P}(X_1 > t) \cdots \mathbb{P}(X_n > t) \\ &= e^{-t(\lambda_1 + \cdots + \lambda_n)}, \qquad t \in \mathbb{R}_+, \end{aligned} \tag{A.8}$$

hence $\min(X_1, \ldots, X_n)$ is an exponentially distributed random variable with parameter $\lambda_1 + \cdots + \lambda_n$.

We also have

$$\mathbb{P}(X_1 < X_2) = \mathbb{P}(X_1 \leq X_2) = \lambda_1\lambda_2 \int_0^\infty \int_0^y e^{-\lambda_1 x - \lambda_2 y} dx dy = \frac{\lambda_1}{\lambda_1 + \lambda_2}, \tag{A.9}$$

and we note that

$$\mathbb{P}(X_1 = X_2) = \lambda_1\lambda_2 \int_{\{(x,y)\in\mathbb{R}_+^2 \ : \ x=y\}} e^{-\lambda_1 x-\lambda_2 y} dxdy = 0.$$

iv) The gamma distribution.

In this case we have

$$f(x) = \frac{a^\lambda}{\Gamma(\lambda)} \mathbf{1}_{[0,\infty)}(x) x^{\lambda-1} e^{-ax} = \begin{cases} \dfrac{a^\lambda}{\Gamma(\lambda)} x^{\lambda-1} e^{-ax}, & x \geq 0 \\ 0, & x < 0, \end{cases}$$

where $a > 0$ and $\lambda > 0$ are parameters and

$$\Gamma(\lambda) = \int_0^\infty x^{\lambda-1} e^{-x} dx, \qquad \lambda > 0,$$

is the Gamma function.

v) The Cauchy distribution.

In this case we have

$$f(x) = \frac{1}{\pi(1+x^2)}, \qquad x \in \mathbb{R}.$$

vi) The lognormal distribution.

In this case,

$$f(x) = \mathbf{1}_{[0,\infty)}(x) \frac{1}{x\sigma\sqrt{2\pi}} e^{-\frac{(\mu-\log x)^2}{2\sigma^2}} = \begin{cases} \dfrac{1}{x\sigma\sqrt{2\pi}} e^{-\frac{(\mu-\log x)^2}{2\sigma^2}}, & x \geq 0 \\ 0, & x < 0. \end{cases}$$

Exercise: For each of the above probability density functions, check that the condition

$$\int_{-\infty}^{\infty} f(x)dx = 1$$

is satisfied.

Remark A.1 *Note that if the distribution of X admits a density then for all $a \in \mathbb{R}$, we have*

$$\mathbb{P}(X = a) = \int_a^a f(x)dx = 0, \tag{A.10}$$

and this is not a contradiction.

In particular, Remark A.1 shows that

$$\mathbb{P}(a \leq X \leq b) = \mathbb{P}(X = a) + \mathbb{P}(a < X \leq b) = \mathbb{P}(a < X \leq b) = \mathbb{P}(a < X < b),$$

for $a \leq b$.

In practice, Property (A.10) appears for example in the framework of lottery games with a large number of participants, in which a given number "a" selected in advance has a very low (almost zero) probability to be chosen.

Given two absolutely continuous random variables $X : \Omega \longrightarrow \mathbb{R}$ and $Y : \Omega \longrightarrow \mathbb{R}$ we can form the $\mathbb{R}^2$-valued random variable (X, Y) defined by

$$\begin{aligned}(X, Y) \quad : \quad & \Omega \longrightarrow \mathbb{R}^2 \\ & \omega \longmapsto (X(\omega), Y(\omega)).\end{aligned}$$

We say that (X, Y) admits a joint probability density

$$f_{(X,Y)} : \mathbb{R}^2 \longrightarrow \mathbb{R}_+$$

when

$$\mathbb{P}((X, Y) \in A \times B) = \int_A \int_B f_{(X,Y)}(x, y) dx dy$$

for all measurable subsets A, B of $\mathbb{R}$. The density $f_{(X,Y)}$ can be recovered from the distribution functions

$$(x, y) \longmapsto \mathbb{P}(X \leq x,\ Y \leq y) = \int_{-\infty}^{x} \int_{-\infty}^{y} f_{(X,Y)}(s, t) ds dt,$$

and

$$(x, y) \longmapsto \mathbb{P}(X \geq x,\ Y \geq y) = \int_{x}^{\infty} \int_{y}^{\infty} f_{(X,Y)}(s, t) ds dt,$$

as

$$\begin{aligned} f_{(X,Y)}(x, y) &= \frac{\partial^2}{\partial x \partial y} \int_{-\infty}^{x} \int_{-\infty}^{y} f_{(X,Y)}(s, t) ds dt \qquad (A.11) \\ &= \frac{\partial^2}{\partial x \partial y} \int_{x}^{\infty} \int_{y}^{\infty} f_{(X,Y)}(s, t) ds dt, \end{aligned}$$

$x, y \in \mathbb{R}$.

The probability densities $f_X : \mathbb{R} \longrightarrow \mathbb{R}_+$ and $f_Y : \mathbb{R} \longrightarrow \mathbb{R}_+$ of $X : \Omega \longrightarrow \mathbb{R}$ and $Y : \Omega \longrightarrow \mathbb{R}$ are called the <u>marginal densities</u> of (X, Y) and are given by

$$f_X(x) = \int_{-\infty}^{\infty} f_{(X,Y)}(x, y) dy, \qquad x \in \mathbb{R}, \qquad (A.12)$$

and

$$f_Y(y) = \int_{-\infty}^{\infty} f_{(X,Y)}(x, y) dx, \qquad y \in \mathbb{R}.$$

The conditional density $f_{X|Y=y} : \mathbb{R} \longrightarrow \mathbb{R}_+$ of X given $Y = y$ is defined by

$$f_{X|Y=y}(x) := \frac{f_{(X,Y)}(x,y)}{f_Y(y)}, \qquad x, y \in \mathbb{R}, \tag{A.13}$$

provided $f_Y(y) > 0$.

Discrete Distributions

We only consider integer-valued random variables, i.e., the distribution of X is given by the values of $\mathbb{P}(X = k)$, $k \in \mathbb{N}$.

Examples:

i) The Bernoulli distribution.

We have

$$\mathbb{P}(X = 1) = p \qquad \text{and} \qquad \mathbb{P}(X = 0) = 1 - p, \tag{A.14}$$

where $p \in [0, 1]$ is a parameter.

ii) The binomial distribution.

We have

$$\mathbb{P}(X = k) = \binom{n}{k} p^k (1-p)^{n-k}, \qquad k = 0, 1, \ldots, n,$$

where $n \geq 1$ and $p \in [0, 1]$ are parameters.

iii) The geometric distribution.

We have

$$\mathbb{P}(X = k) = (1-p)p^k, \qquad k \in \mathbb{N}, \tag{A.15}$$

where $p \in (0, 1)$ is a parameter.

Note that if $(X_k)_{k\in\mathbb{N}}$ is a sequence of independent Bernoulli random variables with distribution (A.14), then the random variable

$$X := \inf\{k \in \mathbb{N} \;:\; X_k = 0\}$$

has the geometric distribution (A.15).

iv) The negative binomial distribution (or Pascal distribution).

We have

$$\mathbb{P}(X = k) = \binom{k + r - 1}{r - 1}(1 - p)^r p^k, \qquad k \in \mathbb{N},$$

where $p \in (0, 1)$ and $r \geq 1$ are parameters. Note that the negative binomial distribution recovers the geometric distribution when $r = 1$.

v) The Poisson distribution.

We have

$$\mathbb{P}(X = k) = \frac{\lambda^k}{k!} e^{-\lambda}, \qquad k \in \mathbb{N},$$

where $\lambda > 0$ is a parameter.

Remark A.2 *The distribution of a discrete random variable cannot admit a density. If this were the case, by Remark A.1 we would have $\mathbb{P}(X = k) = 0$ for all $k \in \mathbb{N}$ and*

$$1 = \mathbb{P}(X \in \mathbb{R}) = \mathbb{P}(X \in \mathbb{N}) = \sum_{k=0}^{\infty} \mathbb{P}(X = k) = 0,$$

which is a contradiction.

Given two discrete random variables X and Y, the conditional distribution of X given $Y = k$ is given by

$$\mathbb{P}(X = n \mid Y = k) = \frac{\mathbb{P}(X = n \text{ and } Y = k)}{\mathbb{P}(Y = k)}, \qquad n \in \mathbb{N},$$

provided $\mathbb{P}(Y = k) > 0$, $k \in \mathbb{N}$.

Expectation of a Random Variable

The expectation of a random variable X is the mean, or average value, of X. In practice, expectations can be even more useful than probabilities. For example, knowing that a given equipment (such as a bridge) has a failure probability of 1.78493 out of a billion can be of less practical use than knowing the expected lifetime (e.g., 200000 years) of that equipment.

For example, the time $T(\omega)$ to travel from home to work/school can be a random variable with a new outcome and value every day; however, we usually refer to its expectation $\mathbb{E}[T]$ rather than to its sample values that may change from day to day.

The notion of expectation takes its full meaning under conditioning. For example, the expected return of a random asset usually depends on information such as economic data, location, etc. In this case, replacing the expectation by a conditional expectation will provide a better estimate of the expected value.

For example, life expectancy is a natural example of a conditional expectation since it typically depends on location, gender, and other parameters.

In general, the expectation of the indicator function $\mathbf{1}_A$ is defined as

$$\mathbb{E}[\mathbf{1}_A] = \mathbb{P}(A),$$

for any event A. For a Bernoulli random variable $X : \Omega \longrightarrow \{0,1\}$ with parameter $p \in [0,1]$, written as $X = \mathbf{1}_A$ with $A = \{X = 1\}$, we have

$$p = \mathbb{P}(X = 1) = \mathbb{P}(A) = \mathbb{E}[\mathbf{1}_A] = \mathbb{E}[X].$$

Discrete Distributions

Next, let $X : \Omega \longrightarrow \mathbb{N}$ be a discrete random variable. The expectation $\mathbb{E}[X]$ of X is defined as the sum

$$\mathbb{E}[X] = \sum_{k=0}^{\infty} k\mathbb{P}(X = k),$$

in which the possible values $k \in \mathbb{N}$ of X are weighted by their probabilities. More generally we have

$$\mathbb{E}[\phi(X)] = \sum_{k=0}^{\infty} \phi(k)\mathbb{P}(X = k),$$

for all sufficiently summable functions $\phi : \mathbb{N} \longrightarrow \mathbb{R}$. Note that the expectation $\mathbb{E}[\phi(X)]$ may be <u>infinite</u> even when $\phi(X)$ is always *finite*, take for example

$$\phi(X) = 2^X \quad \text{and} \quad \mathbb{P}(X = k) = 1/2^k, \qquad k \geq 1. \tag{A.16}$$

The expectation of the indicator function $X = \mathbf{1}_A$ can be recovered as

$$\mathbb{E}[\mathbf{1}_A] = 0 \times \mathbb{P}(\Omega \setminus A) + 1 \times \mathbb{P}(A) = \mathbb{P}(A).$$

Note that the expectation is a <u>linear</u> operation, i.e., we have

$$\mathbb{E}[aX + bY] = a\,\mathbb{E}[X] + b\,\mathbb{E}[Y], \qquad a, b \in \mathbb{R}, \tag{A.17}$$

provided

$$\mathbb{E}[|X|] + \mathbb{E}[|Y|] < \infty.$$

The conditional expectation of $X : \Omega \longrightarrow \mathbb{N}$ given an event A is defined by

$$\mathbb{E}[X \mid A] = \sum_{k=0}^{\infty} k\mathbb{P}(X = k \mid A),$$

whenever the above series converges.

Lemma A.1 *Given an event A such that $\mathbb{P}(A) > 0$, we have*

$$\mathbb{E}[X \mid A] = \frac{1}{\mathbb{P}(A)} \mathbb{E}\left[X \mathbf{1}_A\right]. \tag{A.18}$$

Proof. By Relation (A.5) we have

$$\begin{aligned}
\mathbb{E}[X \mid A] &= \frac{1}{\mathbb{P}(A)} \sum_{k=0}^{\infty} k\mathbb{P}(\{X = k\} \cap A) = \frac{1}{\mathbb{P}(A)} \sum_{k=0}^{\infty} k\,\mathbb{E}\left[\mathbf{1}_{\{X=k\}\cap A}\right] \\
&= \frac{1}{\mathbb{P}(A)} \sum_{k=0}^{\infty} k\,\mathbb{E}\left[\mathbf{1}_{\{X=k\}} \mathbf{1}_A\right] = \frac{1}{\mathbb{P}(A)} \mathbb{E}\left[\mathbf{1}_A \sum_{k=0}^{\infty} k \mathbf{1}_{\{X=k\}}\right] \\
&= \frac{1}{\mathbb{P}(A)} \mathbb{E}\left[\mathbf{1}_A X\right],
\end{aligned} \tag{A.19}$$

where we used the relation

$$X = \sum_{k=0}^{\infty} k \mathbf{1}_{\{X=k\}}$$

which holds since X takes only integer values. □

If X is independent of A (i.e., $\mathbb{P}(\{X = k\} \cap A) = \mathbb{P}(\{X = k\})\mathbb{P}(A)$, $k \in \mathbb{N}$) we have $\mathbb{E}[X\mathbf{1}_A] = \mathbb{E}[X]\mathbb{P}(A)$ and we naturally find

$$\mathbb{E}[X \mid A] = \mathbb{E}[X].$$

If $X = \mathbf{1}_B$ we also have in particular

$$\begin{aligned}
\mathbb{E}[\mathbf{1}_B \mid A] &= 0 \times \mathbb{P}(X = 0 \mid A) + 1 \times \mathbb{P}(X = 1 \mid A) \\
&= \mathbb{P}(X = 1 \mid A) \\
&= \mathbb{P}(B \mid A).
\end{aligned}$$

One can also define the conditional expectation of X given that $\{Y = k\}$, as

$$\mathbb{E}[X \mid Y = k] = \sum_{n=0}^{\infty} n\mathbb{P}(X = n \mid Y = k).$$

In general we have

$$\begin{aligned}
\mathbb{E}[\mathbb{E}[X \mid Y]] &= \sum_{k=0}^{\infty} \mathbb{E}[X \mid Y = k]\mathbb{P}(Y = k) \\
&= \sum_{k=0}^{\infty}\sum_{n=0}^{\infty} n\mathbb{P}(X = n \mid Y = k)\mathbb{P}(Y = k) \\
&= \sum_{n=0}^{\infty} n \sum_{k=0}^{\infty} \mathbb{P}(X = n \text{ and } Y = k) \\
&= \sum_{n=0}^{\infty} n\mathbb{P}(X = n) = \mathbb{E}[X],
\end{aligned}$$

where we used the marginal distribution

$$\mathbb{P}(X = n) = \sum_{k=0}^{\infty} \mathbb{P}(X = n \text{ and } Y = k), \qquad n \in \mathbb{N},$$

that follows from the *law of total probability* (A.4) by taking $A_k = \{Y = k\}$.

Hence we have the relation

$$\mathbb{E}[X] = \mathbb{E}[\mathbb{E}[X \mid Y]], \tag{A.20}$$

which is sometimes referred to as the tower property.

Random sums

Based on the "tower property" or ordinary conditioning, the expectation of a random sum $\sum_{k=1}^{Y} X_k$, where $(X_k)_{k\in\mathbb{N}}$ is a sequence of random variables, can be computed from the "tower property" (A.20) as

$$\begin{aligned}
\mathbb{E}\left[\sum_{k=1}^{Y} X_k\right] &= \mathbb{E}\left[\mathbb{E}\left[\sum_{k=1}^{Y} X_k \Big| Y\right]\right] \\
&= \sum_{n=0}^{\infty} \mathbb{E}\left[\sum_{k=1}^{Y} X_k \Big| Y = n\right] \mathbb{P}(Y = n) \\
&= \sum_{n=0}^{\infty} \mathbb{E}\left[\sum_{k=1}^{n} X_k \Big| Y = n\right] \mathbb{P}(Y = n),
\end{aligned}$$

and if Y is independent of $(X_k)_{k\in\mathbb{N}}$ this yields

$$\mathbb{E}\left[\sum_{k=1}^{Y} X_k\right] = \sum_{n=0}^{\infty} \mathbb{E}\left[\sum_{k=1}^{n} X_k\right] \mathbb{P}(Y = n).$$

Similarly, for a random product we will have

$$\mathbb{E}\left[\prod_{k=1}^{Y} X_k\right] = \sum_{n=0}^{\infty} \mathbb{E}\left[\prod_{k=1}^{n} X_k\right] \mathbb{P}(Y = n).$$

Example:

The life expectancy in Singapore is $\mathbb{E}[T] = 80$ years overall, where T denotes the lifetime of a given individual chosen at random. Let $G \in \{m, w\}$ denote the gender of that individual. The statistics show that

$$\mathbb{E}[T \mid G = w] = 78 \quad \text{and} \quad \mathbb{E}[T \mid G = m] = 81.9,$$

and we have

$$\begin{aligned} 80 &= \mathbb{E}[T] \\ &= \mathbb{E}[\mathbb{E}[T|G]] \\ &= \mathbb{P}(G = w)\,\mathbb{E}[T \mid G = w] + \mathbb{P}(G = m)\,\mathbb{E}[T \mid G = m] \\ &= 81.9 \times \mathbb{P}(G = w) + 78 \times \mathbb{P}(G = m) \\ &= 81.9 \times (1 - \mathbb{P}(G = m)) + 78 \times \mathbb{P}(G = m), \end{aligned}$$

showing that

$$80 = 81.9 \times (1 - \mathbb{P}(G = m)) + 78 \times \mathbb{P}(G = m),$$

i.e.,

$$\mathbb{P}(G = m) = \frac{81.9 - 80}{81.9 - 78} = \frac{1.9}{3.9} = 0.487.$$

Distributions Admitting a Density

Given a random variable X whose distribution admits a density $f_X : \mathbb{R} \longrightarrow \mathbb{R}_+$ we have

$$\mathbb{E}[X] = \int_{-\infty}^{\infty} x f_X(x) dx,$$

and more generally,

$$\mathbb{E}[\phi(X)] = \int_{-\infty}^{\infty} \phi(x) f_X(x) dx,$$

for all sufficiently integrable function ϕ on $\mathbb{R}$. For example, if X has a standard normal distribution we have

$$\mathbb{E}[\phi(X)] = \int_{-\infty}^{\infty} \phi(x) e^{-x^2/2} \frac{dx}{\sqrt{2\pi}}.$$

In case X has a Gaussian distribution with mean $\mu \in \mathbb{R}$ and variance $\sigma^2 > 0$ we get

$$\mathbb{E}[\phi(X)] = \frac{1}{\sqrt{2\pi\sigma^2}} \int_{-\infty}^{\infty} \phi(x) e^{-(x-\mu)^2/(2\sigma^2)} dx.$$

In case $(X, Y) : \Omega \longrightarrow \mathbb{R}^2$ is a $\mathbb{R}^2$-valued couple of random variables whose distribution admits a density $f_{X,Y} : \mathbb{R} \longrightarrow \mathbb{R}_+$ we have

$$\mathbb{E}[\phi(X, Y)] = \int_{-\infty}^{\infty} \int_{-\infty}^{\infty} \phi(x, y) f_{X,Y}(x, y) dx dy,$$

for all sufficiently integrable function ϕ on $\mathbb{R}^2$.

The expectation of an absolutely continuous random variable satisfies the same linearity property (A.17) as in the discrete case.

Exercise: In case X has a Gaussian distribution with mean $\mu \in \mathbb{R}$ and variance $\sigma^2 > 0$, check that

$$\mu = \mathbb{E}[X] \qquad \text{and} \qquad \sigma^2 = \mathbb{E}[X^2] - (\mathbb{E}[X])^2.$$

The conditional expectation of an absolutely continuous random variable can be defined as

$$\mathbb{E}[X \mid Y = y] = \int_{-\infty}^{\infty} x f_{X|Y=y}(x) dx$$

where the conditional density $f_{X|Y=y}(x)$ is defined in (A.13), with the relation

$$\mathbb{E}[X] = \mathbb{E}[\mathbb{E}[X \mid Y]] \tag{A.21}$$

as in the discrete case, since

$$\begin{aligned}\mathbb{E}[\mathbb{E}[X \mid Y]] &= \int_{-\infty}^{\infty} \mathbb{E}[X \mid Y = y] f_Y(y) dy = \int_{-\infty}^{\infty} \int_{-\infty}^{\infty} x f_{X|Y=y}(x) f_Y(y) dx dy \\ &= \int_{-\infty}^{\infty} x \int_{-\infty}^{\infty} f_{(X,Y)}(x, y) dy dx = \int_{-\infty}^{\infty} x f_X(x) dx = \mathbb{E}[X],\end{aligned}$$

where we used Relation (A.12) between the density of (X, Y) and its marginal X.

Conditional Expectation

The construction of conditional expectation given above for discrete and absolutely continuous random variables can be generalized to σ-algebras.

For any $p \geq 1$ we let

$$L^p(\Omega) = \{F : \Omega \longrightarrow \mathbb{R} \ : \ \mathbb{E}[|F|^p] < \infty\} \tag{A.22}$$

denote the space of p-integrable random variables $F : \Omega \longrightarrow \mathbb{R}$. Given $\mathcal{G} \subset \mathcal{F}$

a sub σ-algebra of $\mathcal{F}$ and $F \in L^2(\Omega, \mathcal{F}, \mathbb{P})$, the conditional expectation of F given $\mathcal{G}$, and denoted

$$\mathbb{E}[F \mid \mathcal{G}],$$

can be defined to be the orthogonal projection of F onto $L^2(\Omega, \mathcal{G}, \mathbb{P})$.

That is, $\mathbb{E}[F \mid \mathcal{G}]$ is characterized by the relation

$$\langle G, F - \mathbb{E}[F \mid \mathcal{G}]\rangle = \mathbb{E}[G(F - \mathbb{E}[F \mid \mathcal{G}])] = 0,$$

i.e.,

$$\mathbb{E}[GF] = \mathbb{E}[G\,\mathbb{E}[F \mid \mathcal{G}]], \tag{A.23}$$

for all bounded and $\mathcal{G}$-measurable random variables G, where $\langle \cdot, \cdot \rangle$ is the scalar product in $L^2(\Omega, \mathcal{F}, \mathbb{P})$.

In other words, anytime the relation

$$\mathbb{E}[GF] = \mathbb{E}[GX]$$

holds for all bounded and $\mathcal{G}$-measurable random variables G, and a given $\mathcal{G}$-measurable random variable X, we can claim that

$$X = \mathbb{E}[F \mid \mathcal{G}]]$$

by uniqueness of the orthogonal projection onto the subspace $L^2(\Omega, \mathcal{G}, \mathbb{P})$ of $L^2(\Omega, \mathcal{F}, \mathbb{P})$.

The conditional expectation operator has the following properties.

i) $\mathbb{E}[FG \mid \mathcal{G}] = G\,\mathbb{E}[F \mid \mathcal{G}]$ if G depends only on the information contained in $\mathcal{G}$.

Proof: By the characterization (A.23) it suffices to show that

$$\mathbb{E}[HFG]] = \mathbb{E}[HG\,\mathbb{E}[F|\mathcal{G}]], \tag{A.24}$$

for all bounded and $\mathcal{G}$-measurable random variables G, H, which implies $\mathbb{E}[FG \mid \mathcal{G}] = G\,\mathbb{E}[F \mid \mathcal{G}]$.

Relation (A.24) holds from (A.23) because the product HG is $\mathcal{G}$-measurable hence G can be replaced with HG in (A.23).

ii) $\mathbb{E}[G|\mathcal{G}] = G$ when G depends only on the information contained in $\mathcal{G}$.

Proof: This is a consequence of point (i) above by taking $F = 1$.

iii) $\mathbb{E}[\mathbb{E}[F|\mathcal{G}] \mid \mathcal{H}] = \mathbb{E}[F|\mathcal{H}]$ if $\mathcal{H} \subset \mathcal{G}$, called the tower property.

Proof: First we note that (iii) holds when $\mathcal{H} = \{\emptyset, \Omega\}$ because taking $G = 1$ in (A.23) yields

$$\mathbb{E}[F] = \mathbb{E}[\mathbb{E}[F \mid \mathcal{G}]]. \tag{A.25}$$

Next, by the characterization (A.23) it suffices to show that

$$\mathbb{E}[H\,\mathbb{E}[F|\mathcal{G}]] = \mathbb{E}[H\,\mathbb{E}[F|\mathcal{H}]], \tag{A.26}$$

for all bounded and $\mathcal{G}$-measurable random variables H, which will imply (iii) from (A.23).

In order to prove (A.26) we check that by (A.25) and point (i) above we have

$$\begin{aligned} \mathbb{E}[H\,\mathbb{E}[F|\mathcal{G}]] &= \mathbb{E}[\mathbb{E}[HF|\mathcal{G}]] = \mathbb{E}[HF] \\ &= \mathbb{E}[\mathbb{E}[HF|\mathcal{H}]] = \mathbb{E}[H\,\mathbb{E}[F|\mathcal{H}]], \end{aligned}$$

and we conclude by the characterization (A.23).

iv) $\mathbb{E}[F|\mathcal{G}] = \mathbb{E}[F]$ when F "does not depend" on the information contained in $\mathcal{G}$ or, more precisely stated, when the random variable F is *independent* of the σ-algebra $\mathcal{G}$.

Proof: It suffices to note that for all bounded $\mathcal{G}$-measurable G we have

$$\mathbb{E}[FG] = \mathbb{E}[F]\,\mathbb{E}[G] = \mathbb{E}[G\,\mathbb{E}[F]],$$

and we conclude again by (A.23).

v) If G depends only on $\mathcal{G}$ and F is independent of $\mathcal{G}$, then

$$\mathbb{E}[h(F,G)|\mathcal{G}] = \mathbb{E}[h(x,F)]_{x=G}.$$

The notion of conditional expectation can be extended from square-integrable random variables in $L^2(\Omega)$ to integrable random variables in $L^1(\Omega)$, cf. e.g., [40], Theorem 5.1.

Moment Generating Functions

The characteristic function of a random variable X is the function $\Psi_X : \mathbb{R} \longrightarrow \mathbb{C}$ defined by

$$\Psi_X(t) = \mathbb{E}[e^{itX}], \qquad t \in \mathbb{R}.$$

The Laplace transform (or moment generating function) of a random variable X is the function $\Phi_X : \mathbb{R} \longrightarrow \mathbb{R}$ defined by

$$\Phi_X(t) = \mathbb{E}[e^{tX}], \qquad t \in \mathbb{R},$$

provided the expectation is finite.

In particular we have

$$\mathbb{E}[X^n] = \frac{\partial^n}{\partial t}\Phi_X(0), \qquad n \geq 1,$$

provided $\mathbb{E}[|X|^n] < \infty$. The Laplace transform Φ_X of a random variable X with density $f : \mathbb{R} \longrightarrow \mathbb{R}_+$ satisfies

$$\Phi_X(t) = \int_{-\infty}^{\infty} e^{tx} f(x)dx, \qquad t \in \mathbb{R}.$$

Note that in probability we are using the *bilateral* Laplace transform for which the integral is from $-\infty$ to $+\infty$.

The characteristic function Ψ_X of a random variable X with density $f : \mathbb{R} \longrightarrow \mathbb{R}_+$ satisfies

$$\Psi_X(t) = \int_{-\infty}^{\infty} e^{itx} f(x)dx, \qquad t \in \mathbb{R}.$$

On the other hand, if $X : \Omega \longrightarrow \mathbb{N}$ is a discrete random variable we have

$$\Psi_X(t) = \sum_{n=0}^{\infty} e^{itn}\mathbb{P}(X = n), \qquad t \in \mathbb{R}.$$

The main applications of characteristic functions lie in the following theorems:

Theorem A.1 *Two random variables $X : \Omega \longrightarrow \mathbb{R}$ and $Y : \Omega \longrightarrow \mathbb{R}$ have same distribution if and only if*

$$\Psi_X(t) = \Psi_Y(t), \qquad t \in \mathbb{R}.$$

Theorem A.1 is used to identify or to determine the probability distribution of a random variable X, by comparison with the characteristic function Ψ_Y of a random variable Y whose distribution is known.

The characteristic function of a random vector (X, Y) is the function $\Psi_{X,Y} : \mathbb{R}^2 \longrightarrow \mathbb{C}$ defined by

$$\Psi_{X,Y}(s, t) = \mathbb{E}[e^{isX+itY}], \qquad s, t \in \mathbb{R}.$$

Theorem A.2 *Given two* independent *random variables $X : \Omega \longrightarrow \mathbb{R}$ and $Y : \Omega \longrightarrow \mathbb{R}$ are independent if and only if*

$$\Psi_{X,Y}(s, t) = \Psi_X(s)\Psi_Y(t), \qquad s, t \in \mathbb{R}.$$

A random variable X is Gaussian with mean μ and variance σ^2 if and only if its characteristic function satisfies

$$\mathbb{E}[e^{i\alpha X}] = e^{i\alpha\mu - \alpha^2\sigma^2/2}, \qquad \alpha \in \mathbb{R}. \tag{A.27}$$

In terms of Laplace transforms we have, replacing $i\alpha$ by α,

$$\mathbb{E}[e^{\alpha X}] = e^{\alpha\mu + \alpha^2\sigma^2/2}, \qquad \alpha \in \mathbb{R}. \tag{A.28}$$

From Theorems A.1 and A.2 we deduce the following proposition.

Proposition A.1 *Let* $X \simeq \mathcal{N}(\mu, \sigma_X^2)$ *and* $Y \simeq \mathcal{N}(\nu, \sigma_Y^2)$ *be independent Gaussian random variables. Then* $X + Y$ *also has a Gaussian distribution*

$$X + Y \simeq \mathcal{N}(\mu + \nu, \sigma_X^2 + \sigma_Y^2).$$

Proof. Since X and Y are independent, by Theorem A.2 the characteristic function Ψ_{X+Y} of $X + Y$ is given by

$$\begin{aligned}
\Phi_{X+Y}(t) &= \Phi_X(t)\Phi_Y(t) \\
&= e^{it\mu - t^2\sigma_X^2/2} e^{it\nu - t^2\sigma_Y^2/2} \\
&= e^{it(\mu+\nu) - t^2(\sigma_X^2+\sigma_Y^2)/2}, \qquad t \in \mathbb{R},
\end{aligned}$$

where we used (A.27). Consequently, the characteristic function of $X + Y$ is that of a Gaussian random variable with mean $\mu + \nu$ and variance $\sigma_X^2 + \sigma_Y^2$ and we conclude by Theorem A.1. □

Exercises

Exercise A.1 Compute the expected value $\mathbb{E}[X]$ of a Poisson random variable X with parameter $\lambda > 0$.

Exercise A.2 Let X denote a centered Gaussian random variable with variance η^2, $\eta > 0$. Show that the probability $P(e^X > c)$ is given by

$$P(e^X > c) = \Phi(-(\log c)/\eta),$$

where $\log = \ln$ denotes the natural logarithm and

$$\Phi(x) = \frac{1}{\sqrt{2\pi}} \int_{-\infty}^{x} e^{-y^2/2} dy, \qquad x \in \mathbb{R},$$

denotes the Gaussian cumulative distribution function.

Exercise A.3 Let $X \simeq \mathcal{N}(\mu, \sigma^2)$ be a Gaussian random variable with parameters $\mu > 0$ and $\sigma^2 > 0$, and density function

$$f(x) = \frac{1}{\sqrt{2\pi\sigma^2}} e^{-\frac{(x-\mu)^2}{2\sigma^2}}, \qquad x \in \mathbb{R}.$$

1. Write down $\mathbb{E}[X]$ as an integral and show that

$$\mu = \mathbb{E}[X].$$

2. Write down $\mathbb{E}[X^2]$ as an integral and show that

$$\sigma^2 = \mathbb{E}[(X - \mathbb{E}[X])^2].$$

3. Consider the function $x \mapsto (x-K)^+$ from $\mathbb{R}$ to $\mathbb{R}_+$, defined as

$$(x-K)^+ = \begin{cases} x - K & \text{if } x \geq K, \\ 0 & \text{if } x \leq K, \end{cases}$$

where $K \in \mathbb{R}$ be a fixed real number. Write down $\mathbb{E}[(X-K)^+]$ as an integral and compute this integral.

Hints: $(x-K)^+$ is zero when $x < K$, and when $\mu = 0$ and $\sigma = 1$ the result is

$$\mathbb{E}[(X-K)^+] = \frac{1}{\sqrt{2\pi}} e^{-\frac{K^2}{2}} - K\Phi(-K),$$

where

$$\Phi(x) := \int_{-\infty}^{x} e^{-\frac{y^2}{2}} \frac{dy}{\sqrt{2\pi}}, \qquad x \in \mathbb{R}.$$

4. Write down $\mathbb{E}[e^X]$ as an integral, and compute $\mathbb{E}[e^X]$.

Exercise A.4 Let X be a centered Gaussian random variable with variance $\alpha^2 > 0$ and density $x \mapsto \frac{1}{\sqrt{2\pi\alpha^2}} e^{-x^2/(2\alpha^2)}$ and let $\beta \in \mathbb{R}$.

1. Write down $\mathbb{E}[(\beta - X)^+]$ as an integral. Hint: $(\beta - x)^+$ is zero when $x > \beta$.

2. Compute this integral to show that

$$\mathbb{E}[(\beta - X)^+] = \frac{\alpha}{\sqrt{2\pi}} e^{-\frac{\beta^2}{2\alpha^2}} + \beta\Phi(\beta/\alpha),$$

where

$$\Phi(x) = \int_{-\infty}^{x} e^{-\frac{y^2}{2}} \frac{dy}{\sqrt{2\pi}}, \qquad x \in \mathbb{R}.$$

Exercise A.5 Let X be a centered Gaussian random variable with variance $\alpha^2 > 0$ and density $x \mapsto \frac{1}{\sqrt{2\pi\alpha^2}} e^{-x^2/(2\alpha^2)}$ and let $\beta \in \mathbb{R}$.

1. Write down $\mathbb{E}[(\beta + X)^+]$ as an integral. Hint: $(\beta + x)^+$ is zero when $x < -\beta$.

2. Compute this integral to show that

$$\mathbb{E}[(\beta + X)^+] = \frac{\alpha}{\sqrt{2\pi}} e^{-\frac{\beta^2}{2\alpha^2}} + \beta\Phi(\beta/\alpha),$$

   where

$$\Phi(x) = \int_{-\infty}^{x} e^{-\frac{y^2}{2}} \frac{dy}{\sqrt{2\pi}}, \qquad x \in \mathbb{R}.$$

Exercise A.6 Let X be a centered Gaussian random variable with variance v^2.

1. Compute

$$\mathbb{E}\left[e^{\sigma X} \mathbf{1}_{[K,\infty[}(xe^{\sigma X})\right] = \frac{1}{\sqrt{2\pi v^2}} \int_{\frac{1}{\sigma}\log\frac{K}{x}}^{\infty} e^{\sigma y - y^2/(2v^2)} dy.$$

   Hint: use the decomposition

$$\sigma y - \frac{y^2}{v^2} = \frac{v^2\sigma^2}{4} - \left(\frac{y}{v} - \frac{v\sigma}{2}\right)^2.$$

2. Compute

$$\mathbb{E}[(e^{m+X} - K)^+] = \frac{1}{\sqrt{2\pi v^2}} \int_{-\infty}^{\infty} (e^{m+x} - K)^+ e^{-\frac{x^2}{2v^2}} dx.$$

3. Compute the expectation (A.28) above.

Bibliography

[1] Y. Achdou and O. Pironneau. *Computational methods for option pricing*, volume 30 of *Frontiers in Applied Mathematics*. Society for Industrial and Applied Mathematics (SIAM), Philadelphia, PA, 2005. 144

[2] P. Barrieu, A. Rouault, and M. Yor. A study of the Hartman-Watson distribution motivated by numerical problems related to the pricing of Asian options. *J. Appl. Probab.*, 41(4):1049–1058, 2004. 198

[3] H. Bermin. Essays on Lookback options: a Malliavin calculus approach. PhD thesis, Lund University, 1998. 194

[4] T. Björk. *Arbitrage Theory in Continuous Time*, volume 121 of *Oxford Finance*. Oxford University Press, 2004. 20

[5] T. Björk. On the geometry of interest rate models. In *Paris-Princeton Lectures on Mathematical Finance 2003*, volume 1847 of *Lecture Notes in Math.*, pages 133–215. Springer, Berlin, 2004. 301

[6] D. Bosq and H.T. Nguyen. *A Course in Stochastic Processes: Stochastic Models and Statistical Inference*. Mathematical and Statistical Methods. Kluwer, 1996. 347

[7] A. Brace, D. Gatarek, and M. Musiela. The market model of interest rate dynamics. *Math. Finance*, 7(2):127–155, 1997. 305

[8] D. Brigo and F. Mercurio. *Interest rate models—theory and practice*. Springer Finance. Springer-Verlag, Berlin, second edition, 2006. 282, 304

[9] P. Carr and M. Schröder. Bessel processes, the integral of geometric Brownian motion, and Asian options. *Theory Probab. Appl.*, 48(3):400–425, 2004. 198

[10] R.-R. Chen, X. Cheng, F.J. Fabozzi, and B. Liu. An explicit, multi-factor credit default swap pricing model with correlated factors. *Journal of Financial and Quantitative Analysis*, 43(1):123–160, 2008. 339

[11] R.-R. Chen and J.-Z. Huang. Credit spread bounds and their implications for credit risk modeling. Working paper, Rutgers University and Penn State University, 2001. 339

[12] R. Cont and P. Tankov. *Financial modelling with jump processes.* Chapman & Hall/CRC Financial Mathematics Series. Chapman & Hall/CRC, Boca Raton, FL, 2004. 353, 357, 361, 365

[13] J.C. Cox, J.E. Ingersoll, and S.A. Ross. A theory of the term structure of interest rates. *Econometrica*, 53:385–407, 1985. 278

[14] J.C. Cox, S.A. Ross, and M. Rubinstein. Option pricing: A simplified approach. *Journal of Financial Economics*, 7:87–106, 1979. 35

[15] R.-A. Dana and M. Jeanblanc. *Financial markets in continuous time.* Springer Finance. Springer-Verlag, Berlin, 2007. Corrected Second Printing. 185

[16] J. L. Devore. *Probability and Statistics for Engineering and the Sciences.* Duxbury Press, sixth edition, 2003. 389

[17] L.U. Dothan. On the term structure of interest rates. *Jour. of Fin. Ec.*, 6:59–69, 1978. 278, 285

[18] R.M. Dudley. *Real analysis and probability*, volume 74 of *Cambridge Studies in Advanced Mathematics.* Cambridge University Press, Cambridge, 2002. Revised reprint of the 1989 original. 71

[19] D. Duffie and K.J. Singleton. *Credit risk.* Princeton series in finance. Princeton University Press, Princeton, NJ, 2003. 339

[20] D. Dufresne. Laguerre series for Asian and other options. *Math. Finance*, 10(4):407–428, 2000. 198

[21] B. Dupire. Pricing with a smile. *Risk*, 7(1):18–20, 1994. 144

[22] J. Eriksson and J. Persson. Pricing turbo warrants. Preprint, 2006. 29, 171

[23] G. B. Folland. *Real analysis.* Pure and Applied Mathematics (New York). John Wiley & Sons Inc., New York, second edition, 1999. 64

[24] H. Föllmer and A. Schied. *Stochastic finance*, volume 27 of *de Gruyter Studies in Mathematics.* Walter de Gruyter & Co., Berlin, 2004. 7, 12, 14, 34, 35, 50, 58

[25] J.P. Fouque, G. Papanicolaou, and K.R. Sircar. *Derivatives in financial markets with stochastic volatility.* Cambridge University Press, Cambridge, 2000. 135

[26] M.B. Garman and S.W. Kohlhagen. Foreign currency option values. *J. International Money and Finance*, 2:231–237, 1983.

[27] H. Geman, N. El Karoui, and J.-C. Rochet. Changes of numéraire, changes of probability measure and option pricing. *J. Appl. Probab.*, 32(2):443–458, 1995. 255, 270

[28] H. Geman and M. Yor. Bessel processes, Asian options and perpetuities. *Math. Finance*, 3:349–375, 1993. 199, 200

[29] H.U. Gerber and E.S.W. Shiu. Martingale approach to pricing perpetual American options on two stocks. *Math. Finance*, 6(3):303–322, 1996. 250

[30] P. Glasserman. *Monte Carlo methods in financial engineering*, volume 53 of *Applications of Mathematics (New York)*. Springer-Verlag, New York, 2004. Stochastic Modelling and Applied Probability. 386

[31] X. Guo, R. Jarrow, and C. Menn. A note on Lando's formula and conditional independence. Preprint, 2007. 338, 339

[32] D. Heath, R. Jarrow, and A. Morton. Bond pricing and the term structure of interest rates: a new methodology. *Econometrica*, 60:77–105, 1992. 293

[33] J. Hull and A. White. Pricing interest rate derivative securities. *The Review of Financial Studies*, 3:537–592, 1990. 293

[34] N. Ikeda and S. Watanabe. *Stochastic Differential Equations and Diffusion Processes*. North-Holland, 1989. 74

[35] J. Jacod and P. Protter. *Probability essentials*. Springer-Verlag, Berlin, 2000. 389

[36] F. Jamshidian. An exact bond option formula. *The Journal of Finance*, XLIV(1):205–209, 1989. 317

[37] F. Jamshidian. Sorting out swaptions. *Risk*, 9(3):59–60, 1996. 269, 270

[38] M. Jeanblanc and N. Privault. A complete market model with Poisson and Brownian components. In R. Dalang, M. Dozzi, and F. Russo, editors, *Seminar on Stochastic Analysis, Random Fields and Applications (Ascona, 1999)*, volume 52 of *Progress in Probability*, pages 189–204. Birkhäuser, Basel, 2002. 379

[39] M. Jeanblanc, M. Yor, and M. Chesney. *Mathematical Methods for Financial Markets*. Springer Finance. Springer-Verlag London Ltd., London, 2009.

[40] O. Kallenberg. *Foundations of Modern Probability*. Probability and its Applications. Springer-Verlag, New York, second edition, 2002. 409

[41] Y. El Khatib. Contributions to the study of discontinuous markets via the Malliavin calculus. PhD thesis, Université de La Rochelle, 2003. 195

[42] Y. El Khatib and N. Privault. Computations of replicating portfolios in complete markets driven by normal martingales. *Applicationes Mathematicae*, 30:147–172, 2003. 194

[43] R. Korn, E. Korn, and G. Kroisandt. *Monte Carlo Methods and Models in Finance and Insurance.* Chapman & Hall/CRC Financial Mathematics Series. CRC Press, Boca Raton, FL, 2010. 386

[44] D. Lamberton and B. Lapeyre. *Introduction to Stochastic Calculus Applied to Finance.* Chapman & Hall, London, 1996. 50, 202

[45] D. Lando. On Cox processes and credit risky securities. *Review of Derivative Research*, 2:99–120, 1998. 337, 339

[46] F.A. Longstaff and E.S. Schwartz. Valuing american options by simulation: a simple least-squares approach. *Review of Financial Studies*, 14:113–147, 2001. 240, 243

[47] W. Margrabe. The value of an option to exchange one asset for another. *The Journal of Finance*, XXXIII(1):177–186, 1978. 268

[48] H. Matsumoto and M. Yor. Exponential functionals of Brownian motion. I. Probability laws at fixed time. *Probab. Surv.*, 2:312–347 (electronic), 2005. 197

[49] R. Merton. Theory of rational option pricing. *Bell Journal of Economics*, 4(1):141–183, 1973. 269

[50] G. Di Nunno, B. Øksendal, and F. Proske. *Malliavin Calculus for Lévy Processes with Applications to Finance.* Universitext. Springer-Verlag, Berlin, 2009. 50, 123

[51] C. Pintoux and N. Privault. A direct solution to the Fokker-Planck equation for exponential Brownian functionals. *Analysis and Applications*, 8(3):287–304, 2010. 286

[52] C. Pintoux and N. Privault. The Dothan pricing model revisited. *Math. Finance*, 21:355–363, 2011. 286

[53] J. Pitman. *Probability.* Springer, 1999. 389

[54] N. Privault. Stochastic analysis of Bernoulli processes. *Probab. Surv.*, 5:435–483 (electronic), 2008. 50

[55] N. Privault. *Stochastic Analysis in Discrete and Continuous Settings*, volume 1982 of *Lecture Notes in Mathematics.* Springer-Verlag, Berlin, 309 pp., 2009. 50, 51, 52, 71, 74, 111, 123

[56] N. Privault. *An Elementary Introduction to Stochastic Interest Rate Modeling (Second Edition).* Advanced Series on Statistical Science & Applied Probability, 16. World Scientific Publishing Co., Singapore, 2012. xvi, 293, 301, 302, 303, 304, 305, 318

[57] N. Privault and T.-R. Teng. Risk-neutral hedging in bond markets. *Risk and Decision Analysis*, 3:201–209, 2012. 269, 320, 326

[58] N. Privault and W.T. Uy. Monte Carlo computation of the Laplace transform of exponential Brownian functionals. *Methodol. Comput. Appl. Probab.*, 15(3):511–524, 2013. 286

[59] P. Protter. *Stochastic Integration and Differential Equations. A New Approach.* Springer-Verlag, Berlin, 1990. 259

[60] P. Protter. A partial introduction to financial asset pricing theory. *Stochastic Process. Appl.*, 91(2):169–203, 2001. 123, 270

[61] P. Protter. *Stochastic integration and differential equations*, volume 21 of *Stochastic Modelling and Applied Probability.* Springer-Verlag, Berlin, 2005. 77, 83, 116, 124, 280, 281

[62] D. Revuz and M. Yor. *Continuous Martingales and Brownian Motion.* Springer-Verlag, 1994. 64

[63] L.C.G. Rogers and Z. Shi. The value of an Asian option. *J. Appl. Probab.*, 32(4):1077–1088, 1995. 204

[64] J. Ruiz de Chávez. Predictable representation of the binomial process and application to options in finance. In *XXXIII National Congress of the Mexican Mathematical Society (Spanish) (Saltillo, 2000)*, volume 29 of *Aportaciones Mat. Comun.*, pages 223–230. Soc. Mat. Mexicana, México, 2001. 50

[65] K. Sato. *Lévy processes and infinitely divisible distributions*, volume 68 of *Cambridge Studies in Advanced Mathematics.* Cambridge University Press, Cambridge, 1999. 366

[66] J. Schoenmakers. *Robust LIBOR Modelling and Pricing of Derivative Products.* Chapman & Hall/CRC Financial Mathematics Series. Chapman & Hall/CRC, Boca Raton, FL, 2005. 326

[67] A.N. Shiryaev. *Essentials of Stochastic Finance.* World Scientific Publishing Co. Inc., River Edge, NJ, 1999. 94, 95

[68] S.E. Shreve. *Stochastic Calculus for Finance. II.* Springer Finance. Springer-Verlag, New York, 2004. Continuous-time models. 156, 164, 178, 207, 227, 228, 248, 275

[69] J.M. Steele. *Stochastic Calculus and Financial Applications*, volume 45 of *Applications of Mathematics*. Springer-Verlag, New York, 2001. 237

[70] S. Turnbull and L. Wakeman. A quick algorithm for pricing European average options. *Journal of Financial and Quantitative Analysis*, 26:377–389, 1992. 200

[71] O. Vašiček. An equilibrium characterisation of the term structure. *Journal of Financial Economics*, 5:177–188, 1977. 277, 282

[72] D.V. Widder. *The heat equation*. Academic Press [Harcourt Brace Jovanovich Publishers], New York, 1975. *Pure and Applied Mathematics*, Vol. 67. 102

[73] D. Williams. *Probability with Martingales*. Cambridge Mathematical Textbooks. Cambridge University Press, Cambridge, 1991. 50

[74] H.Y. Wong and C.M. Chan. Turbo warrants under stochastic volatility. *Quant. Finance*, 8(7):739–751, 2008. 29

[75] M. Yor. On some exponential functionals of Brownian motion. *Adv. in Appl. Probab.*, 24(3):509–531, 1992. 197

Subject Index

Author Index